MATERIALS SCIENCE RESEARCH
Volume 5

CERAMICS IN SEVERE ENVIRONMENTS

MATERIALS SCIENCE RESEARCH

MATERIALS SCIENCE RESEARCH • Volume 5

CERAMICS IN
SEVERE ENVIRONMENTS

Proceedings of the Sixth University Conference on Ceramic Science
North Carolina State University at Raleigh
December 7-9, 1970

Edited by

W. Wurth Kriegel

Department of Materials Engineering
North Carolina State University

and

Hayne Palmour III

Department of Engineering Research
North Carolina State University

℗ **PLENUM PRESS • NEW YORK • 1971**

SIXTH UNIVERSITY CONFERENCE ON CERAMIC SCIENCE

This Conference was held at the Hilton Inn adjacent to North Carolina State University at Raleigh, December 7-9, 1970. The Conference was conducted by the School of Engineering through the Division of Continuing Education of North Carolina State University in cooperation with the National Aeronautics and Space Administration, the U.S. Army Research Office–Durham, and the U.S. Air Force Aerospace Research Laboratories.

Library of Congress Catalog Card Number 63-17645

SBN 306-38505-8

© 1971 Plenum Press, New York
A Division of Plenum Publishing Corporation
227 West 17th Street, New York, N.Y. 10011

United Kingdom edition published by Plenum Press, London
A Division of Plenum Publishing Company, Ltd.
Davis House (4th Floor), 8 Scrubs Lane, Harlesden, NW10 6SE, London, England

Printed in the United States of America

PREFACE

This volume constitutes the Proceedings of the Sixth University Conference on Ceramic Science which was held at North Carolina State University in Raleigh, December 7-9, 1970. Previous host institutions and resultant publications in this ongoing series have been:

1. North Carolina State University, November 16-18, 1964.
    W. W. Kriegel and H. Palmour III, Eds., The Role of Grain Boundaries and Surfaces in Ceramics. 631 pp., Materials Science Research, Vol. 3, Plenum Press, New York, 1966.

2. University of Notre Dame, June 21-23, 1965.
    G. C. Kuczynski, N. A. Hooten and C. F. Gibbon, Eds., Sintering and Related Phenomena, 894 pp. Gordon and Breach Science Publishers, Inc., New York, 1967.

3. University of California at Berkeley, June 13-16, 1966.
    R. M. Fulrath and J. A. Pask, Eds., Ceramic Microstructures-Their Analysis, Significance and Production. 1008 pp. John Wiley & Sons, Inc., New York, 1968.

4. State University of New York, College of Ceramics at Alfred University, June 18-23, 1967.
    T. J. Gray and V. D. Frechette, Eds., Kinetics of Reactions in Ionic Solids. 571 pp. Materials Science Research, Vol. 4, Plenum Press, New York, 1969.

5. University of Florida, November 10-14, 1969.
    L. L. Hench and D. B. Dove, Eds., Physics of Electronic Ceramics. M. Dekker, Inc. (In Press).

The 1970 conference directed special attention to *applications* of ceramics. In many instances ceramics are qualified (often uniquely) for service in situations so severe that other classes of materials cannot be reliably employed. This volume, comprising 41 invited papers by a total of 72 contributing authors, examines new findings and presents technologically significant effects (and limitations) relating to ceramic materials, their properties, and their uses which are attributable to chemical, thermal, mechanical, electromagnetic and/or nuclear radiation environments.

Interest is focused on environmental factors, singly or in combination, which are characterized as severe or even extreme in terms of their influence upon materials. Though they challenge materials scientists and engineers and the materials they develop and produce, these severe environments are both characteristic of and critically important to our highly technological society.

The breadth and importance of the field is so large, and the stature of the individual contributing authors so great, that it was not possible to designate any one person or paper to keynote the Conference and its Proceedings. Rather, we suggest that the theme itself--*Ceramics in Severe Environments*--be considered the keynote issue. Ceramics of real technological interest are likely to have their most important advantages and their greatest use potentials in these difficult situations. Future trends in this vital field will be inescapably linked (as have those in the past) with economic and technologic justifications which both stimulate and place practical limits upon all phases (research, design, fabrication, testing, and evaluation) of significant new product developments. Clearly, scientific understanding of constitution-microstructure-environment-property relationships of the kinds considered in this volume is urgently needed in preparing to meet complex engineering requirements for ceramic and composite materials in the future.

One cannot engage in planning, conducting, and editing an educational venture of this magnitude without support and assistance from a great many persons. We emerge from the experience with renewed appreciation for the accomplishments of teamwork and with deep personal gratitude and respect for the dedicated efforts of each of the individuals who contributed so much to making it all possible.

The List of Contributors (P. viii) formally acknowledges the many services rendered by the members of the Conference Staff and the several Session Chairmen, as well as the creative efforts of an international group of distinguished Authors. We extend our special personal thanks to each one for the cooperative attitudes, timely responses, and many helpful suggestions which have characterized all of our working relationships with them.

On behalf of the participants, the University, and the ceramic profession at large, we also gratefully acknowledge financial support provided by the three sponsoring agencies: National Aeronautics and Space Administration, U.S. Army Research Office - Durham, and U.S. Air Force Aerospace Research Laboratories. To have proceeded without adequate funding would have been impractical; to have obtained it in these fiscally difficult times without dedicated supporting effort on the part of the cognizant sponsors' representatives would have been impossible. We are endebted to J. J. Gangler (NASA), H.M. Davis (ARO-D), and H.C. Graham, N.M. Tallan and H.A. Davis (ARL) for their effective and most helpful advocacy.

As in the past, North Carolina State University has contributed very substantially of its time, effort and good will in making

possible this sixth in the series of University Conferences on
Ceramic Science. Frequent assistance was rendered and patient
forbearance was often displayed by our University associates, including students, faculty and staff colleagues, and concerned
administrators, during the many months we have been deeply involved
in planning, organizing, and editing. We are especially grateful
for the active participation in the Conference program by Provost
H. C. Kelly, welcoming attendees to the University; Dean of Engineering R. E. Fadum, presiding at the Conference Dinner; and
Associate Dean of Engineering H. B. Smith, introducing the tour of
the campus research facilities. Chancellor John T. Caldwell was
featured as the dinner speaker, presenting a stimulating and timely
address entitled "The Role of Education in Man's Changing Environment."

In any such undertaking, the prospects of a favorable outcome
really rest upon the skills, enthusiasms, and efforts of a few
persons who work with dedication, but largely behind the scenes.
Thus, we wish to acknowledge with very special personal thanks
Mrs. Janice Mudge for secretarial assistance before, during, and
after the Conference; M. L. Huckabee and C. E. Zimmer for serving
so effectively as Conference projectionists; Mrs. Ann Ethridge,
assisted by Miss Beth Johnson, for skillful, patient, and effective
typing of the entire Proceedings; and Miss Johnson, Mr. Huckabee,
and Mr. Zimmer for major assistance in indexing. Finally, it is
appropriate to acknowledge with real affection the patience, tol-
erance, and tangible and moral support we have been accorded by
our respective families and friends during this period of editorial
servitude.

W.W. Kriegel

Hayne Palmour III

Raleigh
June 24, 1971

CONFERENCE STAFF
NORTH CAROLINA STATE UNIVERSITY

Co-Chairmen
- W. W. Kriegel, Professor and Graduate Administrator, Materials Engineering Department
- H. Palmour III, Research Professor of Ceramic Engineering, Department of Engineering Research

Program and Publication Advisors
- W. O. Doggett, Professor of Physics
- R. A. Douglas, Associate Head, Department of Engineering Mechanics
- H. H. Stadelmaier, Research Professor of Metallurgy
- R. F. Stoops, Acting Director, Department of Engineering Research

Arrangements
- M. E. Shields, Coordinator, Division of Continuing Education

Hospitality and Tours
- W. W. Austin, Head, Department of Materials Engineering
- T. S. Elleman, Professor and Graduate Administrator, Department of Nuclear Engineering
- J. C. Hurt, Assistant Professor, Materials Engineering

Publicity
- M. N. Yionoulis, Engineering Publications and Information, Office of Information Services

CONFERENCE SESSION CHAIRMEN

L. L. Hench
University of Florida, Gainesville, Fla. 32601
G. C. Kuczynski
University of Notre Dame, Notre Dame, Ind. 46556
E. E. Mueller
State University of New York at Alfred University, Alfred, N. Y. 14802
J. I. Mueller
University of Washington, Seattle, Wash. 98105
R. M. Spriggs
Lehigh University, Bethlehem, Pa. 18105
S. M. Wiederhorn
National Bureau of Standards, Washington, D.C. 20234

AUTHORS

A. ACCARY
Centre D'Etudes Nucleaires de Saclay, Gif-sur-Yvette, France

C. B. ALCOCK
University of Toronto, Toronto 5, Canada

M. G. BADER
University of Surrey, Guildford, Surrey, England

J. E. BAILEY
University of Surrey, Guildford, Surrey, England

H. A. BARKER
University of Surrey, Guildford, Surrey, England

J. L. BATES
Battelle Northwest, Richland, Wash. 99352

P. F. BECHER
U.S. Naval Research Laboratory, Washington, D.C. 20390

J. R. BEELER
North Carolina State University, Raleigh, N.C. 27607

I. BELL
University of Surrey, Guildford, Surrey, England

H. S. BENNETT
National Bureau of Standards, Washington, D.C. 20234

S. A. BORTZ
I.I.T. Research Institute, Chicago, Ill. 60616

W. A. BRANTLEY
U.S. Army Materials and Mechanics Research Center, Watertown,
Mass. 02172

W. H. BOYER
Kaiser Aluminum and Chemical Corporation, Pleasanton, Calif. 94566

J. D. BUCKLEY
National Aeronautics and Space Administration, Hampton, Va. 23365

J. E. BURKE
General Electric Research and Development Center,
Schenectady, N.Y. 12301

S. C. CARNIGLIA
Kaiser Aluminum and Chemical Corporation, Pleasanton, Calif. 94566

A. E. CLARK
University of Florida, Gainesville, Fla.   32601

A. H. CLAUER
Battelle Memorial Institute, Columbus, Ohio   43201

J. C. CONWAY
Pennsylvania State University, State College, Pa.   16801

J. A. COOLEY
Martin-Marietta Corporation, Denver, Colo.   80201

A. R. COOPER
Case Western Reserve University, Cleveland, Ohio   44106

D. R. CROPPER
University of California, Berkeley, Calif.   94720

H. H. DAVIS
Wright-Patterson Air Force Base, Ohio   45433

R. H. DOREMUS
General Electric Research and Development Center
Schenectady, N. Y.   12301

L. L. FEHRENBACHER
Wright-Patterson Air Force Base, Ohio   45433

W. J. FERGUSON
U.S. Naval Research Laboratory, Washington, D.C.   20390

R. F. FIRESTONE
Case Western Reserve University, Cleveland, Ohio   44106

D. J. GODFREY
Admiralty Materials Laboratory, Holton Heath
Dorset, (BH16 6JU) England

H. C. GRAHAM
Wright-Patterson Air Force Base, Ohio   45433

G. O. HARRELL
North Carolina State University, Raleigh, N.C.   27607

D. P. H. HASSELMAN
Lehigh University, Bethlehem, Pa.   18015

L. L. HENCH
    University of Florida, Gainesville, Fla.   32601

A. H. HEUER
    Case Western Reserve University, Cleveland, Ohio   44106

W. B. HILLIG
    General Electric Research and Development Center
    Schenectady, N.Y.   12301

S. F. HULBERT
    Clemson University, Clemson, S.C.   29631

R. E. JEAGER
    Bell Telephone Laboratories, Murray Hill, N. J.   07974

R. N. KATZ
    U.S. Army Materials and Mechanics Research Center, Watertown,
    Mass.   02172

J. H. KLAWITTER
    Clemson University, Clemson, S.C.   29631

N. J. KREIDL
    University of Missouri, Rolla, Mo.   65401

I. A. KVERNES
    Central Institute for Industrial Research, Oslo-Blindern,
    Norway (Wright-Patterson Air Force Base, Ohio   45433)

W. J. LACKEY
    Oak Ridge National Laboratory, Oak Ridge, Tenn   37830

T. G. LANGDON
    University of Southern California, Los Angeles, Calif.   90007
    (University of British Columbia, Vancouver 8, Canada)

R. B. LEONARD
    Clemson University, Clemson, S.C.   29631

N. LEVY, JR.
    W. R. Grace and Co., Clarksville, Md.   21029

L. D. LINEBACK
    North Carolina State University, Raleigh, N.C.   27607

C. R. MANNING
    North Carolina State University, Raleigh, N.C.   27607

E. R. W. MAY
  Admiralty Materials Laboratory, Holton Heath,
  Dorset, BH16 6JU England

C. E. McNEILLY
  Battelle-Northwest, Richland, Wash. 99352

N. MIZOUCHI
  Owens-Illinois, Inc., Toledo, Ohio 43607

J. E. NEELY
  Kaiser Aluminum and Chemical Corporation, Pleasanton, Calif. 94566

R. E. NICKELL
  Bell Telephone Laboratories, Whippany, N.J.

H. PALMOUR III
  North Carolina State University, Raleigh, N.C. 27607

J. A. PASK
  University of California, Berkeley, Calif. 94720

J. L. PENTECOST
  W. R. Grace and Co., Clarksville, Md. 21029

N. J. PETCH
  University of Newcastle Upon Tyne, Newcastle Upon Tyne,
  NEI 7RU, England

J. J. RASMUSSEN
  Battelle Northwest, Richland, Wash. 99352

R. W. RICE
  U.S. Naval Research Laboratory, Washington, D.C. 29390

R. C. ROSSI
  The Aerospace Corporation, Los Angeles, Calif. 90045

E. RUH
  Harbison-Walker Refractories Co., Pittsburg, Pa. 15227

J. L. SCOTT
  Oak Ridge National Laboratory, Oak Ridge, Tenn. 37830

M. S. SELTZER
  Battelle Memorial Institute, Columbus, Ohio 43201

W. J. SMOTHERS
  Bethlehem Steel Corporation, Bethlehem, Pa. 18016

J. D. SNOW
  Ferro Corporation, Cleveland, Ohio   44101

N. M. TALLAN
  Wright-Patterson Air Force Base, Ohio   45433

V. J. TENNERY
  Oak Ridge National Laboratory, Oak Ridge, Tenn.   37830

W. C. TRIPP
  Systems Research Laboratory, Dayton, Ohio   45401

J. TULLIS
  University of California at Los Angeles, Los Angeles, Calif. 90024

W. D. TUOHIG
  University of Florida, Gainesville, Fla.   32601

A. M. TURKALO
  General Electric Research and Development Center,
  Schenectady, N. Y.   12301

T. N. WASHBURN
  Oak Ridge National Laboratory, Oak Ridge, Tenn.   37830

B. A. WILCOX
  Battelle Memorial Institute, Columbus, Ohio   43201

S. ZADOR
  University of Toronto, Toronto 5, Canada

CONTENTS

THE CORROSION OF CERAMIC OXIDES BY ATMOSPHERES CONTAINING

SULPHUR AND OXYGEN

C. B. Alcock and S. Zador

University of Toronto

Toronto, Canada

ABSTRACT

The thermodynamics of ceramic oxide corrosion by sulphur and oxygen-bearing atmospheres are considered. The conditions under which condensed sulphates and sulphides are expected to form, and in some cases volatile sulphides, are presented graphically in a chemical potential diagram which makes it possible to determine the total composition of the gas phase with respect to SO, SO_2 and SO_3 for any pair of sulphur and oxygen potentials in a simple and rapid way. Finally, indications concerning the choice of refractory materials in sulphur-bearing furnace atmospheres are obtained from an empirical correlation involving the radius of the cation in the sulphides and oxides.

INTRODUCTION

In a number of industrial operations and others under consideration, ceramic containers must operate at high temperatures under circumstances in which the gas phase contains sulphur. The precise chemical form or forms of sulphur-containing molecules which are present in the gaseous phase obviously depends on the temperature, the total pressure and the concentrations of oxygen and other gases which are also present in the gas. Normally the presence of a number of gaseous compounds of sulphur is to be anticipated, and in this paper we shall consider SO, SO_2, SO_3, together with S_2, as the most important species. The possible reactions which may occur between ceramic oxides and the complex gaseous phase are usually the formation of a metal sulphate, for which the term "sulphation" is generally used, or of a metal sulphide, which might be called "sulphidation", but will here also be referred to as sulphation.

1

The conditions for the formation of metallic sulphates and sulphides at a constant temperature have hitherto been represented on "Stability diagrams"[1]. In these, log p SO_2 and log p O_2 are used as ordinates, and the diagrams show the range of these gas partial pressures over which single phases exist, and have straight lines representing two-phase equilibria. The SO_3 and S_2 pressures which are exerted by coexisting SO_2 and O_2 partial pressures can also be shown on the same isothermal diagram since

$$\log p\ SO_3 = \log p\ SO_2 + 1/2 \log p\ O_2 - \frac{\Delta G^\circ (SO_3)}{4.575T} \tag{1}$$

and

$$1/2 \log p\ S_2 = \log p\ SO_2 - \log p\ O_2 = \frac{\Delta G(SO_2)}{4.575T} \tag{2}$$

where $\Delta G(SO_3)$ and $\Delta G(SO_2)$ are the standard free energy changes for the reaction $SO_2 + 1/2\ O_2 \rightarrow SO_3$ and $1/2\ S_2 + O_2 \rightarrow SO_2$, respectively.

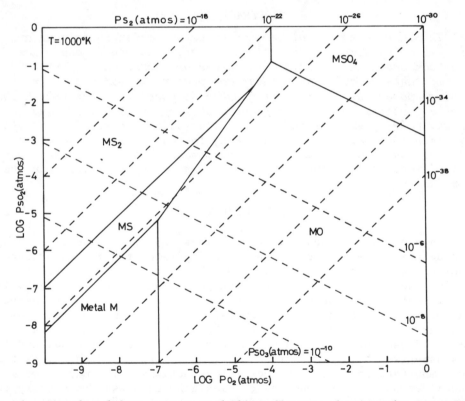

Fig. 1.   Metal-sulphur-oxygen stability diagram showing the range of gas compositions which are in equilibrium with the condensed phases.

Figure 1 shows a schematic diagram for a hypothetical metal-sulphur-oxygen system, which expresses all of these ideas. It can readily be concluded that the representation of the data for more than one system on this single diagram would leave the figure over-filled with lines. Furthermore, if it is required to show results for other temperatures, then either separate diagrams for each temperature must be given, or a three-dimensional figure must be employed.

It is for this reason that the data to be presented in this paper will be given in a novel form which employs the chemical potentials RT ln p, rather than the logarithms of pressures, as ordinates. The following thermodynamic formulae should be applied for understanding the structure of these diagrams. In the binary A-B system where X_A is the atom fraction of component A of chemical potential $\Delta\mu_A$, and X_B, $\Delta\mu B$ are the corresponding values for B, the integral free energy of mixing A and B in these proportions is given per g atom of mixture by

$$\Delta G^M = X_A \Delta\mu_A + X_B\Delta\mu_B. \tag{3}$$

For a gaseous species

$$\Delta\mu_A = RT \ln a_A = RT \ln p_A = 4.575T \log_{10} p_A \tag{4}$$

where the standard state is for the gas at one atmosphere pressure. For the gaseous compound SO_n in the reaction $1/2\ S_2 + n/2\ O_2 \rightarrow SO_n$, the standard free energy of formation, the equilibrium constant, and the chemical potentials of each component and the compound are related thus

$$\Delta G°(SO_n) = -RT \ln K(SO_n) = 1/2\ \Delta\mu(S_2) + n/2\ \Delta\mu(O_2) - \Delta\mu(SO_n). \tag{5}$$

Any pair of chemical potentials on the diagram showing ΔG^M as a function of the atom fractions of sulphur and oxygen are joined by a line which passes the point on the diagram for $1/n+1\ \Delta G°(SO_n)$ by the amount $1/n+1\ \Delta\mu(SO_n)$. Thus when the line joining any two sulphur and oxygen potentials passes through the value of $1/n+1\ \Delta G°(SO_n)$, these potentials are in equilibrium with one atmosphere pressure of SO_n. Note that sulphur and oxygen are represented in the pure state by S_2 and O_2 respectively since these are the commonest gaseous species of these elements in the temperature range which is of immediate interest here.

The sulphur and oxygen potentials for a metal-metal sulphide or metal-metal oxide can be shown by marks at the appropriate ordinates.

In the reaction $MO + 1/2\ S_2 \rightarrow MS + 1/2\ O_2$, sulphidation has the equilibrium constant

$$K = \frac{a_{MS}}{a_{MO}} \cdot \frac{p^{1/2}O_2}{p^{1/2}S_2}$$

It follows that the equilibrium ratio of the partial pressure $p^{1/2}O_2/p^{1/2}S_2$ for the conditions where the oxide and sulphide have unit activity can be obtained from the difference between the corresponding oxygen and sulphur potentials for these M–MO and M–MS equilibrium marks.

The SO_3 dissociation chemical potential, $RT\ \ln p\ SO_3$, of the metal oxide–metal sulphate equilibrium is shown as a point below that for $1/4\ \Delta G°\ (SO_3)$ and is depressed by the amount $1/4\ RT\ \ln p\ SO_3$. The SO_2/O_2 ratio for the given SO_3 partial pressure is obtained by joining this point with that for the selected pO_2 on the $\Delta\mu^{1/2}\ (O_2)$ scale, projecting a line from the corresponding point vertically below $1/3\ \Delta G°(SO_2)$. The length of the projection is then $1/3\ \Delta\mu(SO_2)$. It is clearly possible to make a complete analysis of the metal-sulphur oxygen system at a given temperature. Since the representation of one system only adds markers at the M–MS and M–MO sulphur and oxygen potentials respectively, it is a simple matter to add data for other systems to one diagram.

The Ellingham diagrams for standard free energy of formation of metal oxides[2] and sulphides[3] show a linear relationship between $\Delta G°$ and temperature with changes of slope at metal or metal compound transition temperatures. The data for a system may therefore be represented on one chemical potential diagram over a range of temperatures by putting upper and lower temperature marks; the information for any intervening temperature between the upper and lower limits may then be obtained by linear interpolation. The data for the gaseous sulphur oxides are shown through two curves, one for each temperature limit.

THE SULPHATION THERMODYNAMICS OF LIME AND MAGNESIA

As a first, and most simple, example of the application of the chemical potential diagram to ceramic oxide systems, we will consider the reactions of calcium and magnesium oxides to form sulphides and sulphates. Figure 2 shows the sulphur potentials of Mg/MgS and Ca/CaS at 1273 and 1773°K and the oxygen potentials of Mg/MgO and Ca/CaO at the same temperatures. The data which were used for the construction of the diagram were all taken from the compilation by Kubaschewski et al.[4] It will be noticed that the difference between

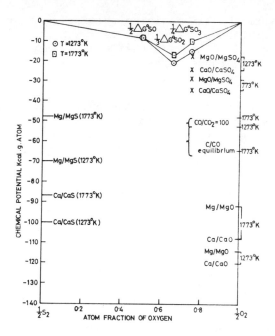

Fig. 2. Chemical potential diagram for the reactions of magnesia and
        lime with gaseous sulphur-oxygen systems.

the sulphur and oxygen potentials for the magnesium compounds is
much greater than that between the calcium compounds. This means
that at any given oxygen pressure CaO is converted to CaS at a lower
sulphur pressure than MgO is converted to MgS.

    The stability of CaO with respect to sulphide formation may be
enhanced by lowering the activity of the oxide by combination with
another oxide. Thus, in a reaction yielding calcium metasilicate,
$CaO + SiO_2 \rightarrow CaSiO_3$, the chemical potential of lime is reduced by
about 20 kcal mole^{-1}

$$\Delta G° \sim \Delta H° = -21.5 \text{ kcal.} \qquad (7)$$

The entropy change of this reaction will be practically zero since
it involves only solid phases. If $a(CaSiO_3)$ = $a(SiO_2)$ = 1 then
$\Delta G°(CaSiO_3)$ = $\Delta\mu(CaO)$; hence in this combination lime is as resistant
to sulphide formation as is pure magnesia.

    On the oxygen ordinate, the oxygen potentials are shown for an
atmosphere which is very reducing, due to the presence of a substantial
partial pressure of CO. Even under these very reducing circumstances,
the sulphur pressure which is required to form MgS is very high
$(\Delta\mu(S_2) \sim -18$ kcal, $\log p$ S_2 = -2.5 at 1500°K). By comparison of the

position of the straight line joining this sulphur potential to the
C-CO equilibrium or $CO/CO_2 = 100$ markers, with the positions for
$1/2 \Delta G°(SO)$, $1/3 \Delta G°(SO_2)$ and $1/4 \Delta G°(SO_3)$, it may be readily
concluded that none of these gases would be present to any significant
extent under these circumstances.

The data for the sulphation of the oxides[5] show that the SO_3
dissociation pressures are quite large and will only be surpassed
in a gas having a high oxygen potential. Sulphation is unlikely,
then, under most industrial conditions with temperatures in excess
of 1000°K, and is made even less probable by a reduction of the
activity of the metal oxide by combination with another oxide such
as silica. The markers on the diagram are for $1/4 \Delta G°(MgSO_4)$ and
$1/4 \Delta G°(CaSO_4)$ at 773 and 1273°K, and these must be raised as the
activities of lime and magnesia are lowered, e.g., in the reaction
$MgO + SO_3 \rightarrow MgSO_4$

$$\Delta G°(MgSO_4) = \Delta \mu(SO_3) + \Delta \mu(MgO) \tag{8}$$

when $a(MgSO_4) = 1$.

REACTIONS INVOLVING ALUMINA AND SILICA

The reactions which are possible between sulphating atmospheres
and alumina and silica are somewhat more limited than in the case
of the Group II oxides. Silica does not form a stable sulphate by
combination with SO_3 and does not react with sulphur to form a
condensed sulphide except under very high sulphur pressures.
Alumina does form a stable sulphate[6] and the heat of formation of
the sulphide is known[4]. The entropy of formation of Al_2S_3 from
aluminium metal and diatomic sulphur is assumed to be the same as
that for the formation of the oxide from metal and oxygen.

Both aluminium and silicon form volatile sulphides AlS and
SiS[7], and although complete data are not available for either
species, these published may be used to calculate approximate,
two-term free energy equations. Those which were used for the
calculations in this paper were:

$Al + 1/2 S_2 \rightarrow AlS(g)$ $\Delta G° = 32,500 - 21.2T \text{cal mole}^{-1}$ \hfill (9)

$Si + 1/2 S_2 \rightarrow SiS(g)$ $\Delta G° = 11,400 - 21.8T \text{cal mole}^{-1}$. \hfill (10)

Since the free energies of fusion of the elements are relatively
small in comparison with the uncertainties of these equations, they
have been ignored in the temperature range 298-1773°K. As might be
expected, the entropy changes are very similar for these formation
reactions, and it can be seen that the silicon compound is always
about 20 kcal mole^{-1} more stable than the aluminium monosulphide
(Fig. 3).

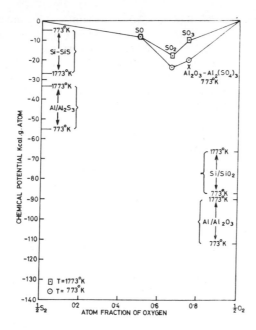

Fig. 3.   Chemical potential diagram for the reactions involving
silica and alumina.

The very large difference in free energies of formation between
Al_2S_3 (s) and Al_2O_3 (s) clearly show that the formation of the solid
sulphide on the oxide would be expected only under extremely reducing
circumstances, and in the presence of high sulphur pressure.   This
is similar to the conclusion which was reached concerning MgO.   The
same information for the C-CO equilibrium can be applied to show
that the sulphur pressure for the formation of Al_2S_3 must be approximately one atmosphere under these very reducing conditions.   On the
contrary, the stability of $Al_2(SO_4)_3$ makes Al_2O_3 quite susceptible
to sulphate formation when the gas phase is oxidizing at temperatures
below 1000°K.

The novel aspect of attack on oxides by sulphating atmospheres
which must now be considered is the formation of the volatile
sulphide SiS.   Due to the sign of the entropy change for the formation
of the gas, the tendency to form this gas is greater the higher the
temperature whilst the stability of the solid oxide SiO_2 is decreased
by increasing temperature.   The standard state which was chosen for
SiS(g) was the normal one atmosphere pressure and the markers on the
diagram are for this standard condition.   Obviously, it is not
necessary for the gas to be formed at one atmosphere for substantial
attack of SiO_2 to take place, and consideration must be given of the
fact that a continuous erosion of silica would occur in a gaseous
environment containing quite a small sulphur pressure.   As an example,

if the partial pressure of SiS is 10^{-3} atmos., the corresponding
marker on the $\Delta\mu 1/2(S_2)$ ordinate for Si-SiS (10^{-3} atmos.) for
1773°K is at -39 kcal. This is 12 kcal below Si-SiS (1 atmos.)
at this temperature. The erosion rate can be reduced by increasing
the oxygen pressure since the formation of the gaseous sulphide must
accord with the equilibrium constant for the formation reaction
$SiO_2(s) + 1/2\ S_2 \rightarrow SiS(g) + O_2$

$$K = p\ SiS \cdot \frac{p\ O_2}{p^{1/2}\ S_2} \quad (a_{SiO_2} = 1). \tag{11}$$

If a choice must be made between silica- and alumina-based
refractories, then clearly alumina is to be preferred under reducing
atmospheres because of the instability of AlS(g). On the other hand,
SiO_2 is to be preferred when the atmosphere is oxidizing, because of
the absence of a stable sulphate and the suppression of the volatile
sulphide formation which is effected by a high oxygen partial pressure.

GENERAL TRENDS IN SUSCEPTIBILITY TO SULPHATION CORROSION

The results which have been presented above were based on
calculations using experimental data. There are, however, a number
of ceramic oxides for which such studies have not been made and
it may be useful to attempt to predict what might be expected for
sulphation thermodynamics by means of an empirical correlation.

Probes which are used for measuring oxygen potentials in
furnaces[8] make use of solid oxide electrolytes containing CaO or
Y_2O_3 mixed with ZrO_2 or ThO_2. How would these probes react to a
furnace atmosphere which contains a small partial pressure of
sulphur compounds? Clearly sulphate formation will be relatively
unimportant if the temperature is in excess of 1000°C, but sulphide
formation is certainly possible when CaO is present. There are no data
for yttrium and zirconium sulphides and so a prediction must be made.

The reaction $MO(s) + 1/2\ S_2(g) \rightarrow MS(s) + 1/2\ O_2(g)$ will have
virtually zero entropy change, because the number of gaseous molecules
is the same on both sides of the equation. The free energy change
is thus practically the same as the heat change for the reaction.
As has been demonstrated with CaO and MgO, the ratio
$\Delta H°(sulphide)/\Delta H°(oxide)$ changes from calcium to magnesium, and it
is useful to inspect the effect of the cation radius on the value of
this ratio (Fig. 4). The results show that as the cation radius
increases the relative stability of sulphide to oxide increases and
thus the possibilities of corrosion by sulphide formation increase
in the same way. It would seem, then, that from the point of view
of resistance to sulphur corrosion the zirconia-based electrolytes are
preferable to the thoria-based ones but that the additions of CaO or
Y_2O_3 to enhance the oxygen ion conductivity of the electrolyte
material are equally hazardous in this respect.

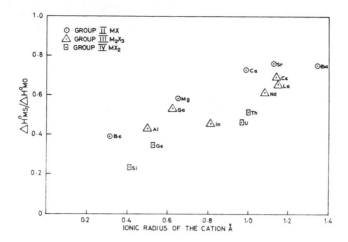

Fig. 4. The correlation between the ratio $\Delta H°$(sulphide)/$\Delta H°$(oxide) and the radius of the metal cation.

Such a correlation between the heat of formation ratio and the cation radius is to be expected when ionic bonding predominates in the compounds. The replacement of oxide ion by sulphide ion will have a greater proportional effect on increasing the internuclear distance, and hence of decreasing the lattice energy, when the cation is small compared to the anion. When the compounds of a large cation are compared, the ratio of $\Delta H°$(sulphide)/$\Delta H°$(oxide) approaches unity. In fact, for the caesium ion (r^+ = 1.69 Å) this ratio achieves the value 1.06.

For the elements of Group IV and higher, the ratio of $\Delta H°$(sulphide)/$\Delta H°$(oxide) for the same valence state is not a very good guide. These elements form lower sulphides even though lower oxides are unstable, e.g., US and UO_2. The tendency to form the lower sulphides will be greater than is indicated in Fig. 4 for these elements. Also consideration must be given to oxysulphides, e.g., UOS.

REFERENCES

1. H. H. Kellogg and S. K. Basu, Trans. AIME 218, 70 (1960).
2. F. D. Richardson and J. H. E. Jeffes, J. Iron & Steel Inst. 160, 261 (1948).
3. F. D. Richardson and J. H. E. Jeffes, J. Iron & Steel Inst. 171, 165 (1952).

4. O. Kubaschewski, E.Ll. Evans and C. B. Alcock, Metallurgical
   Thermochemistry - 4th edn., Pergamon Press, London (1967).
5. E. W. Dewing and F. D. Richardson, Trans. Far. Soc. 55, 611
   (1959).
6. T. R. Ingraham, Trans. AIME 242, 1299 (1968).
7. D. D. Wagman et al., National Bureau of Standards Technical
   Note 270-3, Washington, D. C., (1968).
8. C. B. Alcock and B. C. H. Steele, Science of Ceramics, Vol. 2,
   p. 397 Academic Press, London, (1965).

PROPERTIES OF MOLTEN CERAMICS

J. L. Bates, C. E. McNeilly and J. J. Rasmussen

Battelle Memorial Institute, Pacific Northwest Labs

Richland, Washington

ABSTRACT

The density, volume change on melting, melting point, phase transitions, surface tension, viscosity, compressibility, and thermal diffusivity of molten ceramics have been measured up to 3000°C. The experimental techniques used to make these measurements are described. The properties for a number of oxide systems, e.g., Al_2O_3, Sm_2O_3, UO_2, Cr_2O_3, and basalt, determined as a function of temperature are reviewed. The influences of decomposition and stoichiometry changes, vaporization loss, incompatibility with and choices of containment materials, at these extreme temperatures are considered.

INTRODUCTION

Temperatures which approach or exceed the melting point impose severe conditions on ceramics: kinetic effects involved with compatibility and phase equilibria are enhanced, and experimental problems are compounded. For such reasons, few property measurements have been made of the behavior of liquid refractory ceramics. However, needs for such information now exist in practical applications and in basic understanding of the liquid state of molten ceramics.

Precise measurements of liquid properties are the basis for understanding of the structure of the high temperature liquids. Properties such as expansion, viscosity, compressibility, and thermal diffusivity must depend upon structural parameters. Measurements of thermal expansion, density, surface tension, and compressibility relate to the thermodynamic properties of liquid

11

ceramic materials. Many practical applications for these data are
evident in the technology of ceramics and ceramic based composites,
particularly in crystal growth, fusion casting, and other liquid
phase processing steps.

Except for glass formers, such as SiO_2 and B_2O_3, reliable
data for crystalline ceramics in the liquid state is almost nonexistent. In contrast, a large amount of data is available for
nonoxide ionic salts which melt at relatively low temperatures.[1,2]

This paper is concerned with techniques which have been
developed for measuring these properties of molten ceramic materials
to 3000°C. Many of them are extensions of general methods[3] used
to measure liquids at lower temperatures, with modifications to
reflect the severe conditions imposed by the higher temperatures.

EXPERIMENTAL TECHNIQUES

High Temperature Radiography

The expansivity, density, and volume change on melting of a
ceramic are determined by observing the volume of the encapsulated
material, and surface tension is determined from the shape of a
meniscus or pendant drop. To permit observation through the sealed
refractory metal capsule, X-radiography is employed.

The radiographic system is illustrated schematically in Fig. 1,
employing a 300 kV X-ray source. An X-ray image intensifier and
vidicon camera monitors the sample continuously, and film may be
inserted for a permanent record. A tungsten-mesh resistance
heating element surrounded by split radiation shields allows viewing
under effectively isothermal conditions with a minimum of absorbing
material in the image-forming X-ray beam. The furnace can be
operated with inert, reducing, or vacuum environments.

The large density change on melting for a material such as
Al_2O_3 allows direct radiographic observation of the volume change
on melting, meniscus, and expansion of the liquid (Fig. 2). The
liquid-solid interface of a partially melted sample may be observed
in a capsule which is heated with the midpoint approximately 100°C
hotter than the ends. Melting starts inward from the wall at
the midpoint. The liquid-solid interface is clearly visible, as are
two bubbles which formed in the liquid. Precision film densitometry
could possibly be used to determine the relative density from these
nonisothermal radiographs.

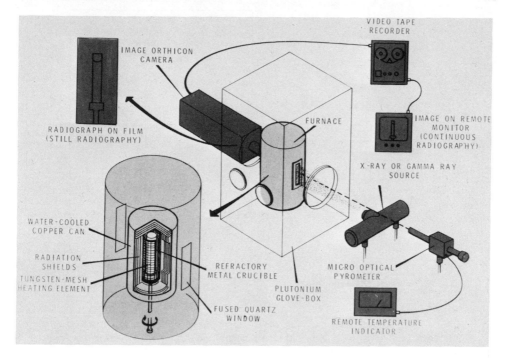

Fig. 1.   Facility for radiographic measurement of liquid density,
          volume change, and surface tension.

Density. The density of the molten ceramic material is calculated from the known weight of the sample and its volume as a
function of temperature. The volume is calculated by measuring the
height of the liquid column from the radiographs and using the
inside diameter of the refractory metal crucible after correcting
for thermal expansion. Corrections also are made for the volume
of the material contained in the meniscus.

Volume Change On Melting. The volume change on melting is
determined by (1) extrapolating the measured volume of the solid
as a function of temperature to the melting point and (2) extrapolating the volume of the liquid as a function of temperature to
the melting point. The difference is the volume change on melting.
Results of measurements on the thermal expansion and volume change
on melting of UO_2 are illustrated in Fig. 2.

Radiographic measurements of volume change on melting and
liquid expansivity of single crystal sapphire were made in
molybdenum and tungsten capsules. In each experiment, hysteresis
in the liquid volume was observed as the specimen was heated and
cooled above the melting point. In one case in which the tungsten
capsule was heated in flowing argon to 2800°C, the volume change on

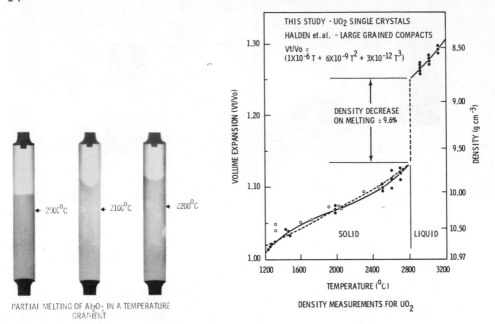

PARTIAL MELTING OF Al₂O₃ IN A TEMPERATURE GRADIENT

DENSITY MEASUREMENTS FOR UO₂

Fig. 2.   Density changes associated with melting of ceramics.
           Left - Radiographs of partially melted Al_2O_3;
           Right - Density-temperature discontinuity associated
           with melting of UO_2.

solidification was approximately 24% compared to 33% on melting and
the expansivity on cooling was correspondingly greater than the
expansivity on heating. These effects were due to the presence
of bubbles in the initial melt. The decrease in specific volume
with time and temperature above the melting point indicates that
the gas can be removed from the liquid. A sufficiently long
refining period results in apparent specific volume values
approaching the true value.

    Surface Tension. Surface tension may be measured by either of
two techniques: (1) pendant drop shape, and (2) equilibrium meniscus
shape. Pendant drops are formed on the end of solid oxide rods
which are suspended in the capsule such that the free end of the
rod coincides with the maximum temperature in the gradient developed
along the capsule. An appropriate temperature is established by
resistively heating a capsule machined with a variable wall
thickness to provide a resistance (and temperature) gradient. In
this manner, only the end of the rod is melted and a stable pendant
drop is formed. Several drops can be formed from each rod by
gradually increasing the temperature. As more material becomes
molten, a drop increases in size, becomes unstable, falls, and a
new drop is formed. The shape of the meniscus which forms in the
refractory metal capsule when the oxide becomes molten can also be
used to obtain the surface tension.

Impurities which are not surface active have a negligible effect in small concentrations; however, a surface active impurity species can markedly change the surface tension value. The pendant drop technique is useful in that the molten drop is in contact only with a solid of the same nominal composition. This removes the possibility of the surface tension being affected by the presence of impurities picked up from a supporting material. Measurement is limited, however, to the value at the melting point. In contrast, the capillary meniscus method yields surface tension values as a function of temperature but these values may be affected by contamination from the crucible material.

From the drop coordinates and the density of the liquid,[4] the surface tension can be calculated utilizing the technique developed by Bashforth and Adams,[5] analyzing the shape of the drop or the meniscus using a computer program. Results of the surface tension and density of molten Al_2O_3, $MgAl_2O_4$, Sm_2O_3, and UO_2 are shown in Table 1. These data for Al_2O_3 agree well with measurements which have been reported utilizing other techniques.[6-11]

TABLE 1.   Surface Tension and Density of Molten Ceramics

Material	σ dynes/cm		ρ g/cm^3
Al_2O_3	574 ± 68	pendant drop	$5.632 - 1.127 \times 10^{-3}$ T
	360 ± 40	meniscus, Mo capsule	for T > 2333°K
	638 ± 100	meniscus, W capsule	
$MgAl_2O_4$	481 ± 102	meniscus	$5.324 - 9.865 \times 10^{-4}$ T
	490 ± 98	pendant drop	for T > 2403°K
Sm_2O_3	962 ± 53	pendant drop	5.9
UO_2	441 ± 35	meniscus	8.7

Viscosity

The kinematic or shear viscosity of molten ceramics can be measured with an oscillating cup viscometer[3,12] (Fig. 3), by determining the time rate of decrease in the amplitude of a free-oscillating, closed capsule containing the liquid, suspended from a torsion wire. The amplitude decrease (or damping) results from both the viscous motion of the liquid and the internal friction of the torsion wire. The wire is chosen such that its damping is small compared to the damping of the system.

Amplitude changes can be measured by a variety of electronic, mechanical, or optical methods. A convenient method is to view the oscillating scale with a television system and replay a video-tape record of the graduated scale in stop action mode. The total period

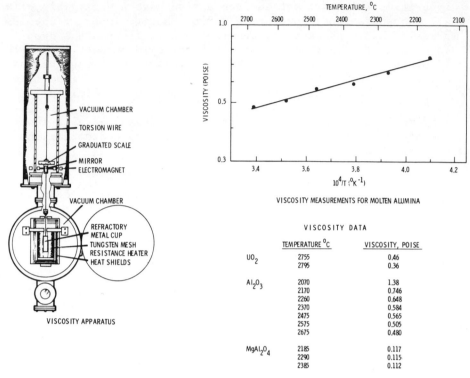

VISCOSITY MEASUREMENTS FOR MOLTEN ALUMINA

VISCOSITY DATA

	TEMPERATURE °C	VISCOSITY, POISE
UO$_2$	2755	0.46
	2795	0.36
Al$_2$O$_3$	2070	1.38
	2170	0.746
	2260	0.648
	2370	0.584
	2475	0.565
	2575	0.505
	2675	0.480
MgAl$_2$O$_4$	2185	0.117
	2290	0.115
	2385	0.112

Fig. 3.   Viscosity apparatus and data for Al$_2$O$_3$, UO$_2$, and MgAl$_2$O$_4$.

of a number of successive oscillations is measured with a stop watch.
Viscosity values are calculated from the measured damping and period,
the known values of the moments of inertia of the sample and
pendulum, the liquid density and the crucible geometry.

For high temperatures, the sample is contained in a vacuum
welded, refractory metal capsule and suspended from a W or W-Re
torsion wire inside a W mesh resistance heated, vacuum furnace.
The system is started oscillating to a predetermined amplitude and
period using an electromagnet and pendulum which is placed above and
outside the hot zone. A mirror and graduated scale attached to the
pendulum is used to measure the damping. The magnetic cocking
system must be precisely balanced to prevent any lateral swing of
the capsule. Both the capsule and the torsion system must be
operated *in vacuo*.

Temperatures are measured with (1) an optical pyrometer sighted
on the rotating capsule and (2) a W-5% Re/W-26% Re thermocouple
positioned near the bottom of the rotating capsule. At these levels
both temperature measurements are difficult; however, if the melting
point of the sample is known, an internal check of temperature is
available.

The volume of the liquid ceramic must not completely fill the capsule nor should any of the meniscus reach the top of the capsule. Consideration is also given to the viscous drag on the bottom. It is essential that the capsule dimensions and sample height be optimally chosen. A very low viscosity requires a larger sample diameter to reduce the damping rate. Thus, a great deal of trial and error is involved in choosing the proper crucible size, the sample height, and type and size of torsion wire.

The viscosity has been measured for a number of oxides including Al_2O_3, UO_2, and $MgAl_2O_4$. The results are shown in Fig. 3. For Al_2O_3, the viscosity decreases linearly from 0.75 poise above the melting point to 0.5 poise at 2700°C. The viscosity of UO_2 was measured just above the melting point. Assuming a liquid UO_2 density of 8.7 g/cm^3, the calculated values at 2755°C and 2795°C were 0.46 and 0.36 poise respectively. The observed melting temperature is 50-100°C less than other reported values, which illustrates the difficulty in measuring absolute temperatures. Other investigators report values from 0.5 to 10 poise.[13,14] The viscosity of $MgAl_2O_4$ was significantly less than the viscosity of Al_2O_3 and exhibited a very small temperature linear dependence from 2185 to 2385°C.

Adiabatic Compressibility

The adiabatic compressibility of molten ceramics is determined at high temperatures by measuring the ultrasonic velocity of longitudinal waves in the liquid.[15] This interference technique, employed extensively for low melting ionic salts,[2] has been modified to measure molten refractory oxides. A rod transmits an ultrasonic pulse from the cold transducer to the melt;[16] the quartz transducer acts both as the transmitter and receiver. The 18 MHz longitudinal waves propagate along a 40 cm long, 1.9 cm dia. tungsten single crystal rod to and from the molten oxide contained in a poly-crystalline tungsten crucible vacuum welded to the end of the crystal, Fig. 4. The crucible cavity is machined with parallel surfaces and contains an expansion plenum. The sample, generally a single crystal, is machined so that at a temperature just below the melting point the crucible is nearly filled. Upon melting, the specific volume increases, filling the cavity and covering the reflecting surfaces. The excess liquid flows into the expansion chamber above and outside the crucible, and does not interfere with the propagation of sound waves. Scattering slots at the top prevent reflection of unwanted echoes. The sealed end containing the sample is heated *in vacuo* in a tungsten mesh resistance furnace. The transmission rod passes through O-ring seals and is externally water cooled. Temperatures are measured with an optical pyrometer sighted in a black-body hole drilled in the side of the crucible. In addition, a W-5% Re/W-26% Re thermocouple is centered directly

above the crucible. The system is calibrated using the observed
melting point of Al₂O₃ (2050°C). The thermocouple is consistently
lower than the melting and pyrometer temperature but is used
conveniently to control the furnace temperature.

Upon melting, the sound waves propagate along the W crystal to
the molten sample, pass through the liquid and reflect from both
surfaces of the tungsten crucible in contact with the molten oxide.
Phase comparisons of the transmitted and reflected waves are
measured, from which sound velocities in the liquid can be
calculated.

The adiabatic compressibility (β) of the liquid is calculated
from the relation

$$\beta = (\rho v^2)^{-1} \tag{1}$$

where ρ is the density and v is the longitudinal sound velocity.
The results for Al₂O₃ are plotted as a function of temperature in
Fig. 5.

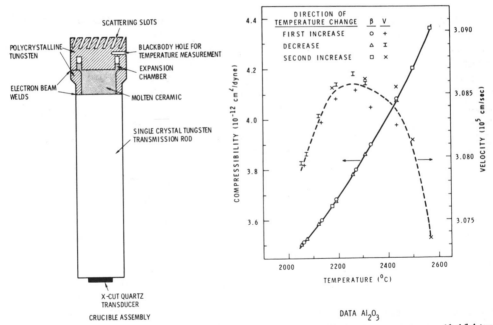

Fig. 4.   Ultrasonic velocity
          measuring apparatus.

Fig. 5.   Adiabatic compressibility
          and ultrasonic velocity for
          molten Al₂O₃.

Thermal Diffusivity

The laser-pulse thermal diffusivity technique, generally used to measure the diffusivity of solids, was modified to measure heat conduction of molten materials. The pulse method [17,18] subjects one surface of a thin sample disc to a short heat pulse from a laser. The pulse passes through the sample and the temperature transient on the opposite surface of the sample is measured (Fig. 6). The temperature change is recorded and the thermal diffusivity is determined from the shape of the temperature-time curve.

To obtain valid results with such a technique, (1) the energy pulse must be uniformly absorbed over the front surface and at the surface only; (2) the laser pulse time must be short compared to the time for the heat pulse to pass through the sample; (3) heat flow must be normal to the heated surface; and (4) no heat must be lost from the sample during the measurements. If these conditions are met, the thermal diffusivity (α) can be determined from the sample thickness and time-temperature curve.

For high temperature measurements, the sample is heated in a tungsten mesh resistance furnace in vacuum, inert, or reducing atmospheres. The specimen, approximately 1.0 cm in diameter and 0.05 to 0.15 cm thick, is supported in a ceramic holder which is positioned in the hot zone of the cavity on a 1.4 cm dia. tungsten tube. Vertical arrangement minimizes loss of essential optical alignment.

The temperature transient on the back surface of the sample is measured with an optical detector such as cooled (liq. N_2) indium antimonide which is focused on the sample surface. The output is amplified, displayed on an oscilloscope and photographed. Temperatures are measured and the furnace controlled with a self supported bare-wire W-5% Re/W-26% Re thermocouple set in a hole in the sample holder. An optical pyrometer is also used to measure the temperature at the back surface.

Such measurements are generally made on small solid discs. To measure liquids, a sample holder is required to contain the liquid in the disc shape (Fig. 6). Two methods have been used, the choice of which depends upon the opacity of the sample, reactions with the sample crucible, the temperatures required, and the properties of the liquid at high temperatures.

Single crystal Al_2O_3 can be used to contain opaque materials. Since sapphire is transparent to the laser beam and infrared detector, even at high temperatures, an opaque sample could be heated directly with the laser (Fig. 6). The crucible is designed with polished surfaces on top and bottom and has expansion plenums to accomodate the increase in sample volume and for gas release on melting. Thin

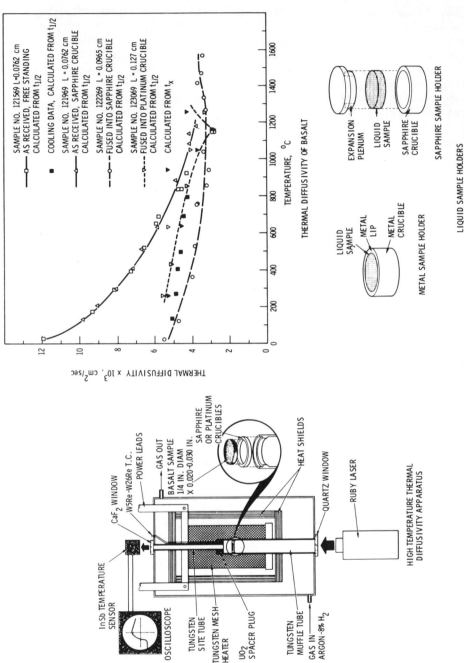

Fig. 6. Thermal diffusivity apparatus, sample holders, and results for Dresser basalt.

melt crucibles of platinum or refractory metals can also be used
(Fig. 6). However, a good thermal contact between the metal and
the sample is necessary. The conductivity of the container must
be large compared to the sample. Conversely, the crucible wall
must be thin to prevent thermal shunting.

At temperatures above the melting point, the sample can wet
the crucible, and may flow over the top edge. This problem can
be remedied by placing a small lip on the rim of the crucible.
However, this does increase the chance of error due to thermal
shunting.

This technique has been used to measure the thermal diffusivity
of basalt. Measurements were made of Dresser and Columbia River
basalt from room temperature to 1600°C; results for Dresser basalt
are illustrated in Fig. 6. The thermal diffusivity of the as-
received rock decreases with increasing temperature similar to that
of oxide insulators. The results for the Dresser basalt were not
reversible, with the diffusivity being significantly lower on cooling
and during subsequent measurements, reflecting a permanent change
from a relatively crystalline basalt to a more vitreous body after
melting. Both basalts show a sudden decrease near the melting
temperature, followed by a slow increase at higher temperatures.

Data for fused basalt using the platinum crucible without the
lip are in good agreement with the measurements made using the
sapphire crucible. However, the addition of a large lip, necessary
to prevent surface movement of the liquid, resulted in apparent
higher values. At temperatures above 1150°C, transparency of the
basalt to the laser and infrared detector again became a problem.
This may have accounted for the apparent sudden increase in
diffusivity above the melting point. Thicker samples appeared to
minimize this problem.

Although this technique appears to provide an easy and rapid
method for measuring the thermal diffusivity of liquids, there are
problems which must be resolved, most of which relate to
(1) maintenance of the thermal boundary restrictions, (2) reactions
between the molten sample and the crucible and (3) unusual
properties of the liquid, e.g., surface tension.

Melting Points and Phase Equilibria

A high temperature differential thermal analysis apparatus was
used to measure melting points and phase equilibria to 3000°C. The
equipment is patterned after that of Rupert.[22] Although essentially
the same as conventional DTA apparatus, unusual features make it
particularly suited for use at these high temperatures (Fig. 7). A
most unusual aspect is the use of reverse-biased, silicon photodiodes

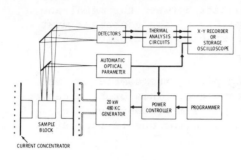

Fig. 7. Block diagram of high
temperature DTA apparatus.

Fig. 8. Sample block, lid, and
heat shield for high
temperature

for temperature sensors. The output is amplified in the thermal
analysis circuit and can be displayed in a number of ways. The
temperature of the sample and/or standard, the temperature derivative, or temperature difference can be plotted against temperature,
time, or derivative of temperature on a recorder or oscilloscope.
Conventional DTA plots can be obtained and ΔT's of a few tenths of
a degree are measured at 2000°C.

The graphite or refractory metal block (1.9 cm dia. and 1.9 cm
high) is heated inductively using a current concentrator (Fig. 8).
Samples ($\sim$0.3 g) are placed in 0.64 cm dia. by 0.64 cm deep cavities.
A lid, with 0.150 cm dia. view holes for the detectors, covers the
block. The block and concentrator are inside a fused quartz
container in vacuum or inert or reducing gas. Greater sensitivity
is obtained *in vacuo*. The choice of a block material depends upon
its reactions with the sample and reference. The ambient temperature is measured and controlled with an automatic pyrometer in
conjunction with a programmer and proportional controller. The
pyrometer is calibrated in place against the melting points of
Au, Ni, Pt, Al_2O_3, Mo, and Ir.

The melting points and phase equilibria have been measured for
some actinide (Cm_2O_3) and lanthanide (Pm_2O_3, Sm_2O_3) oxides.
Reversible, reproducible reactions have been measured. For example,
for Cm_2O_3 the following reactions were observed:

$$1615 \pm 20°C \qquad B \rightleftarrows A$$
$$2000 \pm 20°C \qquad A \rightleftarrows H$$
$$2210 \pm 20°C \qquad H \rightleftarrows X$$
$$2275 \pm 25°C \qquad \text{Melting}$$

Similar phase transformations were obtained for Pm_2O_3 with a melting
point of 2320 $\pm$ 25°C.

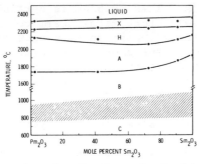

Fig. 9.   DTA results for Pm_2O_3 - Sm_2O_3 binary system.

The Pm_2O_3 - Sm_2O_3 binary system was also investigated (Fig. 9).
The lines separating the different regions should be two-phase
regions; the lower and upper boundaries (<30°C) could not be
distinguished.

EXPERIMENTAL PROBLEMS

There are a number of problems associated with the measurement
and evaluation of high temperature liquid properties of ceramic
materials. These problems often require the use of novel and unusual
methods. Perhaps the most difficult is the incompatibility of the
sample with its container. The elevated temperatures increase the
rates of reactions and enhance the formation of new phases. In most
instances, the choice of container material is narrow and often
restricts the choice of experimental conditions, e.g., refractory
metals require low oxygen pressures. Related to this problem is the
difficulty in capsule fabrication. Closure is complicated by poor
quality refractory metals and machining and welding techniques.

Vapor loss and sample decomposition always must be considered
at these temperatures; loss of sample and compositional change may
alter properties and/or tests. However, appropriate data for liquid
ceramics are very limited. It has been noted that both single
crystal and polycrystalline oxides release significant amounts of
absorbed gas on melting. Bubbles can form in the sample, complicating
property measurements; eventually the released gas, coupled with a
potential high sample vapor pressure, can cause excessive swelling
of the sealed metal container. If a vacuum-sealed capsule is heated
in gas atmosphere, permeation of the gas through the container can
occur. Choice of cover gas can minimize this problem.

Temperature measurement and calibration become increasingly
more difficult as the temperature increases. It is impractical or
impossible to measure the temperature of the liquid directly. Use
of thermocouples is complicated by electrical conductivity in
insulators, diffusion in thermocouples, reactions with the system and

lack of calibrated emf's. This problem is partially solved by using self-supporting, bare refractory metal thermocouples in the hot zone. Optical pyrometer and thermocouple measurements are generally made simultaneously as a consistent check. Temperature calibration is difficult, yet the system can be self-calibrating for samples with known melting points, since melting is easily observed. References and standards at such high temperatures are nonexistent. Methods for evaluating data at lower temperatures often do not apply; suitable high temperature evaluations have yet to be developed.

All these severe conditions impose limitations which reduce the accuracy and precision of the data. However, some data is better than no data at all, and one hopes that new and improved techniques can be developed which will improve the knowledge of the liquid state of refractory ceramic materials.

REFERENCES

1. G. J. Janz, F. W. Dampier, G. R. Lakshminarauanan, P. K. Lorenz, and R. P. T. Tomkins, Molten Salts: Volume I, Electrical Conductance, Density and Viscosity Data, NSRDS-Nat. Bur. Stand. 15, (1968).
2. G. J. Janz, Molten Salts Handbook. Academic Press, New York, 1967.
3. J. O'M Bockris, J. L. White, and J. D. Mackenzie, Physicochemical Measurements at High Temperatures. Butterworths Scientific Publications, London, 1959.
4. J. N. Butler and B. H. Bloom, Surface Sci. 4 [1] (1966).
5. S. Bashforth and J. C. Adams, An Attempt to Test the Theory of Capillary Action. Cambridge University Press, London, 1883.
6. W. D. Kingery, J. Am. Ceram. Soc. 42 [1] 6-10 (1959).
7. H. V. Wartenberg, G. Wehner, and E. Saran, Nachr. Ges. Wiss. Gottingen, Math-Physik, Klasse, (Fachgr. II) 2 65-71 (1936).
8. R. W. Bartlett and J. K. Hall, Am. Ceram. Soc. Bull. 44 [5] 444-448 (1965).
9. O. K. Sokolov, Izv. Akad. Nauk. SSSR Met. i. Gorn. Dela, 4 59-64 (1963).
10. H. V. Wartenberg, G. Wehner, and E. Saran, Nachr. Ges. Wiss. Gottingen, Math-Physik, Klasse, (Fachgr. II) 2 73-75 (1936).
11. A. D. Kirshenbaum and J. A. Cahill, J. Inorg. Nucl. Chem. 14 283-287 (1960).
12. L. S. Priss, Zhur. Tekh. Fiz. 22 [6] 1051-1061 (1952).
13. P. Kozakevitch, Rev. Met. 57 149-60 (1960).
14. A. A. Hasapis, A. J. Melveger, M. C. Panish, L. Reif, and C. L. Rosen, The Vaporization and Physical Properties of Certain Refractories, Pt. II, Experimental Studies, (WADD-TR 60-463 Pt. II), p. 38 (July 1961).

15. O. D. Slagle and R. P. Nelson, J. Am. Ceram. Soc. 53 [11]
        637-38 (1970).
16. H. J. McSkimin, IRE Trans. Ultrasonics Eng., PGUE-5,25 (1957).
17. W. J. Parker, R. J. Jenkins, R. J. Butler, and G. L. Abbott,
        J. Appl. Phys. 32 1679-1684 (1961).
18. J. Lambert Bates, High Temperature Thermal Conductivity of
        Round Robin Uranium Dioxide, BNWL-1431 (1970).
19. G. N. Rupert, J. Rev. Sci. Inst. 36 1629 (1965).

DISCUSSION

D. Shanefield (Western Electric Co.): What atmosphere is suitable
for melting Al_2O_3 in Mo and W capsules?
Authors: Measurements can be made in any inert or reducing
atmosphere with low O_2 pressure, depending upon system and sample;
we find a vacuum ($\sim 10^{-5}$ torr) or a 92% Ar-8% H_2 mixture to be
satisfactory.

R. C. Rossi (Aerospace Corp.): Have you identified the gases
evolved on melting?
Authors: No. They will vary depending upon the ceramic studied,
e.g., Al_2O_3 crystals are grown in N_2 or Ar and these are probably
the gases released.

R. W. Rice: (Naval Research Laboratory): Sulphur and related
species may be outgassed from some of the oxides, e.g., Al_2O_3,
which may be more likely in single crystals. Do you observe
outgassing from single crystals, as well as from polycrystalline
samples and, if so, how do they generally compare?
Authors: We do observe outgassing from single crystal Al_2O_3
and $MgAl_2O_4$. However, more gas appears to be released from polycrystalline materials. Preheating for long periods near but below
the melting point prior to sealing can reduce the quantity in the
capsule.

G. C. Kuczynski (Univ. of Notre Dame): What is the activation
energy of viscosity for Al_2O_3?
Authors: It has yet to be determined.

S. J. Schneider (National Bureau of Standards): The NBS is now
issuing a Standard Reference Material suitable for temperature
calibration (Al_2O_3, m.p. 2054°C). In your experiments what was
the thickness of the container walls, and what magnitudes of
pressures are generated within them?
Authors: The container walls are 0.03 to 0.05 in. thick. The
internal pressures have not been measured, but will vary with
the amount of absorbed gas and the vapor pressure of the sample.

V. J. Tennery (Oak Ridge National Laboratory): What W contamination levels are observed in oxide samples after melting?

Authors: There is very little contamination of Al_2O_3. Some W is observed on the surfaces of UO_2 samples. Spinel is not seriously contaminated, but MgO is. Oxygen activity in the furnace or within the capsule very strongly influences contamination by the container metal.

OXIDATION OF BORON NITRIDE IN AN ARC HEATED JET

John D. Buckley

NASA Langley Research Center

Hampton, Virginia

ABSTRACT

Two grades of hot pressed boron nitride and a boron nitride composite were subjected to oxidation tests in a 2.5 megawatt atmospheric arc jet. The results showed that fabrication and/or composition influenced thermal shock and oxidation resistance. Changes in surface structure and recession due to oxidation suggest correlation with specimen composition. The boron nitride composite reacted with the oxygen in the hot subsonic airstream to produce a glassy coating on the hot face surface.

INTRODUCTION

A continuing need exists for improved materials for use in the severe environments associated with real and simulated aerospace regimes. Boron nitride and boron nitride composites are being used or considered for use in arc jet engines, plasma accelerators, power supplies for space vehicles, components such as microwave antennas, leading edges, nose cones for hypersonic aircraft, and reentry heat- shields.[1-7]

The objective of this investigation was to examine the ablative characteristics of a boron nitride composite and two hot pressed boron nitrides designated as (1) standard and (2) low moisture. Thermal shock and oxidation properties were determined in a high-temperature subsonic free air jet.

MATERIALS AND SPECIMENS

The nominal compositions of the three materials investigated are shown in Table 1. In addition to the constituents normally found in hot pressed material (B, N, O, and C), a significant amount of titanium (as TiB_2) was added in producing the boron nitride composite, while a small amount of calcium (as CaO) was added to produce low moisture boron nitride.[8] The configuration of an oxidation specimen is shown in Fig. 1.

Table 1. Nominal Composition Density and Grain Size of Boron
Nitride and Boron Nitride Composite

Grade	Analysis (wt%)						Bulk Density kg/m^3
	B	N_2	Ti	O_2	C	Ca	Other
Standard Boron Nitride	43.3	53.07		3.3	0.3		0.03	2000
Low Moisture Boron Nitride	42.4	53.07		3.3	0.3	0.9	0.03	1900
Composite Boron Nitride	36.2	27.47	33.1	2.6	0.6		0.03	2800

APPARATUS AND TEST PROCEDURES

Oxidation tests were performed in a 2.5 MW atmospheric arc jet (Fig. 2) described elsewhere[9]. Test specimens were mounted on a 2.2 cm dia. water-cooled sting so that the front of the specimen was located at distances varying from 3.8 to 30.5 cm above the arc jet nozzle. The subsonic oxidation test conditions shown in Table 2 are for a specimen location 3.8 cm from the nozzle. Test

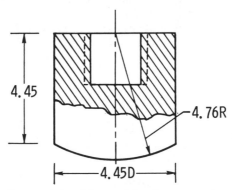

Fig. 1. Boron nitride oxidation specimen (cm).

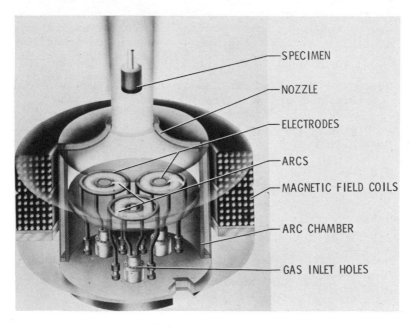

SPECIMEN

NOZZLE

ELECTRODES

ARCS

MAGNETIC FIELD COILS

ARC CHAMBER

GAS INLET HOLES

Fig. 2. Atmospheric arc jet (2.5 MW).

conditions, other than heat transfer rate and specimen surface
temperature, become less precise with increasing distance from the
exit because of the effect of surrounding air mixing with the arc
jet airstream (Fig. 2). The distance from the nozzle to the specimen
was varied to evaluate the effects of varying heating rate, surface
temperature, and surface pressure on thermal shock resistance and
axial recession due to oxidation. Pressure and heat transfer rate
data (from a probe having the same dimensions and shape of the test
specimens) at specific heights above the nozzle are given in
Table 3. The gas enthalpy at the nozzle exit was calculated by
using the method described in reference 9. The surface temperature
of the heated specimens was measured with an optical pyrometer using
an emittance value for boron nitride[10] of 0.5 and 0.9 (extrapolated)
for the boron nitride composite.[11] Specimens were kept in a dry

Table 2. Subsonic Oxidation Test Conditions

Nozzle exit, cm	5.1	Total stream temp., °K	3650
Nozzle throat, cm	5.1	Specimen surface temp., °K	2475
Gas flow, kg/s	0.16	Enthalpy, MJ/kg	6.5
Arc power, kV	2000	Heating rate to specimen	4.0
Velocity, m/s	880	face (Fig.1), MW/m^2	
Mach no.	0.75	Model stagnation pressure, atm.	1.36

atmosphere until immediately before testing. Thermal shock was
accomplished by the rapid insertion and removal of each specimen
from the arc jet airstream. Specimens were exposed from 120 to
300 sec, and were weighed and measured before and after each test.

Table 3. Subsonic Oxidation Test Conditions as a Function of Distance
              (Pressure and Heat Transfer Rate Probe Data)

Specimen Distance Above Nozzle, cm.	Model Stagnation Pressure	Heating Rate, MW/m^2
3.8	1.36	4.00
7.6	1.36	3.25
15.4	1.34	2.42
22.8	1.28	1.89
30.5	1.22	1.30

RESULTS AND DISCUSSION

Arc jet tests under high temperature subsonic airflow
conditions were made to observe the effects of subsonic airflow
on oxidation, erosion resistance, and thermal shock at elevated
temperatures. Table 3, shows the pressure and heat transfer rate
test conditions and Table 4 gives the subsonic test data for
recession along the axis of the specimens as a function of specimen
distance from the nozzle exit. Because of the variation in mass
loss and surface geometry resulting from exposure to the arc jet,
an arbitrary test time of 180 sec was chosen for comparison of
the erosion properties of the materials tested. Figures 3 and 4
show the effect of the high-temperature arc jet air stream on
specimen geometry, 3.80 cm from the nozzle exit. The data in
Table 4 show that standard boron nitride at the various pressures
and heat transfer rates has greater resistance to oxidation and
erosion than the other materials tested.

Table 4. Recession of as a Function of Distance for Specimens
              Subjected to 180 sec in Subsonic Airflow

Distance from Nozzle, cm.	Surface Temp., °K	Axial Recession, cm.		
		Standard BN	Low Moisture BN	BN Composite
3.80	2373	0.751	1.160	1.160
7.60	2314	0.703	1.098	1.090
15.42	2213	0.688	1.010	0.967
22.80	2123	0.646	0.923	0.810
30.50	2013	0.592	0.700	0.695

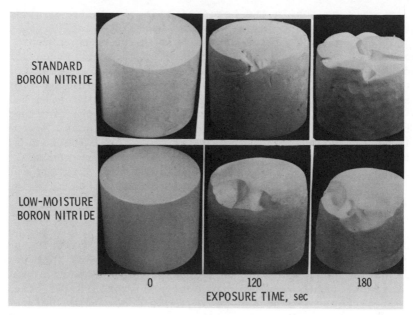

Fig. 3. BN oxidation specimens subjected to subsonic airflow at
hot-face surface temperatures of 2373°K and a heating
rate of 4 MW/m².

The only evidence of thermal shock observed in these tests
was the circular pitlike spalling noted on the sides of a number
of specimens (e.g., Figs. 3 and 4). This type of spalling is
characteristic of hot pressed boron nitride, resulting from the
large difference in thermal expansion parallel and perpendicular
to the pressing direction.[12]

Specimens subjected to subsonic airflow conditions 3.80 cm
above the exit nozzle show deformation of the hot face of the boron

Fig. 4. Composite BN oxidation specimens subjected to subsonic
airflow at hot-face surface temperatures of 2373°K and
a heating rate of 4 MW/m².

nitride compositions, while the boron nitride composite maintained
a relatively uniform shape for test times up to 300 sec (Figs. 3
and 4). Although the hot faces of the boron nitride specimens
showed gouges, measurements made at points of greatest recession
indicated the standard composition suffered less axial recession
for the times tested than did the low moisture material and the
composite. Because of these irregular surfaces, hot pressed boron
nitrides are not considered suitable ablators under the environments
used in this investigation. Newer processing techniques, however,
might eliminate irregular spalling and/or ablation. The composite
boron nitride displayed a more uniform facial geometry and was
tested for up to 300 sec. Figure 5 shows recession data for the
boron nitride composite. The heat of ablation for this material
is 21.6 MJ/kg compared to a nominal value of 30.2 MJ/kg for
graphite.

Visual examination showed the standard and low moisture boron
nitrides to have a surface texture similar to specimens that were
unexposed, while the composite had a glassy phase present on each
specimen tested regardless of exposure time to the arc jet stream.
The lack of coating, or a second phase, on the hot face surface
of the boron nitrides and the presence of a glassy coating on the
composites appear to be related to the sequence of chemical reactions
given in Eqs. 1-4.

$$2BN + 3O_2 \rightarrow B_2O_3 + N_2(g)\uparrow \qquad (\sim1043°K) \qquad\qquad (1)$$

$$B_2O_3(l) \rightarrow B_2O_3(g)\uparrow \qquad (\sim1273°K) \qquad\qquad (2)$$

These reactions[13,14] show that the products would erode away
in the gaseous state. In the composite oxidation of the titanium

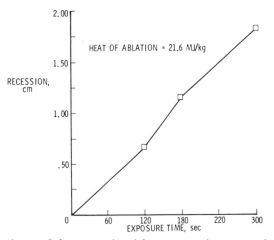

Fig. 5. Recession of boron nitride composite specimens in subsonic
gas flow environments. Surface temperature 2373°K, heating
rate 4 MW/m².

diboride (Eqs. 3 and 4) takes place [in conjunction with Eqs. 1 and 2], ultimately producing the glassy phase present on the surface of the tested composite specimens[15-16]

$$2TiB_2 + 5 \ O_2 \rightarrow 2TiO_2 + 2B_2O_3(l) \qquad (\sim823°K) \tag{3}$$

$$2TiO_2(s) \rightarrow 2TiO_2(l). \qquad (\sim2018°K) \tag{4}$$

There is no loss of titanium oxide due to vaporization since its boiling point is approximately $2793°K$[17].

CONCLUDING REMARKS

Under the test conditions of this investigation the only identifiable evidence of thermal shock was the surface spallation observed on the sides of the hot pressed boron nitride specimens.

The boron nitride composite was the only material that exhibited uniform ablative characteristics.

The glassy coating produced on the surface of the boron nitride composite is believed to result from a series of thermo-chemical reactions between the original components of the boron nitride composite and oxygen present in the arc jet airstream.

REFERENCES

1. J. P. Todd and R. E. Sheets, AIAA Jour., 3 (1) 122-26 (1965).
2. A. F. Carter, G. P. Wood and D. R. McFarland, on a Paper No. 64-699, AIAA Fourth Electric Propulsion Conf., Aug. 31-Sept. 2, 1964.
3. D. F. Swenski and J. R. May, pp. 1161-84 in Progress in Astronautics and Aeronautics, Vol. 16, G. C. Szego and J. E. Taylor Eds., Academis Press, New York, 1966.
4. J. F. Judge Missiles and Rockets, 17 (18) 24-28 (1965).
5. M. L. Hill and J. M. Akridge High Temperature Structures and Materials. John Hopkins University, Res. and Dev. Rept. No. U-ROR/65-1, Jan-Mar 1965.
6. M. Basche and D. Schiff, Materials in Design Engineering, (2) 78-81 (1964).
7. L. Shloss, Government Executive 1 (7) 51-53 (1969).
8. J. Frederickson and W. H. Redany, Metals Progress, 87 (2) 97-101 (1965).

9. R. Brown and B. Fowler, Enthalpy Calculated from Pressure and
      Flow-Rate Measurements in High-Temperature Subsonic Streams.
      NASA TN D-3013, 1965.
10. Southern Research Inst. The Thermal Properties of Twenty-Six
      Solid Materials to 5,000°F or Their Destruction Temperature.
      Tech. Rep. ASD-TDR-62-725, U. S. A. F., Jan. 1963.
11. Y. S. Touloukean, Ed., Thermophysical Properties of High
      Temperature Solid Materials, Vol.4: Oxides and Their Solutions
      and Mixtures. Part 1: Simple Oxygen Compounds and Their
      Mixtures. The Macmillan Co., New York, 1967.
12. Ceramic Data Book 1961-1962. p. 426, Cahners Publishing Co., Inc.,
      Chicago, 1961.
13. A. Munster and   G.Schlamp, Z. Physik. Chem. 116-119 (1960).
14. T. A. Ingles and P. Popper, pp. 144-67 in Special Ceramics.
      Ed. P. Popper, Academic Press, New York, 1960.
15. A. Munster, Z. Electrochem. 63 (7) 807-18 (1959).
16. S. W. Bradstreet, pp. 286-87 in High-Temperature Technology,
      I. E. Campbell and E. M. Sherwood, Eds., John Wiley and
      Sons, Inc., 1967.
17. R. C. Weast, Handbook of Chemistry and Physics. p. B-170.
      50th Edition, 1969-1970. The Chemical Rubber Co., Cleveland
      1969.

DISCUSSION

L. L. Hench (University of Florida): Were the samples pressed
isostatically?
Author: No.

R. W. Rice (U. S. Naval Research Laboratory): What were the reasons
for choosing the BN-TiB$_2$ composite?
Author: The choice was arbitrary.
R. W. Rice: Have you looked at other additives, especially those
that would not degrade dielectric properties, e.g., the BN-SiO$_2$
material now available?
Author: No.

MICROSTRUCTURAL FEATURES OF OXIDE SCALES FORMED ON ZIRCONIUM DIBORIDE MATERIALS

H. C. Graham, H. H. Davis and I. A. Kvernes

Aerospace Research Laboratories

Wright-Patterson AFB, Ohio

W. C. Tripp

Systems Research Laboratories

Dayton, Ohio

ABSTRACT

Both ZrB_2 and $ZrB_2 + 20$ v/o SiC have been oxidized in oxygen over the range 900° to 1500°C. The microstructural features of the oxide scales formed under these conditions are studied as a function of temperature and reaction time. Techniques used to characterize the oxide products include x-ray diffraction, scanning electron microscopy and electronprobe microanalysis. A glassy B_2O_3 is one of the oxidation products at temperatures below 1100°C, resulting in protective behavior. At higher temperatures the formation of an SiO_2-base glass imparts additional protection to $ZrB_2 + 20$ v/o SiC.

INTRODUCTION

Zirconium diboride is of interest because of its excellent properties at very high temperatures, placing it among the leading candidates for both thermal protection and air breathing propulsion materials at use temperatures above 1200°C. For these applications, materials are required that will withstand exposure to high temperature oxidizing environments for long periods of time without drastically changing their configuration.

The diboride materials were originally developed under Air
Force contract F33(615)-3671 and their high temperature thermodynamic properties have been extensively studied by ManLabs Inc.[1]
Included in their study were oxidation recession rate measurements
in the temperature range 1600° to 2200°C.[2]

The materials to be discussed in this paper are ZrB_2 and
ZrB_2 + 20 v/o SiC. Detailed studies of the oxidation kinetics of
these materials up to 1500°C have been accomplished and the results
reported elsewhere.[3-7] In the present study, the microstructural
features of the scales formed on these materials in this temperature
range are examined and related to the oxidation behavior.

EXPERIMENTAL TECHNIQUES

The ZrB_2 and ZrB_2 + 20 v/o SiC materials were supplied by
ManLabs Inc. as hot pressed billets about 75 mm dia. and 25 mm thick.
The major impurity in these materials is ZrO_2 and the B/Zr ratio is
slightly less than two. Small rectangular coupons weighing ~1 g
and about 1.5 x 1.5 x 1 mm in size were cut from the billet with
a diamond saw. Specimens were abraded to a 150 grit finish on a
diamond wheel and ultrasonically cleaned in alcohol.

The oxidation experiments were initiated by inserting the
samples directly into the furnace at the desired temperature and
pressure. Upon completion of the test, the samples were removed
from the hot zone and cooled rapidly to room temperature in the gas
stream in order to retain as much of the vaporizing oxide as possible.
Preparation of the samples for examination was difficult because
of the hygroscopic nature of boron oxide. Therefore, the specimens
were coated with epoxy immediately upon removal from the dry
furnace atmosphere and stored under vacuum. The epoxy also aided
in retaining the glass scales during polishing. It was found
necessary to perform all preparation steps utilizing xylene as the
lubricant in order to protect the boron oxide from moisture.

The scales on the oxidized specimens were characterized by
standard analytical methods. X-ray diffraction was used initially
to identify the reaction products. Scanning electron microscope
(SEM) examination of the scale structure was generally performed
on specimen surfaces and fractured cross sections. In a few cases
it was found beneficial to examine polished cross sections with
the SEM. All specimens for SEM analysis were vapor coated with
Au-Pd to prevent charge build-up in the dielectric portions of the
specimen. The electron microprobe analyzer was used to detect
the elemental distributions of Zr, Si, and when possible B and O.
Optical microscopy was used to correlate the findings of the above
techniques.

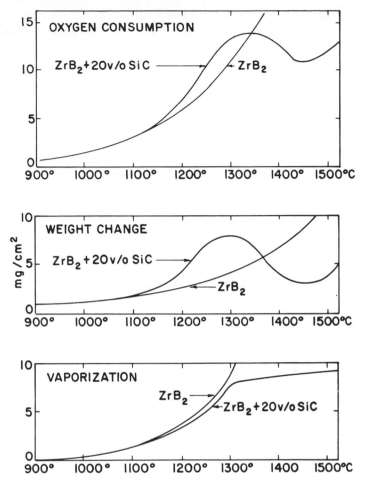

Fig. 1. Thermogravimetric results after 4 hr exposure to O_2 at 250 mm.

RESULTS AND DISCUSSION

Figure 1 gives a summation of the thermogravimetric data for both materials oxidized for 4 hr over the temperature range 900° to 1500°C. These results are presented in the form of three separate measurements: net weight change, total oxygen consumption, and vaporization. The techniques developed to measure these quantities independently have been described previously[3, 6-9]. The net specimen weight change was obtained by conventional weighing techniques where only the products remaining on the sample are weighed. The O_2 consumption represents the total used to form both the volatile and the solid oxides. The vaporization data represents the weight of the volatile oxides formed during the

reaction. The microstructural analysis presented in this study
forms the basis for understanding some of the irregular behavior
observed in the gravimetric results.

ZrB$_2$ Oxidation

It must be stressed that Fig. 1 is a summation of data for
three measurements taken from kinetic curves at the 4 hr point.
These data will vary in both magnitude and relative position if
summarized for shorter or longer oxidation times. It is observed
in Fig. 1 that the O$_2$ consumption for ZrB$_2$ shows a rapid increase
in rate as the temperature exceeds 900°C. Corresponding changes
in the oxide microstructure for 4 hr exposure are illustrated in
Fig. 2. At 900°C (Fig. 2A), the scale consists of two distinct
layers: an external glassy scale of B$_2$O$_3$ which was liquid at

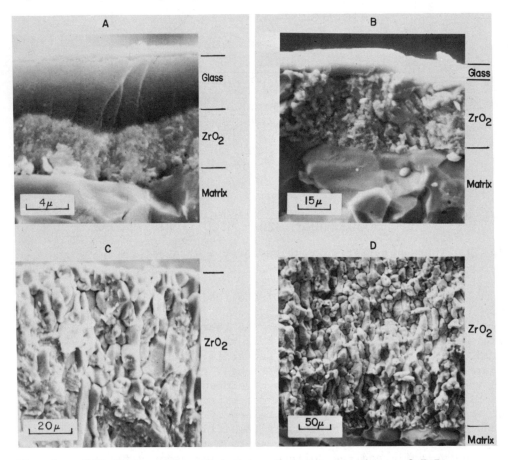

Fig. 2. SEM micrographs of fractured cross sections of ZrB$_2$
         oxidized for 4 hr at: (A) 900°, (B) 1100°, (C) 1300°
         and (D) 1500°C.

temperature, and an internal layer of submicron size particles of
ZrO_2. Though not apparent in the micrograph, it is probable that
the amorphous B_2O_3 extends throughout the inner ZrO_2 layer. A
similar scale structure is observed following reaction for 4 hr
at 1100°C (Fig. 2B). Although the external glassy layer is
approximately the same thickness, the ZrO_2 layer has increased at
least threefold, agreeing with the thermogravimetric data of
Fig. 1. Furthermore, the ZrO_2 particle size is much larger. At
1300° (Fig. 2C) a considerable change has occurred in the oxide
microstructure. Very little glass remains on the outer surface
and the ZrO_2 morphology is becoming columnar in appearance.
Increasing the temperature from 1100° to 1300° has caused a five-
fold increase in the ZrO_2 thickness. However, the corresponding
increase in specimen weight is less than would be expected for
such an increase in condensed oxide. This is explained by the
simultaneous rapid increase in B_2O_3 vaporization over this
temperature span. At 1500° no glassy oxide can be seen and the
columnar structure of the ZrO_2 has become clearly defined throughout
the thickness of the scale. The evaporation rate of the B_2O_3 has
thus increased with temperature so that at 1500° the B_2O_3 evaporates
as fast as it is formed. An enlarged view of the outer surface of
this particular scale is shown in Fig. 3. The absence of glass is
apparent and the surface orientation reveals an end view of the
ZrO_2 columnar grains. This ZrO_2 scale is nonprotective and the
1 mm thick specimen is almost completely oxidized following 4 hr
of exposure.

For comparison, the surface structure of the oxide at 1300° is
shown in Fig. 4. While small patches of glass are found, the

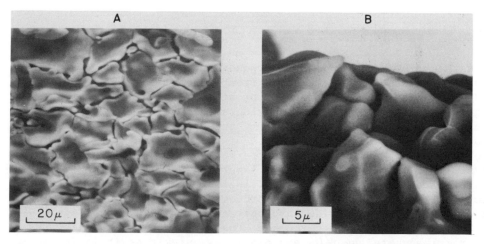

Fig. 3.   SEM micrographs of ZrB_2 oxidized for 4 hr at 1500°C:
         (A) the oxide surface, (B) a fractured cross section of
         the oxide-gas interface.

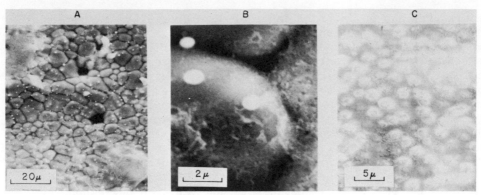

Fig. 4. SEM micrographs of the surface of ZrB$_2$ oxidized at 1300°C
         for 4 hr: (A) overall surface showing both exposed and
         glass covered ZrO$_2$, (B) enlargement of "exposed" area
         showing residual glass around ZrO$_2$ grains, and (C) enlarge-
         ment of a glass covered area showing pores in the glass.

considerable porosity in some portions of the glass indicate the
rapid rate of gas evolution at this temperature. The vapor
pressure of B$_2$O$_3$ at 1500° is 1.5×10^{-3} atm compared to only
10^{-6} atm at 1050°C [10].

     The loss of the glassy B$_2$O$_3$ layer on ZrB$_2$ not only occurs with
increasing temperature, but also may be shown to occur at longer
times at 900°. For short time exposure, the formation of condensed
reaction products imparts protective properties to ZrB$_2$, with the
subsequent oxidation behavior being nearly parabolic. The actual
oxygen consumption has been shown[3,7] to agree with a paralinear
equation of the type

$$moles\ oxygen/cm^2 = at^{1/2} + bt$$

where the constants a and b are exponentially dependent on temperature and the linear term, derived from the vaporization measurements,
predominates at high temperatures or at the lower temperatures for
long times. Figure 5 shows the oxide scales resulting from exposure
times of 1, 24 and 90 hr at 900°. Changes within the glass have
occurred within 24 hr, and the loss of the external glass layer
is observed for the specimen reacted for 90 hr. However, glass
can still be observed in the ZrO$_2$ layer, and some additional
changes have occurred in the ZrO$_2$ itself. At the oxide-matrix
interface, the morphology of the ZrO$_2$ grains has become columnar.
The appearance of the oxide structure near this interface is
similar to that observed throughout the ZrO$_2$ layer at the elevated
temperatures where the B$_2$O$_3$ evaporates as quickly as it is formed.
This columnar structure apparently forms when the loss of the
B$_2$O$_3$ protective layer leads to accelerated growth of the ZrO$_2$.

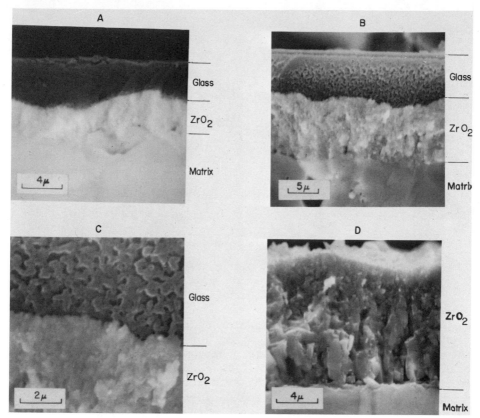

Fig. 5.   SEM micrographs of fractured cross section of ZrB$_2$
oxidized at 900°C for various times: (A) 1 hr, (B) 24 hr,
(C) enlargement of ZrO$_2$-glass interface of 24 hr exposure
sample and (D) 90 hr.

ZrB$_2$ + 20 v/o SiC Oxidation

Up to 1100°C the oxidation behavior of this material after
4 hr of reaction does not significantly differ from that of ZrB$_2$.
However, in the neighborhood of this temperature, where ZrB$_2$
is losing the protective glass layer due to the high evaporation
rate of B$_2$O$_3$, the SiC particles start to undergo appreciable
oxidation.   Figure 6A shows a SiC particle imbedded in the scale
formed at 900°.   The SiC is almost completely surrounded by ZrO$_2$,
yet no apparent reaction has occurred.   At 1100°, intact SiC
particles are still observed in the scale.   Some reaction is observed
at this temperature, as shown in Fig. 6B, but preferential oxidation
of ZrB$_2$ is still predominant.

When the temperature is increased to 1300°C, the SiC particles
undergo appreciable attack.   A particle is shown imbedded in the

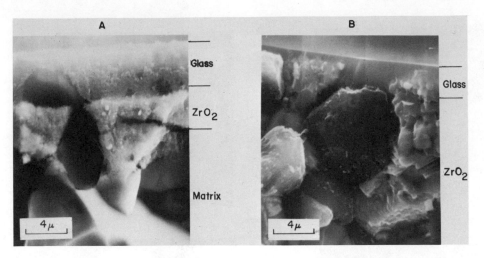

Fig. 6.   SEM micrographs of fractured cross sections of
          ZrB$_2$ + 20 v/o SiC showing SiC particles in oxide
          scale: (A) 5 hr at 900°C and (B) 20 hr at 1100°C.

ZrO$_2$ region (Fig. 7) and the formation of the glassy constituent
can be observed around its edges. Another SiC particle, located
at the oxide-matrix interface, has undergone attack only on the
side adjacent to the oxide region. The presence of the glassy
phase throughout the ZrO$_2$ scale is evident. Surface views of
this sample (Fig. 8) show areas containing thin patches of glass
and areas of exposed ZrO$_2$, analogous to that observed for ZrB$_2$

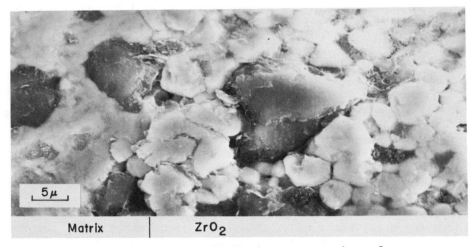

Fig. 7.   SEM micrograph of polished cross section of
          ZrB$_2$ + 20 v/o SiC after oxidation for 24 hr at
          1300°C. Note SiC particles in ZrO$_2$ and at
          ZrO$_2$-matrix interface.

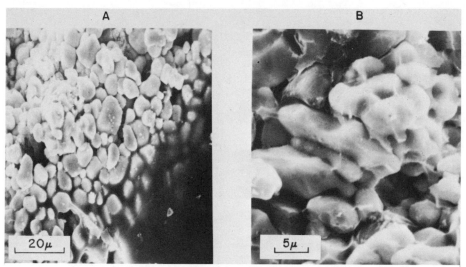

Fig. 8. SEM micrographs of surface of ZrB_2 + 20. v/o Sic after oxidation for 24 hr at 1300°C: (A) overall surface showing both exposed and glass covered ZrO_2 and, (B) enlargement of "exposed" area showing residual glass around ZrO_2 grains.

after only 4 hr exposure. The wetting action of the glass on the oxide particles in Fig. 8B provides a distinct contrast to the surface morphology presented in Fig. 3A where no glass is present.

As the oxidation of the SiC particles increases with temperature, there is an appreciable increase in the amount of the glass phase. At 1400° the amorphous layer has become SiO_2-rich. This can be deduced by recalling that the glass layer formed on ZrB_2 has disappeared above 1300°, while Fig. 9 shows that there exists a thick external glass layer on the SiC-containing material at 1400°. Direct evidence of this is shown by microprobe scans (Fig. 10) taken over the oxide region of a specimen oxidized at 1400°. The glass region is Si-rich, but with some indication of B present. No Zr was detected in the glass layer. The probe scans also reveal measurable quantities of Si and B within the ZrO_2 scale region. These findings correlate with the SEM observation of a glassy matrix surrounding the ZrO_2 grains. These scans were taken on a selected area near the edge of the specimen where an abnormally thick glass was present in order to more easily distinguish the phases.

The change in the glass composition from B_2O_3-rich to SiO_2-rich on ZrB_2 + SiC has an associated decrease in the vapor pressure of the glass. Because the SiO_2 glass phase is undergoing less evaporation, a thick amorphous layer (analogous to the B_2O_3 glass which is formed below 1100°C) begins forming on this material

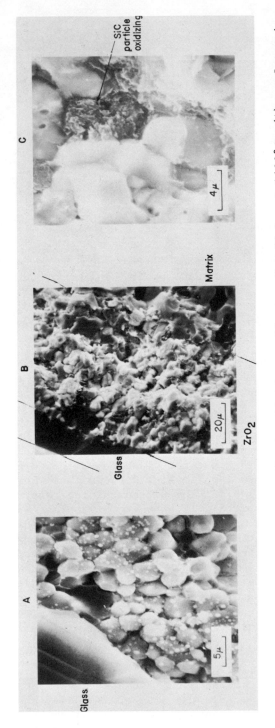

Fig. 9. SEM micrographs of ZrB_2 + 20 v/o SiC after oxidation for 24 hr at 1400°C: (A) surface view showing thick external glass region and exposed ZrO_2 where external glass layer was fractured off, (B) fractured cross section, (C) enlargement of cross section showing SiC particle during oxidation.

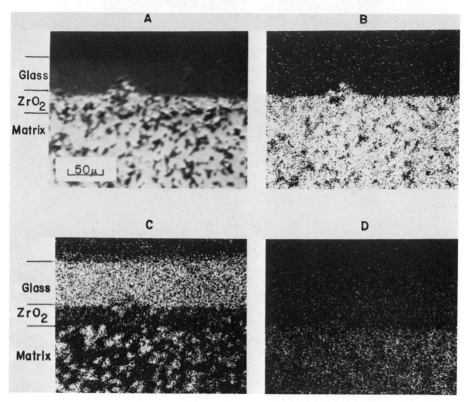

Fig. 10. Electron microprobe images of ZrB_2 + 20 v/o SiC showing
          elemental distributions after 24 hr exposure at 1400°C:
          (A) back-scatter electron image, (B) Zr x-ray image,
          and (D) B x-ray image.

with increasing temperature. The protective characteristics of the
glass starts to become effective at about 1300°, accounting for the
corresponding improvement in both the oxygen consumption and the
vaporization behavior observed for ZrB_2 + SiC in Fig. 1.

Oxidation Mechanism

     The results obtained from examination of the microstructural
features of the scales formed on ZrB_2 and ZrB_2 + 20 v/o SiC were
used to explain some of the anomalous results observed in the
gravimetric data. The correlation of microstructure with thermogravimetric data permits an insight into the mechanism of oxidation
of these materials. Figure 2D shows the columnar structure of the
scale formed on ZrB_2 at 1500°C. The oxidation rate is very rapid
at this temperature and gravimetric data have revealed an almost

Fig. 11.   Optical micrograph of polished cross section of
           ZrB_2 + 20 v/o SiC oxidized for 24 hr at 1400°C showing
           geometry of scale growth.

linear time dependence. This must mean that the structure of the
ZrO_2 scale is very porous and molecular oxygen is freely transported to the boride-oxide interface when no glass is present to
fill the open structure. However, at lower temperatures, glass is
present in the scale and the columnar growth of ZrO_2 is not observed.
A different oxidation mechanism occurs due to the glass on the
surface, and the resulting reaction rate is slower.

It has been shown (Fig. 10) that there is no detectable Zr
in the glass layer. Therefore, Zr does not diffuse to the glass-
oxygen interface. It is possible that the B and/or Si could
diffuse to the outer surface or even react with oxygen in the glass
layer. The presence of glass throughout the ZrO_2 layer suggests that
the rate controlling mechanism is the inward diffusion of oxygen
through the continuous glass phase. Thermogravimetric results
published previously show that the reaction rate is paralinear and
dependent on glass thickness.

Figure 11 shows a typical geometry observed on all specimens
after reaction. The surface of the scale is square at the corner
(upper left) and the corner of the oxide-matrix interface is rounded.
Since the geometry of the scale surface represents that of the
unoxidized sample, this indicates that the oxide has grown in from
this original surface and that the oxidation reaction is controlled
by the inward diffusion of oxygen.

SUMMARY

Based on the microstructural results presented in this paper
and the results of previous thermogravimetric studies, the following
summation is given for ZrB_2 and ZrB_2 + 20 v/o SiC materials at

250 mm of pure oxygen in the temperature range 900° to 1500°C. These two materials oxidize in a similar manner at the lower temperatures where a protective B_2O_3 glass layer remains on the sample. At the higher temperatures, an SiO_2 glass provides similar protection to ZrB + 20 v/o SiC while ZrB_2 is oxidizing at an almost linear rate due to the immediate vaporization of B_2O_3 and the porous nature of the remaining ZrO_2 scale. The behavior of these two materials begins to differ in the range 1100 to 1300°C, i.e., when appreciable amounts of SiO_2 glass are forming due to the oxidation of the SiC particles. All evidence indicates that the oxidation reaction is controlled by the inward diffusion of oxygen through the continuous glass phase. In the absence of glass, the reaction is controlled by transport of molecular oxygen through the open ZrO_2 structure.

ACKNOWLEDGEMENTS

This work was partially supported by USAF Contract F33(615)-69-C-1017. The authors wish to acknowledge the assistance of R. A. Harmer of the University of Dayton Research Institute for the SEM and electron microprobe analyses reported here. We are also indebted to J. R. Fenter, AFML, for supplying the ZrB_2 materials and background information on the diboride systems.

REFERENCES

1. E. V. Clougherty, K. E. Wilkes, and R. P. Tye, Research and Development of Refractory Oxidation-Resistance Diborides. AFML-TR-68-190, Part II Vol. V (Nov. 1969).
2. L. Kaufman and H. Nesor, Stability Characterization of Refractory Materials under High Velocity Atmospheric Flight Conditions. AFML-TR-69-84, Part III Vol I (Sept. 1969).
3. W. C. Tripp and H. C. Graham, Thermogravimetric Study of the Oxidation of ZrB_2 in the Temperature Range of 800° to 1500°C, accepted for publication, J. Electrochem. Soc.
4. W. C. Tripp and H. C. Graham, J. Electrochem. Soc. 117 255 (1970).
5. W. C. Tripp and H. C. Graham, Bull. Am. Cer. Soc. 49 394 (1970).
6. H. C. Graham and W. C. Tripp, p. 9-1 in AGARD Conference Proceedings No. 52, Reactions Between Gases and Solids. 1969.
7. W. C. Tripp, Relationship Between Point Defects and Charge and Mass Transport. Systems Research Laboratories, Inc. Report No. 6390-11-7 (April 1970).
8. H. C. Graham, W. C. Tripp, and H. H. Davis, "Microbalance Techniques Associated with Oxidation Studies", presented at Ninth Conference on Vacuum Microbalance Techniques, West Berlin, June 1970. To be published by Plenum Press.
9. W. C. Tripp and H. C. Graham, Air Force Research Review, 1 (July-August 1970).

10. J. R. Soulen, P. Sthapitanonda, and J. L. Margrave, J. Am.
    Chem. Soc. 59 132 (1955).

DISCUSSION

D. J. Godfrey (Admiralty Materials Laboratory): Have you any
quantitative analyses of the composition of the protective glass
e.g., from microprobe data?
Authors: While it would be informative, we have not yet obtained
any quantitative data concerning glass compositions in the range
where both B_2O_3 and SiO_2 are being formed.

G. C. Kuczynski (Univ. of Notre Dame): I would like to point out
that the micrograph presented in Fig. 5C is an excellent illustration of spinodal decomposition.

S. G. Ampian (U. S. Bureau of Mines): The temperature range of
your studies, 900°-1500°C, encompasses the monoclinic to tetragonal
ZrO_2 transformation. What bearing does this polymorphism have on
the change of the columnar ZrO_2 to a more squarish or cubic habit?
Authors: The columnar morphology apparently has no relation to the
transformation in question. First, columnar growth is observed
on oxidized ZrB_2 (Fig. 5D) following long time exposure at 900°C
as well as above the normal transformation temperature. Second,
the morphology was not columnar on ZrB_2 + 20 v/o SiC oxidized at
1300°-1400°C where SiO_2 glass was formed. It therefore appears
that the columnar growth is associated with the absence of the
glass phase.

TESTING OF GRAPHITE COMPOSITES IN AIR AT HIGH TEMPERATURES

S. A. Bortz

I. I. T. Research Institute

Chicago, Illinois

ABSTRACT

Extreme environments associated with atmospheric re-entry require materials capable of withstanding severe thermal and mechanical conditions. Generally, data available from tests made under ambient conditions of previously oxidized material, or from tests of materials made at elevated temperatures in an inert atmosphere, do not relate well to properties of materials in the actual re-entry environment. Procedures were developed for high temperature testing in air. The loss of material with time was measured and correlated with observed strength data. Shapes of temperature-strength curves for oxidized JTA graphite were similar to those obtained under inert conditions. As anticipated, strengths after oxidation were lower, depending on the length of exposure.

INTRODUCTION

Materials exposed to extreme environments often exhibit considerable variability from handbook property data, which commonly had been obtained under simplified test conditions. Designers who use handbook data often must penalize materials with safety factors large enough to assure that failure will not occur in service situations. Such practices limit material choices for many rocket, re-entry, and nuclear applications. This paper indicates the importance of characterizing and testing a material in the environment in which it will be employed.

A composite, refractory, oxidation-resistant material developed to resist re-entry environments (JTA graphite) was used in this study.

49

The as-formed material is strongly anisotropic, giving rise to property differences when measured in with-grain and across-grain directions. Test specimens were subjected to a series of temperatures and oxidation environments, and the mechanical properties of interest to design engineers were correlated with test environments.

TEST PROCEDURES

The effects of oxidation on the mechanical properties of JTA graphite were observed by two methods. In the first, samples were exposed to three temperatures (1000°, 1750° and 2000°C) in each of four oxygen partial pressures (ambient, 10^1, 10^3 and 10^5 torr), and changes in weight and flexural strength were measured subsequently under room conditions. Two crystallographic orientations (with- and across-grain) were studied. Five replicates at each condition (temperature, atmosphere and orientation) were tested, a total of 120 specimens.

In the second, strength changes were measured in three test modes (flexure, tension and compression) while the specimens were being oxidized in air at ambient pressure at three temperatures (1000°, 1750° and 2000°C). Both orientations (with- and across-grain) were studied. Ten replicates at each test condition (test mode, temperature, and orientation) were tested, totaling 180 specimens. Other temperatures (notably 1500°C) and environments within this experimental range were used in auxiliary and supporting studies. Depths of oxidation, rates of oxidation and nature of the oxidation products were determined. Results were compared with data obtained at elevated temperatures in an inert gas (Ar) atmosphere.

Flexure tests (four point loading) were conducted on 1/4 x 1/4 x 3 in. samples. The distance between supports was maintained at 2.5 in., a ratio of 10:1 with regard to the beam depth.

Tensile strength was measured in a direct uniaxial test. The specimens had a 1/4 in. circular gage section 2 in. long (4 in. total length) and were held in collet grips (Fig. 1). A split ring on the inside fitted the transition radius of the test specimen, and on the outside provided a spherical loading surface. The split ring was held together by a circular collar, and fitted into the grip holder, whose inside surface mated with the spherical surface of the split ring. The assembled grip holder was threaded to the load train of the test machine.

Cylindrical specimens 1/4 in. dia by 3/4 in. long were used for measurements of compressive strength. A 1 in. dia spherical load joint aligned the axis of force with the axis of the specimens.

Fig. 1.   Collet grips used   Fig. 2.   Tension fixture for high
          with tensile samples.   temperature air testing.

Fixtures of Poco graphite were used for all inert atmosphere
high temperature and ambient tests.   Water-cooled stainless steel
fixtures and special test procedures were developed for elevated-
temperature tests in an oxidizing atmosphere.

All mechanical tests were performed with an Instron Testing
Machine (20,000 lb capacity).   Inert atmosphere tests were performed
in a Brew high temperature furnace.   Samples tested in air were

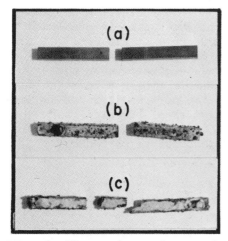

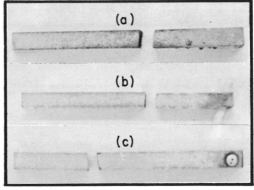

Fig. 3. Flexural specimens of   Fig. 4. Formation of oxide coating
        graphite after 60 min   on JTA samples on exposure
        exposure at various   to air at 1750°C for
        temperatures (a) 1500°,   (a) 20, (b) 40 and (c) 60
        (b) 1750° and (c) 2000°C.   min.

heated by induction; the coils were fabricated for minimum interference with the load train. The tension fixture is illustrated
in Fig. 2; generally similar water cooled fixtures and appropriately
shaped coils were employed for flexure and compression. Temperatures were read directly with an optical pyrometer.

MATERIALS CHARACTERIZATION

    All test specimens were cut from JTA graphite. Manufactured
by the Carbon Products Division of Union Carbide Corporation, JTA
is a hot pressed, multiphase composite material containing 48% C,
35% Zr, 8% B and 9% Si. Under oxidizing conditions, a protective
coating is formed on the surface. Figure 3 illustrates the effect
of increasing temperature on coatings developed after 1 hr exposure.
Figure 4 shows coating build-up as a function of time at 1750°C.
X-ray analyses of coatings formed at that temperature indicated
that zirconia is the dominant crystalline phase, with concentrations
increasing from about 65% after 20 min exposure to more than 95%
after 60 min. The coating build-up was slow at 1500°, but considerably faster at 1750° C. A glassy phase in the form of globules
was observed at the latter temperature. At the highest temperature
(2000°C) a smooth coating was formed, which did not prevent
considerable loss of graphite.

    Figure 5 illustrates oxide surfaces after exposures at 1000°,
1500° and 1750°C. At 2000°C both the coating and the graphite
were too friable for metallographic polishing. A protective
coating did not develop until the material was exposed to temperatures on the order of 1750°C. Coatings showed progressive buildup
with time (Fig. 5c - 5e). Loss of graphite with time is presented
in Fig. 6. More material was lost at 1000° than at either 1500°
or 1750°C. Figure 7 indicates the oxidation loss after 1 hr at
various temperatures. The with-grain loss is greater than the
across-grain loss, an important design consideration. A correlation
of material loss with strength (based on the no-loss inert atmosphere
test data, assuming a reduced cross-section area) revealed that
material in the interior not affected by oxidation retained
its unit strength. All strength data were compared on the basis
of 1 hr exposure.

STRENGTH

    Results of physical testing are presented in Fig. 8 for
flexure, Fig. 9 for tension, and Fig. 10 for compressions. In
flexure, JTA graphite tested at elevated temperatures exhibited
increased strength at approximately 1650°C, although strength
levels in air were considerably below those in inert atmospheres.

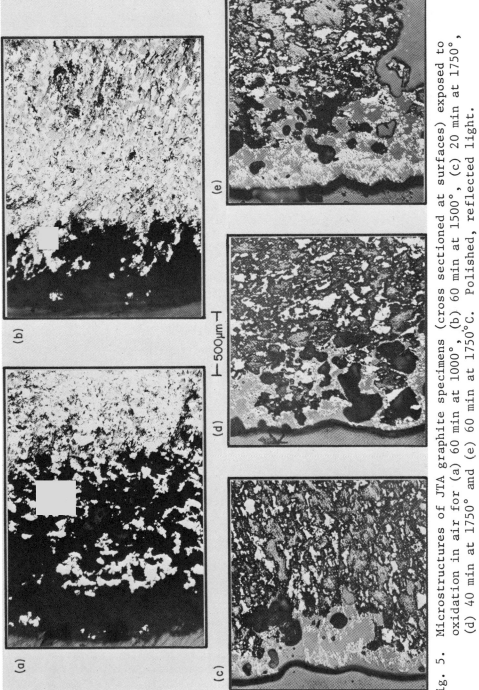

Fig. 5. Microstructures of JTA graphite specimens (cross sectioned at surfaces) exposed to oxidation in air for (a) 60 min at 1000°, (b) 60 min at 1500°, (c) 20 min at 1750°, (d) 40 min at 1750° and (e) 60 min at 1750°C. Polished, reflected light.

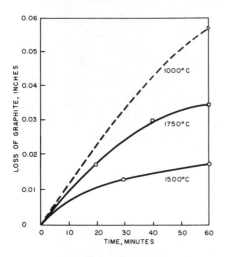

Fig. 6. Oxidation loss for JTA
graphite (with-grain).

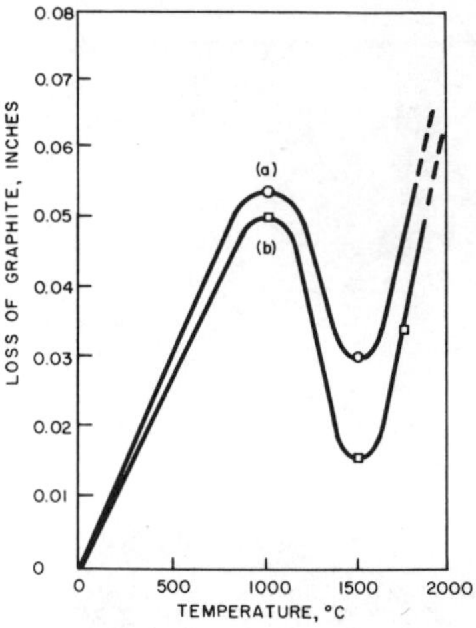

Fig. 7. Oxidation loss (1 hr) for
JTA graphite (a) with-
grain, (b) across-grain.

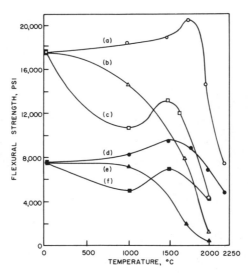

Fig. 8.   Flexural strength of JTA graphite.   Tested with-grain:
(a) at temp. in Ar, (b) at room temp. after oxidation,
(c) at temp. during oxidation. Across-grain: (d) at temp.
in Ar (e) at room temp. after oxidation, (f) at temp.
during oxidation.

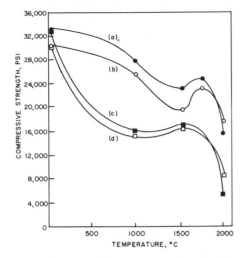

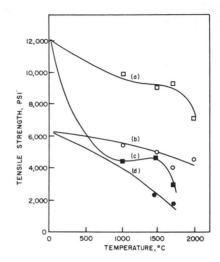

Fig. 9. Tensile strength of JTA graphite. In vacuum: (a) with-grain, (b) across-grain. In air: (c) with-grain and (d) across-grain.

Fig. 10. Compressive strength of JTA graphite. In vacuum: (a) across-grain, (b) with-grain. In air: (c) across-grain, (d) with grain.

For this material, strength is directly proportional to remanent cross-sectional area. Inasmuch as oxidation of JTA material is a surface-related phenomenon, its loss in strength will be directly proportional to the amount of material lost; oxidation losses were indicated in Figs. 6 and 7. Large material losses at 1000° in air account for the marked decrease in strength at this temperature. Strength loss in the across-grain direction is less than in the with-grain orientation, in agreement with the data presented in Fig. 7. Flexure specimens exposed to the elevated temperature air environment and then tested under ambient conditions exhibited a more rapid strength loss with temperature than materials tested at elevated temperatures. Strengths in tension (Fig. 9) and compression (Fig. 10) followed generally similar trends.

CONCLUSIONS

The experiments described clearly demonstrate that to estimate the true behavior of a material in a specific environment, the environment must be simulated. Also, if the environment is corrosive in its effect on the material, loss with time must be determined, since strength changes depend on the amount of good material remaining in the load bearing cross-section. Actual test data can be misleading if this situation is not appreciated. To

truly understand mechanical behavior in an extreme environment,
materials must be (1) fully characterized before and after exposure
to the environment to determine material changes and losses,
(2) tested under conditions that simulate the working environment
and (3) findings from (1) and (2) must be correlated in order to
facilitate design of a successful structure.

ACKNOWLEDGEMENTS

*This work was carried out under Contract AF 33 (615)-3028,
Task No. 738106, Proj. No. 7381, "Properties of JTA Graphite".
The assistance of Mr. M. Knight, Air Force Materials Laboratory,
is gratefully acknowledged.*

DISCUSSION

I. Ahmad (Watervliet Arsenal): Because of the water cooled grips
there must be a considerable temperature gradient along the length
of the specimen. At what position was the temperature measured?
Author: The temperature gradient along the 2 in. gage section
(tension specimen) was less than 30°C. The gradient was very
steep at the grips. The total length of the specimen was 4 in. and
the temperature was measured at the gage section.

S. W. Bradstreet (Consultant, Dayton, O.): What changes were
observed in tensile strain to failure?
Author: It increased with temperature; above 2000°C considerable
plastic deformation was observed.

L. L. Hench (University of Florida): Can the increase in strength
at ~1500°C for samples tested at temperature be explained by
crack healing?
Author: Yes.

MgO REFRACTORIES IN THE BASIC OXYGEN STEELMAKING FURNACE

S. C. Carniglia, W. H. Boyer and J. E. Neely

Center for Technology, Kaiser Aluminum & Chemical Co.

Pleasanton, California

ABSTRACT

The BOF presents an example of a hostile, complex, transient thermomechanical and chemical environment, drawn from a major commercial industry. The most modern and successful refractory linings developed for these furnaces are composed of magnesia and carbon: a ceramic composite. The character of these refractory materials is traced from its origin in processing through interaction with the corrosive BOF service environment, and the bases of durability are discussed. Present performance is illustrated by a case study.

INTRODUCTION

The refining of crude blast furnace iron by rapid pneumatic processes has long threatened to overtake open hearth steel production. The most dramatic growth has been registered by the Basic Oxygen Steelmaking Process (BOSP), spurred by the availability of low cost oxygen. Introduced to the U. S. from Europe in 1954, this process alone in 1970 yields approximately half of all steel made in this country, and its share continues to increase rapidly.

In less than an hour in the BOSP, at metal temperatures starting at about 1300°C and ending at 1600-1700°C, pure O_2 introduced through a water cooled lance over a charge of about 30% scrap and 70% molten pig iron oxidizes C and other impurities. Compounds of Si, P, S, and several other elements are refined out by the addition of suitable slag-forming materials such as burnt lime and fluorspar.

57

Among the many advantages of the BOSP is its substantial saving
in refractory comsumption as compared with the open hearth.
Refractory working linings for the Basic Oxygen Furnace (BOF) in
the U. S. are laid up of brick whose principal constituent is MgO.[1-3]
A number of commercial refractories for the BOF contain also up to
roughly 50% CaO;[1-3] however, this discussion will be concerned only
with lining materials above 80% MgO. For reasons that will become
evident the BOF employs, almost exclusively, brick which also
contain several weight per cent of elemental carbon distributed in
their microstructure.[1-3] This carbon is derived within the BOF
itself by pyrolysis of coal tar pitch, a constituent of the brick
as manufactured. Though refractories of this broad type have been
in steelmaking use since early in the 19th century, only within the
last decade have they become engineered, high performance
composites.[1] Today the scientific content of their material design
and the understanding of their behavior are increasing on an
accelerating scale.   An excellent treatment of these and other
steelmaking refractories has just been published in a volume of
monographs.[4] The present discussion will attempt to highlight and
simplify the extremely complex environmental and material aspects
of these composite refractories and their application.

THE ENVIRONMENT

The Burn-In

The life cycle of the BOF refractory begins with its installation within the back-up lining of the steel vessel. The environments which it will see are all transient. The first of these,
the "burn-in", has the dual purposes of bringing the furnace to
operating temperature and of preparing the lining for its subsequent
use.

The vessel is charged with coke, O_2 is introduced, and the
brick face temperature is raised at a controlled rate ($\sim$50° C/min)
to as least 550°C, in some shops to as high as 1000°C. The rate of
heating may then be decreased until a maximum temperature between
1100° and 1500°C is reached; this temperature is held 1/2 - 2 hr
before the steelmaking "campaign" is begun. The atmosphere throughout this period is principally CO. The residual coke is dumped at
the conclusion of the burn-in.

The Campaign

The "campaign" comprises all of the ensuing cycles (called
"heats") of charging with metal, "blowing" with O_2, sampling, and
tapping, until the spent lining is removed and replaced. The furnace,
mounted on trunions, is characteristically positioned for operation
as shown in Fig. 1.

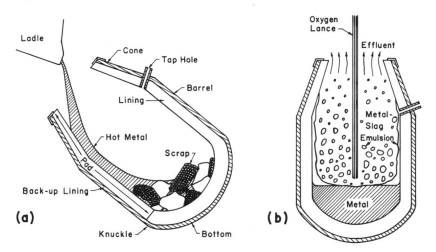

Fig. 1. BOF in characteristic positions: (a) charging, (b) blowing.

Up to 100 tons of scrap steel, in a typical case, is dropped
on the slanted wall and falls to the bottom, producing impact and
abrasion (Fig. 1a). This is followed by the charge of hot metal,
250 tons delivered in minutes at a temperature of ~1300°C. The
"charge pad" which receives this treatment must be thermal shock
resistant, hard, and mechanically sturdy.

After charging, the furnace is positioned upright, the lance
is inserted, and the O_2 "blow" begins (Fig. 1b). CaO, fluorspar,
and mill scale are added (to flux and stabilize the oxidized
impurities), and over the course of the "blow" the bath temperature
rises to 1600°-1700°C. During this period the churning metal is
overlaid with an emulsion of iron in slag which is lifted by the
CO/CO_2 combustion gases and refluxes over the entire lining up
to the conical section. At the exit, particulates entrained in
the gas stream scour the cone. The atmosphere is reducing: mole
ratios of $CO:CO_2$ run generally around 10:1.

When the O_2 flow is interrupted for sampling, the furnace is
tilted nearly horizontal, and the metal and slag lie over the charge
pad and surrounding lining well up into the cone. This position
is reversed for tapping, bathing the opposite wall in metal and
slag. Tapping is concluded with pouring off of excess slag. The
furnace then resumes the position of Fig. 1a for recharging; intense
thermal shock and impact accompany the dropping of cold scrap onto
the 1600°C lining.

Hundreds of heats are conducted on a cycle time of less than
an hour each, for a total lining elapsed lifetime of several weeks.
Conclusion of a campaign is signalled by a dangerous and unrepairable

wear or loss of the refractory lining. Refractory consumption is
5 - 10 lb/ton of steel produced, or between 0.03 and 0.06 in.
average recession per heat. Our principal concern will be with
the nature of this wear and its minimization.

BOF REFRACTORIES

Two types of BOF brick structure will be recognized here.
Both depend upon the production and appropriate sizing of "grain",
which comprises the mineral or ceramic content of the composite,
and both contain coal-tar pitch in their manufacture.[1-4]

In the first and more extensively used type the sized inorganic
material is hot-mixed with a measured quantity of pitch to coat all
particles evenly, and then the mass is warm-compacted into its
final shape. The product is called a "pitch-bonded" or "tar-bonded"
brick; the latter term will be used here. It has become increasingly
widespread practice to subject tar-bonded brick to a final heat-
treatment in manufacture, the purpose of which is to advance the
curing of the pitch in order to diminish the temporary fluidity
which it will subsequently exhibit in the burn-in.[4-6] Products
so treated are called "tempered".

In the second type of brick, the sized inorganic constituents
are cold pressed and then fired to produce a relatively familiar
sintered refractory product. This is then vacuum impregnated with
hot molten pitch, which substantially fills the connected porosity.
The final product is called a "tar-impregnated" brick. It is
evident that the making of the BOF refractory is not in fact
completed until it is installed in the BOF and "burned in": here
the pitch is progressively decomposed, or "coked", to elemental
carbon.[1-4]

For obvious reasons the tar-impregnated brick has found use
in the BOF pad, while tar-bonded and/or tempered brick line part
or all of the remainder of the BOF interior. (Special materials
are used to line the tap hole, and still others for patching,

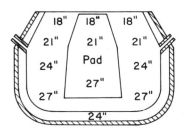

Fig. 2. Fold-out diagram of two zone BOF lining, slit through
         tap hole. (After Ref. 4).

repair, and temporary protection; these will not be discussed here.)
A simple two-zone BOF lining[4] is depicted in Fig. 2. The numbers
in the sketch give the thickness in inches of the lining in various
areas, adopted according to the different rates of recession
experienced. "Balanced" linings of much more complex architecture
than this have been developed.[2,4,7]

   The principal and most vital component of the BOF refractory
is MgO, and the development of this unique type of composite
material rests upon a centuries-old history of deriving refractories
from mineral resources. Nevertheless, it is convenient to achieve
an appreciation of the interworking of C and MgO in the composite
by treating carbon and pitch first: this we shall now do.

Carbon

   All sintered magnesia refractories and refractory grains are
somewhat porous. They also contain from a fraction to a few per
cent of siliceous material which is located predominantly in the
boundaries and/or interstices between MgO crystallites. This
material is thus present in the mineral "bond" within grains, and
additionally in a burned brick, among grains. [The term "grain"
is used as in the refractories industry to designate the sized
particles or fragments of dense inorganic, generally polycrystalline,
material employed in brick manufacture. The term "crystallite" is
common to the vocabulary of physical metallurgy and ceramics,
which often however uses "grain" in the sense of "crystallite".]

   This interstitial material is in communication with the molten
slag of the BOSP; and penetration of the latter into the interior
of the refractory along pore channels and silicate networks can
be rapid and extensive, altering the composition.[4,8,9] In the absence
of carbon this penetration occurs in the liquid state to a depth
corresponding to the melting point of the intruding material,
typically as much as 2-3 in. back from the hot face.[4,8,9] The liquid-
containing mineral structure in front is accordingly weakened, and
becomes susceptible to rapid mechanical wear and erosion if not
to fracture under very low loads.[4-6] Furthermore, the solid-liquid
interface moves back and forth as the hot-face temperature is
cycled, subjecting the material to intergrain and intercrystalline
volume changes and spalling or "peeling".[4,4,8,9] As large depths of
refractory are removed by these processes, the hot face and the
slag penetration limit move back in successive large steps and at
a rapid average rate. To the extent that interior surfaces of the
brick are covered with carbon, penetration by slag is
inhibited,[4,4,8-10] because the liquid does not readily wet
carbon.[2,3,11,12] In Fig. 3 the distribution patterns of pertinent
elements are plotted vs. depth in the absence and presence of
carbon[4,10] and in Fig. 4 the corresponding wear rates of brick in
the BOF are compared.[4]

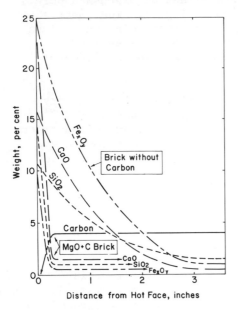

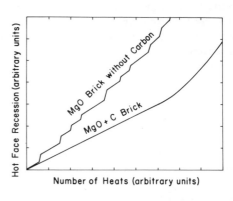

Fig. 3. Influence of carbon on
penetration of slag com-
ponents into MgO brick
(After Ref. 10).

Fig. 4. Schematic representation of
brick wear in the BOF,
showing effect of carbon
(After Ref. 4).

Carbon Coverage

For both tar-impregnated and tar-bonded refractories the wear
rate tends to decrease with increasing contained C,[3-6,11] i.e., with
an increase in the extent of the covering of interior surfaces and
the filling of voids. Ways of accomplishing these ends through
control of the viscosity, mixing, and penetration of pitch into
the structure are obvious. In tar-bonded brick it might be
thought that the pitch fraction could be increased at will to
achieve maximum coverage; but limitations are found here in brick
density and other processing characteristics, as well as in the
subsequent liquefaction of the pitch when the brick is first heated.
Under these limitations, one fruitful area of research remaining
has had as its object the maximizing of the percentage of residual
carbon derived from pitch through control or selection of the
chemical makeup of the latter.

The pitch softening point has the strongest correlation with
residual carbon,[3,4,10] being related to the molecular sizes, complexity, unsaturation, etc., in the pitch (Table 1). Other useful
correlations have been established with the "benzene-insoluble"
fraction and with the "quinoline-insoluble" fraction.[4,9,10] These

are said to be measures of the quantity of nuclei existing for
the deposition and retention of C during pyrolysis. Many of the
other theoretical and practical avenues open to the synthetic
organic chemist are of course barred by economic considerations
here.

Table 1. Carbon Contents as Functions of Pitch Softening Points

Softening Point(°C)	Conradson Coking Value (Wt % C)[4,10]
65-70	38-45
80-85	45-53
95-100	51-57
110-115	56-60
125-130	59-62

Geometrical Aspects of Carbon Coverage

It is logical that the volume of voids in the mineral structure
which require protection by C should be as small as possible.
Interstices and pores among grains are minimized in total volume,
and also in individual dimensions, by optimizing the size distribution, mixing, and compaction of the magnesia grains in the making
of brick.[4,13,14] Inclusion of a milled "fines" fraction (< 100 mesh
to submicron) of up to 30 v/o of the brick mix is instrumental in
this (as well as for providing sinterability in fired brick).
Control of the dimensions of voids is important: in a fired brick,
those below a limiting size, even if connecting, will not fill with
pitch; and large voids in either brick type present a problem of
a more universal kind.

A large space between grains will in all likelihood be filled
with pitch. When the latter is subjected to burn-in, its volume
shrinkage on pyrolysis may cause the resulting carbon to withdraw
from the surrounding mineral structure. We have,

$$\Delta\ell/\ell = \Delta V/3V = \text{constant}$$

from which $\Delta\ell \propto \ell$. Thus the dimensional mismatch $\Delta\ell$ between the
carbon residue and the surrounding magnesia is roughly proportional
to the original interstitial dimensions. While pitch effectively
wets MgO, carbon does not adhere strongly; hence a new void may
be formed during pitch pyrolysis, one of whose walls is bare oxide,
and whose dimensions can be such that liquid slag can intrude in
spite of its not wetting the adjacent carbon.

Although the pitch-filled void dimensions and the total pitch-
filled volume between grains should in principle be examined

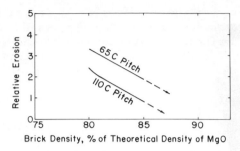

Fig. 5.   Relative erosion in slag test vs density, tar-bonded
          periclase brick (After Ref. 5).

separately, both tend to decrease with increasing brick density,
and so the latter shows an overriding correlation with slag
attack (Fig. 5).[3-5]   Pitch of lower softening point exhibits a
lesser residual carbon and a higher volume shrinkage $\Delta V/V$ on
pyrolysis, thus yielding a more permeable final structure at
equivalent mineral densities.

Carbon Structure

The pyrolysis of pitch at 1100°-1500°C yields a material which
is relatively lacking in long range order as evidenced by x-ray
diffraction,[4,9] but which has nonetheless a particulate character,
i.e., interfaces, as opposed to an indefinitely extended network.
X-ray evidence exists that the degree of graphitic stacking of
atoms in this residue is increased somewhat during BOF service, but
the change is relatively small.[4,9]   The microstructure and distribution of the carbon deposit within the cavities or interstices of
the mineral mass are apparently of greater importance, or at least
capable of much greater variation, than is the degree of crystal-
linity of the carbon.

A scanning electron microscope study of carbon in magnesia
brick has been reported by Gilbert and Batchelor[15] (Fig. 6).
Fig. 6a, a coked-brick fractograph, shows the flaky character of
the layers of carbon adhering to and broken away from particles
of MgO.   Fig. 6b illustrates the "mud-crack" structure resulting
from pyrolysis shrinkage when the carbon lateral dimensions are
too large.   The MgO/C parting discussed previously has undoubtedly
occurred in the plane of the photograph.   The incoherent nature
of some C-C interfaces is illustrated by a pyrolyzed mixture of
carbon black with pitch (Fig. 6c).   Bonding is poor between the
original and the newly-formed carbon.   Characteristic "ball" and
"socket" configurations can be seen where the interfaces surrounding
the spherical carbon black particles have failed in fracture.

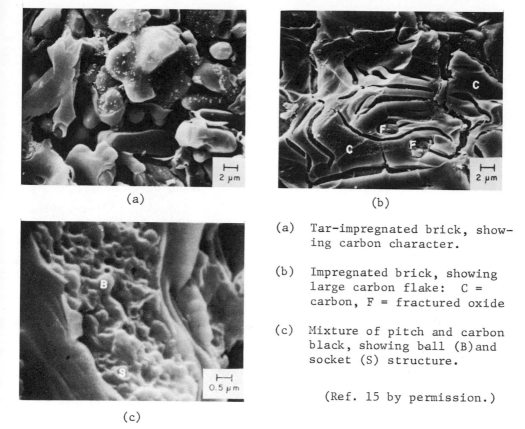

(a)   (b)

(a) Tar-impregnated brick, show-
     ing carbon character.

(b) Impregnated brick, showing
     large carbon flake: C =
     carbon, F = fractured oxide

(c) Mixture of pitch and carbon
     black, showing ball (B)and
     socket (S) structure.

(Ref. 15 by permission.)

(c)

Fig. 6. SEM View of Coked Carbon Microstructure.

The protection afforded to the mineral material in the BOF
brick by C is only transitory, i.e., depends upon kinetic phenomena.
The gaseous environment in the furnace is capable of oxidizing C;
and so is the hot slag, which contains considerable quantities of
iron oxide.[3-5,8,9,12] The depletion of C at the hot face of the brick
is clearly evident in Fig. 3, and the slag penetration depth is
limited to the depleted zone.[3,4,8-10,12] The micrograph of Fig. 7
illustrates this slag reaction boundary. Since erosion, abrasion,
washing, and impact of the slag-enveloped surface grains of magnesia
are principal processes causing recession,[4,6,8,12] it is apparent
that the recession rate can be impeded by impeding the rate of
oxidative loss of C through minimizing of its surface. The microstructural objective for this purpose is identical to that for
maximum protection of the underlying oxide. It is, under the
existing constraints of manufacture, approached most directly by
achieving a uniform, flaw-free, adherent coating of carbon on all
mineral surfaces to which the pitch has access.

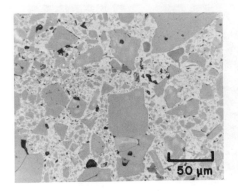

Fig. 7. Sectioned used tar-bonded Fig. 8. Sectioned coked tar-bonded
        periclase brick, showing   periclase brick, showing
        slag penetration to de-   fine-grained carbon (white).
        carbonized boundary.

Apart from the purely geometrical aspects of achieving such
a coating, a number of observations have been made which suggest
that coating uniformity can be improved through controlling the
characteristic dimensions of the carbon particles or layers
deposited in the coking process. Hubble[6] reported that the carbon
in burned and impregnated brick (after coking) was in the form of
large, irregularly distributed grains within the relatively coarse
pores; while in coked tar-bonded brick the structure of the carbon
was much finer and its distribution more uniform. Books, et al.[14]
observed the same trend with pore size, viz. coarse-grained
carbon in 10-20μm dia pores, fine-grained carbon in 1-4μm dia pores.
Herron and Runk[16] reproduced these findings, both in laboratory
experiments using a variety of pitches and in their own examination
of coked brick. They further related the presence of fine-grained
C to the presence of finely subdivided MgO, concluding that the
latter nucleated (or, interrupted) carbon deposition in the coking
of the pitch. Similar observations of carbon structural variations
in brick had been made earlier.[8] Fig. 8 illustrates the fine
carbon inclusions in coked tar-bonded periclase brick.

Evidences of superior performance of fine-grained carbon (or,
of mineral fines that produce it) are numerous, though these do not
correlate readily with improved protection against slag on the one
hand and improved resistance to oxidative loss on the other. These
evidences include, in addition to decreased pitch mobility in the
course of pyrolysis,[14,16] an increased carbon yield during coking[14]
and increased efficiency of a given percentage of C in resisting
slag advance.[12,16]

Carbon Corrosion Reactions

The destruction of carbon may occur to some extent through washing or eroding of large particles (which bond poorly either to other carbon or to oxides), but in an optimum structure it proceeds primarily by oxidation. The role of O_2 and of CO_2 in the BOF in oxidizing C is apparent, and these agents dominate in the cone where slag coverage is incomplete. But in the barrel section under continual washing of the lining with metal and slag, little direct oxidation of C by gases occurs during a heat. The primary oxidizing agent here is Fe_xO_y contained in the slag, and behind the immediate hot face metallic iron is usually found accompanying carbon depletion.[3-5,8,9,12] It has also been noted that original iron oxide contamination present in the brick will result in C loss and in the appearance of Fe metal, far deeper into the refractory microstructure than the penetration depth of slag, and even in the absence of slag.[8,9,16]

Very recent studies have identified two more potential oxidizers of carbon, both inherent in the refractory material. These are SiO_2[17] and finally MgO itself.[18] Reactions of C with these components are minor at temperatures much below 1600°C, but surface reactions of this type have long been postulated in connection with the low wetting qualities of carbon with oxides and silicates. The papers cited conclude that these reactions will set temperature limits on the serviceability of pitch-containing refractories in the BOF that are only a little above the 1700°C maximum of present practice.

Magnesia Grain

Study of Figs. 3 and 7 shows that the carbon and the mineral content of BOF brick work cooperatively in the process environment, each protecting the other. Carbon inhibits the intrusion of liquid slag and its consequences, while the exposed MgO at the hot face protects the immediately underlying C from oxidizing agents. Thus resistance of the exposed surface grain to slag attack and removal is a vital part of the dynamic chain of refractory life.[3,4,6,8,19] It is important to maximize both the amount (accomplished by sizing and other brick processing parameters) and durability (discussed below) of the MgO standing between the carbon and the hot face.

Grain Density

Since voids in the inorganic structure would be the primary avenues of slag penetration, the first object of grain design for durability is to minimize pores. Substantial advances in BOF brick performance have been made by increasing the density of the grain. In largest part this has been accomplished in manufacture by adopting a two-step firing process.[13,20] The mineral source material

is fired first as powder. This firing removes some impurities,
converts the major constituent ($MgCO_3$ or $Mg(OH)_2$) to MgO, and
begins the diffusion processes which perfect the lattice structure
of the MgO crystallites hence beginning their densification. This
fired powder is then compacted at high pressure, usually in
briquetting rolls, to yield "beetles" or pellets of high green
density, characteristically sized ~1/2 - 1 in. These are larger
than the largest grains that will be subsequently derived, but
small enough to permit effective heat transport and mass transport
processes in the second firing. The final firing may remove other
volatiles of concern in particular cases,[21] but principally it
completes the densification and bonding within the grain. Coarse
commercial grain made in this manner ranges between 5-10% porosity.[4,13]
As is familiar in sintering, the powder crystallite sizing, compaction, temperature of prior calcination, and detailed powder
chemistry as well as firing conditions all affect the result. High
grain densities are being achieved in tonnage quantities by the
use of gas-fired shaft kilns, in which the peak temperature exceeds
2000°C.[4,13]

That dense grain should be more resistant to disintegration
by slag than porous grain is evident; the same qualitative correlation is expected as with the overall mineral density of brick
(see Fig.5). This is a matter of minimizing the surface area
and the channel volume exposed by the solid system to chemical
attack. The remaining degree of freedom in refractory raw material
design is in the grain chemistry.

Grain Chemistry

Commercial magnesia grain is always comprised of more than one
phase. The MgO crystallites of coarse grains constitute the most
chemically stable and refractory component in the BOF: their dissolution in slag is slow and can be safely ignored as a primary
contributor to brick wear. It is the impurity system in the exposed
MgO grain which is most susceptible to chemical attack, and the
principal mechanisms of MgO loss follow the liquefaction of grain
and crystallite boundaries.[4,8,9,19,20] The effect is clearly visible
in the micrograph of Fig. 9.

If this impurity network is an avenue of slag penetration, we
would expect to find a correlation between the overall impurity
level of BOF refractories and slag erosion rates. Figure 10 plots
laboratory slag test data vs. impurity level for tar-impregnated
and for tar-bonded, tempered brick.[4,5] Considering the number of
other variables of brick constitution which can affect the result,
the correlation must be regarded as strong and favoring the highest
chemical purity achievable. As is the case for other refractory
oxides, most of the impurity can be found in the crystal boundary

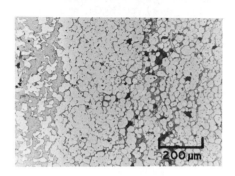

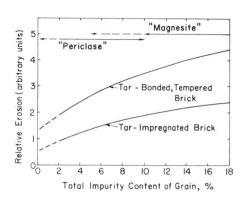

Fig. 9.Slag penetration along
crystallite boundaries
within MgO grain at hot
face.

Fig. 10. Relative erosion in slag
test vs impurity content
of MgO grain (After Ref. 5).

and interstitial regions; hence this correlation is simply with
the volume fraction of the non-MgO phases present, irrespective of
their own chemical variability.[4,6,8]

High-MgO grain is made commercially from several sources. With
due allowance for overlap arising from source selection and
processing differences, it is generally true that the natural
minerals magnesite ($MgCO_3$) and brucite ($Mg(OH)_2$) are the least pure
and of the most variable chemistry, while synthetic magnesia
precipitated as $Mg(OH)_2$ from brine or seawater is of the highest
and most uniform purity. [4,12,22] Refractories made from the first
sources are usually called "magnesite" refractories; these span
the approximate range of 80-90% MgO, with higher levels achieved by
beneficiating. The aqueous sources yield "periclase" refractories,
generally spanning the range 90-98% with a potential for exceeding
99%.

The resistance to slag reaction is, of course, by no means so
simple as Fig. 10 might suggest. It is necessary to investigate
individual impurities (as well as other characteristics) at depth
in order to explain the behavior of different refractory products.
Typical ranges of impurity contents in commercial MgO grains are
given in Table 2. The $CaO:SiO_2$ ratio, included in the data, will
be shown to be a chemical characteristic of very great importance.
In the refractories and steel industries this ratio is always
expressed by weight; however, for the present discussion, the
corresponding mole ratio will be used.

To examine grain chemistry intelligently, one should first
be aware of the chemical nature of the BOSP slag and its entrain-
ments. The principal components are Fe, SiO_2, CaO, Fe_xO_y, P_xO_y,
and CaF_2. Iron and its oxides come from the charge, its

Table 2.   Chemistry of Some Magnesite and Periclase Grains+

Type	% MgO	CaO	SiO$_2$	Fe$_2$O$_3$	Al$_2$O$_3$	Cr$_2$O$_3$	B$_2$O$_3$	ΣSiO$_2$ +R$_2$O$_3$	CaO:SiO$_2$ (moles)
M*	87.0	4.1	3.5	4.0	1.2	0.0	0.0	8.7	1.3
M	87.2	5.5	4.3	1.3	1.0	0.0	0.0	6.6	1.4
M	91.3	3.2	3.1	1.1	0.6	0.0	0.0	4.8	1.1
M	92.1	3.5	1.4	3.1	0.2	0.0	0.0	4.7	2.7
M	94.3	3.3	1.8	0.5	0.1	0.01	0.0	2.4	1.9
P*	90.8	5.5	2.8	0.3	0.3	0.2	0.1	3.7	2.1
P	93.2	2.0	1.8	1.0	1.1	·0.02	0.16	5.0	1.2
P	94.4	2.1	0.9	1.4	0.9	0.15	0.10	3.5	2.5
P	95.3	3.0	1.0	0.4	0.3	**	**	1.7	3.2
P	95.5	2.4	1.4	0.3	0.3	0.0	0.05	2.1	1.8
P	97.0	1.3	1.0	0.2	0.2	**	**	1.4	1.4
P	97.2	1.7	0.5	0.1	0.2	0.3	0.05	1.2	3.6
P	98.0	0.8	0.5	0.2	0.2	0.0	0.3	1.2	1.7
P	98.0	1.0	0.3	0.2	0.1	0.2	0.15	1.0	3.5
P	99.6	0.2	0.1	0.0	0.0	**	**	0.2	2.1

*M = Magnesite: Mineral origin   +Data from various sources
*P = Periclase: Seawater or brine origin in the literature and from
** = Data not available   this laboratory

oxidation, and added mill scale; SiO$_2$ and compounds of phosphorus
result from oxidation of major impurities in the iron. CaO (largely
solid initially, the dissolved fraction increasing toward 100% with
increasing "blowing" time and temperature) and CaF$_2$ (melting and
dissolving immediately) are added chemicals.[1,4] In the remaining discussion metallic Fe, free CaO$_{(s)}$, and the phosphorus and fluorine
compounds will be all but disregarded.

Again Kappmeyer and Hubble[4] include valuable data, from which
Fig. 11 is derived. The basicity of molten slag, as measured by the

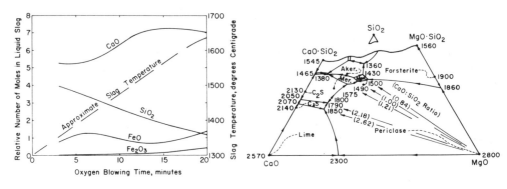

Fig. 11.   Chemical composition   Fig. 12.   CaO-MgO-SiO$_2$ phase
           of liquid BOSP slag   system (After Ref. 24).
           (After Ref. 4).

CaO:SiO$_2$ mole ratio, increases markedly with time from about 4:3 very early in the "blow" to about 5:2 midway through the period and to almost 6:1 after 20 min. The FeO content peaks early, then dips, and finally rises together with the Fe$_2$O$_3$ content as the temperature increases and C and Si in the iron become exhausted. With the iron oxides taken into account and allowing for the fluxing effect of fluorspar, the CaO content of the melt (by dissolution of the added lime) appears to stay fairly near to the saturation level as a function of temperature until late in the "blow" when the lime supply becomes exhausted.

Treffner[9] has reported CaO:SiO$_2$ ratios in a number of slags, that begin a little higher than the above and end significantly lower: no ratio above 4:1 was listed. Both ferrous and ferric oxide contents reported were generally higher than those above. P$_2$O$_5$ was also reported in the range of 1-3 weight per cent, with comparable MnO, and minor amounts of Al$_2$O$_3$ and TiO$_2$. Petrographic examinations disclosed what was probably CaF$_2$ among the many solid phases present in the cooled slag. Although Treffner's petrography confirmed expectations for high-lime compositions, e.g., Ca$_2$SiO$_4$, Ca$_3$SiO$_5$, and Ca$_2$Fe$_2$O$_5$, considerable glassy material and other evidences of nonequilibrium states (e.g., free CaO) were in early and mid-period slags. Numerous other slag analyses[4,12,19] generally agree on time trends but rarely on details of composition.

A safe conclusion is that BOSP slags are of quite variable composition according to the origin of the iron and to local shop practices as well as with time. It is useful for designing brick, however, to note that the slag is coolest when it is most acidic, and that by the time its temperature reaches 1500°C it is fairly reliably basic (i.e., into the Ca$_2$SiO$_4$, Ca$_2$Fe$_2$O$_5$ fields of the CaO-SiO$_2$-Fe$_x$O$_y$ system). Since it is impossible to maximize chemical resistance to both acids and bases in the same material, it would appear well to design the chemical content of BOF brick to be most resistant to basic slag. One desirable characteristic of that chemical content will obviously be that it should be as high-melting as possible.

For a simplified view, then, we consider the high temperature slag to be a source of dissolved CaO at high chemical activity, of SiO$_2$, and of Fe$_2$O$_3$ and FeO. The last named is readily soluble in MgO, and to simplify matters further we can imagine the slag which penetrates along magnesia crystal boundaries and interstices in the decarbonized zone to be depleted in this component, leaving principally the first three.

All commercial grains for BOF refractories also contain the same three impurity components, CaO, SiO$_2$, and Fe$_2$O$_3$, though in varying amounts. Iron oxide has long been recognized as depressing the melting point of silicate systems and as decreasing the interfacial tension between the silicate and the oxide;[22,23] pains have been taken to control this impurity. We can therefore start with the MgO–CaO–SiO$_2$ phase system[24] as approximately describing refractory-grade magnesites and periclases. A simplified, partial view of this system is presented in Fig. 12. The impurities CaO and SiO$_2$ must freeze first on cooling to give crystalline species that coexist in binary equilibrium with the MgO phase. In order of increasing CaO:SiO$_2$ mole ratio, these species are listed in Table 3.

Table 3. Magnesium and Calcium Silicates in Binary Equilibrium
with MgO

CaO:SiO$_2$ Ratio	Mineral Name	Nominal Formula	Nominal M.P.°C	Minimum M.P.°C	CaO:SiO$_2$ Range
0	Forsterite	2MgO·SiO$_2$	1900	1860–1430	0.00–0.84
1	Monticellite	CaO·MgO·SiO$_2$	1490	1500–1430	0.84–1.00
1.5	Merwinite	3CaO·MgO·2SiO$_2$	1575	1575–1435	1.00–1.21
2	Di-Ca Silicate	2CaO·SiO$_2$	2130	1800–1575	1.21–2.18
3	Tri-Ca Silicate	3CaO·SiO$_2$	2140	1850–1790	2.18–2.62

The minor phases existing at room temperature in grain or brick are usually identifiable petrographically or by x-ray diffraction, and in some writings the melting points of the pure compounds have been associated with these. But if the eutectic and peritectic valleys are traced out, depending on the total amount of SiO$_2$ present, the impurity system may freeze last on cooling, or first commence melting upon heating, anywhere in the "Minimum M.P." range given above. The "CaO:SiO$_2$ Range" gives the limits (in mole ratio) between which the indicated mineral should predominate in the MgO interstices. Monticellite and merwinite, especially, are capable of wide variations in composition as solid solutions.

When small quantities of other impurities are accounted for, e.g., Al$_2$O$_3$, Fe$_2$O$_3$, etc., the minimum melting point ranges given above must be still further depressed (in fact, thirteen solid phases are listed as able to coexist at equilibrium with MgO in various regions of the MgO–CaO–SiO$_2$–Fe$_2$O$_3$–Al$_2$O$_3$ system).[25] It is easily seen, then, that if the interstitial phases are not to be permitted to melt at 1650°C, the mole ratio CaO:SiO$_2$ must be adjusted either well below 0.5 or else close to 2. (Ca$_3$SiO$_5$ and free CaO, corresponding to CaO:SiO$_2$ > 2.2, are susceptible to hyration). Furthermore, if a ratio below 0.5 obtains, reaction with CaO-rich slag will produce a low-melting inter-diffusion zone by raising the ratio, while the lime-rich interstitial material is not subject to this cause of increasing reaction rate. (The reverse is true for an acidic slag,[9,12,19] but this argument was set aside, above.) Consequently, a mole ratio of ~2 is preferred in the impurity fraction of theBOF grain; and this ratio becomes the more important to maintain, the higher the content of SiO$_2$. Higher values are actually desirable to account for the solubility of CaO in MgO,[26,27] and for the presence of impurities such as Al$_2$O$_3$, Cr$_2$O$_3$, and Fe$_2$O$_3$ which combine with CaO in the interstitial system.[9,22,23,25]

If Fe_2O_3 or FeO is present in *significant* quantity together with
MgO, CaO, and SiO_2 as above, melting at BOF hot face temperatures is
certain.[20,22-24] Since this condition is precisely that of the slag-
refractory interface, the interstitial material in the refractory will
necessarily melt there. The resistance to slag attack is thus clearly
a matter of kinetics, and adjustments of grain chemistry such as that
just described can only diminish the rate of attack, not prevent it.

Research is bringing similar reasoning to bear on other components of brick and slag. For example, B_2O_3 is a characteristic
impurity of seawater and brine periclases,[4,20,28] and this component
depresses low-calcium silicate melting points and increases wetting
even more severely than does Fe_2O_3.[4,20,28,29] Taylor, et al.[29] have
shown that in part its melting point effect can be nullified by
increasing the CaO (or Ca_2SiO_4) content of the system; while other
work[20] has been devoted to its elimination from periclase in the
course of synthesis.

By contrast with B and Fe, Cr_2O_3 raises the minimum melting point
of low-calcium silicate systems perceptibly,[24] and decreases the
wetting tendency of the liquid. For these and other reasons Cr_2O_3 is
found as an additive (rather than impurity) in some synthetic BOF
periclases. As was mentioned, such an additive requires increasing
the CaO content if the impurity system is to remain in the Ca_2SiO_4
field.

Whether or not a portion of the intercrystalline material is glass
at low temperatures, and what this may mean in brick behavior, have
long intrigued investigators. Tighe and Kreglo[30] recently brought
transmission electron microscopy, selected area electron diffraction,
and the electron beam microprobe together to focus on this question.
Their study was of a periclase of $CaO:SiO_2$ ratio ~1.8 and B_2O_3 content ~0.3%. Several views they obtained of an intercrystalline glass
are reproduced in Fig. 13. The role of glass, if any, in magnesias of
low B_2O_3 content or high $CaO:SiO_2$ ratio has yet to be determined.

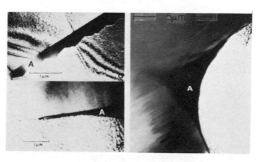

Fig. 13.   Transmission electron micrographs showing intercrystalline
           glass (A) in a periclase refractory (Ref. 30, by permission).

Meanwhile, Jones and Melford[22] have used the electron beam microprobe together with optical microscopy in an extended study of the partition of elements between major and minor phases. Table 4 combines their findings with the influence of lime:silica ratio observed in this laboratory. In characterizing selected magnesites and periclases, Jones and Melford found evidence of nonequilibrium states in the latter such that loss of CaO from the silicate system into the MgO phase could be expected in BOF service. This finding may relate to both long term storage behavior (i.e., hydration) and high temperature slag resistance of periclase refractories.

Table 4. Influence of $CaO:SiO_2$ Ratio on Phase Distributions
of Elements

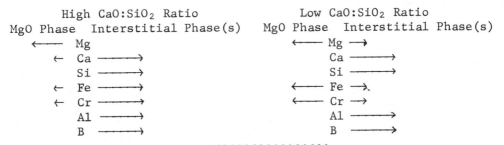

	High $CaO:SiO_2$ Ratio		Low $CaO:SiO_2$ Ratio	
	MgO Phase	Interstitial Phase(s)	MgO Phase	Interstitial Phase(s)
Mg	←—		←— Mg —→	
Ca	← Ca ——→		Ca ——→	
Si	Si ——→		Si ——→	
Fe	← Fe ——→		←— Fe —→.	
Cr	← Cr ——→		←— Cr —→	
Al	Al ——→		Al ——→	
B	B ——→		B ——→	

Numerous investigators[4,8,9,12,13,19,20,22,28,29] have noted microcstructural features of magnesite and periclase grain and fired brick that are consistent with the foregoing discussion and that bear on brick properties of importance. They are summarized in Table 5 in terms of factors which encourage angular MgO crystallite character and "direct"MgO-MgO bonding, with secondary phases relatively isolated in crystallite interstices; and factors which encourage rounded MgO crystallite character and a "silicate bond", i.e., a relatively continuous silicate network tending to surround the MgO crystallites.

Table 5. Influence of Composition and Microstructure on
Crystallite Bonding

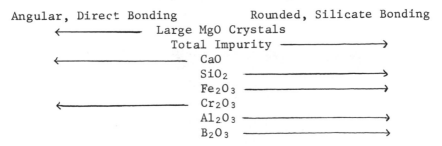

Angular, Direct Bonding   Rounded, Silicate Bonding
        ←——————— Large MgO Crystals
                      Total Impurity ——————→
        ←——————— CaO
                      SiO_2 ——————→
                      Fe_2O_3 ——————→
                      Cr_2O_3
        ←——————— Al_2O_3 ——————→
                      B_2O_3 ——————→

Fig. 14 illustrates microstructural features ranging from one of these extremes to the other. Clearly, not only minimization but also isolation of the silicate system will impede chemical attack by molten slag.

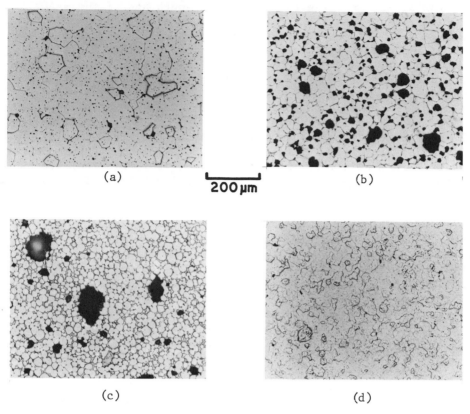

Fig. 14. Sections of selected MgO grains showing variations in
          bonding of crystallites: (a) direct-bonded, angular
          periclase; (b) intermediate periclase structure;
          (c) silicate-bonded, rounded periclase; (d) direct
          bonded, angular magnesite.

    Knowledge of the impurity chemistry and the microstructure of
MgO refractory grain can thus give considerable insight into the
slag resistance of BOF brick made from this grain and pitch. It
is quite evident that this same knowledge concerning a fired brick
should relate well to the high temperature mechanical properties
of that brick, because both slag resistance and high temperature
stiffness and strength depend to some extent on common features.

        Mechanical Behavior of Burned, Tar-Impregnated Brick

    The mechanical demands of BOF service are extremely difficult
to translate into quantitative terms. They consist essentially
of abrasion resistance, modest low temperature strength, high
resistance to thermal cycling damage in the elastic range, and

high temperature creep resistance and creep-rupture life. The
maximizing of low and high temperature strength and of thermal
shock resistance present a conflict, e.g., extremely dense, direct-
bonded MgO should excel in both strength and corrosion resistance
but is poor under thermal cycling. Economic constraints are also
severe, e.g., some ways of improving hot strength which are recognized and technically achievable are not economically competitive.

      Much has been written concerning these mechanical demands and
the corresponding brick properties,[4,6,10,13,20-22,28,29,31-34] and the
steel industry has adopted certain mechanical test practices and
specifications for product acceptance as BOF lining. Rather than
to repeat this considerable literature here, just three selections
are presented below which illustrate the applicability of chemical
principles.

      First, consider the compressive creep data reported by Van
Dreser[13] (Fig. 15). Specimens of fired periclase brick of various
purities were subjected to static load at 1600°C in air and the
compressive strain after 72 hr was recorded. Carbon was not
present in these specimens. Brick density was not constant in
the series, and undoubtedly affected the results, though not the
conclusions. Two specimens exhibited early shear failure, i.e.,
intergranular parting, and a third, though it survived, showed
very high creep in view of its overall purity of ~98%. These three
were of $CaO:SiO_2$ mole ratio < 1.8; it is evident that in the two
which failed the impurity phase was literally molten, while there
could be some doubt about the third. Recalling the influence
of amount of impurity on the degree of direct MgO-MgO bonding (i.e.,
isolation of the impurity phase), the specimen which survived may

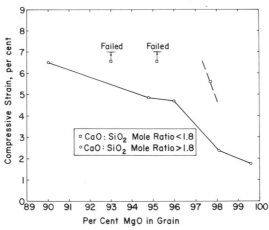

Fig. 15.   Creep of periclase refractories vs grain purity: 1600°C,
           72 hr, 25 psi (After Ref. 13).

have contained a sufficient network of MgO-MgO bonds to prevent its
rupture in spite of the presence of a liquid phase. The remaining
five specimens (connected by the solid line) were of $CaO:SiO_2$
ratio >1.8. There is, accordingly, less liquid present at 1600°C.
But the influence of the $CaO:SiO_2$ ratio on direct bonding is strong
and must explain survival of even the 90%-MgO specimen. These two
factors in combination would yield the greater stiffness of the whole
group, with the extent of direct MgO-MgO bonding increasing with
overall purity.

There are implications that creep measurements over a range of
temperatures might be highly informative. A very few of these
are on record;[21,34] these have been reserved for discussion in a
companion paper (this volume, page 387).

Next, consider hot modulus of rupture data on periclase brick,
burned and burned-impregnated, as reported by Kappmeyer and Hubble[4]
(Fig. 16). Two different periclases are represented which yield
quite different strength-temperature curves. Refractory qualities
of the impurity phases can be immediately inferred. In both cases
the bricks are appreciably strengthened by the presence of carbon,
but (a) the effect is nil at low temperatures and (b) the general
strength-temperature characteristics of the two types of brick are
nearly unchanged. The improvement of strength in each case can be
attributed to the prevention of liquid silicate movement by carbon,

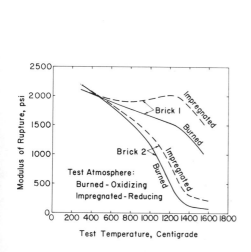

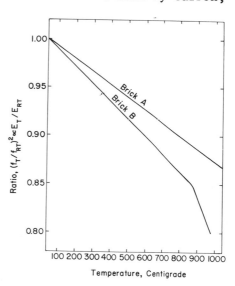

Fig. 16. Modulus of rupture of
two periclase bricks,
as-fired and tar-impreg-
nated (After Ref 4).

Fig. 17. Temperature dependence of
sonic modulus of elasticity
of two periclase bricks
(Ref 35, by permission).

as was suggested in Ref. 4; but carbon bonding of MgO does lend an
increment of strength, which may have become increasingly evident
as the mineral bond became more plastic with increasing temperature.

Finally, with the value of making measurements over a wide
temperature span clearly in evidence, consider the sonic modulus
of elasticity which is a sensitive index of grain (or crystallite)
boundary sliding. Unpublished data obtained recently at Bethlehem
Steel Corporation[35] is shown in Fig. 17. Again, carbon was absent
from these structures. Repeated measurements made on specimens
in the vicinity of 3×10^3 cps were said to give quite reproducible
results. The square of the ratio of the fundamental flexural
frequency at temperature to that at room temperature is plotted vs.
temperature: the ordinate is proportional to the ratio E_T/E_{RT}
(where E = Young's modulus).   The data for material
A indicate no curvature up to the highest test temperature, while
curve B shows a sharp knee at about 880°C as well as a less-stiff
quality over the whole elastic (linear) range. Whether material
B is less stiff in an absolute sense is not evident, because the
plots do not distinguish room temperature values of E. The slopes
do not relate simply to the quantity of impurities present: the
stiffer material A actually contains about twice the $(SiO_2 + R_2O_3)$
of the other. Knowing that fact, and assuming the impurity systems
in the two cases to have comparable elasticity at low temperatures,
we could only conclude from the difference in initial slopes that
there is greater *isolation* of the impurity into interstices in
material A, i.e., there is a greater extent of MgO–MgO bonding, a
lesser breadth of impurity phase available to strain in shear
between MgO crystals, and a lesser cooperative strain in the
impurity network because of its interruption. If we knew further
that $(E_A/E_B)_{RT} > 1$,[35] this conclusion would be confirmed; and that
is actually so.[35]

The knee in curve B reflects a lower melting constituent in
this material than exists in material A. The sharp change in slope
could signal incipient melting of a very minor component of the
system (which seems unlikely in view of the data discussed under
"Grain Chemistry" above), or it could signal greatly enhanced
solid state diffusion in a component occupying a more significant
volume fraction of the system. The latter interpretation is
tentatively preferred. From the authors' experience elsewhere,
such a knee would be found at between 2/3 and 3/4 of the absolute
melting point of the impurity, from which limits of about 1250°
and 1450°C are determined for that melting point. The impurity
melting point in material A must be at least 150°C higher, by the
same reasoning.

The most probable differences between materials A and B are
therefore in CaO:SiO_2 ratio and/or in levels of such impurities as

Fe_2O_3 and B_2O_3. From the range of melting points estimated for the impurity in B, it seems certain that Fe_2O_3 and/or B_2O_3 are present; and from that of A, its $CaO:SiO_2$ ratio should be near 2:1. These deductions are supported by chemical analyses of materials A and B[35] (Table 6). The actual minimum melting points of the impurity systems are estimated to be, respectively, ~1500°C and ~1200°C.

Table 6. Compositions of Periclase Bricks
A and B (After Ref. 35; See Fig. 17)

Constituents	A, w/o	B, w/o
MgO	95.5	98.0
CaO	2.4	0.8
SiO_2	1.4	0.5
Fe_2O_3	0.3	0.2
Al_2O_3	0.3	0.2
B_2O_3	0.05	0.3
$CaO:SiO_2$ (moles)	1.8	1.7

Mechanical Behavior of Coked Tar-Bonded Brick

When magnesia grains are not mineral bonded, but are joined by a network of carbon (see Fig. 8), the mechanical properties are greatly altered from the above. In general, the intercrystalline silicate system within each grain assumes far less importance, while the MgO grain sizing, overall density, and the quantity, distribution, and bonding or adhesive qualities of the carbon between grains become dominant.

Assuming optimum grain size distribution and optimum compacting, the larger MgO grains in the system are virtually in contact, separated by only a thin film of C. The strength would be expected

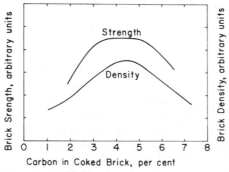

Fig. 18. Strength and density as schematic functions of carbon content of coked tar-bonded BOF brick.

to peak at this point, decreasing with either increasing or
decreasing carbon content about the optimum value as illustrated
in Fig. 18. The MgO-C bond being less strong than the MgO-MgO
bond, the low temperature strength of coked tar-bonded refractories
is of course considerably below that of burned brick, e.g., compare
cold crushing strengths of 6,000 and 12,000 psi. On the other
hand, the microstructure is characteristically resilient and hence
exceptionally thermal shock resistant, and curves of strength or
creep vs. temperature should be considerably flatter than for burned
brick up to the point at which the carbon bond is destroyed by
chemical reaction (>1500°C). Consequently, once burned-in as a BOF
lining, tar-bonded brick display very satisfactory bulk thermomechanical properties.

The resistance of surface grains to impact and abrasion is
necessarily inferior to that of tar-impregnated brick. Destruction
of the carbon bond by oxidation at the hot face in the BOF (visible
in Figs. 3 and 7) leaves the exposed MgO grains vulnerable to
mechanical removal. It is principally for this reason that the
more costly burned and impregnated brick has been preferred for the
charge pad and sometimes for other high wear areas such as at the
exit end of the cone, or "nose", of the BOF. But the thermal shock
sensitivity of such brick nullifies some of its wear advantage.
Conversely, either sintering or slag reaction occurring in the hot,
decarbonized zone of tar-bonded periclase refractories can develop
strength in their surface layers that would not otherwise be
expected. With utilization of residual slag coatings or other
temporary protective measures to decrease abrasive damage during
charging, it is possible to operate a BOF economically with an all
tar-bonded lining.

The MgO-C bond has been subjected to but little detailed physical
and mechanical investigation to date. It is possible that future
research into carbon microstructures and C-MgO interfaces in tar-
bonded periclase refractories will lead to still further improvement
of their mechanical characteristics.

A CASE STUDY

A subject not discussed at all in the preceding pages is the
tremendous contribution made to the survival of BOF linings by the
steel industry's study and adjustment of its own operating
procedures. These will not be examined here; suffice it to say
that a major part of the improvement of BOF refractory service life
has been directly attributable to the competence and diligence of
the refractory user. Nevertheless, it is equally true that the art
of manufacture of these materials has improved steadily over the
past decade.

Whatever the combination of reasons, new shop records for BOF lining life (measured in heats per campaign) are appearing at great frequency across the U.S.A. To close the discussion on a note of current reality, let us examine one BOF lining material (tempered, tar-bonded periclase) from a 1970 record campaign. The brick in question made up the barrel and cone sections of a 200 ton vessel.

The periclase was derived from seawater, using calcined dolomite (CaO·MgO) as precipitant; the basic process has been described elsewhere.[36] The grain was of the "double-pass" type, i.e., hearth furnace calcined, briquetted, and shaft kiln fired, followed by crushing, screening, and blending to optimum sizing specifications including use of ball-milled fines. Maximum grain size was ~3/8 in. For the manufacture of brick,[37] the coarse grain fractions were preheated to 200°-250°C, and these were mixed together with the fines and with preheated high softening-point pitch in a double shaft batch pug mill. A jacketed holding mixer and jacketed feeder maintained the mix above about 100°C for delivery into the cavity of the mechanical brick press. The warm bricks emerging from the press were palleted on cars, and immediately entered a gas fired tunnel tempering oven containing a final air cooling zone. Following completion of quality assurance procedures, the product was packaged for shipment.

The plant identification of the product brick (partial list) is given in Table 7, together with data obtained independently in this laboratory on specimens taken from this shipment. This material is seen to be generally in the middle of the plant production range. It has a $CaO:SiO_2$ mole ratio of 4.0, desirable in view of the R_2O_3 content and the B_2O_3 in particular. The grain microstructure would be well represented by Fig.14a, and the carbon deposit after coking appears as in Fig. 8. The residual carbon is below the plant average, but in range.

A multiple zoned, "balanced" lining was installed in the BOF, including the above brick as noted. Thicknesses of the barrel and cone linings were principally 27 and 24 in., respectively.

Minor spalling of the tempered, tar-bonded brick was noted in the cone and low in the barrel during burn-in to 1300°C. The two small lower spalls (1 1/2 - 2 in. deep) spread in width but not in depth during the first hundred heats, after which they gradually disappeared as wear rate became even throughout the area. Operation from then on was apparently routine, employing the regular lining maintenance procedures of the particular shop. At the conclusion of the campaign, a pattern of uneven wear had developed which was usual and characteristic of this shop operation.

Table 7. Properties of Tempered, Tar-Bonded Pericalse BOF Brick: Comparison of Catalog and Actual (Single Shipment) Values

	As Packaged		Coked ($1000°C$, N_2)		Ignited ($1000°C$, Air)	
	Catalog	Actual	Catalog	Actual	Catalog	Actual
Bulk Density, lb/ft^3	190. – 193.	190.4	186. – 189.	187.7	178. – 181.	179.1
App. Sp. Gr., g/cm^3	3.17 – 3.20	3.18	3.29 – 3.33	3.31	3.42 – 3.45	3.44
App. Porosity, %	3. – 6.	5.0	9.5 – 11.5	9.8	16.5 – 17.5	17.2
Ignition Loss %	6.	6.2				
Residual C, %			4.4 – 4.7	4.4		
Chemistry (plant data), %:						
MgO					98.	97.9
CaO					0.8 – 1.2	1.12
SiO_2					0.2 – 0.4	0.30
Fe_2O_3					0.2	0.25
Al_2O_3					0.1	0.11
Cr_2O_3					0.2	0.16
B_2O_3					0.1 – 0.2	0.15

Fig. 19.   Used tempered tar-bonded periclase BOF bricks (a)[l. to r.,
           back] new brick, no. 3, no. 4, [front], no. 2, no. 6,
           no. 1; (b) no. 1; (c) no. 2; (d) no. 3.

      Fig. 19 shows five of the seven worn brick recovered from this
campaign, together with a specimen representing the as-installed
external condition. Owing to the manner of BOF lining removal, it
is rarely possible to identify a brick remnant, via records, with
its original location in the furnace. The locations of the seven

Table 8. Microscopic Observations of Used Tempered, Tar-Bonded Periclase BOF Brick.

Location	Slag Layer	Decarbonization	Brick Alteration	Wear Mode
Cone	Discontinuous, splatter; variable composition from early to finishing.	Gaseous (O_2, CO_2): gradational, without evidence of Fe, to several mm depth.	No slag penetration; sintering in the decarbonized zone.	Repetitive (early) peeling of decarbonized and sintered zone; abrasion.
Upper Barrel	Transition from cone to lower barrel character: low CaO, FeO, some Fe; some flow features; local variations.	Variable between gaseous (graded) and slag reaction (sharp boundary, evidence of Fe).	Variable: some sintered areas without slag penetration, some slag-intruded areas with reaction to the carbon line.	Combined processes: erosion (principal), corrosion, peeling at hot face; locally variable.
Lower Barrel	Continuous, uniform; finishing composition. Principally Ca_2SiO_4, $Ca_2Fe_2O_5$, MgO-FeO solid solution (both grain residues and slag precipitate). See Figure 20 a.	Slag reaction: fine Fe inclusions, Fe_xO_y depletion, sharp decarbonized zone boundary. Variable depth, averaging several mm. See Figures 20b and c.	Slag penetration throughout decarbonized zone, via boundaries and matrix area. No sintering. Slag at C boundary contains Ca_3SiO_5, Fe, no precipitated MgO. See Figure 20c.	Corrosion (principal), erosion. Even and uniform hot face recession.

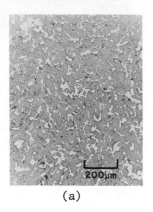 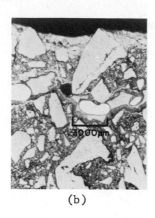 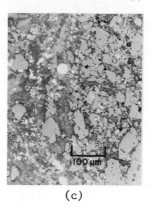

(a)   (b)   (c)

Fig. 20.   Microstructures of used tempered tar-bonded BOF brick
           from lower barrel: (a) slag layer, Ca_2SiO_4 rich area;
           (b) brick section, decarbonized zone; (c) brick section,
           slag (left) - carbon (right) interface.

specimens deduced from microscopic examinations were Nos. 1 and 2
in the cone, No. 4 in the upper barrel, and Nos. 3, 5-7 in the
lower barrel. Figures 19b, c, and d show closeups of specimens
Nos. 1, 2, and 3, illustrating the different surface conditions
corresponding to different locations and exposures.

    Sections were taken of bricks 4 and 5 (1 in. increments) from
the hot face to 8 in. back. The average carbon content of each
1 in. section was determined: results were quite constant at about
5.5+0.2%, which is considerably higher than the preinstallation
(coking test) value. The higher carbon content of the used brick
is attributed to distillation of pitch constituents from the
original hot face toward the cold face during burn-in. These
analyses, averaging the carbon content of large depth intervals,
do not indicate the location of the decarbonized zone. The latter,
together with important characteristics of slag and brick, were
identified by optical microscopy. Table 8 gives these characteristics as functions of the brick locations, and refers to the
illustrative optical micrographs of Figs. 20a, b, and c.

    A significant feature of the kinetic protection afforded the
refractory by slag is the variation in composition of the latter
with depth of penetration, whereby the slag becomes more refractory
as its iron oxide content decreases toward the carbon boundary.
This causes a stiffening of the slag, which helps to bond the
surface MgO grains in place, as well as a decrease in interdiffusion
rates which helps protect the underlying carbon against oxidation.

    Considering all features of preinstallation brick character,
a knowledge of both general BOF exposure and the specific practices

of this particular steelmaking shop, and the normal features found
above in these brick specimens after service, it can be said that
this lining was typical of the advanced technology of today's tar-
bonded periclase refractory products and deserving of a share in
the BOF campaign records of 1970. Research, meanwhile, continues
toward advancing performance levels and reliability apace with the
needs of the steelmaking industry.

REFERENCES

1. B. L. Dorsey, Am. Ceram. Soc. Bull. 39 (5) 261 (1960).
2. R. N. Ames, Iron and Steel Eng. Sept., 1963.
3. R. W. Limes, J. Metals 18 (7) 865 (1966).
4. K. K. Kappmeyer and D. H. Hubble, Ch. 1 in Refractory Materials.
   Allen M. Alper, ed., Academic Press, New York, 1970.
5. D. H. Hubble and K. K. Kappmeyer, Am. Ceram. Soc. Bull. 45 (7)
   646 (1966).
6. D. H. Hubble, Am. Ceram. Soc. Bull. 47 (2) 170 (1968).
7. R. C. Padfield and C. R. Beechan, J. Metals 19 (4) 17 (1967).
8. J. Martinet, W. H. Boyer and J. E. Allen, "Wear Mechanism is
   Basic Oxygen Steel Process Refractories", presented at
   15th Pacific Coast Regional Meeting, Am. Ceram. Soc., San
   Francisco, Oct. 17, 1962.
9. W. S. Treffner, Am. Ceram. Soc. Bull. 44 (7) 546 (1965).
10. D. H. Hubble, W. H. Powers and J. A. Lamont, Am. Ceram. Soc.
    Bull. 44 (3) 226 (1965).
11. R. H. Herron, C. R. Beechan and R. C. Padfield, Am. Ceram. Soc.
    Bull. 46 (12) 1163 (1967).
12. P. T. A. Hodson & H. M. Richardson, Trans. Brit. Ceram, Soc.
    69 (3) 45 (1970).
13. M. L. Van Dreser, Am. Ceram. Soc. Bull. 46 (2) 196 (1967).
14. W. C. Books, R. H. Herron, and C. R. Beechan, Am. Ceram. Soc.
    Bull. 49 (7) 643 (1970).
15. V. Gilbert and J. D. Batchelor, Am. Ceram. Soc. Bull. 50 (2)
    156 (1971).
16. R. H. Herron and E. J. Runk, Am. Ceram. Soc. Bull. 48 (11)
    1048 (1969).
17. B. Brezny and R. A. Landy, "Effects of Heat and Silicate Phase
    on the Oxidation of Carbon in Burned Impregnated Brick", pre-
    sented at Refractories Division Meeting, Am. Ceram. Soc.,
    Bedford Springs, Pa. Oct. 9, 1970.
18. G. D. Pickering and J. D. Batchelor, "The Stability of Magnesium
    Oxide in BOF Refractories", presented at 72nd Annual Meeting,
    Am. Ceram. Soc. Philadelphia, May 5, 1970.

19. A. J. Owen, D. R. Shepherd, and G. Bull, "Mode of Wear of Lime-Magnesia-Carbon Refractories in L-D Vessels", presented at XI th International Ceramic Conference, Madrid, Spain, Sept. 22, 1968.
20. W. C. Gilpin, Refractories J. 45 (3) 68 (1969).
21. T. S. Busby and M. Carter, Trans. Brit. Ceram. Soc. 68 (5) 205 (1969).
22. D. G. Jones and D. A. Melford, Trans. Brit. Ceram. Soc. 68 (5) 241 (1969).
23. B. Jackson, W. F. Ford, and J. White, Trans. Brit. Ceram. Soc. 62 (7) 577 (1963).
24. E. M. Levin, C. R. Robbins, and H. F. McMurdie, Phase Diagrams for Ceramists, The American Ceramic Society, Inc., Columbus, Ohio, 1964.
25. J. R. Rait, Basic Refractories, Interscience Publishers, New York, 1950.
26. R. C. Doman, J. B. Barr, R. N. McNally, and A. M. Alper, J. Am. Ceram. Soc. 46 313 (1963).
27. T. Hatfield, C. Richmond, W. F. Ford, and J. White, Trans. Brit. Ceram. Soc. 69 (2) 53 (1970).
28. H. M. Richardson, M. Lester, F. T. Palin, and P. T. A. Hodson, Trans. Brit. Ceram. Soc. 68 (1) 29 (1969).
29. M. I. Taylor, W. F. Ford, and J. White, Trans. Brit. Ceram. Soc. 68 (4) 173 (1969).
30. N. J. Tighe and J. R. Kreglo, Jr., Am. Ceram. Soc. Bull 49 (2) 188 (1970).
31. H. J. S. Kriek and B. B. Segal, Trans. Brit. Ceram. Soc. 66 (2) 65 (1967).
32. R. L. Coatney, "Effect of Boron on the Compressive Strength of Periclase Grain", presented at Refractories Division Meeting, Am. Ceram. Soc., Bedford Springs. Pa., Oct. 5, 1968.
33. R. L. Coatney, "Compressive Strength of Polycrystalline Magnesia at 1500°C", presented at 6th Annual Symposium on Refractories, St. Louis Section, Am. Ceram. Soc., St. Louis, Mo., April 17, 1970.
34. J. R. Kreglo and W. J. Smothers, J. Am. Ceram. Soc. 50 (9) 457 (1967).
35. J. R. Kreglo, Jr., and W. J. Smothers, Bethlehem Steel Corp., Bethlehem, Pa., unpublished data.
36. C. R. Havighorst and S. L. Swift, Chem. Engr., Aug. 2, 1965, p. 84.
37. D. H. Taeler, Ceramic Age, 84 (8) 18-21 (1968).

DISCUSSION

H. M. Davis (Army Research Office - Durham): At 1650°C and higher, MgO is reducible by carbon. If graphite is taken as the standard state, the activity of carbon at the surface of the MgO grains

may be even greater than unity. What, then, is the kinetic limitation which prevents the destruction of the MgO? Is the carbon
spending itself in reaction with more readily reducible substances,
i.g., oxides of iron?
Authors: Two factors are probably involved in avoiding extensive
MgO-C reaction in BOF brick. The first is the temperature
profile, under which more readily reducible species (Fe_xO_y, CO_2
O_2, and even SiO_2) are present at the hot face but in particular
the temperature drops at roughly 50°C/in. through the brick,
such that the ΔF becomes positive not far behind the hot face.
The second is that the reaction is between two solids, with
gaseous products; in the absence of a mass transport medium (e.g.,
a liquid phase), reaction of MgO with carbon causes immediate
separation of the two phases and is hence self-limiting. If a
liquid (siliceous) phase is present, then its SiO_2 should be more
readily reduced than MgO, although both may react at temperatures
above 1600°C. Laboratory (isothermal) experiments have clearly
indicated reaction of MgO with carbon in basic brick specimens
at BOF hot face temperatures.

W. J. Smothers (Bethlehem Steel Corp.): We have some evidence
on used brick from the BOF that reduction of MgO took place.
Authors: Such evidences would appear to be in accord with the
findings covered by Refs. 17 and 18.

S. K. Dutta (Army Materials and Mechanics Research Center): Did
you measure the dihedral angles and solid-solid to solid-liquid
interfacial ratios for all the oxides in addition to MgO?
Authors: Many interaction effects are involved in wetting, so
that statements of the effects of individual components are
really an over-simplification. The available literature on the
subject is fairly well represented in the References.
S. K. Dutta: Did you observe any grain size effect by carbon?
Authors: Beyond the rough correlations cited in the paper, there
appears to be little or no more detailed information available at
this time.

THERMAL STRESS CRACK STABILITY AND PROPAGATION IN SEVERE THERMAL ENVIRONMENTS

D. P. H. Hasselman

Lehigh University

Bethlehem, Pennsylvania

ABSTRACT

A fracture-mechanical analysis is presented for stability criteria and propagation behavior of thermal stress cracks in brittle ceramics in environments so severe that initiation cannot be avoided. It is based on a mechanical model consisting of a rigidly constrained, uniformly cooled thin flat plate with a uniform distribution of microcracks; results are qualitatively similar to those obtained for a three-dimensional body with penny-shaped cracks. High stability of thermal stress cracks is attained in materials with high values of surface fracture energy, and low values of thermal expansion and Young's modulus. On catastrophic propagation of an initially short crack, the final crack is subcritical and has a length which is independent of material properties but depends only on the initial crack length and the crack density. It is suggested that materials with very high thermal shock resistance can be developed by synthesizing materials with high densities of microcracks.

INTRODUCTION

Ceramics because of their chemical inertness and superior high temperature mechanical behavior often are the only materials available for high temperature environments. Such applications, however, very frequently involve high heat fluxes and rapid temperature variations which will result in high levels of thermal stress. Under these conditions ceramic materials, because of their general brittleness, are susceptible to catastrophic failure, often rendering the material unsuitable for further use.

Two criteria currently are in vogue in the selection of materials with high resistance to thermal stress fracture. The first applies to crucibles, nozzles, laboratory and white ware, kiln-furniture, spark-plug insulators, etc., which generally are used in applications where thermal stress fracture and resulting cracking cannot be tolerated. For these conditions materials are selected with maximum resistance to the initiation of thermal stress fracture. The second criterion is applicable to severe thermal environments in which thermal stress fracture initiation cannot be avoided, e.g., refractory linings of blast furnaces, Bessemer converters, cyclicly operated kilns and ovens. For these applications materials are selected which show a minimum time-dependent change in strength, permeability and/or weight, i.e., minimizing recession of the furnace wall.

Theoretical studies[1-4] have shown that ceramic materials with high resistance to the initiation of thermal stress fracture should have high values of tensile strength, thermal conductivity and diffusivity and low values of coefficients of thermal expansion, Young's modulus of elasticity and Poisson's ratio. In addition, a low value of the effective viscosity[5] in creep can lead to rapid thermal stress relaxation. Experimental studies[2,3,6-10] to verify theoretically predicted behavior have shown that for well-characterized materials in well-defined thermal environments, thermal stress fracture behavior can be calculated with a good degree of reliability.

Selection of ceramic materials on the basis of minimum changes in physical behavior due to thermal stress fracture obviously is based on minimizing the extent of crack propagation rather than avoiding the initiation of thermal stress fracture. Some years ago, the writer[11] suggested that the extent of crack propagation in thermal stress fracture would be proportional to the elastic energy at fracture and inversely proportional to the surface fracture energy required to create the new crack surfaces. On this basis it was suggested that superior materials should have high values of the "thermal stress damage resistance parameter"

$$R''''= EG/S_t^2(1-\nu) \tag{1}$$

where E is Young's modulus, G is the surface fracture energy, S_t is the tensile strength and ν is Poisson's ratio. It is of interest to note that high values of R'''' require high values of Young's modulus and low values of tensile strength, directly opposite to the requirement of high strength and low Young's modulus to avoid the initiation of thermal stress resistance. Nakayama and Ishizuka[12] investigated the validity of this parameter. For a variety of materials they found a positive correlation of R'''' and

thermal shock behavior as determined by the number of cycles
required to produce a given percentage loss of weight. Conversely,
a negative correlation was found between thermal shock resistance
and the parameters which determine the resistance to thermal stress
fracture initiation. It is apparent, however, that the validity
of R"" must be limited as $S_t \rightarrow 0$.

    More recently, thermal stress crack stability and propagation
has been analyzed[13] using a fracture mechanical approach which
followed Berry's[14] analysis of crack propagation under constant
deformation, based on the concept of an effective modulus of
elasticity of material containing cracks. This concept describes
the change in compliance of the material due to the presence of
the cracks. The mechanical model consisted of a three-dimensional
solid containing penny-shaped cracks, rigidly constrained from
deforming when cooled through a temperature difference.[13]

    This paper presents a similar analysis for a flat plate more
closely approximating thermal stress fracture nucleated in the
surface. Analytical results for this model are shown to be
qualitatively similar to the results for the three-dimensional solid.
However, for the two models a quantitative difference between crack
length and critical temperature difference is shown to exist. The
theory is discussed with emphasis on crack propagation behavior
in severe thermal environments in which thermal stress fracture
initiation cannot be avoided.

                              THEORY

    Figure 1 illustrates the flat plate mechanical model which
has a crack density of N cracks/unit area. All cracks are assumed

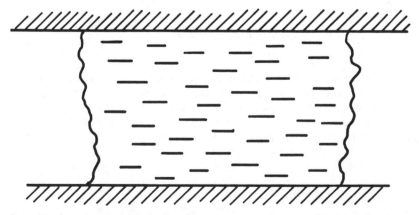

Fig. 1. Mechanical model for analysis of thermal stress crack
                          stability.

to be of equal length. The plate is uniaxially constrained with the cracks oriented perpendicularly to the direction of constraint and is cooled through a temperature difference ΔT.

After Berry[14], the effective Young's modulus is

$$E_{eff} = E(1 + 2\pi N\ell^2)^{-1} \tag{2}$$

where ℓ is the half-length of the cracks.

The total energy (W) per unit area is the sum of the strain and surface energies of the cracks,*

$$W = \frac{\alpha^2 (\Delta T)^2 E}{2(1 + 2\pi N\ell^2)} + 4G\ell N \tag{3}$$

where α is the coefficient of thermal expansion, ΔT is the temperature difference and G is the surface fracture energy required to produce unit area of new crack surface.

After Griffith[15, 16], the cracks are unstable whenever

$$dW/d\ell \leq 0 \tag{4}$$

which is satisfied at a critical temperature difference (ΔT_c) such that

$$\Delta T_c = (2G/\pi\ell\alpha^2 E)^{1/2} (1 + 2\pi N\ell^2). \tag{5}$$

Data calculated from Equation 5 are shown in Figure 2 by solid curves. It may be noted that for a given value of ΔT_c crack instability occurs between two values of crack length. ΔT_c is a function of N at long crack lengths only. This also can be shown by approximating ΔT_c (Eq. 5) for short and long crack lengths. For short crack length, $2\pi N\ell^2 \ll 1$,

$$\Delta T_c \approx (2G/\pi\ell\alpha^2 E)^{1/2} \tag{6}$$

which is independent of N.

*For simplicity, transverse strains are taken equal to zero.

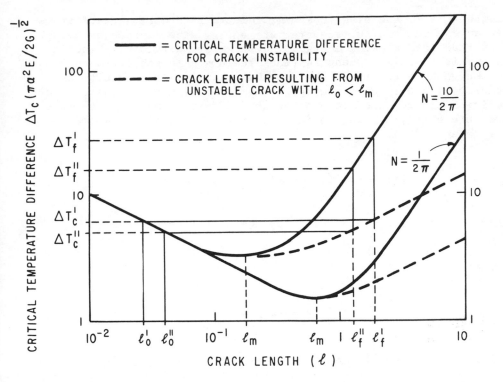

Fig. 2. Thermal stress crack stability and propagation behavior for rigidly held flat plate with N cracks per unit area, cooled through a temperature difference ΔT.

For long crack length, $2\pi N \ell^2 \gg 1$,

$$\Delta T_c \sim (8\pi G N^2 \ell^3 / \alpha^2 E)^{1/2} \qquad (7)$$

which shows ΔT_c to be directly proportional to N. For short crack lengths the present model shows an identical relationship between ΔT_c and crack length ℓ as for the three dimensional model with penny-shaped cracks. However, for long crack lengths, the model shows that ΔT depends on $\ell^{3/2}$ whereas for the penny-shaped crack a $\ell^{5/2}$ dependence is found. The minima in the stability curves (Fig. 2) can be calculated to occur at a crack length ℓ_m where

$$\ell_m = (6\pi N)^{-1/2} \qquad (8)$$

Equation 5, however, is not sufficient to completely describe crack propagation behavior. For unstable cracks with initial length $\ell_o < \ell_m$ the elastic energy release rate during crack propagation exceeds the surface fracture energy, the difference being converted into kinetic energy. As shown by Berry, this kinetic energy is a maximum when the crack reaches a length corresponding to the upper boundary of the unstable region. As a result, the crack will continue to propagate when all the kinetic energy and additionally released strain energy are transformed into surface fracture energy. This condition is met at a final crack length (ℓ_f) when

$$\alpha^2 (\Delta T_c)^2 E\{[(1 + 2\pi N \ell_o^2)]^{-1} - [(1 + 2\pi N \ell_f^2)]^{-1}\} = 2G(\ell_f - \ell_o) \qquad (9)$$

which is indicated in Figure 2 by the dotted curves.

For short crack lengths, with the aid of Equation 5, Equation 9 yields

$$\ell_f = (4\pi N \ell_o)^{-1}. \qquad (10)$$

Interestingly, equation 10 shows that the final crack length which results from an initially short crack $(\ell_o \ll \ell_m)$ is independent of material properties and is a function only of the crack density and original length. For the penny-shaped crack only a dependence on Poisson's ratio was found. Substitution of the Griffith solution for the critical fracture stress (S_t) for a crack with length ℓ_o yields

$$S_t = (GE/\pi\ell_o)^{1/2}. \qquad (11)$$

Substitution of Equation 11 into Equation 10 yields

$$\ell_f = S_t^2/4NGE. \qquad (12)$$

Equation 8 shows that the area A over which the cracks propagate $(A \propto \ell_f$ for the present model) with the exception of the factor $(1-\nu)$ is inversely proportional to the thermal shock damage resistance parameter R'''' (Eq. 1). The present theoretical approach suggests that the parameter R'''' is restricted to initially short crack lengths only when $\ell_o < \ell_m$. In contrast, long cracks with $\ell_o < \ell_m$ will not attain kinetic energy, but on instability will propagate in a stable manner and increase in length uniformly with increasing ΔT, as described by Equations 5 and/or 7.

This fracture-mechanical approach permits one to obtain from a single expression (Eq. 3)embodying the sum of elastic and surface energies three solutions: a) the thermal condition for fracture initiation, b) the extent of crack propagation and c) the thermal condition for instability of the newly formed cracks. The conventional engineering approach,based on the criterion that failure will occur when the maximum thermal stress reaches the tensile strength, results only in the condition for fracture initiation.

III. DISCUSSION

These results have a number of practical implications for the selection of materials for severe thermal environments in which thermal stress fracture cannot be avoided. The following discussion will be limited primarily to the effect of crack propagation on strength. Figure 2 illustrates the behavior of two cracks with initial length $\ell_o' < \ell_o'' < \ell_m$. It is assumed that (at instability) cracks can propagate rapidly, so that during the time required for propagation, ΔT is essentially constant. When the cracks are unstable ($\Delta T = \Delta T_c$) they will propagate to a length ℓ_f. This length is subcritical with respect to ΔT_c and ΔT must be increased to ΔT_f before the cracks will continue to propagate. The relative propagation behavior for the two cracks can be written

$$\ell_o' < \ell_o'' < \ell_m \quad \rightarrow \quad \Delta T_c' > \Delta T_c'', \ell_f' > \ell_f'' \text{ and } \Delta T_f' > \Delta T_f''. \tag{13}$$

The cracks are subcritical over the temperature difference intervals

$$0 < \Delta T < \Delta T_c', \Delta T_c'' \tag{14a}$$

$$\Delta T_c', \Delta T_c'' < \Delta T < \Delta T_f', \Delta T_f'' \tag{14b}$$

As suggested by Equation 11, the strength of a solid is an inverse function of crack length. Equation 13 suggests that for two different materials the initially stronger will require higher ΔT_c to initiate fracture. Crack propagation behavior is such, however, that after fracture the originally weaker material will be the stronger. Also strength will show an instantaneous decrease at ΔT and will be invariant over the temperature ranges given by Equation 14. Only when $\Delta T \geq \Delta T_f$ will strength decrease uniformly with increasing ΔT.

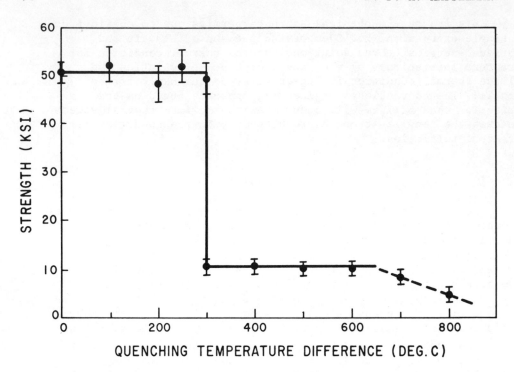

Fig. 3.   Strength behavior of Coors AD-94 alumina rods (0.187 in. dia.)
as a function of quenching temperature difference (after
Hasselman[17]).

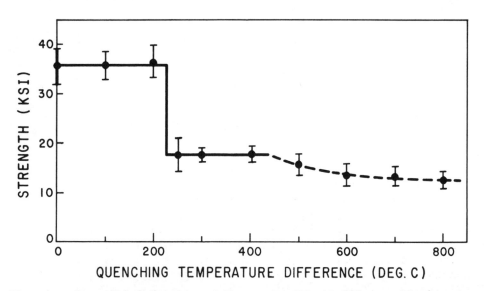

Fig. 4.   Strength behavior of Wesgo AL-300 (0.195 in. dia.)
alumina rods as a function of quenching temperature
difference (after Hasselman[17]).

Figures 3 and 4, respectively, show recent data[17] illustrating this general behavior for the thermally shocked AD-94* and AL-300** alumina rods which had been quenched into room temperature water from various temperatures. Strength was measured in 4-point bending. It is clear that for mild thermal shock where thermal stress fracture is not expected, the AD-94 material is preferred because of its high original strength. In severe thermal environments, however, where fracture cannot be avoided, the AL-300 alumina is the logical choice because of its superior strength behavior after thermal stress fracture.

Strength of shocked materials having long cracks ($\ell_o \geq \ell_m$) is expected to differ significantly from that of materials with $\ell_o \ll \ell_m$: (1) it is not expected to show an instantaneous decrease at ΔT_c, (2) it will be invariant over only one range of temperature difference ($0 < \Delta T < \Delta T_c$) and, (3) it is expected to show a continuous decrease with increasing ΔT for $\Delta T > \Delta T_c$. This behavior is illustrated in Figure 5 for strength of a chrome-alumina cermet material ("chromal") as determined by Tacvorian[18]. Figure 6 shows strength behavior[18] expected from a material with initial crack length only slightly shorter than ℓ_m. For severe thermal environments the general behavior of strength as shown in Figures 5

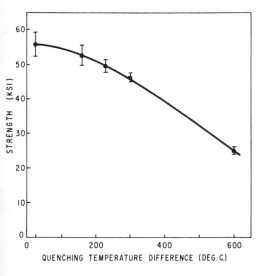

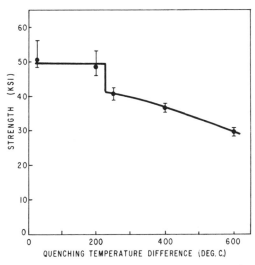

Fig. 5. Strength behavior of "chromal" ABF-5 (after Tacvorian[18]).

Fig. 6. Strength behavior of "chromal" AB-7 (after Tacvorian[18]).

---

*   Coors Porcelain Company, Golden, Colorado.
**   Western Gold and Platinum Company, Belmont, California.

and 6 is to be preferred to the behavior shown in Figures 3 and 4.

The evaluation and development of new insulating refractories generally appears to be aimed in this direction. Figure 7 shows strength differentials resulting from thermal shock for fire clay-refractories as determined by Morgan[19]. It may be noticed that the higher the initial strength the greater is the loss in strength. This behavior contradicts the theories of thermal stress fracture initiation where the weakest material should have the lowest thermal shock resistance. However, this behavior is in agreement with the present theory which predicts that the weakest material (i.e. the material with the longest cracks) should exhibit a minimum extent of crack propagation and should retain most if not all of its original strength.

The preference of large initial crack size to small initial crack size can be inferred further from the curves in Figure 2. Suppose that for a crack density $N = 10/2\pi$, a crack length ℓ_f' is the maximum which can be tolerated. This crack length for instance may correspond to the cross-sectional area (i.e., zero strength) of the specimen. For strong material with short cracks having lengths $\ell_o << \ell_m$, ΔT_c can only be increased by decreasing the initial crack length ℓ_o to the minimum permissible value (ℓ_o') which gives a maximum $\Delta T_c = \Delta T_c'$. If however, a material is selected with

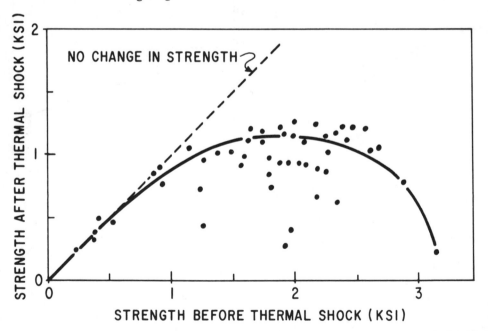

Fig. 7. Strength after thermal shock as a function of initial strength for refractory clay bodies (after Morgan[19]).

initial crack length $\ell_o = \ell_m$, on attaining crack instability, the crack will propagate along the solid line with the result that strength will reach zero (i.e. $\ell_f = \ell_f^!$) when $\Delta T = \Delta T_f^!$ which is well in excess of $\Delta T_c'$. In this manner materials with high densities of long cracks can be subjected to considerably more severe thermal environments before complete failure than strong materials with short cracks. This phenomenon is offered as an explanation for the observations of Rossi[20] for the high thermal shock resistance of MgO-W composites with high densities of microcracks which result from the thermal expansion mismatch between the MgO and W phases. It is expected that the heavily microcracked magnesium dititanate and β–eucriptite materials investigated by Bush and Hummel[21] will exhibit similar behavior. It may be pointed out that no advantage is attained by introducing cracks with $\ell_o > \ell_m$. This will only lead to a weakening prior to thermal shock without increasing the temperature difference required for complete failure.

That the deliberate introduction of cracks can increase ΔT_c can also be ascertained from the data in Figure 3. The rods subjected to a quenching temperature $\Delta T < 300°C$ can be considered as a material with a high density of surface cracks. The crack density for this material was observed[17] to be $N \approx 60$ cracks cm^{-2}. As indicated by the strength results the cracked material must be subjected to a $\Delta T \approx 650°C$ for the cracks to become unstable, whereas for the original uncracked material a value of ΔT of only $300°C$ is required for crack instability.

RECOMMENDATIONS

It becomes clear that the problem of thermal shock resistance of brittle solids can be approached in three ways:
1. For relatively mild thermal environments the incidence of thermal stress fracture can be avoided by selecting and developing materials with high values of strength, thermal conductivity and thermal diffusivity, and low values of coefficient of thermal expansion, Young's modulus, Poisson's ratio, emissivity and viscosity.

2. In severe thermal environments in which fracture cannot be avoided even in the best materials, Equations 10 and 12 suggest that the extent of crack propagation and the resulting reduction in strength can be minimized by selecting materials with relatively large cracks and by enhancing the number of cracks. In terms of engineering properties this implies the selection of materials with moderate values of strength and high values of surface fracture energy and Young's modulus of elasticity. It may be noted here that efforts to avoid thermal stress fracture initiation by increasing strength, unless successful, become more and more disastrous. Only if materials with tensile strengths approaching 1×10^6 psi are

developed can the problem of thermal shock resistance be solved.
This, however, is not expected to occur in the near future.

3. The most promising approach is to select and/or develop
materials with low values of thermal expansion and Young's modulus,
and a high value of surface fracture energy, combined with high
densities of microcracks. The material should be tailored such
that the original crack length l_o exceeds l_m (Eq. 8). This
assures that when thermal stress fracture is initiated, crack
propagation will occur in a stable manner. Catastrophic failure
is thereby avoided.

In the opinion of the writer this approach is most feasible
at the present time to overcome many immediate engineering
problems involving severe thermal shock. As evidenced by the work
of Rossi[20], this method should be economically feasible and present
a minimum of technical difficulties.

Many methods can be suggested for introducing microcracks in
solids. Since the grain size and Griffith crack size are considered
to be related, large crack size can be attained simply by over-firing
the material during fabrication in order to enhance grain growth.
Also, the introduction of pores increases the "effective crack length"
of existing cracks, as shown theoretically by Bowie[22]. It is well-
known that a pore phase enhances the resistance to thermal shock
damage[23-25]. Artificial cracks can be introduced in composite
materials with individual components with widely different
coefficients of thermal expansion. This approach was followed by
Rossi[20]. This method also gives close control over crack density.
Another method is to develop materials with dispersed phases of
soft particles or platelets as was followed by Rossi et al.[26] and
Hasselman and Shaffer[27] in the development of highly thermal shock
resistant carbide-graphite composites. For thermal stress fracture
nucleated in the surface, high crack densities can be achieved by
heavy abrasion and/or by deliberately cutting grooves in the surface.
Cutting grooves gives the additional advantage of giving control
over the plane of crack propagation. Specimen fragmentation by
crack intersection is thereby minimized.

It is always desirable to maximize strength for a given crack
size. This can be accomplished by developing materials with high
values of fracture surface energy which may be achieved by
incorporating plastic or viscous second phases in the material.
Strong continuous fibers can also be used[28,29]. This procedure
was followed in the development of metal-fiber reinforced
refractory oxides which show excellent resistance to crack
propagation from thermal cycling. The surface fracture energy also
appears[30] to increase with grain size which constitutes a further
reason for overfiring to enhance grain growth. It is strongly

recommended that further research be carried out on the variables
which affect the value of surface fracture energy of brittle solids.
It is also recommended that in the development and production of
refractory solids measurements of the surface fracture energy be
carried out routinely in addition to the usual standard tests such
as those for tensile and compressive strength, elasticity, thermal
expansion and thermal conductivity.

CONCLUSIONS

The present fracture mechanical approach based on a simple
mechanical model throws new light on thermal stress fracture and
crack propagation. The role of the individual physical properties
is established; this should aid in the proper interpretation of
experimental data and lead to the development of new materials
with superior thermal shock resistance.

REFERENCES

1. W. D. Kingery, J. Am. Ceram. Soc., 38 [1] 3-15 (1955).
2. W. B. Crandall and J. Ging, J. Am. Ceram. Soc., 38 44 (1955).
3. D. P. H. Hasselman, J. Am. Ceram. Soc., 46 [5] 229-34 (1963).
4. W. R. Buessum, Sprechsaal, 93 137-41 (1960).
5. D. P. H. Hasselman, J. Am. Ceram. Soc., 50 [9] 454-57 (1967).
6. B. Schwartz, J. Am. Ceram. Soc., 35 [12] 325-33 (1952).
7. E. Glenny and M. G. Royston, Trans. Brit. Ceram. Soc., 57
    [10] 645-77 (1958).
8. R. L. Coble and W. D. Kingery, J. Am. Ceram. Soc., 38 [1]
    33-37 (1955).
9. R. L. Coble and W. D. Kingery, J. Am. Ceram. Soc., 39 [11]
    377-83 (1956).
10. W. B. Crandall and J. Ging, J. Am. Ceram. Soc., 38 [1]
    44-54 (1955).
11. D. P. H. Hasselman, J. Am. Ceram. Soc., 46 [11] 535-40 (1963).
12. J. Nakayama and M. Ishizuka, Bull. Am. Ceram. Soc., 45
    [7] 666-69 (1966).
13. D. P. H. Hasselman, J. Am. Ceram. Soc., 52 [11] 600-7 (1969).
14. J. P. Berry, J. Mech. Phys. Solids, 8, 206-17 (1960).
15. A. A. Griffith, Phil. Trans. Roy. Soc. (London) A221 [4]
    163-98 (1920).
16. A. A. Griffith, pp. 55-63 in Proc. First Intern. Congr. Appl.
    Mech., Delft, 1924.
17. D. P. H. Hasselman, J. Am. Ceram. Soc. (to be published).
18. M. S. Tacvorian, Soc. Franc, Ceram. Bull., 29 20-40 (1955).
19. W. R. Morgan, J. Am. Ceram. Soc. 14 [12] 913-23 (1931).
20. R. C. Rossi, Bull. Am. Ceram. Soc., 48 [7] 736-37 (1969).
21. E. A. Bush and F. A. Hummel, J. Am. Ceram. Soc., 41 [6] 189-95
    (1958); ibid. 42 [8] 388-91 (1959).

22. O. L. Bowie, J. Math. Phys., 35 [1] 60-71 (1956).

23. C. W. Parmelee and A. E. R. Westman, J. Am. Ceram. Soc.
     11 [12] 884-95 (1928).

24. O. Bartsch, Ber. Deut. Keram. Ges., 18 [11] 465-89 (1937).

25. S. Kato and H. Okuda, Nagoya Kogyo Gijutsu Shikensko
     Hokoku 8 [5] 37-43 (1959).

26. R. C. Rossi and R. D. Carnahan, in Ceramic Microstructures,
     R. M. Fulrath and J. A. Pask (eds), John Wiley and Sons,
     Inc., (1968), pp. 620-635.

27. D. P. H. Hasselman and P. T. B. Shaffer, WADC-TR 60-749,
     (April 1962)

28. Y. Baskin, C. A. Arenberg and J. H. Handwerk, Bull. Am.
     Ceram. Soc., 38 [7] 345-49 (1959).

29. J. R. Tinklepaugh, in Cermets, Reinhold Corp. (1960),
     pp. 170-180.

30. P. L. Gutshall and P. E. Gross, Eng. Fracture Mechanics,
     1, 463-71 (1969).

DISCUSSION

D. D. Briggs (Coors Porcelain Co.): Have you abandoned your
earlier advocation of crack capture by microporosity?
Author: In general, the effect of porosity is to make it easier
to nucleate fracture as shown by Coble and Kingery.[89] However,
porosity also causes fracture to occur at a lower stress level,
i.e., having lower elastic energy at fracture, which limits the
extent of crack propagation. Thus in severe environments, porous
ceramics are preferred over dense ones, the latter spalling
catastrophically.

D. J. Godfrey (Admiralty Materials Laboratory): This thesis is
beautifully confirmed by recent work of the British Ceramic
Research Association on zirconia, a ceramic not usually famed
for its thermal shock resistance. Low density materials with a
strength ~5000 psi has performed very well when fabricated into
crucibles used for casting superalloys. Also, work at the National
Gas Turbine Establishment and at the AML with reaction bonded
silicon nitride containing appreciable void fractions, low density
materials (2.1 g/cm^3) are marginally better than higher density
forms (2.5 g/cm^3), although fully dense hotpressed silicon nitrides
are better than the porous (2.1) materials.
Author: It is gratifying to learn of these further confirmations
of the theory presented here.

J. C. Conway, (Pennsylvania State University): Could you not expect
branching of the crack at a critical energy release rate?
Author: Yes, this would be desirable. In effect, branching
increases the number of propagating cracks, which should reduce the
distance over which they propagate.

S. A. Bortz (I.T.T. Research Institute):   How do we get designers
to use a cracked body?
Author: Generally the designer is concerned with supporting an
applied load.   However, in thermal shock the major problem is to
accommodate thermally induced strains. Thermal stresses arise (as
dependent variables) whenever these thermal strains cannot be
accommodated.   High densities of large cracks are very effective
in lowering Young's modulus without an equivalent reduction in
strength.   The strain at fracture is thereby greatly increased so
that large thermal strains can be accommodated before the cracks
propagate.   Indeed, this has been the approach of the refractories
industry.   However, at present I see no solution to requirements
which combine high load-bearing capabilities with resistance to
severe thermal shock unless we develop materials with tensile
strengths on the order of 10^6 psi.

S. C. Carniglia (Kaiser Aluminum andChemical Corp.):   The refractories industry has employed the Hasselman principles successfully
for many decades, albeit empirically.   They and their customers
do not need the motivation of this message, though they do need
its science.   In referring to the kinetic energy of crack growth,
do you mean an elastic wave in the solid medium accompanying the
moving tip; or if not how is it defined?
Author: The fracture-mechanical theory presented here should
allow quantitative interpretations and it is hoped better understanding of the material parameters which affect the thermal shock
resistance of refractory materials.   Clearly, the theory points
out the need for fracture-toughness testing in addition to those
usually made.   The kinetic energy of a crack refers to that of the
material adjacent to the crack plane which moves as a result of
the crack opening during propagation.   Cracks will be arrested
only when all the kinetic energy of propagating cracks is transformed
into fracture surface energy.

S. W. Wiederhorn (National Bureau of Standards):   As noted, the
elastic energy of the crack arises from motion of material away
from the crack plane.   A good review article containing the
analytical treatment of this problem is given by Anderson
[O. L. Anderson in Fracture.   B. L. Averback, *et al.*, Eds., John
Wiley & Sons, Inc., New York. 1959].

OPTIMUM PROPERTIES OF ZIRCONIA CERAMICS FOR HIGH PERFORMANCE
STORAGE HEATERS

L. L. Fehrenbacher

Aerospace Research Laboratories

Wright-Patterson AFB, Ohio

ABSTRACT

The simulation of true supersonic flight conditions for the development of hypersonic air breathing propulsion systems is a current objective of the Air Force. Heat transfer, aerodynamic heating and propulsion problems associated with re-entry and hypersonic flight within the atmosphere have led to the use of ceramic storage heaters for the source of high temperature air in hypersonic "blow down" wind tunnels. Described are: (1) typical operating conditions, (2) role of thermal, mechanical, chemical and microstructural properties of zirconia ceramics in heater design optimization, (3) engineering evaluation tests on bed and insulation materials; and (4) results of in-service performance in operating heaters.

INTRODUCTION

The development of Air Force ground facilities capable of testing air-breathing propulsion and aerodynamic systems in the high supersonic and hypersonic Mach number range requires a reliable source of dust free, high temperature air at high pressures and mass flow rates. True stagnation pressures and temperatures must be produced since factors affecting mechanical and thermal loading on structures must be mutually compatible with combustion and aerodynamic parameters. Scaling or similitude techniques cannot simultaneously account for these parameters. The type of facility currently under consideration is a blowdown tunnel using regenerative ceramic storage heaters for heating high

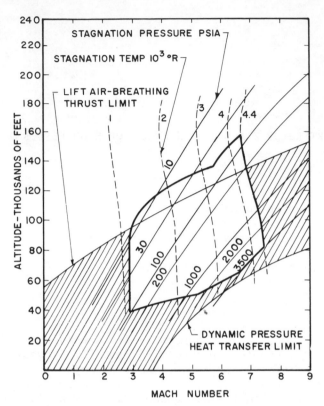

Fig. 1. Typical flight corridor for advanced propulsion systems
         showing wind tunnel true simulation capability.

pressure air. The storage heater "blowdown" principle offers
marked economic advantages over continuous electrically heated
tunnels in the high velocity regions (Mach 4 - 8).

    The potential capability of ceramic storage heaters for
providing true stagnation temperatures and pressures at high Mach
numbers is shown in Fig. 1. For example, a hypersonic vehicle
cruising at an altitude of 80,000 ft at Mach 7 velocity would be
exposed to temperatures from 3800-4000°R (~2100°-2200°K) at approximately 2000 psi. At the low altitude limit of the true simulation
envelope, the ceramics heater must provide large flows of 4000°F
(~2205°C), 3500 psi air upon demand. The chemical, thermal and
mechanical stability requirements on the ceramic heat storing media
truly represent state-of-the-art limits of ceramic material technology. Prior to discussing the interdependence of material properties
and operating conditions on heater design, a brief description
of the principles of wind tunnel air heater operation is given as
background for the materials engineer.

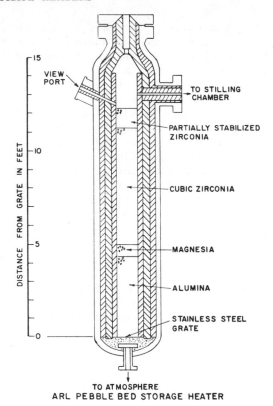

Fig. 2.   Aerospace Research Laboratories (ARL) pebble bed storage
heater.

Storage Heater Operation

A ceramic air heater consists of a steel pressure vessel containing the heat storage bed (matrix) and several layers of ceramic
insulation which minimize heat losses and thermally protect the
vessel shell.   Previous tunnel heaters have generally used composite beds of ceramic spheres (pebbles) made from Al_2O_3, MgO,
and/or ZrO_2 surrounded by hot face insulation of similar compositions
(Fig. 2).   Less expensive firebrick was used as backup insulation
in these "pebble bed" air heaters.

The bed is heated by passing combustion gases of methane-
and/or propane- air (and O_2) mixtures from the top to the bottom
of the matrix at a low mass flow rate.   The heat energy is extracted
by the ceramics resulting in an axial temperature distribution as
illustrated by the idealized profile in Fig. 3.   The nearly uniform
temperature region in the top portion is known as the plateau and
and the linear slope from the plateau to water cooled steel grate
at the bottom is called the ramp.   After the plateau and ramp profile desired for a specific tunnel test is reached, the reheat

cycle is complete and the burners are shut off.   Then, cold air at
the desired stagnation pressure is introduced at the bottom of the
vessel and is heated as it passes through the hot ceramics in the
counterflow direction out through a nozzle and into the test
chamber.   The heat energy which was stored in the bed at low rates
over long times is thus extracted by the cold air at much higher
rates during the test run.   The amount of energy released from the
storage heater is proportional to the area between the before and
after bed temperature distributions.   During blowdown stagnation
conditions exist for a "useful run time" varying from seconds to
minutes depending on the mass flow and pressure requirements.
Useful run time is defined as that period of time during which
stabilized conditions exist in the test stream at the desired
Mach number, pressure and temperature.

CERAMIC PROPERTY-HEATER DESIGN RELATIONSHIPS

Although "pebble bed" heaters have been used extensively in
several "blowdown" wind tunnels, the development of a hexagonal
cored brick element (Fig. 4) was a necessity for high capacity

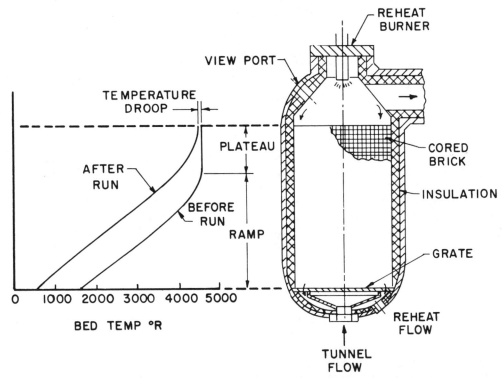

Fig. 3.   Storage heater schematic showing before run and after run
          temperature distributions in the bed.

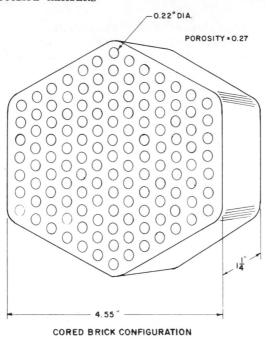

Fig. 4.   Configuration of cored brick used in ARL storage heater.

hypersonic tunnels. Dust contamination, low mass flow rates, and
maximum temperature stability are significant limitations of pebble
bed heaters. Bed floatation due to clumping (sintering of pebbles
into large masses), high point contact loading and phase decomposition of partially stabilized ZrO_2 pebbles have been the major
causes of dusting in zirconia pebble bed air heaters.[1] Bed
floatation or lifting of the ceramic bed occurs if the force
resulting from the pressure drop across the bed exceeds the bed
weight. The cored brick shape offers not only the obvious advantages of increased clumping and dusting resistance in comparison
to ceramic pebbles but significantly lowers the pressure differential through the bed length minimizing bed floatation problems as
well as improving the mass flow flux and heat transfer capability
markedly.

     Since dust formation is the most critical criterion in hyper-
sonic test facilities, fracturing of the cored brick resulting
from severe thermal stress gradients or bed floatation impacts
must be avoided at all costs. Thus, these parameters may dictate
the entire heater design, i.e., the brick geometry, properties, and
quantities of ceramic matrix and insulation are all related to
thermal stress and bed floatation limitations imposed by the most
severe operating conditions of the heater. The cored brick and
overall heater design must also be balanced by the interdependent
criteria of fabricability, cost and engineering performance.

Following sections describe the application of simplified mathematical relationships relating thermal stress and bed floatation characteristics to the elastic, physical, and thermal properties of the ceramic, the brick geometry, the thermal properties of the gas, and heater operating conditions in arriving at a practical heater design. The analytical development of these relationships is reported elsewhere.[2]

Thermal stresses arise in a homogeneous, unconstrained body due to a temperature gradient. Since the cored brick matrix assembly is composed of independent expansion free columns, only temperature gradients will impose thermal stresses. Of practical interest, of course, is whether the maximum tensile stress developed exceeds the fracture strength (usually, tensile strength) of the ceramic. Hence, all thermal stress equations are concerned with the maximum temperature difference necessary to cause fracture of the ceramic under specified thermal shock conditions. Calculations of the temperature distributions developed in a solid from heat flux equations are relatively simple for steady state conditions and simple shapes. For a transient (non-steady state) thermal environment, temperatures can be readily determined as a function of time and position (assuming various boundary conditions) by numerical or analogue methods on a computer.

Thermal Stresses in Storage Heater Brick

Temperature gradients arising in ceramic cored brick during blowdown place the hole surfaces in tension. Thus, thermal stress fracture is most likely to initiate at these locations. To facilitate thermal stress design calculations, the cored brick can be treated as a series of infinitely long cylindrical tubes as seen in Fig. 5.

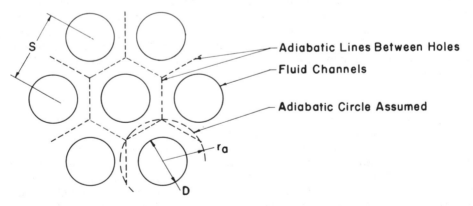

Fig. 5.   Cored brick cross section showing assumed cylinders arranged in an equilateral triangular array.

Assuming only radial temperature gradients, the tangential (circumferential) stress (σ_θ) and axial (σ_z) stress   at the hole surface are equal and given by

$$\sigma_\theta = \sigma_z = \frac{E\alpha}{1-\nu} (T_m - T_s) \qquad (1)$$

where T_m is the mean temperature of the solid material (See Table 1, List of Symbols). The radial stress is zero at the surface. Using the tensile fracture strength of the material in the equation, the

Table 1. List of Symbols

σ_θ = tangential stress in hollow cylinder

σ_z = axial stress in hollow cylinder

σ_r = radial stress in hollow cylinder

E = Young's modulus of solid

α = thermal expansion coefficient

ν = Poisson's ratio

T_m = mean temperature of cylinder wall

T_s = surface temperature of inner cylinder wall

T_g = gas temperature

β = Biot's modulus = $h\, r_o/k_s$

h = heat transfer coefficient

r = hole radius

k_s = thermal conductivity of solid

G_1 = geometrical constant relating to brick geometry and S/D ratio

S = hole center-to-center distance

D = hole diameter

$\dot{m}$ = mass flow rate of air

C_p = specific heat of air at constant pressure

z = bed length

A = cross section area of bed

$\frac{\partial T}{\partial z}$ = axial temperature gradient

G_2 = cored brick geometry constant

f = friction factor

R = gas constant

g = gravitational force

p = bed porosity, area of holes divided by A

W_B = weight density of bed

P = pressure

C = constant related to frictional forces and brick geometry

difference in the hole surface (T_s) and mean temperatures becomes
the allowable maximum and can be used as a design parameter. The
difference maximum is related to the brick geometry and heater
operating conditions. Assuming purely radial heat conduction in
the solid and quasi-steady state heat transfer (ignoring initial
transient conditions) through the solid and convective transfer
to the flowing air, two useful equations can be derived which
couple the thermal stress (ΔT) maximum to the maximum solid-to-fluid
gradient, the maximum axial temperature (ramp) gradient, the brick
geometry, the mass flow rate of air and the thermal properties of
the ceramic and air. The relationship between the induced temperature gradient in the solid and the gas-surface temperature
difference and the axial temperature gradient is given by

$$T_m - T_s = \beta \, 1/G_1 \, (T_s - T_g) \tag{2}$$

and

$$T_m - T_s = G_2 \, \frac{\dot{m} \, C_p}{Ak} \, \frac{\partial T}{\partial z} \quad \text{respectively.} \tag{3}$$

These equations assume that $\frac{\partial T_g}{\partial z} = \frac{\partial T}{\partial z}$. As can be seen from these
equations, the thermal gradient in the solid web portion of the
cored brick, and hence the thermal stress, varies functionally
as follows: (1) increases with (a) heat transfer coefficient,
(b) mass flow rate of air, (c) axial temperature gradient, and
(d) hole-to-hole spacing (web thickness) and (2) decreases with
thermal conductivity of solid.

For a given material, brick geometry, initial bed temperature
distribution and operating conditions, the radial temperature
gradient in the solid at any axial location as a function of time
can be calculated from these equations. In turn, the resultant
thermal stresses can be derived and compared to the thermal stress
limit of the ceramic for the hole surface temperature at that
position in the bed. For design purposes the highest energy
extraction rate operating conditions (largest mass flows, pressures,
and temperatures) should be used for various brick properties and
configurations to determine the minimum bed volume and length that
brings the gas-to-solid temperature difference at the inlet (Eq. 2)
and the ramp slope profile (Eq. 3) within the allowable thermal
stress gradient.

The other important design criteria for assessing heater size
and brick dimensions are the total run time for a specified temperature drop at the most demanding blowdown conditions. The discharge
temperature drop (usually 10% of maximum) and run time combined
with the related thermal stress (Eqs. 2 and 3) define the total bed

volume and length. The temperature drop and run time, for example, would indicate the minimum volume of solid ceramic needed for heat storage purposes. If the mean-to-hole surface temperature difference, however, is too great at the inlet (bottom) end, the bed length and consequently the matrix volume would have to be increased to lower this thermal gradient to an acceptable level. Of course, the brick geometry and properties can be changed to reduce thermal stress gradients. However, too small a hole spacing-hole dia (S/D) ratio (thin webs) would exceed cored brick fabrication capability and require a single storage heater of impractical size coupled with increased refractory costs.

Bed Floatation of Brick Matrix

The pressure drop across a cored brick matrix assuming slight hole misalignment and frictional effects can be represented by

$$\frac{dP}{dz} = \frac{fRT}{2gPD} \left(\frac{\dot{m}}{A}\right)^2 \frac{1}{\sigma} = W_B \tag{4}$$

for a specific height of bed.[3] If the pressure drop exceeds the weight of any portion of the bed, lifting will result. The pressure differential and thus, lifting is more likely to occur during blowdown in the upper, higher temperature portions of the bed as reflected by Eq. 4. Application of these thermal stress and floatation expressions to high capaity storage heater design revealed that the thermal stress limit of the ceramic brick is the governing parameter for matrix size.

For converting existing pebble bed tunnel air heaters to cored brick, Eqs. 3 and 4 can be rearranged to determine the optimum heater operating conditions for a fixed material and geometry. Equation 3 becomes

$$\dot{m} = \frac{1}{G_2} \left(\frac{k}{C_p}\right) A(T_m - T_s) \frac{\partial z}{\partial T} \tag{5}$$

and Eq. 4,

$$\dot{m} = C \, W_B^{1/2} \left(\frac{P}{T}\right)^{1/2} \tag{6}$$

Thus, the allowable mass flow rates for specific bed temperature profiles at thermal stress limited cooling rates and floatation limited pressures can be calculated.

Optimum Properties of Cored Brick

Since the thermal stress criterion appears to control the heater design, improving stress resistance properties or changing the brick geometry for a given material could yield a substantial reduction in bed volume or alternatively, increased run times. In terms of the optimization of matrix element properties and dimensions to provide the most efficient heater and tunnel performance, a cored brick of minimum web thickness and highest obtainable density would be the best choice. Increased density optimizes heat storage capacity per unit volume, thermal conductivity, fracture strength, surface abrasion resistance, creep and hot load resistance while the small S/D ratio provides large heat transfer surface area and low pressure loss characteristics. Although the elastic modulus is increased with density, the thermal stress restriction is relaxed by the enhancement of fracture initiation strength, thermal conductivity, and minimum web thickness.

MATERIAL TESTING AND EVALUATION PROGRAM

Since ground facility development for hypersonic flight testing within the atmosphere imposed new maxima on the operating capability of ceramic air heaters, an extensive testing and evaluation program of commercial zirconia refractories was undertaken. The zirconia material development program was separated into two phases: (1) subscale engineering tests to assess critical properties and (2) in-service performance in operating heaters. The subscale tests were conducted under contract by Fluidyne Engineering Corporation of Minneapolis. Since the cyclic and high temperature stability and thermal stress resistance of partially and fully magnesia-and calcia-stabilized zirconia materials were questionable, the subscale engineering tests were designed to measure (1) permanence of stabilization-cycling between 500°F and 220°F; (2) high temperature stability from 3700°F to 4200°F and (3) thermal stress cycling to failure with increasing cooling rates.

Engineering Test Results

The results of an initial 262 hr-3700°F thermal soak test indicated that MgO-and CaO-stabilized ZrO_2 compositions might be unstable at maximum bed temperatures. Therefore, several yttria-and yttria, mixed rare earth oxide zirconia compositions were investigated [The yttria, rare earth oxide (YRE_2O_3) stabilizer consisted of 90% Y_2O_3 with the balance being the other heavy (type "C") rare earth oxides]. These Y_2O_3-ZrO_2 and YRE_2O_3-ZrO_2 refractories, along with the more conventional CaO-and MgO-ZrO_2 types, were exposed to a flowing combustion gas-oxygen-propane atmosphere for 240 hr at 4200°F. This soak duration is the equivalent of several months of actual storage heater operation.

The performance of Y_2O_3- and YRE_2O_3- stabilized ZrO_2 cored bricks was exceptional in comparison to the CaO-and MgO- stabilized ZrO_2 specimens. Their chemical and structural stability was verified by the constancy of crystalline phase concentrations, the lattice parameters of the cubic stabilized phases and the bulk densities of the Y_2O_3-ZrO_2 samples. The CaO-and MgO-ZrO_2 specimens destabilized due to vapor loss of the stabilizing agent as confirmed by emission spectrographic concentration values and cchanging lattice parameters of specific CaO-or MgO-ZrO_2 compositions. The mechanism of this structural degradation is thus in contrast to the low temperature phase destabilization that is induced by the monoclinic-tetragonal phase inversion.

The permanence of stability tests confirmed the destabilization phenomenon reported by Buckley and Wilson.[4] Only fully cubic (5% monoclinic or less) stabilized zirconia insulation and matrix samples remained insensitive to cubic phase destabilization. These results suggested that all ZrO_2 refractories in the air heater should be fully stabilized for optimum structure stability and dusting resistance.

Thermal stress cycling tests from 3500°F revealed that Y_2O_3-and YRE_2O_3-ZrO_2 cored brick specimens could withstand steep cooling rates (50°F/sec) without limiting optimum heater performance. Thermal stress damage resistance of insulation materials was confined to in-service heater operation. Dense Y_2O_3-ZrO_2 insulation samples exhibited remarkable dusting resistance as they survived pressurization-depressurization rates greater than 1000 psi/sec at a maximum pressure of 3500 psi.

Zirconia Cored Brick Specifications

The results of the subscale tests and compilations of property data on the favorable candidate ZrO_2 specimens permitted specifications to be made for composition (% stabilizer), purity, and density of matrix and insulation components for a 14 in. dia bed storage heater at Arnold Engineering Development Center (AEDC) at Tullahoma, Tennessee. The refractory orders included (1) 90% dense (minimum) fully stabilized (less than 3% monoclinic) ZrO_2 cored brick of 8.4 w/o and 9.25 w/o YRE_2O_3-ZrO_2 compositions, (2) fully stabilized Y_2O_3-ZrO_2 and CaO-ZrO_2 hot face insulation for the upper half of the storage heater and (3) Al_2O_3 cored brick and insulation for the lower regions of the heater.

Selected quality control samples of the commercial zirconia refractories were subjected to the"permanence of stability" cycling test and analyzed prior to installation in the pilot heater. All the 8.4 w/o YRE_2O_3-ZrO_2 cored brick were rejected because of cubic phase destabilization and strength loss. These brick were replaced by 9.25 w/o elements.

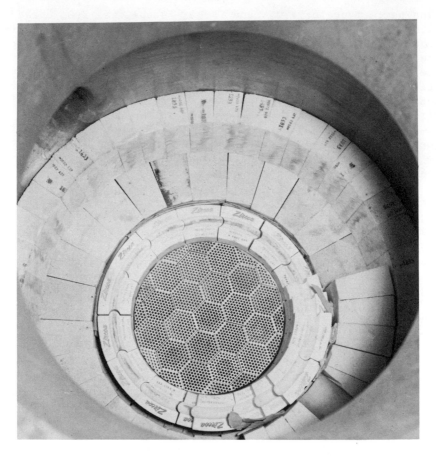

Fig. 6.   Arrangement of ceramic cored brick matrix and insulation
          in AEDC storage heater.

     A top view of the 14 in. dia bed and surrounding insulation
as installed in the AEDC heater is shown in Fig. 6.   The bricks
are 2 7/8 in. across hexagonal faces, with 0.2 in. hole dia and
0.085 in. web thickness. The hole alignment of the vertical
stacked columns is maintained during thermal expansion by tongue
and groove keys on the ends of the variable length brick.

     A 50 run heater shakedown and checkout program under increasing
pressure, temperature, and mass flux conditions was planned.
Pt-6Rh/Pt-30Rh thermocouples were placed throughout the bed length
and insulation to provide the data necessary to determine air
mass flow and heat transfer characteristics of the bed as well as
overall heater efficiency.   The checkout program is currently in
progress.

In-Service Performance

The storage heater can provide valuable information on the effects of actual operating conditions on full scale ceramic matrix and insulation shapes. Operation of AEDC and ARL air heaters to date have produced the following behavioral data on zirconia refractories:

(1) AEDC pilot heater: (a) Inspection of the 9.25 w/o YRE_2O_3-ZrO_2 cored bricks after 27 runs (3000°F max) of varying pressure which included 6 cycles to ambient revealed extensive cracking and strength reduction. X-ray analysis indicated that cubic phase destabilization had occurred as the monoclinic concentrations increased from 5 to around 15 percent. (b) Refiring of the pilot heater to 4000°F for 96 hr eliminated the monoclinic phase, restored the strength, and healed many of the microcracks of the 9.25 w/o YRE_2O_3-ZrO_2 brick. (c) New cored bricks of 10.4, 12, 14, and 16 w/o YRE_2O_3-ZrO_2 compositions were added to the reconstituted bed and the heater shakedown program was continued. (d) No monoclinic phase or friability was detected in 9.25, 10.4, 12, 14, and 16 w/o YRE_2O_3-ZrO_2 brick that were examined after five heater blowdown runs each at 3000°F and 3500°F. (e) A new autoclave "permanence of stability" test exposed these same YRE_2O_3-ZrO_2 variable compositional elements to an air-water atmosphere for 250 hr at 600°F and 2200 psi. All these samples remained fully cubic. However, any brick originally containing monoclinic phase experienced destabilization in this test.

(2) ARL heater: (a) Cored brick of 8.0 and 9.0 w/o Y_2O_3-ZrO_2 compositions have seen continuous use for two years at temperatures to 4200°R with approximately 20 cycles to room temperature. No evidence of strength deterioration or destabilization has been found. (b) Cracking of some partially stabilized CaO-ZrO_2 and fully stabilized 8.5 w/o Y_2O_3-ZrO_2 hot face insulation brick was also attributed to cubic phase destabilization.

Destabilization Mechanism in Zirconia Refractories

A destabilized 9.25 w/o YRE_2O_3-ZrO_2 cored brick taken from the AEDC heater after 27 runs and samples from it which have subsequently been annealed for 50 hr at 1600°C, 750 hr at 1300°C and 1100 hr at 1000°C were analyzed by microstructure-electron microprobe techniques. Results showed that the destabilization process is dependent on the presence of monoclinic second phase particles which, in turn, act as nuclei for the cubic phase decomposition reaction. The amount of monoclinic solid solution phase increased with decreasing annealing temperature although the composition of monoclinic (4 w/o YRE_2O_3) and cubic (10.4 w/o YRE_2O_3) zirconia phases remained constant. The concentration profile (2 μm) step scan of Fig. 7 typifies the relative concentrations of Y and Zr

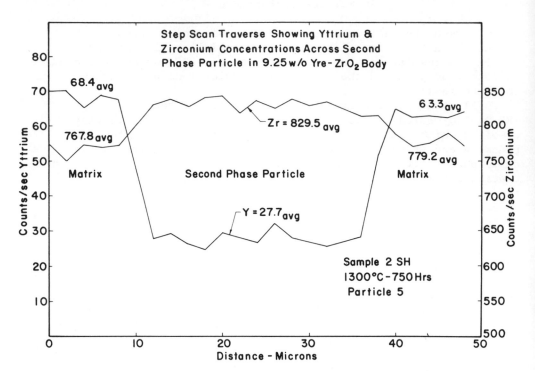

Fig. 7.   Step-scan traverse showing Y and Zr concentrations across
          second phase particles in 9.25 w/o YRE₂O₃-ZrO₂ brick.

in the two phases. The low temperature (2500°F or less)-high
pressure combustion gas cycling conditions used in the initial
stages of the AEDC pilot heater program enhanced the kinetics
of the destabilization reaction much more than isothermal treatments
and/or low temperature cycling.

    This destabilization phenomenon can be prevented by
    (1) total elimination of the monoclinic phase in the as-fabri-
cated cored brick through use of higher concentrations of the
YRE₂O₃ stabilizer, and
    (2) proper heater operation which requires a 4000°F (2205°C)
initial soak and maintenance of the zirconia portion of the bed
above 1500°C the majority of the time.

                        RECENT DEVELOPMENTS

    The creep behavior of 10.4 w/o YRE₂O₃-stabilized ZrO₂ brick
specimens was measured over the range 1400°C-1535°C, 800-7500 psi

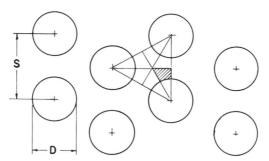

Fig. 8.   Axisymmetric segment of cored brick used for thermo-
           structural analyses.

in order to make credible predictions of creep rates at various
operating conditions in a ceramic storage heater.[5] From the
experimental data, stress-time to failure curves were plotted
and correlated with idling conditions (2500-2800°F) and stresses
(10-50 psi). These time to failure (3% creep strain) extrapolations
indicated creep of 95% dense cored brick in the 2500-3000°F range
to be negligible. Calculations of creep rates and times to failure
in the 4000°F temperature-low stress (10-20 psi) regions of a heater
were based on an experimentally determined deformation equation
and a modified Nabarro-Herring equation, the latter providing
an estimation of grain size influence. These results suggested
that fairly high creep rates and hence short service life might
be expected at the maximum operating conditions. Hence, microstructural changes were recommended to improve the creep resistance
of the zirconia matrix elements. Further testing is planned to
critically assess the influence of differences in purity, grain
size, and composition on creep behavior. Present equipment is
being modified to make hot load deformation and creep measurements
on cored brick samples at 4000°F in a combustion gas environment.

     State-of-the-art thermal analysis and finite element elastic
thermostructural analysis computer programs will be used to
quantitatively calculate thermal stress distributions in zirconia
cored brick for a variety of heater operating conditions and
brick geometries. A transient thermal analysis on a axisymmetric
solid segment (accounting for non-uniform wall thickness, see Fig.8)
of the cored brick will produce accurate time-dependent temperature
distribution profiles for critical locations in the heater.
Elastic thermal stresses will then be computed for a selected number
of time points and compared with fracture strengths to estimate
the possibility of thermal stress failure. The effect of brick
design on thermal stress cracking will also be assessed.

     Fully stabilized zirconia cored brick containing a minimum of
6 mole % pure (99.6%) Y_2O_3 has been ordered to replace the zirconia
refractories in the AEDC pilot heater.

CONCLUSIONS

The development of the cored brick shape for the matrix element and the use of YRE$_2$O$_3$ fully stabilized zirconia have resulted in a regenerative ceramic wind tunnel storage heater that should be vastly superior to the old pebble bed predecessors. Air temperatures to 4000°F and higher at greatly increased mass flow rates with minimal dust contamination can now be realized. The ARL and AEDC storage heaters will continue to provide valuable information on present bed and insulation refractories in addition to serving as test beds for the exploitation of new design and material concepts of the future.

ACKNOWLEDGEMENTS

The writer is grateful to G. Arnold (Chief, Facility Development Division, AEDC), D. Hagford, D. Decoursin and J. Plunkett (Fluidyne Engineering) for their close cooperation and encouragement throughout this program. The support of Col. W. Moran (Commander, Aerospace Research Laboratories) and N. Tallan, (Chief, Metallurgy and Ceramics, ARL) is sincerely appreciated. A special thanks is rendered to B. Ruh (AFML) for assisting in the quality control analyses; to F. Bailey, N. McKinnon and J. James (C.S.I.R.O., Australia) for their experimental assistance in performing the creep measurements; to D. Kohler (Battelle, Columbus) for the electron microprobe results; D. Franks (ARL), for the x-ray diffraction analysis; and J. Koenig (AFML) for financial and technical support of the finite element thermostructural cored brick analysis.

REFERENCES

1. C. H. Weissinger and G. W. Barnes, "Experience with Pebble Beds and Air Heaters", presented at Air Heater Conference, Worcester, Massachusetts, May 1963.
2. W. Hedrick, F. Larsen, B. Lindahl, and D. DeCoursin, "Storage Heater Design Study for the Hypersonic True Temperature Tunnel", AEDC-TDR-64-48, 185 pp. Arnold Air Force Station, Tennessee, July 1964.
3. D. DeCoursin, D. Hagford, G. Arnold, and D. Male, "Recent Development of Storage Heaters to Provide Flight Simulation for Air Breathing Propulsion Systems", Paper presented at the AIAA Third Propulsion Joint Specialist Conference July 17-21, 1967, Washington, D. C.
4. J. Buckley and H. Wilson, J. Am. Ceram. Soc. 46 (10) 510 (1963).
5. L. Fehrenbacher, F. Bailey, and N. McKinnon, SAMPE Quarterly Journal, 2 (2), January 1971.

DISCUSSION

A. R. Cooper, Jr. (Case Western Reserve University): Failure due to phase separation of Y_2O_3-stabilized ZrO_2 is interesting: it would seem to be a type of spinodal decomposition since it can be avoided by going to high-Y_2O_3 compositions, presumably outside the two phase region. However, your comments about the avoidance of phase separation by elimination of every particle of the low-Y_2O_3 phase suggest that nucleation of this phase is difficult, and the phase separation is controlled by nucleation.

A. H. Heuer (Case Western Reserve University): The question of whether destabilization occurs by spinodal decomposition or by nucleation and growth is confused because this phase transformation (tetragonal→monoclinic) is a *shear* transformation, and hence very sensitive to the stress condition. The main role of monoclinic impurities may be as stress raisers - due to differential and anisotropic thermal expansion - enhancing the destabilization. This view would also be consistent with the differences observed between "static" and "dynamic" thermal cycling, in that internal stresses would be expected to be greater in the latter type of cycling.

S. W. Wiederhorn (National Bureau of Standards): Since high stresses are involved, could pressure-induced polymorphic transformations account for destabilization of the ZrO_2?

H. Palmour III (North Carolina State University): If Dr. Heuer's comments are correct in diagnosing these destabilization phenomena as due to shear processes, wouldn't strain rate effects arising from thermal (strain) cycling be important?

Author: Several factors, no doubt, contribute to the rate of the destabilization reaction in cubic stabilized ZrO_2. First, the necessity of monoclinic solid solution nuclei of nearly constant composition in a polycrystalline body certainly suggests that a time-dependent nucleation and growth mechanism is responsible for the decomposition process. The wide variance in destabilization rates of cubic ZrO_2 compositions have confused the peculiar role of several independent kinetic-influencing parameters. Based on considerable experience with these materials, I feel the growth rate of the monoclinic solid solution phase is functionally dependent as follows: (1) The magnitude of internal and/or residual stresses and associated strains that, in turn, are related to (a) the thermal expansion coefficient difference between the monoclinic and cubic phases and (b) the large, polyaxial stresses and strains associated with the reversible monoclinic-tetragonal phase inversion and (c) the amount of original monoclinic phase present and the rate of cyclic heating and cooling. Thus, the kinetics of the decomposition process are stress rate (strain rate) dependent. (2) The presence

of moisture (H_2O) in a high pressure atmosphere, as emphasized by
the autoclave air-water vapor environment (600°F, 2200 psi).
Since this temperature is well below the temperatures of the
monoclinic→tetragonal transformation (typically 850° - 1020°F,
as determined by hot stage x-ray analysis) for cubic stabilized
ZrO_2 materials of the 3-5% monoclinic as-fabricated composition,
the destabilization process does not require the stresses induced
by cycling through the monoclinic-tetragonal phase change. Slow
cycling from 2000°F to ambient in air atmosphere furnaces will
eventually produce formation of more monoclinic phase, but its
effect appears less severe than the high pressure air-water
vapor autoclave test.

The air storage heater environment combines all these kinetic
factors, and consequently the high pressure air-combustion gas-
cycling conditions of the early AEDC pilot heater runs would be
expected to enhance the diffusion rate of cations and the growth
of the monoclinic phase markedly.

THERMAL-SHOCK-RESISTANT MATERIALS

R. C. Rossi

The Aerospace Corporation

El Segundo, California

ABSTRACT

Aerospace applications which take advantage of the refractory nature of ceramic materials also impose upon them severe conditions of thermal shock. The severity of this thermal environment exceeds that for which it is possible to prevent crack nucleation; therefore, the design of thermal-shock-resistant materials is based on the concept of preventing crack propagation. More explicitly, composite materials with either carbide or oxide matrices were developed with the capability of sustaining thermal strains without generating the thermal stresses that could lead to catastrophic failure. Thermal simulation tests revealed a resistance to thermal shock far superior to that which could be realized by more conventional ceramic materials. Improved quality is related to the role of the individual phases within the microstructures.

INTRODUCTION

For specific aerospace applications such as rocket nozzles, nose tips, and leading edges of reentering vehicles, there are few materials refractory enough to meet the thermal requirements. Ceramics have been candidate materials for these applications, and they offer the additional advantage of relative resistance to erosion by either oxidation or abrasion. However, the ionic-covalent ceramic bond that provides their resistance to high temperature is also responsible for their brittle fracture nature and their susceptibility to failure from thermal shock; the technical limitation of ceramics in these critical aerospace applications is the direct consequence

123

of an intrinsic material property. ("Thermal shock" as used in
this text is taken to mean the materials' response to a sudden
change in ambient temperature. Specific reference is made to
fracture, failure, or damage as applied to the case in point.)

THEORETICAL CONCEPTS

Among the analytical theories for thermal shock behavior,
there exist but two fundamental concepts. One relates the parameters that affect crack initiation by thermal shock;[1] the other
relates those parameters responsible for thermal crack propagation.[2]
In the first case, high strength and low Young's modulus of
elasticity provide resistance to thermal shock by preventing crack
nucleation. In the second case, low strength and high elastic
modulus provide resistance to thermal shock damage by decreasing
stored strain energy at fracture. In the latter case, it is
assumed that the thermal shock environment is too severe for the
prevention of crack nucleation, therefore, the prevention of
catastrophic failure by limitation of the damage caused by propagating cracks is most feasible.

These theories do not immediately suggest a technique for
designing materials with a resistance to thermal shock. For most
ceramic materials, those microstructural parameters that affect
the elastic modulus also affect strength similarly; the range of
strength-to-modulus ratios normally obtainable through variation
of these parameters is insufficient to satisfy the microstructural
design criteria for the thermal shock environment experienced in
aerospace applications.

An alternate approach is possible, however, that does not
necessarily depend upon a relationship between strength and
modulus. It is necessary to differentiate the effect of thermal
shock upon the mechanical behavior of materials from the more
commonly experienced effect of mechanical loading. In the latter
case, the strain of a material is a response to the stress applied
and is determined by the appropriate elastic modulus. In thermal
shock, stress is the material's response to thermal strain, and it
is determined by both elastic modulus and thermal expansion
gradients. In this case, stress is the dependent variable rather
than the independent variable. Therefore, a candidate material
must be designed to sustain the required thermal strain without
generating sufficient stress to cause failure. The essence of this
argument is presented in the schematic stress - strain curves in
Fig. 1. The curve with the higher slope represents a typical
ceramic material; the curve with the lower slope represents a
material designed to resist thermal shock. The modulus has been
decreased such that the elastic strain increases by a factor of

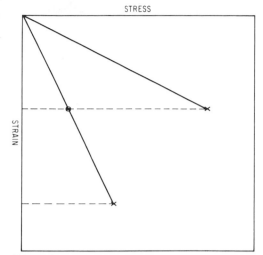

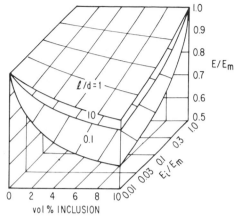

Fig. 1. Schematic stress-strain
diagram for two materials.
That with the lower modu-
lus has an improved re-
sistance to thermal shock
fracture at equal strengths
and equal strain energy
and improved resistance
to thermal shock damage at
equal strength-to-modulus
ratios.

Fig. 2. Effect of volume concentration of inclusion, in-
clusion modulus-to-matrix
modulus ratio, and shape
of the included phase on
the composite modulus-to
matrix modulus ratio. (A
randomly oriented spheroi-
dal inclusion was consid-
ered a generalized particle
shape; $\ell/d>1$=prolate sphe-
roid, $\ell/d<1$= oblate sphe-
roid.)

four if strength should remain unchanged. If the strength should
be reduced such that the strain energy at failure remains unchanged,
the elastic strain is doubled. In these cases, increased strain can
be directly translated into proportional increases in thermal
shock resistance. On the other hand, if the strength-to-modulus
ratio is held constant, the stored strain energy at fracture is
reduced, and the thermal shock damage resistance is improved by a
factor of four. In all cases, the thermal-shock resistance of the
material can be improved by reduction of the elastic moduli of the
material; fracture strength becomes less important in design
considerations.

Techniques for reducing elastic moduli, e.g., increasing
porosity, alloying, and use of a glass-phase matrix, are not desirable for the intended applications. The alternative, therefore, is
the fabrication of a material in composite form in which control of
the properties can be attained.

According to the various theories of micromechanical behavior
of composite materials, the two most influential parameters
affecting the composite modulus are the ratio of inclusion modulus
to matrix modulus and the particle shape of the included phase.[3]
The effects of these parameters on composite modulus are shown in
Fig. 2. Over a limited range of the included phase, the greatest
reduction in the composite modulus occurs at low values of E_i/E_m
ratios and for particles with low aspect ratios (disc shaped).

CARBIDE COMPOSITES

Material Fabrication

Among all ceramic materials, the refractory metal carbides have
the greatest potential for aerospace applications because they are
the most refractory of materials and they offer a high abrasion
resistance. Carbides are conventionally fabricated by hot-pressing
techniques. As single-phase materials, carbides do not have a
notably high resistance to thermal shock. However, when they are
fabricated as composite materials with graphite particles as the
included phase, a marked improvement in thermal shock resistance
is realized. In one case[4] in which graphite particles were added
to zirconium carbide in a hot-pressed composite, the temperature
differential necessary to initiate fracture increased from 1500° to
1900°C at a graphite content of 50 vol %. While this increase
represents an improvement of more than 25% in thermal shock resistance, it is insufficient for many aerospace applications. Furthermore, at such high concentrations of graphite, the resistance to
abrasion and erosion are severely sacrificed.

While graphite is an ideal second-phase additive to carbides
because it provides a large modulus mismatch between phases,
graphite particles offer an aspect ratio (ℓ/d) of only 1:3, and these
particles tend to align during hot-pressing, thus producing an
anisotropic composite. An alternative method for fabrication of
carbide - matrix composites is the fusion-casting process, in which
graphite is added in excess of stoichiometry.[5] The refractory metal
carbides form simple eutectics with graphite, and the eutectic temperatures are sufficiently refractory to withstand the aerospace
environment. The eutectic structure is lamellar, with graphite
lamellae dispersed uniformly, randomly, and with an aspect ratio of
about 1:5. An example of the eutectic structure is shown in Fig. 3
for an alloy fabricated by arc-melting under a pressure of 1/2 atm
Ar.[6] Graphite added in excess of the eutectic composition produces
a structure typical of hypereutectic alloys shown in Fig. 3, in
which the primary graphite appears as flakes with an aspect ratio of
1:50. If graphite is added in concentrations less than the eutectic
composition, a hypoeutectic structure is formed that contains
dendrites of the carbide phase, as shown in Fig. 3.

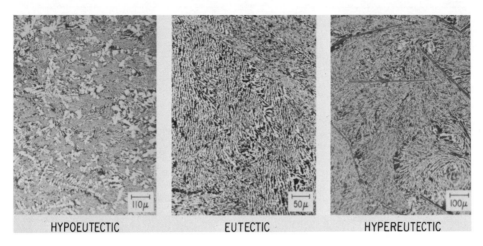

HYPOEUTECTIC   EUTECTIC   HYPEREUTECTIC

Fig. 3.   Typical microstructures of fusion-cast carbide - graphite
          alloys.

The hypereutectic carbide - graphite alloy can be considered
a "natural" composite that meets the ideal microstructural design
conditions for thermal shock resistance; it has a high ratio of
matrix modulus to inclusion modulus and also has disc-shaped inclusions with a very low aspect ratio.

Properties

The mechanical behavior of the fusion-cast carbide - graphite
alloys were determined, the strengths by 4-point bend[7,8] and the
moduli by dynamic resonance techniques.[9]   Typical results are
shown for zirconium carbide - graphite alloys as a function of
composition in Fig. 4.   The rapid and nearly discontinuous change
of properties in the vicinity of the eutectic composition has been
shown to be the direct consequence of the changes in microstructure
that occur near the eutectic composition.   Identical behavior was
observed for all fusion-cast carbide - graphite alloys as a
function of microstructure and independent of absolute composition.
These results show that, while the inclusion of graphite as lamellae
or flakes effectively decreases modulus, these inclusions are even
more effective in decreasing strength.

From mechanical properties, the relative thermal shock resistance can be predicted by either of the methods described.   Fig. 5
presents the calculated strain at failure and the stored elastic
energy at failure as a function of composition.   On the basis of
the arguments previously   presented, the hypereutectic alloys
would have little resistance to the thermal initiation of cracks,
but they would contribute little stored elastic energy for the

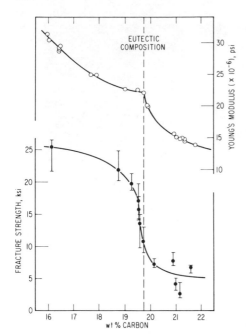

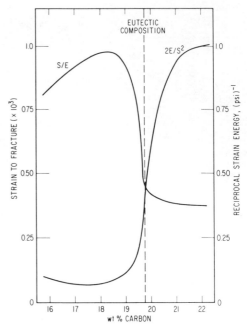

Fig. 4. Young's modulus and fracture strength of fusion-
cast—zirconium carbide –
graphite alloys. Rapid
property changes in the
vicinity of eutectic com-
position are typical of
all similar fusion-cast
carbide – graphite alloys.

Fig. 5. Calculated strain to fracture and reciprocal strain
energy of fusion-cast Zr
carbide – graphite alloys
as functions of composition.

propagation of cracks. In contrast, the hypoeutectic alloys show a
much greater resistance to crack initiation, but a crack, once
initiated, would be reinforced with sufficient strain energy to
assure its propagation to failure. The absolute extent to which
these alloys can withstand a severe thermal environment is not
adequately defined by this analysis.

A more quantitative measure of thermal shock resistance was
obtained through a series of experiments in which zirconium carbide-
graphite alloys were subjected to a nitrogen arc plasma at a heat
flux of 2400 Btu/ft^2 sec for 30 sec. The samples were right
circular cylinders with one end exposed to the arc, the other end
held in a water-cooled grip. Maximum thermal gradients along the
3/4 in. sample length were established in approximately 20 sec,
and maximum apparent front face temperatures reached 3400°C. High-
speed photography was used for observation of the tests.   Thermal

shock failure was taken as that time in which a macroscopic crack could be visually observed. For the purposes of this analysis, the temperature differential between front and rear face at the time of failure was assumed to be the measure of the thermal shock resistance. (In a strict sense, the thermal diffusivity must be included in the analysis.) The results are plotted in Fig. 6 as a function of composition.

The failure profile of these alloys closely resembles the stored elastic energy criterion for minimal thermal shock damage. Reduction of the elastic modulus by the inclusion of graphite, either as eutectic lamellae or as flakes, causes an even more drastic reduction in strength. These results imply that, in the hypereutectic alloys, thermal fracture initiates at low thermal stress, but the elastic energy released on fracture is inadequate to propagate the cracks and early failure is prevented. In contrast, the greater strengths of the hypoeutectic alloys prevent early crack initiation, but once they are initiated, cracks are able to propagate to an earlier failure.

Examination of thermal shock specimens of the hypereutectic alloys revealed the presence of many hairline cracks, which propagated largely along the boundaries between the graphite flakes and the eutectic matrix. This examination, therefore, suggests that the flakes and lamellae may act as gross flaws that initiate early fracture; but once this fracture is initiated, thermal stress

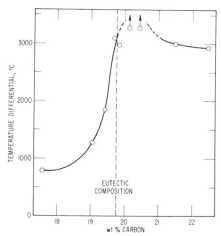

Fig. 6. Thermal shock failure of fusion-cast Zr carbide-graphite alloys represented by temperature difference between front and back faces at the time of macro-crack development.

Fig. 7. SEM fractograph of a fractured hypereutectic alloy revealing structural incoherency between the eutectic matrix and the proeutectic graphite flakes.

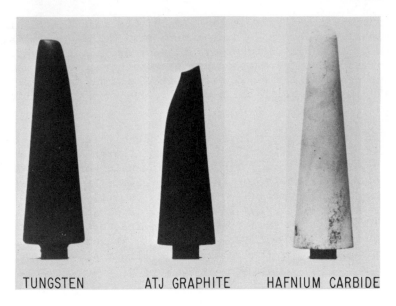

TUNGSTEN   ATJ GRAPHITE   HAFNIUM CARBIDE

Fig. 8.   Comparative results of W, ATJ graphite, and fusion-cast
          hafnium carbide - graphite alloy nose tip models exposed
          for 15 min at 700 Btu/ft^{2}sec to an air plasma arc (General
          Electric arc facilities, after Kendall *et al.*[11]).

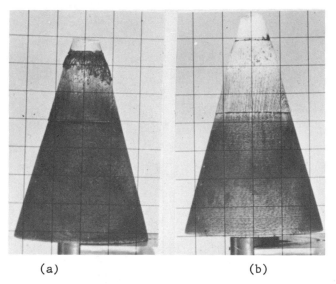

          (a)   (b)

Fig. 9.   Comparative results of (a) sintered W and (b) fusion-cast
          Hf carbide - graphite alloy nose tip models exposed to an
          air plasma for 6 sec at 4080 Btu/ft^{2}sec and just preceding
          the onset of HfO$_2$ melting (Cornell Aeronautical Lab wave
          superheater facility, after Kendall *et al.*[11]).

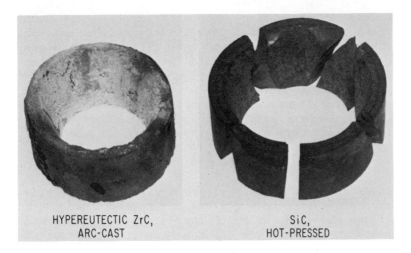

HYPEREUTECTIC ZrC,   SiC,
ARC-CAST   HOT-PRESSED

Fig. 10. Comparison of a fusion-cast carbide - graphite alloy and
         dense, hot-pressed silicon carbide rocket nozzles fired
         in a N_2O_4 - Aerozine 50 fueled engine; gas temperatures
         reached 2600°C. The zirconium carbide alloy survived
         two 60 sec firings; the silicon carbide cracked within
         the first seconds of a single firing (NASA, Lewis Research
         Center, LEM test engine, after Kendall *et al.*[12]).

is relieved and much greater thermal strain can be accomodated
before failure occurs. (An independent evaluation of the thermal
stress behavior of hypereutectic alloys revealed that crack initiation took place within the first second of heating at heat flux
as low as 1000 Btu/ft^{2}sec[10].) Further support for the contention
that graphite flakes are incipient flaws is presented in the
fractograph, Fig. 7, of a mechanically fractured hypereutectic alloy.
A distinct separation between the c-plane face of the graphite and
the carbide matrix can be seen. On the basis of the relative thermal
expansion of the two phases, a typical 10 μm-wide graphite flake
would be expected to shrink away from the carbide matrix as much as
0.5 μm upon cooling from fabrication temperature. Structural discontinuities of this magnitude between carbide and graphite have
been consistently observed by SEM techniques.

Performance

     The results of the experimental work strongly suggest the cap-
ability of the hypereutectic alloys to withstand the thermal shock
environments of aerospace. Therefore, hardware was fabricated for
testing in simulated flight environments and was compared with other
candidate materials. Results of some of these tests are shown in
Figs. 8, 9 and 10. While failure by thermal shock in these very
severe environments has not been observed for the hypereutectic
alloys, hardware failed by erosion as a result of significant shape

change whenever the stagnation temperature exceeded the melting
temperature of the metal oxide. The results clearly establish,
however, that at temperatures up to this melting temperature, the
hypereutectic alloys remain structurally sound at extremely high
heating rates, and furthermore, they do not experience the
relatively high recession rates of tungsten or graphite. At the
higher temperatures, however, the sublimation and ablation rates
of tungsten and graphite are less than the mass loss experienced
by the oxidation products of the carbide - graphite alloys.

Discussion

The successful development of the hypereutectic carbide - graphite alloys for severe thermal shock environments and the understanding of the mechanism by which they perform suggests microstructural design criteria for the development of thermal-shock-
resistant composite materials with brittle matrices. Reduction of
the elastic moduli of a material appears to be a fundamental property
requirement for an increase in the resistance to thermal shock failure
provided by an increased strain to failure. Furthermore, the development of strength beyond a minimal level appears undesirable.

For the severe thermal shock environment experienced in aerospace applications, ceramic materials cannot be made to prevent
crack initiation. Therefore, control of crack propagation is the
predominant method for prevention of catastrophic failure. From
the behavior of the carbide - graphite alloys, it is inferred that
a microstructural system that easily nucleates many cracks simultaneously and allows thermal strain from these cracks, without
allowing the catastrophic growth of any one of them, will produce
the required resistance to failure from thermal shock. Moreover,
these criteria imply that the strain energy released by a nucleating
or propagating crack will be inadequate to continue its propagation
beyond that necessary to accommodate the thermal strains developed
in heating.

OXIDE MATRIX COMPOSITES

The foregoing discussion defines the criteria for the development of oxide composites for severe thermal shock environments.
The more obvious design would incorporate flakes of graphite in an
oxide matrix. Flakes of boron nitride, made by mechanical shearing
of pyrolytically deposited BN, can be used with equal success. An
example of such a material is shown in the microstructure, Fig. 11,
of a composite containing BN flakes in a matrix of beryllium oxide
fabricated by hot-pressing. A composite containing 10 vol % flakes
has sufficient thermal shock resistance to be considered a candidate
for a radome material on a hyper-velocity reentering vehicle.

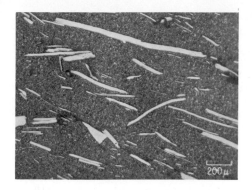 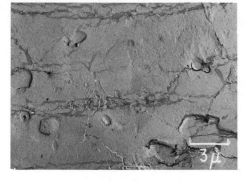

Fig. 11. A microstructure of hot-pressed composite of BeO containing 10 vol % flakes of pyrolytic BN, simulating the hypereutectic structure in the carbide-graphite alloys.

Fig. 12. A microstructure of hot-pressed composite of MgO containing 3 vol % W particles, producing a three-dimensional micro-crack network (After Rossi[13]).

The concept need not be restricted to inclusions having a flake morphology. A composite of magnesium oxide containing as little as 3 vol % tungsten particles has vastly improved thermal shock resistance relative not only to magnesium oxide but also to the best performance of the more thermal-shock-resistant oxides. A microstructure of such a composite hot-pressed by standard techniques and containing a uniform dispersion of 1-2 μm particles of tungsten is shown in Fig. 12. The microstructure contains a network of microcracks that pass through the included particles. Origination of these cracks is the consequence of the differential thermal expansion of the matrix and inclusion. During cooling from fabrication temperature, local stresses around each included particle lead to localized cracking and the development of the microcrack network.

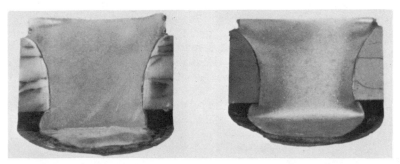

Fig. 13. Cross section of rocket nozzles of BeO (left) and composite of BeO containing 15 vol % 7 μm SiC particles (right) fired in LEM test engine, conditions as in Fig. 10.

A further example of the resistance of the oxide-particle composites is shown in Fig. 13. A cross section of a rocket nozzle of beryllium oxide containing 15 vol % SiC particles is compared with one of beryllium oxide identically tested. Although some cracking is apparent, the extent and severity of the cracking are far less in the composite than in the oxide, which demonstrates the improvement in thermal shock resistance that can be obtained from a microcrack network.

CONCLUSION

The development of thermal-shock-resistant ceramic materials for severe aerospace environments has been achieved through microstructural design. The greatest resistance to damage from thermal shock was found in those materials that contain a very high density of incipient flaws or cracks and accommodate thermal strain through pseudoductility, by the controlled propagation of these cracks. These materials have, by intention, low elastic moduli and low strength; therefore control of crack growth is possible through the minimal release of strain energy from cracks propagating in response to thermal strain.

Thermal-shock-resistant materials with both carbide and oxide matrices in composite form have been successfully developed. Application of microstructural design criteria has permitted the use of brittle ceramics in thermal environments that are so severe that catastrophic damage from thermal shock in conventional materials could not be avoided.

REFERENCES

1. W. D. Kingery, J. Amer. Ceram. Soc. 38 (1) 3-15 (1955).
2. D. P. H. Hasselman, J. Amer. Ceram. Soc. 46 (11) 535-40 (1963).
3. R. C. Rossi, J. Amer. Ceram. Soc. 51 (8) 433-39 (1968).
4. D. P. H. Hasselman and P. T. B. Shaffer, "Factors Affecting Thermal Shock Resistance of Polyphase Ceramic Bodies," WADD-TR-60-749 (Part II), April 1962.
5. E. G. Kendall, R. D. Carnahan, and E. L. Foster, p. 240 in Trans. Vac. Met. Conf. Pap., 8th, New York (Ed. L. M. Bianchi), (1966).
6. R. D. Carnahan, K. R. Janowski, and R. C. Rossi, Metallography, 2 (1) 65-77 (1969).
7. R. C. Rossi and R. D. Carnahan, Ch. 29 in Ceramic Microstructures. Edited by R. M. Fulrath and J. A. Pask, John Wiley and Sons, Inc. N. Y. (1968).
8. R. C. Rossi and E. G. Kendall, J. Amer. Ceram. Soc. 53 (8) 476-7 (1970).
9. R. D. Carnahan, K. R. Janowski, and R. C. Rossi, J. Amer. Ceram. Soc. 52 (9) 475-77 (1969).

10. J. R. Bohn, K. R. King, C. H. Ernst, and K. R. Janowski,
    "Nonsteady-State Thermal Stress Behavior of Refractory
    Materials," AFML-TR-67-315 (Nov. 1967).
11. E. G. Kendall, J. I. Slaughter, and W. C. Riley, AIAA J. 4 (5)
    900-905 (1966).
12. E. G. Kendall, R. D. Carnahan, and R. C. Rossi, Space/Aeronautics
    47 (1) 132 (1967).
13. R. C. Rossi, Amer. Ceram. Soc. Bull. 48 (7) 736 (1969).

DISCUSSION

S. A. Bortz (I.I.T. Research Institute): What size plasma torch was
used for thermal shock tests?
Author: We used a 200 kW subsonic plasma arc facility with a 1/4
in. dia exhaust nozzle at nominal current loads of 1000 A.

S. A. Bortz: Have you considered metal reinforced ceramics to
obtain controlled crack structures?
Author: Yes, but we haven't performed any experiments yet. There
are many problems with such systems, many caused by the thermal
expansion differences between metals and ceramics. Work performed
more than 10 years ago by J. R. Tinklepaugh, et.al. (WADC Technical
Report 58-452, Pt. III) demonstrated this principle. We have not
yet determined how to solve the problems they had encountered.

S. W. Bradstreet (Consultant - Dayton, Ohio): Presently UTC
(Sunnyvale, Calif.) has developed a W-wound, W-sprayed composite
that illustrates the principles enunciated in this talk.

A. H. Heuer (Case Western Reserve University): Can you get nodular,
rather than flaky, precipitates of carbon and do these give any
improvement in performance?
Author: Yes. The inclusion of Mg_2Si in the melt charge will
provide graphite nodules; examples can be seen in Ref. 6. No change
in properties were observed, but the matrix is brittle (rather than
ductile, as it is in cast iron) so no change should be expected.

L. D. Lineback (North Carolina State University): In regard to the
ZrC-graphite and HfC-graphite composites which you have mentioned,
we made several similar to the ones you have described and shocked
them thermally as you have with a nitrogen plasma. The composites
did not exhibit thermal shock degradation in the form of one or
more macrocracks. They were, however, so weak after thermal shock
exposure that they could be crumbled with the fingers. Was that
your experience?
Author: Yes, a nitrogen plasma can impose a sufficiently severe
thermal environment to structurally degrade the composite without
causing a catastrophic failure and weaken it so that it can be
easily destroyed.

L. D. Lineback: Of what value are they? Are they really thermal
shock resistant?

Author: Yes, they are valuable if it is necessary to have a material
that will not fail from thermal stress. However, when using a
material to withstand a very severe thermal shock environment, it
is necessary not to impose a structural requirement on the
material as well. For the nose tips and nozzle applications for
which these composites were designed, it is only necessary that
the material not significantly change shape during operation; they
are not required to sustain structural load besides.

D. P. H. Hasselman (Lehigh University): Regarding the loss of
structural integrity of HfC-graphite and ZrC-graphite alloys in
thermal shock, one can circumvent the problem by artificially
inducing a greater number of sites for crack nucleation. Grooving
by diamond sawing or other techniques for introducing stress
concentrations will solve the problem.

R. W. Rice (U. S. Naval Research Laboratory): Following up on one
of Dr. Hasselman's points, one can get a high density of fine
cracks and thus good thermal shock behavior if you make a fine
grained body from a material with a high degree of thermal expansion
anisotropy. Another example besides those he mentioned is hafnium
titanate, which also offers the advantage that compositions can
be made having predetermined, near-zero macroscopic thermal
expansion.

FACTORS AFFECTING THE THERMAL SHOCK BEHAVIOR OF YTTRIA   STABILIZED

HAFNIA BASED GRAPHITE AND TUNGSTEN COMPOSITES

L. D. Lineback and C. R. Manning

North Carolina State University

ABSTRACT

   *Hafnia-based composites containing either graphite or
tungsten were investigated as rocket nozzle throat inserts in
solid propellant rocket engines. The thermal shock resistance
of these materials is considered in terms of macroscopic thermal
conductivity, thermal expansion, modulus of elasticity (E), and
compressive fracture stress (σ_E). The effect of (a) degree of
hafnia stabilization, (b) density, and (c) graphite or tungsten
content upon these parameters is discussed. The variation of the
σ_E/E ratio with density and its effect upon thermal shock resistance
of these materials is discussed in detail.*

INTRODUCTION

   Solid propellant rocket nozzle throat areas are subjected to
a severe environment which produces thermal shock, mechanical
erosion and chemical corrosion. In the past a number of materials
including graphite, tungsten, alumina, and polymers have been used
in this area but all are unsatisfactory in one or more respects.
In the present study, hafnia and hafnia-based composites containing
either graphite or tungsten were investigated in an attempt to
optimize resistance to thermal shock as well as mechanical and
chemical erosion.

   Thermal shock resistance is not an inherent property of any
material. It is instead the result of the interaction of several
measurable parameters relating to the material, its environment,

137

and its geometry. The material parameters are thermal conductivity, thermal expansion, fracture stress, modulus of elasticity, and Poisson's ratio. Temperature and coefficient of surface heat transfer are important environmental considerations while size, shape, and areas of surface heating are geometrical considerations of importance.

All solid materials can have stresses induced in them by the introduction of thermal gradients. For a semi-infinite body these stresses are given by [1,2]

$$\frac{\partial \sigma_i}{\partial x_i} + \frac{\partial \tau_{ij}}{\partial x_{xi}} + \frac{\partial \tau_{ik}}{\partial x_k} = \left(\frac{\alpha E}{1-2\nu}\right) \frac{\partial T}{\partial x_i}$$ (1)

where α is the coefficient of linear thermal expansion, E is the modulus of elasticity, ν is Poisson's ratio, and T is the temperature on any plane parallel to the heated plane, $x_i = 0$. If the material is heated by a fluid obeying Newton's Law of Cooling the thermal gradient is given by

$$\frac{\partial T}{\partial x_i} + h\ (T-T_o) = 0$$ (2)

where h is the coefficient of surface heat transfer.

These equations have been solved previously for a number of finite geometrical shapes such that the minimum temperature difference to which a body may be exposed and not fail by thermal shock is given in the general form [3]

$$T_{max} = \frac{\sigma_F}{E}\ \left(\frac{1-\nu}{\alpha}\right)\ \left(A + \frac{ck}{Lh}\right)$$ (3)

where σ_F is the fracture stress, k is thermal conductivity, A and C are geometrical constants, and L is the "half thickness". Therefore, for a given environment and geometry, the thermal shock resistance of a material may be increased by increasing σ_F and/or k, or by decreasing E, ν and/or α.

EXPERIMENTAL PROCEDURE

The base material for this study was −325 mesh reactor-grade hafnia stabilized with reactor-grade yttria. Composites were fabricated with additions of grade TRA Poco Graphite (−400 mesh)

or 0.005 in. dia W-3% Rh alloy wire chopped to a aspect ratio of
25:1 as the dispersed materials.

Specimens were fabricated by hot pressing at pressures of 1000,
2000, 4000 and 8000 or 10,000 psi in graphite dies at a temperature
of 2100°C *in vacuo* at 10^{-2} torr or below. Stabilization of the
hafnia was accomplished during hot pressing taking up to 8 w/o Y_2O_3
into solid solution. The degree of stabilization was determined by
x-ray methods.[4,5]

Thermal expansion was measured by the dilatation interferometer
method.[6] Thermal conductivity was measured by a semiabsolute steady-
state method[7] in an apparatus designed by one of the authors (LDL).
The values given were measured in the vicinity of 350° K.

Modulus of elasticity values were obtained by measuring displacements with SR-4 type strain guages attached to specimens with
height to diameter ratios of 1:2 at static compression load levels
corresponding to stresses from 1000 to 3000 psi. Compressive
fracture stresses were obtained by carrying these specimens to
failure. Poisson's ratio was not measured due to limited specimen
sizes.

Thermal shock resistance was determined by exposing a circular
portion of one end face of a cylindrical specimen to a nitrogen
plasma arc at 9000°C. Stress distributions for this method of
heating are given elsewhere.[8,9] The maximum compressive stress is
created on the heated surface and the maximum shear stress is
located at an angle of 45° to this surface and intersects this
surface at the edge of the circular heated zone. Thermal shock
resistance was measured qualitatively by observing the extent of
specimen degradation produced by a 5 sec exposure to the plasma arc.

DISCUSSION OF RESULTS

The linear thermal expansion of the composites did not vary
extensively from that of the base material. In general, expansion
increased with increased degree of stabilization and/or graphite
content and decreased with increased tungsten content.

The effect of increased stabilization of the HfO_2 base material
is shown (Fig. 1) to linearly increase the thermal expansion. The
broken lines represent the scatter of data due to density differences
and uncertainties in measuring the degree of stabilization.
Increasing the hot pressing pressure from 1000 to 10,000 psi was
found to increase the coefficient of linear thermal expansion by
approximately $0.1 \times 10^{-6}°C^{-1}$. Addition of powdered graphite to the
base material increased thermal expansion such that it approached

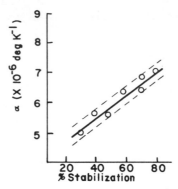

Fig. 1.   Linear thermal expansion coefficient as a function of the
          degree of stabilization of the body-centered tetragonal
          Y_2O_3-HfO_2 phase, hot pressed at 1000 psi, 2100°C.

that of pure graphite (Fig. 2) while addition of tungsten fibers
decreased the expansion such that it approached that of pure W
(Fig. 3).

In general, the results obtained are consistent with those
which may be expected from composite materials. The changes in
coefficient of linear thermal expansion are felt to be secondary
in importance to other changes to be discussed below. Because of

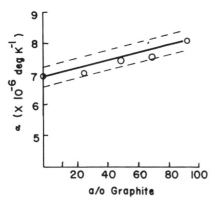

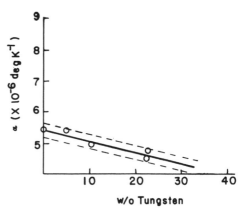

Fig. 2. Linear thermal expansion   Fig. 3. Linear thermal expansion
        coefficient of Y_2O_3-HfO_2 coefficient of Y_2O_3-HfO_2
        based composites as a func-   based composites as a
        tion of graphite content   function of W content
        (hot pressed at 10,000 psi,   (hot pressed at 8000 psi,
        2100°C; 55% stabilized).   2100°C; 70% stabilized).

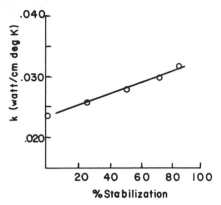

Fig. 4.   Thermal conductivity of Y_2O_3-HfO_2 as a function of degree
          of stabilization (hot pressed at 10,000 psi, 2100°C).

the elevated temperature (~2000°C) phase inversion of hafnia, the
degree of stabilization of the base material is probably of greater
importance to thermal shock resistance.

Thermal Conductivity

     Thermal conductivity of high density hafnia increased only
slightly with large changes in the degree of stabilization (Fig. 4).
Introduction of graphite to high density materials also resulted
in only modest increases in thermal conductivity which did not
appear to be approaching that of graphite (Fig. 5).

     Tungsten additions to the stabilized base material resulted in

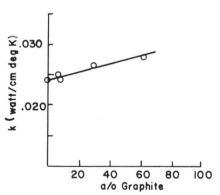

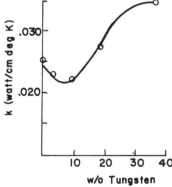

Fig. 5. Thermal conductivity of   Fig. 6. Thermal conductivity of
        Y_2O_3-HfO_2 based composites Y_2O_3-HfO_2 based composites
        as a function of graphite   as a function of W content
        content (hot pressed at   (hot pressed at 8000 psi,
        10,000 psi, 2100°C, 85%   2100°C; 70% stabilized).
        stabilized.

the nonlinear behavior shown in Fig. 6. With small additions of
tungsten the conductivity initially decreased and then recovered
and increased as would be expected with the larger additions of
the metal. However, the conductivities are lower than might be
expected for the larger additions of high thermal conductivity W
metal. Anamolies in thermal conductivity at small W concentrations
can be attributed primarily to interactions between the method of
measurement of conductivity and process-related changes in density.
The density changes from approximately 96% of theoretical for base
material with no tungsten to approximately 88% of theoretical for
composites containing 5-10 w/o W. With larger amounts of tungsten
the density approaches 90-95% of theoretical. It has been established that values of thermal conductivity of porous materials
measured *in vacuo* can be considerably lower than if made in a
fluid medium.[10] The low values observed in conductivity measurements
are attributed to this phenomenon. Other data which support this
hypothesis are presented below. Thermal conductivity variations
in the materials studied appear to be secondary in their influence
on thermal shock behavior.

Modulus of Elasticity and Compressive Fracture Stress

No graphical data is presented relating either modulus of
elasticity or compressive fracture stress of the base material to
the degree of stabilization because of their strong dependence on
density. Young's modulus of elasticity was found to vary from
6×10^6 to 67×10^6 psi as density increased from 88% to 96% of
theoretical. No detectable dependency on degree of stabilization
was found. Additions of graphite to composites formed at high
pressures (Figs. 7 and 8) significantly lowered both the modulus
of elasticity and compressive fracture stresses. Additions of
tungsten initially lowered both the modulus of elasticity and
compressive fracture stress (Figs. 9 and 10). Large additions,
however, increased both such that they approached the values
expected for tungsten. Similar behavior was observed for thermal
conductivity of the tungsten composites and was attributed to a
drop in density at the lower W concentrations. The σ_F/E ratio
(Fig. 11) shows a maximum value near 10 w/o W which is approximately
an order of magnitude greater than that of the base material hot
pressed at the same temperature and pressure. Low density
base material has a σ_F/E ratio of approximately 6.6×10^{-3} as
compared to a ratio of approximately 1.7×10^{-3} for high density
material. The addition of 10 w/o of W-wire increases the ratio
to approximately 1.2×10^{-2}. An addition of graphite in any
amount decreases the ratio to approximately 1.5×10^{-4}.

Thermal Shock Resistance

For maximum thermal shock resistance of the base material, the
optimum degree of stabilization value lies between 50-55% for all
densities. Deviation in either direction resulted in center cracking

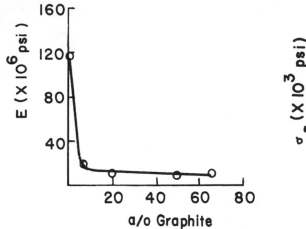

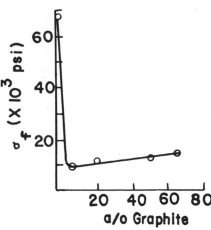

Fig. 7. Modulus of elasticity of
Y$_2$O$_3$-HfO$_2$ based composites
as a function of graphite
content (hot pressed at
10,000 psi, 2100°C; 85%
stabilized).

Fig. 8. Compressive fracture
stress of Y$_2$O$_3$-HfO$_2$ based
composites as a function
of graphite content (hot
pressed at 10,000 psi,
2100°C; 85% stabilized).

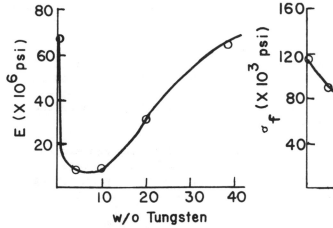

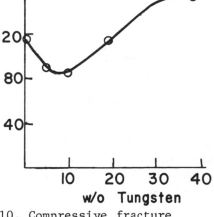

Fig. 9. Modulus of elasticity of
Y$_2$O$_3$-HfO$_2$ based composites
as a function of W content
(hot pressed at 8000 psi,
2100°C; 70% stabilized).

Fig. 10. Compressive fracture
stress of Y$_2$O$_3$-HfO$_2$
composites as a function
of W content (hot pressed
at 8000 psi, 2100°C; 70%
stabilized).

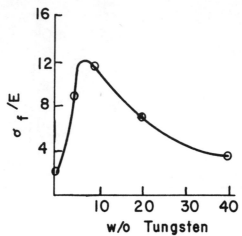

Fig. 11. Ratio of compressive fracture stress, σ_F, to modulus of
elasticity, E, for Y_2O_3-HfO_2 based composites as a function
of W content (hot pressed at 8000 psi, 2100°C; 70%
stabilized).

as predicted by the expression of Sharma.[8] Large deviations also
resulted in ring cracking around the heated surface. It was also
observed that the lower density material was less susceptable to
thermally induced cracking for all degrees of stabilization. This
is attributed to the increase in the σ_F/E ratio for lower density
material.

The addition of graphite in any amount resulted in catastrophic
failure which may be described as explosive. Here again, the
σ_F/E ratio is low.

The addition of tungsten in general improves the thermal shock
resistance. The optimum thermal shock resistance obtained corresponded to the maximum σ_F/E ratio shown in Fig. 11. Composites
containing >20 w/o W yielded central cracks although none of the
composites showed any shear cracks.

Composites containing 10 w/o W fabricated by hot pressing at
8000 psi and 2100°C into 1 in. dia, 1/2 in. thick discs were
exposed to 10 cycles in a nitrogen plasma, followed by quenching
from 2100°F into ice water ten times, without any indication of
cracking.

CONCLUSIONS

1. The maximum thermal shock resistance of HfO_2 can be obtained by
stabilizing (to 50%) with Y_2O_3 and by decreasing the bulk density.

2. Composites based on Y_2O_3-HfO_2 containing graphite have less thermal shock resistance than the base material.
3. Composites based upon Y_2O_3-HfO_2 containing tungsten have greater thermal shock resistance than the base material. The optimum addition of tungsten fibers was 10-20 w/o.
4. For this group of materials the ratio (σ_F/E) of the compressive fracture stress to the macroscopic modulus of elasticity is the major factor affecting thermal shock resistance, while macroscopic thermal conductivity and macroscopic thermal expansion play minor roles.
5. It is concluded that fracture strength and modulus of elasticity for these materials are interrelated parameters and their combined effects rather than each individual effect must be considered when investigating thermal shock resistance.
6. In general, it is concluded that the thermal shock resistance of such composites can be optimized by maximizing the σ_F/E ratio.

ACKNOWLEDGEMENTS

The authors wish to thank the National Aeronautics and Space Administration for financial assistance in the form of Grant NGR 34-002-108 supervised by James P. Howell for the SCOUT Office, NASA-Langley. Special thanks go to Mitchell Haller for his aid in the Young's modulus determinations.

REFERENCES

1. H. Carslow and J. Jaeger, Conductivity of Heat in Solids, Claredon Press, Oxford, England. 1959.
2. B. Boley and J. Weiner, Theory of Thermal Stresses, John Wiley and Sons, New York, 1960.
3. S. S. Manson, in Mechanical Behavior at Elevated Temperatures, J. E. Dorn, ed. McGraw-Hill Book Co., New York, 1961.
4. P. Duwez and F. Odell, J. Am. Ceram. Soc. 32 (5) 180 (1949).
5. J. D. Buckley, "Stabilization of the Phase Transformation in Hafnium Oxide" Ph.D. Thesis, Iowa State University, Ames, Iowa. 1967.
6. American Society for Testing Materials "Standard Method of Test for Linear Expansion of Whiteware" ASTM Designation C 327-56.
7. G. Haacke and D. P. Spitzer, J. Sci. Instrum. 42 702 (1965).
8. R. Sharma, J. Appl. Mech. 23 (4) 527 (1956).
9. Ya. Fridmar, Strength and Deformation in Nonuniform Thermal Fields, Consultants Bureau, New York, 1964.
10. D. R. Flynn, in Mechanical and Thermal Properties of Ceramics. J. B. Wachtman Ed., National Bureau of Standards Special Publication 303. 1969.

DISCUSSION

S. K. Dutta (Army Materials and Mechanics Research Center): What
is the die material and the size of the fabricated compact?
Authors: Specimens ranged from 1/2 to 1 in. dia and 1/4 to 3/4 in.
high; they were fabricated in dies of Poco Hf-1 graphite.
S. K. Dutta: Did you detect quantitatively any evidences of altered
carbon content in the finished compacts due to diffusion of C from
the die material at 2100°C? Also, did you observe reaction of W-
filaments with C from the die at such temperatures?
Authors: First, specimens were protected by tantalum shields from
direct contact with the graphite ram surfaces. Second, exposed
surfaces were ground to eliminate as much carbon contaminated
material as possible. X-ray and microprobe examinations did not
indicate any reaction between HfO_2 and C. In earlier work with HfC,
we observed that the carbide decomposes to the oxide in the presence
of O_2 at high temperatures; this was a factor in the selection of
the oxide as the base material for this study. As for the tungsten,
microprobe analyses failed to reveal substantial O_2 or C levels in
the W fibers, and microhardness data did not indicate serious
embrittlement attributable to them.

N. N. Ault (Norton Company): It is very likely that HfC formed in
the hot pressing of the HfO_2-C composites. Its presence could
explain the poor behavior of the material in the thermal shock
environment.
Authors: Again, that is not what we observe in X-ray and microprobe
studies.

S. Schneider (National Bureau of Standards): What do you mean by
"degree of stabilization"? Is this the quantity of tetragonal phase
present?
Authors: Yes.
S. Schneider: Why not start with equilibrium phases obtained by
solid state sintering?
Authors: Initially, we did. All the base material used in the
hafnia-graphite composites was stabilized in that manner. However,
we found that we could satisfactorily stabilize HfO_2 in the hafnia-
tungsten composites during hot pressing, thus eliminating a process
step and reducing impurities picked up in ball milling of pre-reacted
material.

S. A. Bortz (I.I.T. Research Institute): Have you applied composite
theory to your HfO_2-W composite.
Authors: No. However, it may explain the loss of strength at low
fiber concentrations and the subsequent increase in strength for
larger additions. The effect of fiber additions on density and
hence on strength until minimum values of σ_F and E are reached
is discussed on p 142. The increase in strength to a value above

that of the base material with further additions of fibers, as well
as an increase in modulus approaching that of pure W does suggest
that the fibers are strengthening the composite, in general agreement with composite theory. The favorable aspect ratio (25:1) was
selected for that reason. However, I don't think composite theory
will explain the behavior observed in the hafnia-graphite system.

R. C. Rossi (Aerospace Corporation): Did you determine whether the
HfO_2-W fiber composites contained cracks within the matrix?
Authors: Although we did not specifically look for microcracks
during this phase of the investigation, we are reasonably sure that
they are present in the HfO_2-W materials. Some cracks were observed
in HfO_2 grains by light microscopy. Also deviation from 100% density
in fine grained material suggests that pores may act either as cracks
or as crack initiators. The hafnia-graphite composites, on the
other hand, did not show visible microcracks, and in general,
appeared to have good structural integrity.
R. C. Rossi: About 10 years ago, J. R. Tinklepaugh, et al [J. R.
Tinklepaugh, et al, W.A.D.C. Technical Report 58-452, Parts I, II,
and III, 1958-1960] studied the behavior of an alumina and mixed
oxide matrix composites containing either W or Mo fibers; composites
almost identical to yours. Their composites were excellent in thermal
shock resistance, had low modulus and high strength, but were per-
meated with microcracks which were not always resolved with light
microscopy. (Evidence for the existence of the microcracks was
best revealed during oxidation studies.) The low modulus was a
result of the cracked matrix; the high strength was the consequence
of the fibers pulling out from the matrix which had tightly shrunk
around them. The thermal shock resistance they experienced was
probably a consequence of the cracked matrix, and I suspect this
explanation may be appropriate to your work as well.

THE RESISTANCE OF SILICON NITRIDE CERAMICS TO THERMAL SHOCK

AND OTHER HOSTILE ENVIRONMENTS

D. J. Godfrey and E. R. W. May

Admiralty Materials Laboratory

Poole, Dorset, England

ABSTRACT

The properties of Si_3N_4 ceramics relevant to thermal shock are described, and results of testing and hardware trials discussed. The technology of Si_3N_4 materials, and their potential for other hostile environments, is reviewed.

INTRODUCTION

In the past decade there has been a great awakening of interest in ceramics based on silicon nitride. Their most fundamental advantage is a relative lack of susceptibility to thermal stress problems, due to their very low coefficient of thermal expansion. At a value of 2.5×10^{-6} m/m/°C (20-1000°C) this is only bettered among common ceramics by vitreous silica, and lithium aluminium silicates such as eucryptite and spodumene. However, silica materials soften progressively with rising temperature and also are liable to recrystallization, with deleterious volume changes such as that of the high/low quartz transformation at 573°C. Lithium aluminium silicate glass ceramics can have excellent thermal stress properties, but sometimes suffer from a weakening effect in low expansion formulations, caused by the anistropy of crystal expansion leading to microtesselated thermal expansion mismatch stresses; and have a maximum use limit of 1200°C due to a tendency to revitrification. Silicon nitride has therefore been the object of study for applications in which thermal stress is a serious consideration, such as gas turbine components.[1] With closer study, many other advantageous properties have been discovered or developed,

including great strength[2,3] and unusual fabricability.[4,5] Also, the
material has shown good refractoriness and oxidation resistance.

Two different fabrication routes for the production of Si_3N_4
ceramics have evolved. In the first, silicon powder, suitably
compacted by ceramic, powder metallurgical or thermal spraying
processes, is then reacted with nitrogen at 1200-1450°C to yield
a "reaction-bonded" silicon nitride (RBSN) which contains upwards
of about 12% void. In the second, Si_3N_4 powder is hot pressed with
the aid of an additive, usually a few percent of MgO, at 1600-1850°C
to yield fully dense (3.18 g/cm^3) material (HPSN). Although the
best properties are obtained from HPSN (except, perhaps, hot
strength), the far simpler and cheaper reaction bonding of Si has
considerable fabrication advantages. The volume change on con-
solidation from soft, machineable partially nitrided silicon compacts
to the fired hard ceramic is less than 0.1%, large shapes are easier
to fabricate than in sintered contraction-processed oxide ceramics,
and materials may be joined with unusual ease during fabrication
by silicon thermal spray joining or powder glue techniques.[5]

THERMAL STRESS AND SHOCK BEHAVIOUR

Whilst the comparative thermal stress and thermal shock behaviour of materials are determined largely by thermal expansion
(α) properties, other factors such as strength (σ), elasticity (E)
and thermal conductivity (λ) are involved. The parameters for
instantaneous thermal shock ($E\sigma/\alpha$) and thermal stress ($E\sigma/\alpha\lambda$) for
silicon nitride material compare well with those of other
ceramics.[1] Their potential for gas turbine blade applications,
where high gas velocities and pressures make thermal shock conditions
particularly severe, has been examined by Glenny and Taylor.[6,7]
These authors used fluidized powder heating and quenching beds to
conduct realistic thermal shock tests on several ceramic materials,
and made a particular study of the repeated thermal shock cycling
of several Si_3N_4 ceramics. Metals were found to fail by cracking
in such "thermal fatigue" tests. Plastic deformation appears to
be a basic component of this effect in metals, and identical behaviour would not be expected from brittle materials. In 1958[6]
they reported on our (AML)(2.1 g/cm^3) and British Ceramic Research
Association (BCRA) material (2.5 g/cm^3), in tests cycling between
1020 and 20°C, the specimens being 0.875 in. dia edge tapered discs
with a central hole. The AML material gave a mean life of 1,430
shocks, and the BCRA 55 shocks. Both materials survived quenching
from 1200°C. Test with a 1.625 in. disc proved more severe and
only the 2.1 density was tested, and 3 specimens gave a mean life
of 15 shocks at 1020°C. Vitreous silica gave a similar result,
but SiC materials were slightly inferior, their better conductivity
being less important at the high heat transfer rates of the
fluidized bed. Further work was reported in 1961[7] on Si_3N_4 from

four sources: type A - AML, densities 2.1 and 2.5; B - BCRA, density
2.5; C - Union Carbide, Kokomo, density 2.3-2.5; and D - Plessey Co.,
hot pressed 5% MgO, densities 3.0-3.1 (Si purity 98%) and 3.18
(Si purity 99.9%). Experiments with 1.625 in. discs of type A
(2.1 density) indicated that, although variable, it had a life of
about 150 cycles, and was superior to 2.5 density material which
rarely survived a single quench from 950°C. Type B failed at about
a 750°C quench, but types C and D survived single quenches from
1020°C, and 2 type D specimens (density 3.19) endured a 1320°C
quench. Subsequent National Gas Turbine Establishment work, with
HPSN type D materials showed variable lives varying from 1 to 100
shocks. Plessey workers found a lithia treatment to be beneficial
in improving thermal shock performance.[8] The variations in performance appeared to have some correlations with compositional
variables, including impurity content, and hot pressing processing
parameters, but despite studies on the possible machining damage
to surfaces, the factors determining performance were never
elucidated satisfactorily. The strength of HPSN was also subject
to inexplicable variation, for which there was no consistent cause
in terms of void content, composition or processing, and strengths
were sometimes only comparable with alumina 29,000-43,500 psi
(200-300 MN/m²) but ranging upwards to values over 116,000 psi
(800 MN/m²). Optimisation of Si₃N₄ powder character and hot pressing
procedures finally improved strength and consistency, but no thermal
shock testing was carried out. However, statistical studies show
the material still to be more variable than RBSN.[5] Another group
working on HPSN material made from 98% Si, have demonstrated consistently strength above 100,000 psi (690 MN/m²), but have not
worked extensively on thermal shock. It would appear however that
HPSN material has considerable potential for thermal shock
resistance.

In our work on the thermal shock of RBSN, a water quench
technique was used instead of the National Gas Turbine Establishment
fluidized bed. Although film boiling interferes with heat transfer,
water penetration of the specimen may make the test more severe.
Test discs 0.875 in. dia of 2.3 density were found to sustain
1000°C shocks 300-500 times, but nine discs of 2.54 density averaged
only five 1000°C shocks, and the best endured 15. Glenny's data
also indicated better performance at the lower densities of RBSN.
The larger disc also had a poorer performance in this test, and
suggested performance correlated inversely with RBSN density. Six
specimens of 2.32 density survived an average of five 1000°C shocks
(best 15); six of 2.51 density survived only 800°C quenches, and of
2.60 density discs, three failed after 850, 850 and 900°C quenches,
with a further six surviving an average of one 1000°C shock (best 2);
and seven of 2.66 density survived an average of two shocks (best 4).
The water quench variation of the NGTE test appears similarly
sensitive to variations in material quality, and the indications of

better performance at lower densities suggest that thermal shock
resistance is not simply connected with strength, which over the
density range 2.1-2.65 varies between 11,600-26,100 psi
(80-180 MN/m^2).

PERFORMANCE IN ENGINEERING ENVIRONMENTS

The thermal shock testing already described was oriented
towards the use of ceramics in gas turbines. In order to obtain
a wider spectrum of information about severe environments in the
contect of engineering, several practical engineering studies have
been carried out with RBSN components. These have included thermal
shock cycling of gas turbine nozzle guide vanes, a combustion
structure, an oxygen-propane flame-driven fan, and internal combustion engines of the Wankel rotary piston and conventional
cylindrical piston types.

Gas turbine stator blade shapes[9] of 2.55 density RBSN were
subjected to thermal shock conditions by alternate heating and
cooling in an oxy-propane jet to   ~1500°C and a jet of air at
50 psi delivered through a 3.3 mm nozzle 20 mm away from the leading
edge of the aerofoil. Although the conditions were less severe
than in a gas turbine, it was encouraging that 500 shocks were
endured by the blades without failure. Another gas turbine environment for which ceramics are under consideration is the combustion
chamber.[10] Several tests were carried out on a silicon nitride
slotted dome combustion stabiliser in a Rolls-Royce ram jet combustion test facility. The first type of shape tested failed each
time by thermal stress cracking, the dome end separating,[9] but in
a final test the dome end was slit as shown in Fig. 1, and only
one "finger" then failed, despite a 10 min. test at 1637°C. This
failure was almost certainly due to the "finger" in question being
deeply affected by corrosion during nitridation, arising from
the melting of a kaolin wool spacer used inadvertently to separate
a stack of domes during firing of the sprayed silicon compact
shapes. In order to investigate further the potential of Si_3N_4
materials in high temperature bearings,[11] a small fan was constructed
in RBSN by the silicon flame spraying technique, the blades being
joined by a flame-sprayed fillet deposit of porous silicon to a
cylindrical tube. The tube was fitted on to a central solid powder-
metallurgy route RBSN shaft, drilled centrally to pass low pressure
air, and the journal region with eight 406 μm (0.016 in.) holes.
At an air pressure of 5-10 psi, with an annular spacing of 62 μm,
it became an almost frictionless air bearing, and the fan could be
spun at several thousand rpm by the impingement of a propane-
oxygen torch (Fig. 2). The assembly survived heating of the fan
blades to 1100°C, and the flame being extinguished many times,
demonstrating very well the good thermal shock resistance and air-
bearing capabilities of RBSN.

Fig. 1. Slotted dome combustion stabiliser shape (8 in., 0.25 in.
         wall) after testing at 1637°C. Note slits cut in dome
         cap to relieve thermal strain.

The good performance of RBSN materials in the explosive and
thermal environment of the internal combustion engine, demonstrated
by the successful use of RBSN rotors and tip seals in two types of
rotary piston engine,[9] has been extended by experiments with a
conventional internal combustion engine. A commercial Villiers
50 mm bore single cylinder 75 cc, 1.2 HP gasoline engine was fitted
with a piston, split (1/3) piston rings, and gudgeon pin in 2.55
density RBSN, suitably designed to accommodate the strong but
brittle character of the ceramic, and as shown in Fig. 3 survived
operation at speed and under load without fracture or chipping,
even of the rather delicate piston rings.

Fig. 2. Si$_3$N$_4$ air bearing 2.5 in. dia fan driven at 3000 rpm by
         oxygen-propane torch (electronic flash photograph).

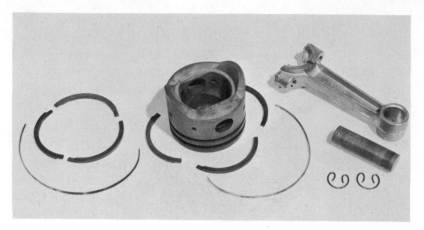

Fig. 3.   Silicon nitride 50 mm dia piston, segmented piston rings
          and gudgeon pin, after running in 1.2 HP Villiers single
          cylinder engine. Metal piston ring backup springs,
          gudgeon pin circlips, and connecting rod are also shown.

PROPERTIES RELEVANT TO PERFORMANCE IN SEVERE ENVIRONMENTS

    Thermal expansion behaviour is probably the factor most
responsible for differences in performance between various materials.
The value of 2.46×10^{-6} m/m (20-1000°C) for α given by Collins
and Gerby in 1955[12] has been confirmed by Popper and Ruddlesden,[13]
and other workers.[14]   There appears to be no evidence of anisotropy,
although single crystals of Si_3N_4 are not available in convenient
sizes and have not been studied. The theoretical reasons for the
unusually low value are not well understood, but in comparison
with SiC the more open structure of Si_3N_4 may allow more vibrational
energy to be absorbed by the greater contribution of less simple
vibrational modes.

    Conductivity (λ) is the other thermal property involved in
thermomechanical stress. Powell and Tye[15] give a value of 18.5
$Wm^{-1} °K^{-1}$ for 3.16 density HPSN (larger than the preliminary value
of Deeley, et al.[2] of 12 $Wm^{-1} °K^{-1}$) and 5 $Wm^{-1} °K^{-1}$ for 2.34 RBSN,
at room temperature. Here the deleterious effect of porosity on
λ may be offset by the high thermal conductivity of unreacted
metallic phase usually present in RBSN, and obvious in microsections.
This yellowish highly reflective phase, which has often been
assumed to be unreacted silicon, has in 98% Si derived material
been shown by electron probe microanalysis to contain less silicon
than the surrounding Si_3N_4, and to consist largely of iron, and
occasionally titanium, alloyed with silicon. Unreacted silicon
appears light grey. These two phases are shown in Fig. 4. Over
0.5 Fe is often present. Aluminium is present in equivalent amounts

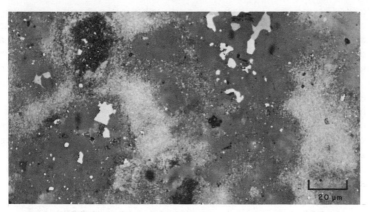

Fig. 4. Reflected partly polarized light micrograph of RBSN
material (2.6 density), showing iron-rich (white) and
silicon-rich (light grey) metallic phases, and low
density region (granular light grey) surrounding high
density (dark grey) regions. Voids appear black.

but does not segregate in this way. These high conductivity phases
(Si 150 $Wm^{-1}°K^{-1}$, steel $\sim$45 $Wm^{-1}°K^{-1}$) are especially important in
high strength RBSN, made by the flame-spray route which allows
compaction to high densities,[16,17] which are more difficult to
nitride thoroughly, and may even improve thermal stress capability
at the expense of hot strength and creep resistance. The Powell
data show a serious effect of porosity on conductivity, and considerably improved values are to be expected from a rise in density
to 2.8. An application of the Moore relation[18]

$$\lambda = \lambda o \ (1 - P)/(1 + P) \tag{1}$$

suggests a value of 14 $Wm^{-1}°K^{-1}$, although this relation probably
holds closely only to void fractions of $\sim$0.08.

The modulus of elasticity of HPSN, given by Deeley[2] as
31.5 x 10^{-6} psi, from our measurements with Stanford Ring tests is
about 39 x 10^{-6}, and this agrees well with Lumby's value[3] of
40 x 10^{-6}.   The figure of 55 x 10^{-6} for whiskers is now thought to
be an overestimate, and a revised figure of 43.5 x 10^{-6} [19] suggests
some effect from the MgO additive in HPSN. Our value of 24.6 x 10^{-6}
for 2.4 density material is slightly lower than the 27.5 x 10^6
calculated from the MacKenzie[20] equation

$$E = Eo \ (1 - 1.9P + 0.9P^2) \tag{2}$$

Strength is the most complex property affecting thermomechanical performance, since it is subject to great variation, especially in early development of Si_3N_4 materials. Lumby and Coe[3] have shown that the strength of HPSN can be made consistently very high, by increasing pressing time at 1700°C, when although grain size increases, large structural features seen in electron micrographs tend to be eliminated. These could be a vitreous phase, since X-ray techniques are unable to detect the MgO additive as a crystalline phase, the existence of which has been postulated by Jack et al.[21] The softening of such a phase may be responsible for the marked loss in strength of HPSN materials at around 1100°C. RBSN materials have been variously described as modestly increasing or decreasing in strength up to 1500°C. Reaction-bonded materials are often low in strength or variable when made by workers inexperienced with the material. But as Davies has demonstrated[5,22] consistent material can be made. A number of 2.56 density RBSN specimens prepared at AML in two sizes (4.5 x 4.5 x 63.5 and 7.55 x 7.55 x 63.5 mm) were tested in 4-point flexure. The smaller ones have a mean strength of 25,500 psi (174 MN/m^2) (123 specimens, Weibull "m" exponent 15.6), the larger 24,000 psi (166 MN/m^2) (54 specimens, m = 14.0). In a second test, commercial material (density 2.41) gave in 4-point flexure 27,500 psi (190 MN/m^2) (47 specimens, m = 16), and in 3-point flexure 31,200 psi (215 MN/m^2) (49 specimens, m = 19). Specimen size was 4 x 4 x 50 mm.

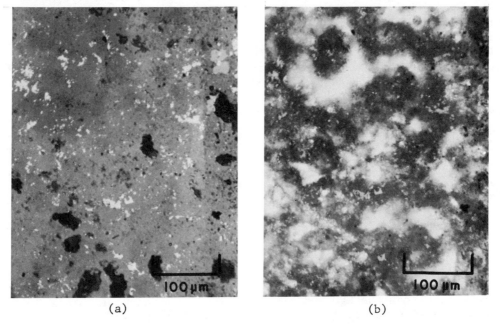

(a)   (b)

Fig. 5.   Microstructure of (a) flame-spray RBSN material with strength of 44,100 psi (303.4 MN/m^2); (b) same area in fully polarized light, white regions are high in porosity.

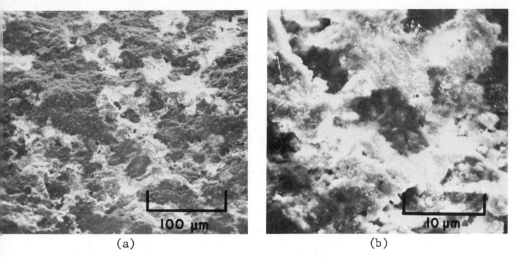

(a)   (b)

Fig. 6.   Scanning electron fractographs of high strength (44,100
           psi) specimen, (a) showing network of porous regions and
           (b) details of porous region in (a).

Strength variability is related to void size by the classical
Griffiths' relationship and to microstructural and compositional
features.   Davidge and Evans[23] have argued the Griffiths flaw
approach persuasively. As shown in Figs. 4-6, void fraction is
divided between large voids up to 100 µm in size, and a reticulated
distribution of very finely porous areas.   A further problem is that
the exothermic reaction of Si with N$_2$ may melt the Si compact
internally, without altering external appearance, as shown in Fig. 7,
a component which showed an exceptionally poor resistance to thermal
shock.   A further consideration is oxygen contamination, caused by
moisture in the nitriding furnace.   Four experimental flame-sprayed
materials nitrided together showed strengths only ~20-30% of
expected, but there was nothing abnormal in weight gain, final density

Fig. 7.   Fracture surfaces of thermally shocked 33 mm thick RBSN
           shape, showing unnitrided core caused by internal melting
           during nitridation.

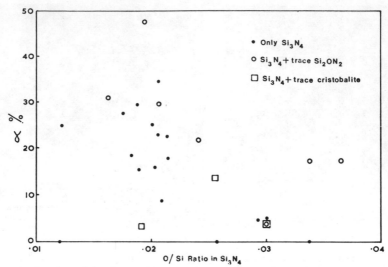

Fig. 8.   Relationship between α-Si₃N₄ (X-ray) and oxygen content
           (neutron activation) of flame-sprayed RBSN.

or microstructural appearance. X-ray examination showed an α/β
Si₃N₄ ratio of 32/68, with a trace of β cristobalite present.
Electron probe micro-analysis suggested 10 times more oxygen than
normal in the microstructure however, and vitreous silica was
probably present. Jack, et al.[24] have shown that α-Si₃N₄ is an
oxynitride, containing a few percent of oxygen, whereas the β-phase
does not contain oxygen. An investigation of oxygen content before
and after nitridation in flame-sprayed silicon route Si₃N₄ showed

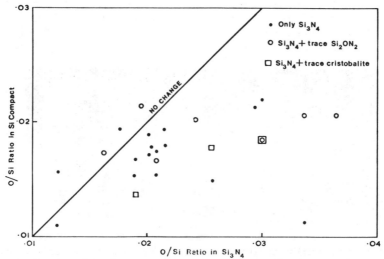

Fig. 9.   Relationship between oxygen content before and after
           nitriding of flame-sprayed Si compacts.

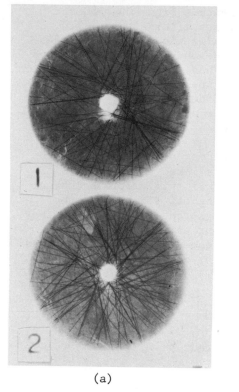
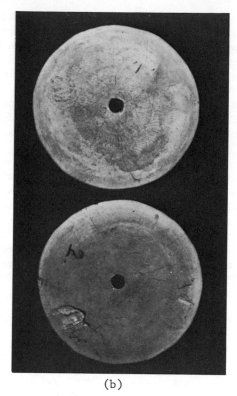

(a)   (b)

Fig. 10.   Reinforced 1.625 in. dia Si₃N₄ disc   (a) radiographs before
           test showing distribution of 0.3 v/o SiC fibre addition,
           and (b) after first thermal shock from 850°C.

no correlation between O and α content (Fig. 8), although it was
noteworthy that oxygen contents often had increased after the
nitriding reaction, as shown in Fig. 9. The role of oxygen in
Si₃N₄ ceramics is still a complex problem.

FIBRE-REINFORCED SILICON NITRIDE

A very significant recent development in Si₃N₄ ceramics has been
the demonstration that reinforcement with fibres is unusually
facile.[5,16] Very significant improvements in crack arrestment have been
reported for low density Si₃N₄ matrix composites reinforced with
10-20 v/o SiC-sheathed W-core fibres, toughness improvements by a
factor of at least 50 being possible.[25] This approach has been
briefly explored for thermal shock capability with 1.625 in. discs
with a small amount of reinforcement. Although they did not withstand
a single thermal shock without cracking, the fibres prevented to an
appreciable extent the brittle fragmentation usually experienced in
this test. The appearance of the discs after a single water quench
from 850°C, and radiographs before testing, are shown in Fig. 10.

The thinner disc (0.125 in.), after quenches from 900, 950 and 1000°C, spalled a large portion of surface after a second 1000°C quench; the thicker (0.18 in.) endured a little more, the surface being detached after the sixth 1000°C quench. However, there was evidence that cracking of the Si matrix had occurred in fabrication during laying up of the composite. The specimens had only 0.3% volume fraction of fibres in a matrix of density 1.24, and further experimentation appears justified in view of the reduction in fragmentation afforded by the reinforcement.

The presence of SiC in the material may act beneficially in three ways: thermal expansion mismatch is likely to put the Si_3N_4 matrix into compression; the presence of SiC will improve thermal conductivity; and the work done in pulling out fibres from the matrix significantly impedes crack propagation. It is likely that fibre reinforcement will have an important part to play in the development of ceramics for severe environments, since it promises some hope of ameliorating the worst defect of ceramic materials, lack of ductility.

ACKNOWLEDGMENTS

Grateful acknowledgment is made of much helpful guidance, advice and assistance from N. L. Parr, M. W. Lindley, D. P. Elias, P. G. Taylor, R. L. Brown and M. J. Leonard, and other colleagues at AML. This paper is published by permission of the Ministry of Defence.

REFERENCES

1. D. J. Godfrey, Metals & Mater., 2 (10) 305 (1968).
2. G. G. Deeley, J. M. Herbert and N. C. Moore, "Dense Silicon Nitride"; Powder Met. (8) 145-151 (1961).
3. R. J. Lumby and R. F. Coe, Proc. Brit. Ceram. Soc.,1970, No. 15, pp. 91-101.
4. N. L. Parr, G. F. Martin and E. R. W. May, p. 102-135, in Special Ceramics. Ed. by P. Popper. Haywood & Co., London, 1960.
5. D. J. Godfrey, Brit. Interplanetary Soc. 22 (5) 353, Oct. 1969.
6. E. Glenny and T. A. Taylor, Powder Met. (1/2), p. 189-226, 1958.
7. E. Glenny and T. A. Taylor, Powder Met. (8), p. 164-195, 1961.
8. British Patent 1117788.
9. D. J. Godfrey and P. G. Taylor, Eng. Mater. Design, p. 1339-1342, Sept. 1969.
10. D. J. Godfrey and N. L. Parr, p. 379-395, in Combustion in Advanced Gas Turbine Propulsion Systems. Ed. by I. E. Smith. Pergamon Press, London, 1968.

11. D. J. Godfrey and P. G. Taylor, p. 265-273, in Special Ceramics
    4. Ed. by P. Popper. British Ceramic Research Association,
    Stoke-on-Trent, 1968.
12. J. F. Collins and R. W. Gerby, J.Metals (8) 612-615 (1955).
13. P. Popper and S. N. Ruddlesden, Trans. Brit. Ceram. Soc. 60,
    603-623 (1961).
14. D. J. Godfrey, Engineering, 126 (8 Aug. 1967).
15. R. W. Powell and R. P. Tye, p. 272-273 in Special Ceramics 1962.
    Ed. by P. Popper. Academic Press, London, 1963.
16. R. L. Brown, D. J. Godfrey, M. W. Lindley and E. R. W. May,
    Special Ceramics 1970, British Ceramic Research Research
    Assoc, Stoke-on-Trent. (in press).
17. R. L. Brown, Metal Constr. 1 (7) 317-320, (July, 1969).
18. J. P. Moore, T. G. Kollie, R. S. Graves and D. L. McElroy,
    "Thermal Conductivity Measurements on Solids Between 20 and
    150°C"; ORNL-412 (June 1967).
19. J. Cooke, Composites, 1 176-180 (Mar. 1970).
20. J. K. MacKenzie, Proc. Phys. Soc. (London), B63, 2, 1950.
21. S. Wild, P. Grieveson, K. H. Jack and M. J. Latimer, in Special
    Ceramics 1970, British Ceramic Research Assoc., Stoke-on-
    Trent. (In press).
22. D. G. S. Davies, Private communication.
23. A. G. Evans and R. W. Davidge, J. Mater. Sci., 5, 314-325 (1970).
24. P. Grieveson, K. H. Jack and S. Wild; p. 237 in Special Ceramics
    4. Ed. by P. Popper BCRA, Stoke-on-Trent, 1968, p. 237; and
    BCRA Fifth Special Ceramics Symposium 1970, in course of
    publication.
25. D. J. Godfrey and M. W. Lindley, Nature 229, 192 (1971).

DISCUSSION

A. H. Heuer (Case Western Reserve University): Is it true that the
hot pressed (i.e., dense) high strength Si₃N₄ shows the very best
thermal shock resistance?
Authors: Yes.

R. C. Rossi (Aerospace Corporation): You use Weibull's "m" parameter
for design purposes. Theoretical work which we have not yet reported
has revealed that this parameter relates to density rather than
volume. Furthermore, values greater than 2.5 explicitly require flaw
interaction. High "m" values like those you report suggest a great
deal of flaw-flaw interaction and may in some way account for the
excellent thermal shock resistance you observe.
Authors: Without greater detail about the work cited, it is difficult to comment, but the hypothesis seems unconvincing in view of
high "m" values which are characteristic of materials of lower
variability, such as metals. Experiments on the volume effect have
shown agreement with theoretical prediction to within 1% (D.C.S.
Davies, unpublished work). The high void fraction of RBSN means an
abundance of flaws, facilitating uniformity in strength properties.

J. J. Gangler (National Aeronautics and Space Administration): Can you comment on the impact and oxidation resistance of Si_3N_4?
Authors: Si_3N_4 has about the same rather low impact strength of other ceramic materials. Our reinforced materials may be better, but we have not yet done impact tests on them. Oxidation resistance is good, being comparable with SiC. A life of at least 1000 hr was found at 1400°C in a flue gas atmosphere. Long lives at 1200°C have been demonstrated[4] and recent work has shown no appreciable recession of surface after 80 hr at 1500°C.

P. W. Heilman (Detroit Diesel-Allison, General Motors Corp.): Could you describe in more detail how the integral turbine wheel (Fig. 2) was fabricated?
Authors: An annular steel sheet was slotted radially, and the segments bent with pliers to angle the blades. Silicon powder was compacted onto this former by flame spraying of Si. A silicon compact tube was inserted axially and joined with a fillet deposit of Si on each side, and the whole nitrided to a monolithic Si_3N_4 component.

B. A. Wilcox (Battelle Memorial Institute): Have you examined the creep behavior of Si_3N_4, and if so, could you comment on the results?
Authors: No further creep work has been carried out on RBSN superceding the data of Ref. 4. Additions of SiC (5–10%) were shown to give secondary creep rates similar to superalloys. The creep performance of HPSN above 1000°C has been studied by workers at Plessey Co. and at N.G.T.E., showing pure material (99.9% Si_3N_4 powder) to be better than that of 98% purity, although strength deterioration above 1100–1200°C can be a problem.

I. Ahmad (U. S. Army Watervliet Arsenal): What is the process of nitridation, and what rates of nitridation are achieved?
Authors: We are speaking of the reaction of Si with N_2 in the temperature range 1200–1450°C, in two or more stages, over 1–5 days. The rate retards parabolically with time, so that in the early stages melting of Si by exothermic heat has to be avoided, and in latter stages less accessible Si has to be reached by the nitrogen.

D. D. Briggs (Coors Porcelain Co.): In general, what are the electrical and chemical resistance properties of silicon nitrides?
Authors: Si_3N_4 is a good insulator. It resists most corrosive media extremely well, with the exception of some molten metals and glasses, and pressurized alkaline media.

THERMAL SHOCK RESISTANT ZIRCONIA NOZZLES FOR CONTINUOUS COPPER CASTING

R. E. Jaeger[*] and R. E. Nickell[**]

Bell Telephone Laboratories, Inc.

[*]Murray Hill and [**]Whippany, New Jersey

ABSTRACT

A partially stabilized zirconium oxide nozzle to be used in a continuous copper casting process has been designed to adequately resist thermal shock and satisfactorily tested in the operating environment. The zirconia has an optimized stabilizer content in terms of free thermal expansion characteristics and high temperature tensile strength. The optimization is achieved through control of composition, microstructure, and crystallographic transformation during the heating cycle. Thermal and stress analyses of the design indicate a positive safety margin subsequently borne out by production testing.

INTRODUCTION

The operating companies of the Bell System amass enormous quantities of scrap metal from equipment replacement. The reprocessing of a large percentage of this scrap, approximately one million pounds per day, is the responsibility of a wholly owned subsidiary - Nassau Smelting and Refining Corporation. The reclaimed metal, primarily lead and copper, is then resold to Western Electric and the operating companies. It is expected that by 1980 the daily consumption of scrap at Nassau will double.

For the past several years, Nassau has been experimenting with a continuous casting and rolling process that seems to have the dual advantages of economy and higher quality rod. It is anticipated that the new process will reduce the cost of producing copper rod

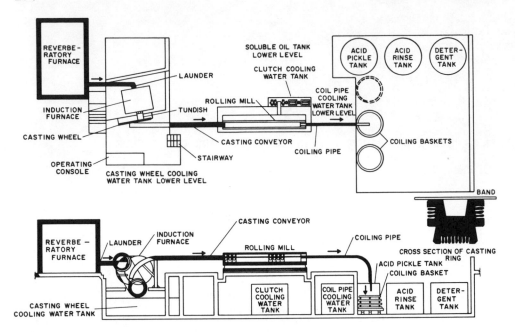

Fig. 1.   Schematic of continuous casting and rolling process.

from scrap copper by half.   A schematic of the process is shown in
Fig. 1.   Two of the 75-ton reverberatory furnaces for refining the
scrap copper are now in operation at Nassau.   When the molten
metal meets purity standards with respect to oxygen and metallic
impuries it is allowed to flow into an induction-heated holding
furnace which serves as an intermediate control on the flow between
the reverberatory furnace and a refractory funnel called the
tundish.   From the tundish the molten copper flows through an
opening into a nozzle which directs the stream into the mold cavity
in the casting wheel at the rate of 20,000 lbs/hr.   A cross section
of the mold and the steel mold closing band is included in Fig. 1.

     As the casting wheel rotates, cooling water is sprayed on the
mold and the steel mold closing band to solidify the copper and
enable the casting to develop adequate strength (while maintaining
appropriate ductility) in order to be conveyed into the rolling
mill.   The continuous cast bar has a 2.4 in.^{2} trapezoidal cross
section and enters the 14-stand rolling mill traveling at 40 ft/min.
The mill rolls the bar into a 5/16 in. dia rod which is wound at
a rate of 1,200 ft/min in 8,000 lb coils.   A more complete description
of the process is given in Ref. 1.

     Several technical problems of a thermo-mechanical nature
inhibit the continuous casting process - such as the service life
of the casting mold and thermal stress fatigue of the closing band.

The most serious of these problems, however, and the limiting factor
in determining the duration of a casting run, is the lifetime of
the graphite nozzles which are currently used to direct the molten
copper from the tundish into the mold cavity. At the pouring
temperature (1130–1180°C) the graphite undergoes erosion by two
mechanisms. The physical size of the nozzle is gradually reduced
by rapid oxidation at the surfaces exposed to the air atmosphere
while the inside bore grows in size as a result of constant abrasion
by the molten copper stream which combine to give a useful life of
about 3 hr for this type of nozzle. The erosion rates imply that
two graphite nozzles are required to empty each of the two reverberatory furnaces per day with all of the attendant problems associated
with starting a casting run four times each day. In addition, the
down times represent a loss of actual casting time which ultimately
reduces the maximum attainable production by 25–30%.

MATERIAL SELECTION

     In an effort to alleviate this problem, a ceramic nozzle
development program was initiated. Its purpose has been to attempt
to find materials to replace the graphite which are capable of
performing adequately as a nozzle under the conditions set by the
current operating procedure.

     Materials for experimentation were limited to those which are
commercially available in the required geometry as well as by
information in the literature concerning: (1) their stability in
contact with molten copper; (2) thermal shock characteristics;
(3) oxidation resistance and (4) refractoriness. Whenever possible,
promising data found in the literature were checked by experiment.
Generally speaking the most refractory ceramic materials are the
carbides, borides, nitrides and single oxides. Materials in these
basic categories were screened for further testing using the above
criteria.

     As a result of this initial screening, the following limited
number of materials were selected for further examination: ZrC,
Al_4C_3, ZrB_2, AlN, Al_2O_3, BeO, MgO, ZrO_2. The stability of these
materials in contact with molten copper is determined to some
extent by their wetting characteristics and also by any chemical
reactions that may occur at the temperatures involved. The wetting
and flow properties in a solid-liquid-vapor system are controlled
by the surface and interface energies involved and can be determined
by considering the shape that a liquid droplet assumes on a solid
substrate.

     A spreading coefficient can be defined, $S = \gamma_{SV} - (\gamma_{LV} + \gamma_{SL})$
where S is the spreading coefficient and γ_{SV}, γ_{LV} and γ_{SL} are the

solid-vapor, liquid-vapor, and solid-liquid interface energies.
When S is positive, spreading occurs. It follows from this that a
necessary condition for spreading is that $\gamma_{LV} < \gamma_{SV}$. The reverse
is true for the materials we have selected in all cases where information on surface energies is available, indicating that copper will
not wet them.

The possibility of chemical reaction between molten copper and
any of these materials is slight, but reaction will occur with any
copper oxide that may be present. Also the presence of CuO will
cause the copper to wet the ceramic more readily.[2] As a check on
the data reported in the literature, and in an attempt to simulate
the conditions on the inside bore of the nozzle, the various
materials were heated in contact with copper to 1150°C in a wet N_2
atmosphere. In addition, any material suspected of questionable
oxidation resistance was heated to the same temperature in air.

Both ZrC and ZrB_2 were subsequently eliminated because of their
extremely poor oxidation resistance, contrary to previously reported
data in the literature. BeO and MgO exhibited the poorest wetting
characteristics (lowest contact angle) and were eliminated on that
basis. In addition, MgO appeared to have a tendency to be reduced
by the molten Cu which resulted in a further decrease in wetting
angle. Al_2O_3 and Al_4C_3 (carburized surface layer on Al_2O_3) were
perhaps the most readily available materials, however, they had
relatively poor thermal shock resistance as will be discussed in
a later section. While meeting all other criteria AlN was unavailable commercially in the quality and size required. The ZrO_2 was
not as readily available as most of the other materials but met all
the chemical stability requirements and had the highest thermal
shock figure of merit.

THERMAL SHOCK CRITERIA

The resistance to thermal shock can be defined as the ability
of solid materials to withstand failure (usually tensile failure
in brittle materials) when subjected to sudden heating or cooling.
A quantitative measure of this resistance is the thermal shock
figure of merit - a product of the *material* heat resistance factor
and the *design* factor.[3] The material heat resistance factor is
computed from

$$R = \frac{\sigma_f(1-\nu)}{\alpha E} \tag{1}$$

where σ_f is the tensile failure strength, ν is Poisson's ratio, α
is the coefficient of linear thermal expansion, and E is the elastic
modulus, and carries the units of temperature. The design factor,

Q, is a function of the Fourier number $F_0 = \kappa t/L^2$, where κ is the thermal diffusivity, t is the time, and L is a characteristic dimension; the Biot number $Bi = hL/k$, where h is the coefficient of heat exchange with the surroundings and k is the thermal conductivity; and the use temperature T_u. The design factor must, in general, be determined from experiment and/or analysis.

Because of the significantly greater values of thermal conductivity and diffusivity for graphite, as compared to those of ceramics, geometrically similar parts will be orders of magnitude different with respect to thermal shock resistance. These differences can be minimized by proper design if the material heat resistance factor is comparable to that of graphite – about 600°C for CS grade graphite. The candidate materials – Al_2O_3 and ZrO_2 – were found to have R = 100°C and 600°C, respectively. On the basis of these calculations, a decision was made to pursue the design of a zirconium oxide nozzle, in this case a MgO-stabilized ZrO_2 composition (Zircoa 1706) supplied by Zirconium Corporation of America.

The thermal conductivity of this material is so low (1.125 Btu/hr ft°F at 2000°F) that the major design effort centered around a thermal conditioning environment which would minimize thermal gradients in the nozzle during initial exposure to the molten copper. The attractive features of Zircoa 1706, on the other hand, were its relatively high strength-to-modulus ratio (which remained to be firmly verified at high temperatures) and the thermal expansion data, which will be discussed in some detail subsequently.

NOZZLE-HEATER DESIGN

The obvious solution to the thermal conditioning problem was to design a furnace capable of preheating the nozzle to the molten copper temperature and which would remain on the nozzle during the casting run. The heater unfortunately did not eliminate the thermal stress problem entirely. The amount of space between the front face of the tundish and the mold closing band is very limited and places rather severe physical restrictions on the overall geometry of the furnace. In reality this resulted in a shorter, less efficient furnace which was incapable of providing a uniform temperature distribution over the entire length of the nozzle.

A cross-sectional detail of the nozzle-heater configuration is shown in Fig. 2. The entrance end of the nozzle is flared with a 90° included angle in order that the exit hole in the tundish be completely covered and a good seal maintained between the nozzle and the tundish. The close proximity of the heater winding to this opening in the tundish is such that even a small copper leak would

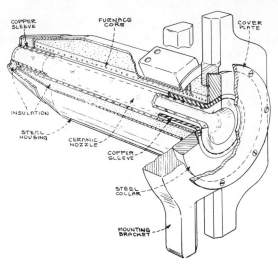

Fig. 2. Sectional view of nozzle-heater design.

result in imminent failure of the furnace. The nozzle is supported
in a stainless steel collar, having a matching 45° taper on its
inside wall, which in turn is recessed into a large steel mounting
bracket. The top surface of the nozzle is allowed to protrude
slightly beyond the surface of the steel collar. During the
mounting procedure, therefore, the nozzle is wedged into the collar
and, as a result, that part of the nozzle in contact with the collar
is put in compression. This detail was required in order that the
nozzle be allowed to survive the mechanical abuse of the mounting
process and subsequent "rodding" during the casting run.

     The heater winding starts at a position coinciding roughly
with the beginning of the flared portion of the nozzle and extends
to a point about 1 in. from the exit end. The air gap between
the O.D. of the nozzle and the I.D. of the heater core allows for
any axial camber resulting from the forming and firing process for
the ceramic and facilitates replacement of the nozzle at the completion of a casting run. This approach was thought to be superior
to a design in which the heater is an integral part of the expendable
nozzle. The exit end was allowed to extend beyond the heater for
several reasons: (1) this part of the nozzle has to be positioned
partially within the 1 $\frac{7}{8}$ in. wide mold cavity of the casting wheel
in order to effectively direct the stream of molten copper; (2) at
the same time, the operator's line of sight view of the liquid
puddle in the mold must not be blocked; and (3) the rather vigorous
splashing of molten copper in this area would be detrimental to the
life of the furnace.

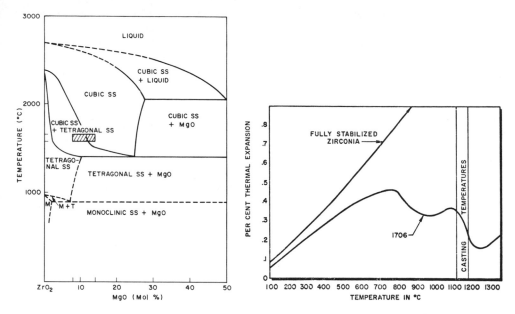

Fig. 3. MgO-ZrO₂ phase diagram. Fig. 4. Thermal expansion of fully
                                          stabilized ZrO₂ and Zircoa
                                          1706.

MICROSTRUCTURE AND THERMAL EXPANSION OF ZIRCOA 1706

A brief discussion of the zirconia-magnesia phase diagram is in
order at this point to achieve a better understanding of the influence
of the structure of these materials on their thermal behavior.
Figure 3 shows a portion of the phase diagram of this system according
to Viechnicki and Stubican.[4] A variety of MgO stabilized ZrO₂ compositions having markedly different properties can be manufactured
in the general region bordered by the two vertical lines which
straddle the 10 mole % MgO composition. These ceramics are fired
in the temperature range bordered by the two horizontal lines around
1600°C. The phase boundary separating the cubic solid solution
region and the tetragonal - cubic region also cuts through our area
of interest. Both monophase (fully stabilized) and biphase (1706)
ceramics can be made by changing the composition of the mix. The
monophase compositions to the right and above the phase boundary are
stronger and more stable dimensionally with thermal cycling. Compositions below and to the left of the boundary are weaker but nevertheless have superior thermal shock resistance resulting from a
reduction in elastic modulus and a low but somewhat anomalous thermal
expansion characteristic.

Figure 4 compares the thermal expansion of a fully stabilized material with that of the #1706 composition from 100-1350°C. The fully stabilized zirconia, containing 4.5 to 5 w/o MgO is fired in the cubic solid solution region and retains the cubic structure upon cooling to room temperature. The 1706 composition, on the other hand, contains only 2.8 w/o MgO and is fired in the cubic + tetragonal solid solution region. At room temperature the structure of this composition is characterized by a discontinuous small grained monoclinic phase dispersed in a large grained cubic matrix. The ratio of cubic to monoclinic determined by x-ray is 60% to 40% by weight, however, the amount of monoclinic phase that can be resolved optically as discrete grains is around 20-25%. The remaining mono-clinic phase exists as submicroscopic domains within the cubic phase. Figure 5a is an electronmicrograph of an as-fired surface at ~4500 X which already begins to show a clear indication of these domains in some of the cubic grains. At higher magnification (Fig. 5b) they appear to be tiny crystals embedded in the cubic grains. Their average diameter is 2000 to 5000 A.

The complex shape of the thermal expansion curve of 1706 is due to the combined effects of the cubic, intergrown monoclinic and discrete monoclinic phases present in the microstructure of this material. The rather large volume contraction occurring in the temperature range from 1100-1250°C is a result of the transformation of the discrete monoclinic phase to tetragonal. This transformation appears to be completed at 1250°C. Inasmuch as the intergrown mono-clinic phase might well be left in a stressed condition within the

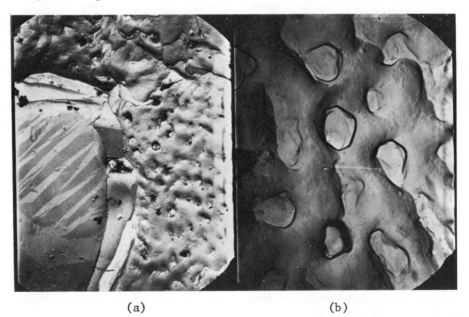

(a)   (b)

Fig. 5.   Electron micrographs of as-fired surface replica illustrating multiphase microstructure (a) ~4500 X, (b) ~18,000 X.

cubic crystals, one might expect that this portion of the monoclinic phase would convert at a lower temperature. The anomalous negative expansion in the temperature range from 800–950°C is evidence that this may indeed be the case.

Considering the nozzle heater previously described, it is clear from Fig. 4 that for temperatures associated with a normal axial gradient, both the magnitude and the sign of the slope of the expansion are changing radically. As this point it became evident that one needed to have additional information regarding the location and magnitude of the thermal stresses in the nozzle in order to evaluate the probability of survival of a given design.

PREHEATING EXPERIMENT

An important ingredient in the rational design process is the mathematical model. Because of the complexity of the nozzle and heater design, a computational model of the thermal shock problem, including both heat transfer and thermal stress aspects, was developed. A preheating experiment, using an available Al_2O_3 nozzle, was then conducted in order to verify the predictive accuracy of the computational model.

Before carrying out the analysis, however, several basic questions with regard to the physics of the problem had to be resolved. The most important were: (a) calculating the volumetric heat generation rate corresponding to the resistance heater wire; (b) determining heat transfer coefficients for the various surfaces of the device corresponding to free convection and radiation; and (c) assessing the heat transfer via radiation between internal boundaries (e.g., across the air gap between the heater core and the nozzle).

Heat is provided by a resistance element composed of 30-mil Pt-Rh wire embedded in a porous aluminum oxide (Alundum) furnace core which is grooved for wire windings. The average winding diameter for the wire is $1\frac{7}{8}$ in. and the core is grooved for 84 full turns of wire in a core length of $9\frac{1}{4}$ in. If all turns are used, a total wire length of 12.57 m would be required and a uniform heating rate over the axial length of the furnace would be achieved. When the heater was operated at 150 V, a current of 9.5 A was drawn, implying a resistance per unit length of wire of 1.25 Ω/m. This number corresponds well with the electrical resistance per unit length for Pt-Rh wire at temperatures near 1500°C. For a uniform distribution, the heat generated per unit axial length per unit time through Joule heating is

$$\dot{Q}_o = \frac{(150 \text{ V}) (9.5 \text{ A})}{(9.25 \text{ in.}) (2.54 \text{ cm/in.})} = 60.6 \text{ J/cm/sec} \qquad (2)$$

As estimate for the effective conductivity across the air gap, taking thermal radiation into account, can be found from the expression for radiation between two infinitely long concentric cylinders:[5]

$$q = \frac{\sigma}{\frac{1}{\varepsilon_1} + \frac{A_1}{A_2}(\frac{1}{\varepsilon_2}-1)} (T_1^4 - T_2^4) \tag{3}$$

where q is the heat flux per unit area; A_1 and A_2 are the inner and outer surface areas of the concentric cylinders, respectively; ε_1 and ε_2 are the respective surface emissivities; T_1 and T_2 are the respective absolute surface temperatures; and σ is the Stefan-Boltzmann constant. For our geometry, the area ratio is about 0.8 and the surface emissivities can be assumed to be 0.9. Then

$$q = \frac{5}{6}(T_1^4 - T_2^4) \tag{4}$$

and an effective conductivity can be calculated from

$$k_{eff} = \frac{q\ r}{(T_1 - T_2)} = \frac{10}{3}\sigma\ \Delta r(T_{avg})^3 \tag{5}$$

where Δr is the radial gap and T_{avg} is the average absolute air temperature in the gap. For average air temperatures on the order of 1650°K, the effective conductivity due to radiation is approximately 20 times the nominal conductivity at that temperature, indicating that the dominant heat transfer mechanism across the gap is radiation.

The last question to be resolved concerns the free convection and radiation from the device surfaces to the surrounds. For free convection at the exterior of a heated cylinder in a vertical position, the formula

$$\overline{Nu} - 0.555(GrPr)^{1/4} \tag{6}$$

is considered satisfactory for laminar flow and

$$\overline{Nu} = 0.0210(GrPr)^{2/5} \tag{7}$$

is suitable for turbulent flow, where $\overline{Nu} = h_c L/k$ is the average Nusselt number based on the length, L, of the nozzle, the thermal conductivity of the air is k, and GrPr is the product of the Grashof and Prandtl numbers.[6] Evaluating the properties at the mean temperature (estimated) between the wall and the free stream, GrPr = 6×10^7. This implies that the free convective flow is laminar and that the

mean transfer coefficient for free convection for the nozzle exterior is $\bar{h}_c$ = 1.4 Btu/hr ft^2 °F. If radiation is taken into
account, then

$$\bar{h} = \bar{h}_c + \bar{h}_r \tag{8}$$

where $\bar{h}_r$ can be estimated from the equation for radiation from a
gray body into a black environment:

$$\bar{h}_r = \frac{\varepsilon\sigma(T_w^4 - T_\infty^4)}{(T_w - T_\infty)} . \tag{9}$$

Here ε is the emmissivity of the stainless steel jacket, T_w is the
absolute wall temperature and T_∞ is the absolute temperature of the
black surroundings. For an estimated wall temperature of 500°C,
the surroundings at 20°C, and an emissivity for polished stainless
steel of about 0.2, $\bar{h}_r$ is approximately 1.5 Btu/hr ft^2 °F. Therefore, the combined heat transfer coefficient is chosen to be
$\bar{h}$ = 3.0 Btu/hr ft^2°F.

A reasonable model for heat transfer from the exit end of the
nozzle is to neglect convection and treat the radiation by considering
the ceramic to be a small gray body in black surroundings. Then,
as before,

$$\bar{h} - \bar{h}_r = \frac{\varepsilon\sigma(T_s^4 - T_\infty^4)}{(T_s - T_\infty)} = \varepsilon F_T \tag{10}$$

where T_s is the nozzle exit end surface temperature and F_T is
referred to as the temperature factor. For an estimated tip temperature of 600°C radiating to black surrounding at 20°C, F_T = 8.3.
If the emissivity is chosen to be ε = 0.8, then the film coefficient
is about 6.5. For the computer simulation, the film coefficient
was assumed to vary linearly from 10.0 at the inside wall to 5.0
at the outside wall.

Free convection on the interior wall of the nozzle can be
neglected since the entrance end was sealed during the preheating

experiment. Now consider the problem of radiation from the inside
surface of the nozzle. The wall temperature is a function of axial
position and depends on four heat flux contributions; (a) heat
conducted across the nozzle wall from the resistance heater;
(b) radiative flux to the environment through the entrance and
(c) exit openings in the nozzle; and (d) radiation to other points
of the inside wall surface. This problem has been addressed[7] and
an expression derived which can be modified in order to calculate
an effective film coefficient for radiation:

$$q_w(z) = \bar{h}_r(T_w - T_\infty) = \varepsilon\sigma T_w{}^4 - \sigma T_1^4\, F(z) - \sigma T_2^4\, F(L-z)$$

$$- (1-\varepsilon)\sigma \int_o^z T_w^4\,(\eta)K(z-\eta)\, d\eta$$

$$- (1-\varepsilon)\sigma \int_z^L T_w^4\,(\eta)K(\eta-z)\, d\eta \qquad (11)$$

and

$$F(z) = \left[\frac{\left[\dfrac{z}{D}\right]^2 + 1/2}{\left[\dfrac{z^2}{D^2} + 1\right]^{1/2}} - \frac{z}{D} \right] \qquad (12)$$

is the shape factor for radiation from the inside wall of a cylinder
of length L and diameter D through an end opening to a black environment at temperature T_1; $F(L-z)$ is the shape factor for radiation
through the other end opening to a black environment at temperature
T_2; and

$$K(z) = 1 - \frac{\left[\dfrac{z}{D}\right]^3 + \dfrac{3z}{2D}}{\left[\dfrac{z^2}{D^2} + 1\right]^{3/2}} \qquad (13)$$

is the shape factor for radiation between differential areas of the
cylinder wall. If the emissivity is assumed to be independent of
temperature and the temperature profile as a function of axial
position is relatively flat in the middle of the heater, an estimate
can be obtained for $\bar{h}_r$. In the middle of the heater, radiation
through the end openings can be neglected and, if T_w is assumed
nearly constant, then

$$\bar{h}_r = \sigma T_w^3 \left[\varepsilon - (1-\varepsilon) \int_o^L K(|z-\eta|)\,d\eta \right]. \qquad (14)$$

For a wall temperature of 1000°C, $\bar{h}_r$ = 10.0; for a wall temperature of 1200°C, $\bar{h}_r$ = 17.0; and, for a wall temperature of 800°C,
$\bar{h}_r$ = 7.0. For the computer simulation, the effective heat transfer
coefficient was selected to be 5.5 Btu/hr ft^2 °F - uniform along
the entire inside wall.

Using this accumulated information, a transient heat conduction
analysis of the nozzle and resistance heater was performed using a
finite element heat conduction computer code.[8] This code has the
capability to treat two-dimensional (plane or axisymmetric) problems
having arbitrary cross-sectional geometry and includes the effects
of temperature dependent thermal properties, among other features.
The heater was subdivided into 1237 finite elements with a total
of 1308 nodal points; 50 rows of nodal points were used in the layout. Ten elements were used in each row to characterize the temperature distribution across the nozzle wall. One element was used
to model heat transfer across the air gap and eight elements were
used to describe the volumetric heat generation of the resistance
wire and the heat conduction across the Alundum core. Up to 10
elements characterized the temperature drop across the insulation
layer. One element was used to model the stainless steel jacket.

The thermal conductivity of the Pt-Rh wire and the stainless
steel jacket were assumed to be constant with temperature, having
values of 40.2 and 9.4 Btu/hr ft °F respectively. All other materials
(Coors AD-995 alumina, air gap and Fiberfrax insulation) were assumed
to have thermal conductivities which depend significantly on temperature at the operating range of the heater. As discussed previously
the dominant mode of heat transfer across the air gap was radiation;
thermal conductivity versus temperature curves for the other three
materials are shown in Fig. 6. Volume specific heats for the

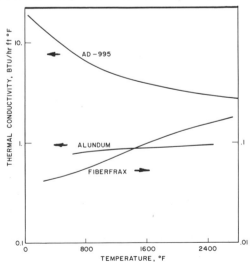

Fig. 6.   Thermal conductivity vs temperature for nozzle-heater
materials.

various materials are shown in Table 1. A uniform heat generation
rate corresponding to 60.6 J/cm sec was selected for the resistance
wire.

Table 1. Specific Heats

Material	AD-995	Air	Alundum	Platinum	Insulation	Steel
Specific Heat (Btu/ft^3 °F)	50.1	.008	20.0	43.0	9.0	54.0

When a very large time step is used in the finite element
analysis, the steady state solution is computed. A slight modification in this procedure enables the nonlinear thermal conductivities
and the radiation across the air gap to be taken into account
correctly. Several very large time steps, each one hour long, are
used; at the end of each time step, the thermal transport properties
are recomputed based on the current temperature distribution. The
temperatures at the end of another hour are then found, and iteration
continues until convergence is achieved. For this analysis, four
iterations were sufficient to assure convergence.

A Zircoa 1706 nozzle was not available at this stage of the
development program; therefore an alumina nozzle (Coors AD-995) of
the same configuration was used for the experiment. The device was
preheated to a steady state with the heater operating at 150V and
9.5 A. Thermocouple and I.R. probe locations where the steady state
temperatures were measured are shown in Fig. 7; for I.R. measurements,

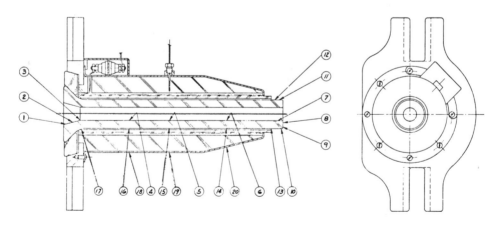

STEADY STATE	T.C. POSITION	1	2	3	4	5	6	7	8	9	10	11	12	13	14	15	16	17	18	19	20	MAT'L.	
1350°C	TEMP°C	680	720	947	1472	1538	1350	720	535	450	465	485	570	565	1374	1568	1430	909	475	480	460	AD 995	EXPERIMENTAL
1393°C		587	619	780	1414	1559	1297	577	525	510	520	531	547	590	1330	1593	1447	806	409	455	575		CALCULATED

Fig. 7. Comparison of experimental and calculated steady-state
         temperature distribution.

emissivity was assumed to be 0.8. In addition, the comparison between experimentally measured and analytically computed temperatures is shown in tabular form on the same figure.

The comparison appears favorable at most probe locations considering the complexity of the modeling problem. Notable exceptions are the locations (1 and 7) at the inner wall near the entrance and exit end openings. The probable source of error at these areas is the assumption of constant film coefficient along the entire inside wall. Re-radiation to these cooler areas from hotter spots in the center of the heater - plus the ability of these points to "see" a cool, black environment through the end openings, thereby drawing heat to the inside wall surface - would likely account for the differences. The discrepancies at locations 17 and 20 can probably be traced with similar reasoning. If the radiation boundary conditions at the inside wall near the exit end of the nozzle were treated more precisely, heat would be drawn away from location 20 toward the inside wall. Similarly, more precise boundary values at the entrance end would draw heat toward location 17.

From an engineering standpoint, however, the results are conservative, since lower temperatures than actual are predicted and they are sufficiently accurate to verify the mathematical model.

THERMAL STRESS ANALYSIS OF ZIRCOA NOZZLE

With the confidence engendered by the experimental confirmation of the mathematical model, a complete heat transfer and thermal stress analysis for a ZrO_2 nozzle under actual operating conditions was undertaken. Two cases were analyzed: (1) a nozzle with a bore of 11/16 in. and a constant wall thickness of 13/32 in.; and (2) a nozzle with a bore of 5/8 in. and a tapered wall thickness at the exit end, having copper sleeves fitted on both ends to draw heat to relatively cool regions.

Both analyses consisted of three parts: (a) the steady-state temperature distribution due to preheating was found, using the finite element heat conduction code, and the nodal point temperatures written on a magnetic tape for subsequent analysis; (b) the transient temperature distribution during copper efflux was then computed, again using this code, with the initial conditions supplied from the magnetic tape mentioned above; nodal point temperatures in the 1706 material (and in the copper sleeves, for the alternate design) at selected times of interest in the transient regime are written on another magnetic tape for use in the stress analysis; and (c) a thermal stress analysis (neglecting inertial effects) is carried out, using a finite element stress analysis code,[9] at these selected times, with the temperatures supplied as input from the second magnetic tape.

In the first case (without the copper sleeves) the grid layout
was essentially identical to that used for the steady-state
experiment (see Fig. 7). A total of 1330 nodal points and 1262
elements were used in the thermal analysis. The film coefficient
for the stainless steel heating jacket was chosen to be 3.0, for
the inside wall a uniform value of 5.0 and for the exit end, values
between 5.0 and 10.0 Btu/hr ft^2 °F were used. The entrance end
was assumed to be in contact with the tundish, which was maintained
at a temperature of 550°C. A total time of two hours was allowed
in order for the system to reach a steady-state during preheating.
The temperature profile for the inside wall at the end of this time
is shown in Fig. 8, where it is compared to a similar profile for
the alternate design having the copper sleeves.

Using these steady-state temperatures as initial data, transient
temperatures during the first few seconds of copper efflux were
found. Boundary conditions for the problem were kept the same with
the exception that the nodal points on the inside wall were all
prescribed to be at a temperature of 1150°C (a mean value for the
molten copper flowing from the tundish). Thermal stresses were
then computed at three times after the beginning of copper flow, i.e.,
0.1, 0.5, 1.5 sec.

From an examination of the computed stresses, together with
the knowledge that the tensile failure stress of the material is
about 18,000 psi at the pour temperature of 1150°C, two regions of
likely failure were seen. Critical hoop stresses occur on the
inside wall approximately two inches from the exit end of the nozzle
and critical axial stresses are found on the outside wall slightly
below the collar region (see Fig. 2 for visual reference). These
stresses are the result of large radial and axial temperature gradients
combined with the large temperature difference between the molten
copper temperature and the cooler entrance and exit ends of the nozzle.

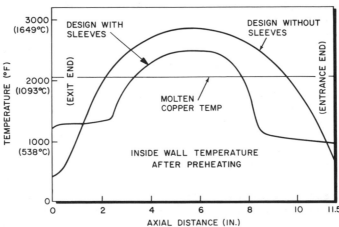

Fig. 8.   Temperature profile on inside bore of nozzle.

The modified design was modeled with 1304 nodal points and 1236 elements for the preheating and transient copper efflux thermal analysis. Once again two hours were used to achieve the steady-state preheating condition. The comparison between preheating temperature profiles at the inside wall for the two designs is shown in Fig. 8. Transient temperatures, using the preheat temperatures as initial conditions, were then found for the first few seconds of copper pour and the thermal stresses were calculated at 2.0 sec. Comparison of these stresses to those mentioned previously showed that critical stress regions had been eliminated, with the possible exception of hoop and axial stresses .at the inside wall near the termination of the copper sleeves.

In an effort to determine whether the hoop and axial stresses in the neighborhood of the sleeve terminations were critical, a heat transfer and stress analysis of the exit end of the nozzle was made. The mesh was refined near the upper end of the copper sleeve in order to improve the computation of temperature gradients and stresses. These temperatures were essentially identical to those computed from the full-size simulation, with the added definition in the neighborhood of the copper sleeve termination.

Transient temperatures during initial copper efflux were then calculated and stresses were computed, based on these temperatures, at three times, 0.1, 0.5, and 2.0 sec. From an examination of these stresses, no cirtical radial or hoop stress regions were seen to exist in the modified design. The highest tensile stresses are at the inside wall - approximately 1/2 in. below the termination of the copper sleeve. These stresses, which are of the order of 17,000 psi (sufficiently close to the tensile strength of the material to give marginal confidence in the design), are caused by the large temperature differences in the axial and radial directions combined with an adverse material property. These large differences are occurring at temperatures where the free thermal expansion of the stabilized zirconia is making a rapid change (in this case, decreasing). Because of the addition of the copper sleeves, however, the magnitude of these thermal stresses has been reduced substantially and positive safety margins exist everywhere.

An informative comparison between the state of thermal stress in the original and modified designs is shown in Figs. 9 and 10. The hoop and axial stresses at points very near the inside wall are plotted as a function of axial positions, with z = 0 indicating the exit end of the nozzle and z = 11.5 in. indicating the entrance end. These comparisons illustrate graphically the advantages which accrue to the modified design.

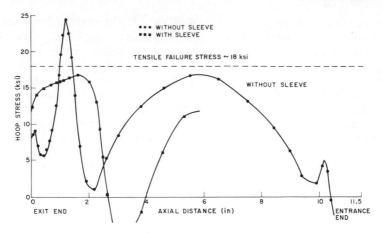

Fig. 9. Inside wall hoop (tensile) stress (2 sec after Cu efflux).

PRODUCTION TESTING

To provide information verifying the predictive accuracy of
the stress analysis, an attempt was made to conduct a controlled
experiment under actual production conditions. The configuration
shown in Fig. 2 was tested with a nozzle fabricated from Zircoa
#1706 with the modification that the copper sleeve at the entrance
end was eliminated while the exit end remained as shown (i.e., the
experiment was intended to simultaneously evaluate both designs).
The furnace was brought to the maximum preheat temperature slowly
and controlled automatically from a portable console designed for
this purpose.

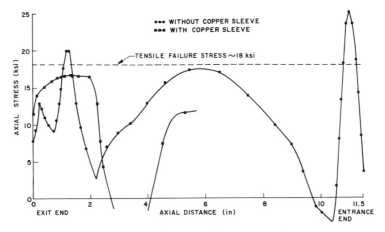

Fig. 10. Inside wall axial (tensile) stress (2 sec after Cu efflux).

This nozzle experienced failure at some time during the first half hour of the casting run. Circumferential and radial cracks appeared at a position approximately $1-1\frac{1}{2}$ in. from the extreme entrance end as predicted. This location corresponds to the point at which the heater winding is terminated and as such would be the area of greatest axial and radial gradients during the period of copper flow. In addition, as mentioned previously, the thermal expansion characteristic of the 1706 composition experiences a rather large negative expansion in the temperature range at which the molten copper is poured. At the extreme ends of the nozzle this behavior has the effect of allowing the cooler outside portions of the nozzle to experience a greater thermal displacement than the hotter ones on the inside. Coincident with this condition at the ends, opposing displacements are occurring just a short distance away in the heater region where the outside wall is hotter than the inside. While this peculiar expansion characteristic leads to an overall low expansion it nevertheless results in severe stress concentrations in these critical areas.

Upon removal of the copper sleeve from the exit end of this same nozzle, a visual inspection of this critical area revealed that no failure had occurred, indicating that the copper sleeve was successful in reducing the stresses below the failure stress of the material; again, as predicted by the stress analysis.

Confident that a successful design could be arrived at, a nozzle was fabricated from Zircoa 1706 and fitted with copper sleeves at both entrance and exit ends. The nozzle wall thickness is 3/16 in. except for the tapered section at the exit end. The copper sleeves have a 1/8 in. wall thickness and extend approximately 3 in. in from the ends of the nozzle. The exit sleeve is held against the nozzle by a Fiberfrax ring which also serves to protect the heater core and winding from the splashing of molten copper.

At the entrance end, a 1/8 in. stainless steel plate was inserted between the front face of the tundish and the nozzle. Serving as an intermediate seal between the tundish and the nozzle, it enabled the the included angle at the entrance of the nozzle to be reduced from 90° to 60°. This change all but eliminated a manufacturing problem associated with the hang-firing of the ceramic. The use of the steel cover plate also allowed a rigid positioning of the nozzle in the heater which greatly facilitated mounting of the assembly on the tundish.

This design was successful in casting a 120,000 lb reverberatory furnace charge in $5\frac{2}{3}$ hr on a continuous basis. There were no signs of imminent failure at the completion of the run. The nozzle was subsequently sectioned to determine the extent of physical and/or chemical degradation resulting from prolonged exposure to the molten metal.

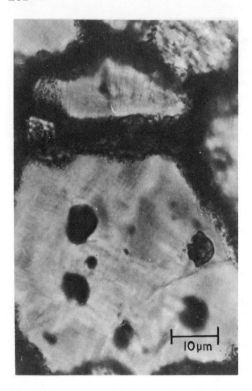

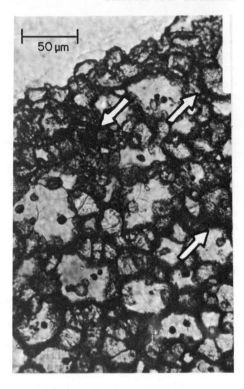

Fig. 11. Photomicrograph of material
        near maximum penetration of
        copper oxide.

Fig. 12. Photomicrograph of
        altered material near
        the copper-refractory
        interface.

Fig. 11 depicts a thin section of the nozzle prepared in a radial
plane parallel to and including the axis of the nozzle. The area
shown is near the maximum penetration of the copper oxide (~0.040 in.
from the inside surface). The grain boundaries in this region are
wider than in the unaltered material near the O.D. of the nozzle.
Small monoclinic grains of secondary origin can be seen in the
boundaries, which occur as a chemical alteration product from the
cubic grains. The presence of these crystals indicates that MgO is
being removed from the cubic solid solution by a chemical species
(copper oxide) which has diffused in along grain boundaries.

The structure of the material near the copper refractory interface
is shown in Fig. 12. The reddish copper oxide shows as dark material
along grain boundaries (see arrows). The total amount of material which
has penetrated is small and in no case was copper metal seen to
penetrate the refractory structure. The grain size of the cubic phase
in this region averages about 35 μm compared to about 55 μm in the

unaltered material. The degree of alteration is moderate with about 25% of the original cubic ZrO_2 being altered to monoclinic. In the region of copper oxide penetration the monoclinic phase is distributed along grain boundaries and becomes the continuous phase, while in the original unaltered structure cubic ZrO_2 was the continuous phase. The structure is not disrupted in any way, however, and appears to remain coherent. The remaining cubic ZrO_2 does not appear to be altered within the grains, and the changes seem to be associated only with internal surfaces.

In the region characterized by a continuous monoclinic phase, thermal cycling over the crystallographic transformation will tend to disintegrate the material. The useful life of the refractory will depend therefore upon not only the total amount of time in contact with the hot copper oxide but equally upon whether the nozzle is thermally cycled over the transformtion temperature range. From the amount of alteration and coherence of the used Zircoa 1706 nozzle it is estimated that it could have remained in service at least several times longer than the test run. Ultimately the useful life of the refractory will be limited by the destabilization reaction observed in Figs. 11 and 12.

Future work involving compositional modification of the Zircoa 1706 material will be centered around the effect of stabilizer content (MgO) on the thermal expansion characteristic as well as its high temperature flexural strength and elastic modulus. The minimum preheat temperatures occurring at the ends of the nozzle are about 600-700°C while the temperature of the molten copper stream is 1100-1180°C. If a compositional modification could be achieved which would exhibit a relatively flat expansion characteristic over this temperature range, all other things being equal, then the thermal stresses in the nozzle would be minimal. This is the direction in which we are currently moving.

REFERENCES

1. J. Cole and H. Moss, The Engineer (WECo) 11 (3) July 1967.
2. H. Okamoto, "High Temperature Reactions of Silica-Glass with Several Metals in a Vacuum", 1961 Transactions of the Eighth Vacuum Symposium and Second International Congress, Ed. by L. E. Preuss, Pergamon Press, New York, 1962.
3. High-Temperature Strength of Materials, Ed. by G. S. Pisarenko, Translated from the Russian by the Israel Program for Scientific Translations, Jerusalem, 1969, NASA TT F-508, TT68-50308.
4. D. Viechnicki and V. S. Stubican, J. Amer. Ceram. Soc., 48 (6) 292-97 (1965).
5. F. Kreith, Principles of Heat Transfer, p. 204, International Textbook Company, Scranton, Pennsylvania, 1961.

6. Ibid, pp. 297-330.
7. C. M. Usiskin and R. Siegel, Trans. ASME, J. Heat Transfer, 82,
   (4) 369-374, (1960).
8. E. L. Wilson and R. E. Nickell, Nuclear Eng. Design, 4 (3)
   276-86 (1966).
9. G. L. Goudreau, R. S. Dunham and R. E. Nickell, "Plane and
   Axisymmetric Finite Elements Analysis of Locally Orthotropic
   Elastic Solids and Orthotropic Shells", Report 67-15,
   Structural Engineering Laboratory, University of California,
   (Berkeley) August 1967.

DISCUSSION

L. M. Gold (U. S. Army, Frankford Arsenal): Were the points shown
in the stress analysis cross-section at the center of each element
or did they represent points for finite difference computation?
Authors: These points represent nodal points for finite element
solution and are the vertices of the quadrilateral or triangular
elements at the element centroids which implies that the stress
profiles shown (Figs. 9 and 10) are *near* , but not *at*, the inside
surface of the nozzle.

FRACTURE MECHANISMS OF VERY STRONG SOLIDS

N. J. Petch

University of Newcastle upon Tyne

Newcastle upon Tyne, England

ABSTRACT

The use of high strength solids is limited by their brittleness. Superfically, many of them are completely brittle and their fracture is widely treated as a purely elastic process. This concept requires reexamination in terms of the relative ease of an atomic shear movement and of an atomic pulling-apart at a crack tip. Theoretical treatments of this problem are considered. Truly brittle fracture emerges as a process that is probably quite rare. The fracture of alumina is taken as a detailed example. Compositional, heat treatment and chemical environmental effects on fractures may arise from their influence on the critical shear stress.

INTRODUCTION

Many very strong solids appear to be completely brittle and their fracture is commonly considered as a purely elastic process. This concept is re-examined in terms of the relative ease of an atomic shear movement and of an atomic pulling-apart at a crack tip. Experimental evidence on the fracture of alumina is considered in detail.

THEORETICAL STRENGTHS

The use of very strong solids is often limited by their brittleness. Macroscopically, many of them appear to be completely brittle, except at high temperatures, and their fracture is widely treated as a purely elastic process. In recent years, the concept of complete

brittleness has received more critical attention and the need for
this is clear when the problem is considered on the basis of interatomic forces.

The tensile stress required to pull atoms apart is roughly
~E/5, but the plastic shearing of atoms past one another is a less
drastic process and commonly requires a shear stress of only
~G/10 or ~E/30. Here, E is Young's modulus and G is the rigidity
modulus. On this basis, one would expect that plastic shearing
should occur, in general, more readily than true brittle fracture
The balance between the two processes may be partly restored by the
stress system. Thus, in uniaxial tension, the maximum shear stress
is only half the tensile stress and at the tip of a crack is a
half to a fourth of the maximum tensile stress depending on the
stress system and the value of Poisson's ratio. Thus, if the
theoretical shear strength τ_{max} is less than one half to one fourth
of the theoretical tensile strength σ_{max}, true brittleness is improb-
able.

The relationship between σ_{max} and τ_{max} depends on the relationship between E and G and, since $G = E/2 (1 + \nu)$, Poisson's ratio ν
is the important quantity. Table 1 gives some elementary calculations
of τ_{max} and σ_{max} along with values of ν. Clearly, with high ν, non-
directionally bonded solids (f.c.c. metals), there is no doubt that
plastic shearing is easier than tensile separation of the atoms.
With b.c.c. metals, the balance between the two processes is closer
and it becomes very fine with MgO and Al_2O_3. On the other hand,
with low ν diamond and NaCl, true brittleness seems to be possible.
In Table 1, τ_{max} refers to the shear of the perfect solid; if dislocations are present and their movement is thermally aided, this
will further favour the plastic shear process.

Table 1.   Estimates of theoretical tensile and shear strengths.

Substance	E 10^6 psi	σ_{max} 10^6 psi	τ_{max} 10^6 psi	ν
Ag	18	3.4	0.1	0.38
Cu	18	3.6	0.2	0.34
Fe	30	6.0	1.0	0.28
W	50	12.5	2.4	0.29
MgO	35	5.4	2.1	0.18
Al_2O_3	60	6.7	2.5	0.20
C (diamond)	150	30.0	18.0	0.10
NaCl	6.4	0.6	0.4	0.20

Thus, on the basis of these simple calculations, the belief that inorganic covalent and ionic solids commonly experience true brittle fracture before atomic shearing is possible does not appear to rest on too firm a foundation. Even more complex calculations (Kelly, Tyson and Cottrell[1])leaves the problem more or less in that situation and an appeal has to be made to experimental evidence.

THE FRACTURE OF ALUMINA

As an example of the experimental evidence, consider the strength of a plate of polycrystalline alumina broken from a central, elliptical, ultrasonically-drilled crack, Fig. 1. It does not seem possible to understand the observed temperature-dependence of the fracture stress in terms of pure elastic breaking, since, from the Griffith equation, the only parameters involved would be the crack size, Young's modulus and the intrinsic surface energy. The measurements were carried out *in vacuo* after outgassing at 350°C, so static fatigue does not appear to be involved.

An explanation of the observed temperature-dependence has been offered in terms of fracture by a plastic shear process (Congleton, Petch and Shiels[2]). Starting from the low temperature end, it is suggested that the initial fall in fracture stress with rise in temperature reflects the increasing ease of the plastic deformation *necessary* for fracture. However *accompanying* plastic deformation will also occur around a growing crack and eventually, as dislocation movement becomes easier, the total fracture energy associated with

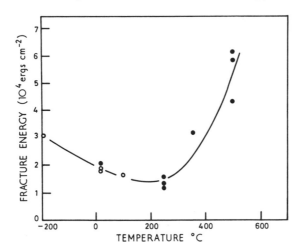

Fig. 1.   Effective fracture energy versus temperature for poly-
          crystalline alumina. The effective fracture energy is
          proportional to the (fracture stress)^2 at constant initial
          crack length.

crack extension may begin to increase as the temperature rises. The fracture stress will then pass through a minimum value and begin to rise again as the temperature is further increased.

This view that plastic deformation is involved in the very brittle fractures of alumina below ~1000°C is supported by the observation that flakes collected from the fracture surface, which are sufficiently thin to be examined by transmission in the electron microscope, show extensive twinning and dislocation generation. [In the case of the brittle fracture of glass, it has also been argued that fracture arises by plastic flow (Marsh[3])].

Above about 1000°C, single crystal alumina experiences general yielding, the yield stress falling as the temperature rises. Fracture then occurs by a dislocation or twin process associated with general yielding and the fracture stress required also falls as the temperature rises.

This high temperature fall occurs rather earlier (~700°C) with polycrystals. The fracture becomes increasingly intergranular, and it is probable that the polycrystals are affected by the inability of alumina to satisfy the von Mises-Taylor and other conditions for compatability of plastic deformation at the grain boundaries. Thus, increasing plastic deformation generates intergranular stresses.

The view advanced here about the fracture of alumina at the lowest temperatures can be expressed more precisely. Consider a cross-sectional area A in a body that is deforming plastically at a stress σ. During an increment in the deformation, the load L required for deformation decreases because of area contraction by an amount $dL = \sigma dA$. At the same time, the load required for deformation increases because of work-hardening by a amount $dL = Ad\sigma$. Thus, the deformation is only stable if $Ad\sigma > -\sigma dA$. Since the increment in the strain is $d\epsilon = -dA/A$, the plastic deformation is stable only if

$$d\sigma/d\epsilon > \sigma. \tag{1}$$

This immediately highlights the difficulty that is faced by very strong solids. If these plastically deform at the theoretical shear stress τ_{max} ~G/10, σ is about G/5, but work-hardening processes associated with the accumulation of dislocations only give $d\sigma/d\epsilon$ ~G/30. In this circumstance, it immediately follows from Eq. 1 that plastic deformation will be unstable. Thus, as soon as plastic deformation begins at the tip of a crack, unstable necking and fracture will ensue.

It is suggested that this is what happens in the first part of Fig. 1. The fracture appears extremely brittle, but it arises from unstable plastic deformation. The fracture stress falls

as the temperature is raised because the yield stress falls.   The
fracture process may be entirely this unstable plastic necking or
it may be that dislocation or twin interactions within the plastic
zone produce cleavage.   The limited data on the yield stress of
alumina available from microhardness neasurements (Westbrook[4]) indicate that the fall in fracture stress as the temperature is raised
up to 250°C (Fig. 1) parallels the fall in the yield stress.

Eventually, as the temperature is raised

$$\sigma = d\sigma/d\epsilon. \tag{2}$$

It seems quite probable that Eq. 2 gives the condition for minimum
fracture strength.   At higher temperatures, plastic deformation at
the yield stress is stable, so additional plastic work may occur at
a crack tip prior to fracture propagation.

The data for alumina appear consistent with this conclusion.
From microhardness measurements (Westbrook[4]), the yield stress of
alumina at the temperature of the minimum strength is ~500 kg mm^{-2},
whereas $d\sigma/d\epsilon$ at $G/30$ is ~550 kg mm^{-2}.   This agrees with determination
of the minimum strength by Eq. 2.

The conclusion that the fracture strength of alumina at the
lowest temperatures is determined by the yield stress indicates that
increasing the difficulty of dislocation movement should increase
the fracture strength at these temperatures.   This is borne out by
preliminary work on the fracture of alumina doped with Cr or Ti
(Congleton, Petch and Shiels[2]).

Although an increase in the yield stress σ will increase the
fracture stress at the lowest temperatures, it should be noted
that there will not necessarily by any increase in the value of the
minimum strength, if this is determined by Eq. 2.   An increase in
the rate of strain hardening $d\sigma/d\epsilon$ is required to increase the
minimum strength.

These conclusions on the fracture mechanisms in strong solids
appear significant in considering the influence of the chemical
environment.   Unless the action of this environment is so vigorous
that removal of material and alteration in the geometry of the
specimen occurs, the only way in which an effect on the fracture
stress can arise, at temperatures below that of the minimum strength,
is through an effect on the yield stress.   That such an effect is
possible is indicated by the observation that adsorbed water films
can significantly lower the microhardness of nonmetallic solids
(Westbrook[4]).   Marsh[3] has also suggested that environmental effects
in the fracture of glass arise from an alteration of the flow stress.

At temperatures above that of the minimum strength, where the local stress at a crack tip exceeds the yield stress when fracture occurs, the environment may again affect fracture via the flow stress. Additionally, if, after local yielding at a crack tip, a definite stage of cleavage crack formation is involved in determining the fracture stress, then an effect on the crack formation stage via surface energy and related properties may also be significant.

CONCLUSIONS

1. With solids of low Poisson's ratio, there is a fine balance between the occurrence of tensile separation of atoms or of atomic shearing.

2. Four temperature regimes are distinguished in the fracture of alumina and this is probably fairly typical of a number of strong solids. Starting from the lowest temperature, these regimes have the following characteristics:

Regime 1. At yield, $\sigma > d\sigma/d\epsilon$. Fracture occurs by unstable plastic yielding from a crack tip. The fracture stress falls as the temperature rises. The minimum strength is probably reached when $\sigma = d\sigma/d\epsilon$.

Regime 2. At yield, $d\sigma/d\epsilon > \sigma$. Plastic deformation at a crack tip is now stable and more work is done prior to crack propagation. The fracture stress rises with temperature.

Regime 3. In polycrystals that do not fulfill the conditions for compatability of plastic deformation, the fracture stress begins to fall again because of the development of intergranular stresses.

Regime 4. General plastic deformation occurs, fracture involves dislocation or twin interactions and the fracture stress falls as the temperature rises.

3. The influence of alloying and of environment on the fracture stress has to be examined in the light of their influence on the yield stress.

REFERENCES

1. A. Kelly, W. R. Tyson and A. H. Cottrell, Phil. Mag., 15, 567-86 (1967).
2. J. Congleton, N. J. Petch and S. A. Shiels, Phil. Mag., 19, 795-807 (1969).

3. D. M. Marsh, Proc. Roy. Soc. (London), A282, 33-43 (1964).
4. J. H. Westbrook, The Temperature Dependence of Hardness of Some
   Common Oxides, G. E. Research Report, No. 65-RL-4017 M.,
   1965, 17 pp.

DISCUSSION

S. M. Wiederhorn (National Bureau of Standards): Prof. Petch has
pointed out an interesting anomaly in the strength behavior of
sapphire. The strength decreases with increasing temperature and
the decrease cannot be explained by the Griffith theory. Prof.
Petch has ascribed the strength behavior to necking of the material
at the root of the crack due to the high stress fields there. One
objection to this mechanism is based on the limited amount of flow
that can occur at a crack tip in sapphire. Accepting the value of
6.7×10^6 psi as the cohesive strength of sapphire[1], the maximum
shear stress at the crack tip is of the order of 3.4×10^6 psi.
At room temperature the yield stress of sapphire, obtained from
hardness indentation data, ranges between 1 and 2.5×10^6 psi
depending on the theory used to calculate the yield strength.
Because the stress fields near a crack tip decrease as one over the
square root of the distance from the crack tip, the maximum distance
that dislocations can move from the crack tip is ~10 to 20 Å. Since
dislocations in sapphire have Burgers vectors that are greater
than approximately 5 Å, only a limited number of dislocations can
be generated from the crack tip before dislocation motion ceases.
Consequently, it is hard to understand how a sufficient amount of
plastic deformation can occur at room temperature to account for
necking at the root of the crack.

Other possible mechanisms may account for the observed decrease
of sapphire strength with increasing temperature. These include
dislocation feeding mechanisms as have been observed in magnesium
oxide[2], static fatigue due to water vapor in the environment[3], and
the possibility of an activated fracture process such as that
observed in glass in the absence of water vapor.[4] Objections may be
raised to the first and second of these suggested mechanisms based
on the limited mobility of dislocations in sapphire and on the fact
that the strength decrease is observed in sapphire even when tested
in vacuum conditions. The last process has not been studied in
sufficient detail to draw any firm conclusions as to its possibility.

It is concluded that none of the suggested mechanisms to explain
the behavior of the strength of sapphire have been adequately proven.
Although Prof. Petch may be right in his interpretation, additional
experimentation is necessary for an unequivocal interpretation of
the strength data of sapphire.

DISCUSSION REFERENCES

1. A. Kelly, Strong Solids, Clarendon Press, Oxford (1966).
2. F. J. P. Clarke, R. A. J. Sambell and H. G. Tattersall, Phil.
   Mag. 7, 393-413 (1962).
3. R. J. Charles, p. 225 in Fracture. Ed. by B. L. Averback, D. K.
   Felbeck, G. T. Hahn and D. A. Thomas. John Wiley & Sons, Inc.,
   New York. 1959.
4. K. Schonert, H. Unhauer and W. Klenn, Influence of Temperature
   and Environment on Slow Crack Propagation in Glass. To be
   published in the Proceedings of the Second International
   Conference on Fracture, Brighton, April 13-18, 1969.

A. H. Heuer (Case Western Reserve University): I suppose that one
has to take sides during this discussion -- my own view is that as
an explanation of the temperature dependence of the strength of
sapphire, plastic deformation contains too many unreliable features
to be tenable. In other words, the low temperature fracture of
sapphire is a more complex phenomenon than would be inferred from
Prof. Petch's paper.

One problem resides in demonstrations that plastic flow can
indeed occur under a microhardness indenter. In such a test, the
accompanying hydrostatic confining pressures enhance plastic flow
and suppress cleavage. On the other hand, the stresses at a crack
tip are such that cleavage is enhanced relative to plastic flow,
and it is not clear at all that plastic flow must precede fracture.
Furthermore, one must say in detail *how* this occurs, inasmuch as
many of the current mechanisms (crack feeding, twin intersections,
etc.) are quite specific.

Author: I should like to acknowledge the benefit I have derived
in the last few years from sharpening my ideas on the fracture of
alumina against the friendly disbelief of Dr. Wiederhorn and
Professor Heuer.

In the present instance, Dr. Wiederhorn's treatment of the
yielding at a crack tip appears too approximate to be conclusive.
Using an elastic calculation, he argues that when the applied stress
generates the theoretical cohesive stress at the crack tip the yield
stress is only exceeded within 10-20 Å of the tip. But, this is not
an accurate estimate of the size of the plastic zone at fracture.
Plastic deformation begins at the crack tip when the yield stress
is reached, redistribution of stress into the regions more remote
from the crack then occurs and the plastic zone spreads. Additionally,
when $\sigma > d\sigma/d\epsilon$, the instability of plastic extension at the crack
tip also causes the plastic zone to spread across the specimen.
When σ is high, the longitudinal spread of the neck will be small.

Dr. Wiederhorn refers to the thermally activated stage that
is observed, even in the absence of water vapour, in the extension
of a crack in glass. On the present microplasticity view of
fracture, this thermal activation of fracture can be simply thermal
activation of yield.

I agree with Professor Heuer that it is not absolutely clear
from theory alone that plastic shear is preferred to cleavage at
a crack tip in alumina, but my contention is that the experimental
evidence favours this conclusion. When $\sigma > d\sigma/d\varepsilon$, fracture must
occur because plastic extension is unstable; the details of specific
twin or dislocation interactions are then only of secondary
importance.

R. W. Rice (U. S. Naval Research Laboratory): I would like to
make three points:
(1) Cracks are often at grain boundaries. Because of the elastic
anisotropy of Al_2O_3, stress concentrations will result that may
increase local stresses by ~50%, thus increasing the region of
possible slip.
(2) The strengths achieved in sapphire cannot, in my opinion,
be cited as clear evidences against microplastic processes in
ceramics. When single crystal specimens are flame polished, one
not only melts out surface flaws, but also (a) melts out a considerable population of mobile dislocations and twins, and (b)
subjects much, if not all, of the bulk specimen to a high temperature anneal. Now if one takes MgO crystals, which clearly have
their normal strength determined by slip (plasticity), first removing
surface flaws and mobile surface dislocations and finally annealing
them to pin internal dislocations, the result is an increase from
the normal 10,000-20,000 psi slip-determined strength to brittle
strengths of 100,000-400,000 psi. Further, testing of such
specimens at elevated temperatures will again allow the pinned dislocations to become active, giving ductile behavior, just as flame
polished sapphire does. Therefore, the behavior of flame polished
sapphire is quite analogous to that of similarly treated MgO
crystals, and thus is not to be considered as sound evidence
against microplasticity in sapphire.
(3) One should not exclude the possibility of twinning, which can
cause many of the same effects associated with slip. Twinning
clearly occurs at ambient temperatures. Recently, Becher and Rice
have shown that if twins are left in specimens after flame polishing,
strengths will be no better than that of mechanically finished
specimens for those cases where twins are intersecting. It turns
out that round rods are not only easier to polish than those of
rectangular section, but are much less likely to develop twins
that can cause failure.
Author: The support from Dr. Rice for a microplasticity theory
of fracture is much appreciated.

In addition to Dr. Rice's discussion of flame-polishing, the
detailed effects of annealing are also of interest. Davies,
[L. M. Davies, Proc. Brit. Ceram. Soc., 6, 29-37 (1966).] has
studied the strengths of sapphire crystals in two conditions
(a) surface-ground and (b) surface ground and annealed. The
variation of fracture strength with temperature is similar to

Fig. 1.   Annealing produces a considerable rise of the low temperature portion of strength-temperature curve, but the high temperature branch of the curve is unaffected. An explanation can be given in terms of the ideas in the present paper if annealing produces pinning of the dislocations. On the low-temperature strength curve, fracture occurs when initial yielding takes place at a notch tip, so the strength should be raised by annealing, as is observed. However, at temperatures above that of the minimum strength, a stable plastic zone forms at a notch tip, and fracture occurs when a critical value of the flow stress is reached in this zone. The conditions for fracture should then be largely unaffected by any initial pinning of the dislocations at the yield point after the annealing treatment, i.e., the flow stress is largely unaffected by initial dislocation pinning, so the fracture stress should be unaffected by annealing, as is observed.

One point in my paper was perhaps not too clear. At low temperatures, fracture is bound to follow when yielding occurs at a crack tip because plastic elongation is unstable, but the fracture may actually be cleavage arising from dislocation or twin interaction. It was said that the minimum strength was probably reached when $\sigma = d\sigma/d\epsilon$ and this perhaps needs amplification. Fracture may continue to occur at yield to somewhat higher temperatures than given by this relationship provided the fracture is by the dislocation or twin-interaction mechanism and that this is still possible at yield.

THE COMPRESSIVE STRENGTH OF CERAMICS

R. W. Rice

Naval Research Laboratory

Washington, D. C.

ABSTRACT

The literature on compressive strengths of crystalline ceramics, especially at room temperature, suggests that microplasticity may be the mechanism of much compressive failure since (1) the yield stress (microhardness/3) is the upper limit of both ambient and elevated temperature compressive strengths and (2) data on grain size dependence are consistent with the Petch equation. A failure theory is developed which is in accord with (1) experimentally observed variations in microhardness and compressive strength data, (2) stress concentrations due to thermal and mechanical anisotropies, impurities, pores and flaws, and (3) twin-induced premature fracture. Applicability of the theory is evaluated for crystalline ceramics in terms of conventional flaw theories, and relevance for non-crystalline ceramics is discussed.

INTRODUCTION

It is well known that the compressive strengths of ceramics are generally severalfold greater than their tensile strengths, which is why ceramics are used in compression wherever feasible. Applications of Griffith theory predict that the compressive strengths of ceramics will exceed their tensile strengths eightfold due to local tensile stresses at the tip of pre-existing flaws. Other theories such as the maximum tensile strain theory also predict compressive strengths that are several times the tensile strength (in this case, the ratio is inversely proportional to Poisson's ratio).[1] Since these theories generally give the right order of

195

magnitude in relating compressive to tensile strengths, they have accentuated the study of tensile strengths. For example, very little study or analysis of the microstructural effects on compressive strengths of ceramics has been undertaken, which is rather surprising in view of possible new applications such as in deep submergence vehicles. The orientation to tensile-sensitive flaws as the sole cause of compressive failure leaves considerable uncertainty about the limit of ceramic compressive strengths as the qualities of ceramic bodies are improved.

In recent years there has been growing recognition that the strength of ceramics may not be controlled exclusively by flaws, but that microplastic* effects may also be a factor, at least in some materials. Such studies have been almost exclusively restricted to tensile strengths. However, since much higher stresses normally are reached in compression, it appears even more important to consider microplastic effects in the compressive mode.

While no explicit study has been conducted, there is a fair amount of data in the literature. By combining these findings with the concept of hardness being related to yield stress (by a factor of about one-third, as in metals) it is shown that microplastic yielding appears to set an upper limit to compressive strengths. Several factors which can reduce compressive strengths below the yield stress are discussed. Analysis of the limited microstructural data is presented in support of this theory.

MICROHARDNESS AND MICROPLASTICITY

Microhardness values determined for metals are known to be about three times their respective yield stresses. It is of importance to assess the role of microplasticity associated with hardness indents in ceramics. The occurrence of microplasticity during indentation of soft materials of simple, symmetric structures such as that of the alkali halides and MgO is fairly well known.[2-5] Though less extensive, similar sound evidence also exists for other moderately soft materials such as other halides and tellurides, some of which have less symmetric structures (e.g., MgF_2 has the tetragonal rutile structure [6-10]). Though less work has been done on hard ceramic materials (in part because of greater difficulty of detection), there

---

* The term "microplasticity" includes both slip and twinning and possibly other nonelastic deformations (e.g., possible densifications or rarefractions in glass). Its use emphasizes that regardless of extent, it normally only results in microscopic changes and does not lead to macroscopic ductility unless temperatures or possible confining pressures are high enough to permit generalized plastic flow.

Table 1. Microplasticity Around Indents in Harder Materials

Material	Structure	Indent	Load (g)	Hardness (kg/mm²)	Microplasticity Evidence	Ref.
Boride						
TiB$_2$	H	K,V	100-1000	(2860)	Surface markings, slip	11
Carbides						
TaC	C	V,K		(1500)	Etched slip bands	12
UC	C	V		(700)	Surface markings, slip	13
WC	H	V	30,90	(1800)	Surface markings, slip	14
WC	H	V		(1800)	Surface markings, slip	15
WC	H	V		(1800)	Surface markings, slip	16
Oxides						
Al$_2$O$_3$	H	K		(1900)	Surface markings	17
Al$_2$O$_3$	H	V		(2360)	Transmission Electron Microscopy	18
Al$_2$O$_3$	H	K		(1900)	Surface markings	19
BeO	H	V		(1130)	Etched slip bands	20
BeO	H	V		(1130)	X-ray analysis	21
SiO$_2$ (qtz)	H	V		(1090)	Surface markings, slip	22
TiO$_2$	T	V		(1280)	Surface markings, slip	23
UO$_2$	C	V,K		525K	Surface markings, etched slip bands	24
BaTiO$_3$	T	V			Etched slip bands	25
Mica	M	S			Surface deformation and slip	26
Silicides						
CoSi$_2$	C			792	Surface markings, slip	
CoSi	C	V	100-1000	1001	Surface markings, slip	27
Co$_2$Si	O			834	Surface markings, slip & twinning	
WSi$_2$	T	K	25-1000	1350	Surface markings	28

C: Cubic; H: Hexagonal; M: Monoclinic; O: Orthorhombic; T: tetragonal
K: Knoop; V: Vickers; S: spherical
Values in parentheses are typical literature values, others are as measured by investigator

is substantial evidence of microplastic deformation around hardness
indents, as shown in Table 1.

To what extent is microplasticity responsible for displacements
at and near the indent? Is plasticity the dominant mechanism?
Cracking is normally not observed at the light loads considered here,
nor is it easy to conceive of cracking being dense and fine enough,
especially under compressive loads, to give the smooth, symmetrical
indents observed. In order for microplasticity to be the sole
mechanism of forming indents, it must at least approach homogeneous
operation on five independent systems (though complete satisfaction
of homogeniety and independence of microplasticity may not be
required, since the indented surface is unconstrained). As will be
discussed later, the small volume of deformation is a key factor
in achieving the necessary homogeniety. The high stresses and their
partially hydrostatic nature indicate that a sufficient number of
independent systems may be operative, as corroborated by studies on
several materials. Stresses of 100,000 - 200,000 psi are adequate to
activate the {100} <100> slip systems which complement the much
more easily operated {110} <110> systems to give the necessary five
independent shear modes in MgO crystals.[29] Similarly, the corresponding slip systems in LiF have been found to operate at stresses
of the order of 12,000 and 800 psi, respectively.[30] The high stresses
needed for all slip systems to function in such materials are clearly
reached in hardness testing (Table 2). Similarly, Westbrook[31]
observed that the yield stress of ionic single crystals (NaCl
structure) oriented for activation of only the easiest operable slip
systems are about 3% of the hardness rather than over 30%. The
author has corroborated this result and indicated that it may apply
to several other materials of other crystal structures.[32] Twinning,
which by itself cannot result in homogeneous deformation as required
for a hardness indent, is observed in bending of sapphire crystals
fracturing at stresses well below 100,000 psi.[33] Thus the microplasticity associated with hardness indents normally represents the
general yielding of crystalline ceramics. (Glasses will be discussed
later.)

MICROHARDNESS AND COMPRESSIVE STRENGTH LIMITATIONS

The microhardness of a wide variety of ceramics, the calculated
yield stresses (H/3), and the maximum observed compressive strengths
are compiled in Table 2. Some strengths have been corrected for
porosity, as discussed later; the averaged hardness data[49] were
obtained by indentations, typically under 100 g loads at room
conditions. It is observed that in all cases, within reasonable
scatter, the hardness-derived yield stress is the upper limit to the
compressive strengths, and that well fabricated materials often reach
half or more of this limit.

While reasons for the compressive strengths falling below the yield stress will be discussed later, it is worthwhile to examine yielding as the limiting strength of a "brittle" material such as Al_2O_3. Soltis[50] has reported evidence of dislocations and resultant tensile yielding of sapphire whiskers at stresses from 4×10^5 to over 2×10^6 psi. The lower stresses were for c-oriented whiskers, and the higher stresses (most commonly 1.0 to 1.5×10^6 psi) were for a-oriented whiskers. Bayer and Cooper[51] have shown similar strengths for chemically polished sapphire whiskers. They attributed failure to dislocation pile-ups and interactions in polished (as well as the strongest unpolished) whiskers. Considering size effects, it was shown that these strengths extrapolated to those of flame polished rods. Mallinder and Proctor[52,53] studied the bend strength of small flame polished sapphire rods, and found that the maximum bend strength was about 1×10^6 psi. Specimens with internal flaws (generally bubbles) typically failed at 3×10^5 - 7×10^5 psi. Similarly, Hurley[54] observed tensile strengths of sapphire filaments to be typically 5×10^5 psi, and found internal bubbles limited strengths. Since the stress concentration due to a bubble will nominally be about two, these values all generally agree with Mallinder and Proctor's limiting strengths of about 1×10^6 psi, in reasonable agreement with the predicted yield stress.

MEASUREMENT ERRORS

Measurements of both compressive strength and hardness are commonly in error, with observed compressive strengths generally being low and hardnesses high. Corrections for these known errors brings compressive and yield strengths closer together.

Compressive Strength Measurement Errors

The two most important problems in measuring compressive strength are (1) nonparallel specimen ends or loading heads and (2) specimen-interface (end constraint) problems.[1,55] While the resultant bending (tensile) stresses from nonparallel conditions can be quite high, careful machining and use of self-leveling heads can minimize these effects so that resultant errors should not exceed a few percent. However, interfacial effects which arise from the combined effect of differential lateral (Poisson) expansion and friction between loading heads and the specimen, have usually not been dealt with as effectively. These effects can result in substantial tensile stresses,[56] e.g., use of samples of square cross section (which is fairly common due to ease of preparation or other experimental considerations) can result in compressive strengths being of the order of 15% lower than with comparable circular cross-sectional areas. Use of a simple cylinder instead of one with a reduced gauge section also can reduce strengths, reported by Brace[57] to be as much as 30% lower. The combined effects of test-related errors will tend to make measured values fall below true compressive strengths.

Table 2.   Hardness-Derived Yield Stresses and Polycrystalline Compressive Strengths

Material	Calculated Yield Stress				Compressive Strength		References
	Vickers ($H_v/3$)		Knoop ($H_k/3$)		Measured	Corrected*	
	kg/mm^2	$psi \times 10^3$	kg/mm^2	$psi \times 10^3$	$psi \times 10^3$	$psi \times 10^3$	
Borides							
TiB$_2$	1130	1600	960	1360	344	–	34
ZrB$_2$	680	970	620	890	247	310 (3,7)	34
						360 (5,7)	34
Carbides							
B$_4$C	1660	2360	1000	1420	414	–	35
Cr$_3$C$_2$	550	780			600		36
SiC	1100	1560	800	1140	150		†
TaC	600	850	490	690	100	122 (5,4)	††
TiC	1130	1600	960	1360	300	348 (5,3)	††
					800 – single crystal		37
WC	590	830	650	920	200	232 (5,3)	††
WC–4.5% Co	600	850			900		38
Nitrides							
AlN	400	570	410	580	300	330 (5,2)	39
Si$_3$N$_4$					500		40
TiN			660	940	140		35

Table. 2 (continued)

Oxides

Material							
Al_2O_3	790	1120	640	900	650	360 (8,2)	41
BeO	380	540	430	610	300		42
MgO	220	310			200		
$MgAl_2O_4$	550	780	400	570	400		43
SiO_2 (crystal)	360	520	250	360	360		44
SiO_2 (fused)	180	260			190		†††
ThO_2	250	360	260	370	400		45
UO_2	200	290	190	260	140	230 (7,8)	46
ZrO_2 (+CaO)	470	660	430	620	290		38
$ZrSiO_4$	270	380	380		210		38

Other Materials

Material							
$MoSi_2$	480	680			350		38
C (polycrystal-line diamond)	3000	4260			850	910 (8,1)	47
C (glassy carbon)	67	95			128		48
	55	78			85		48
NaCl	7	10			6		44
CaF_2	58	83			50		††

* Strength corrected using the equation $S = S_0 e^{-bP}$ where S = strength at volume fraction porosity P, S_0 = strength at zero porosity, and b = an empirical constant. Numbers in parentheses indicate values of (b, % P) employed in a specific correction.

† Data from the Carborundum Company.

†† Data from National Beryllia Corporation.

††† Data from Amersil Quartz.

Hardness Variations

Many factors affect hardness values, e.g., indentation of carefully dried surfaces often gives values about 15% higher than
normal.[58] Machining can work harden the surface[59,60] and thus
increase measured values in comparison to the bulk material which
will control compressive strengths. Brace[61] has observed that hardness values of NaCl will generally be high by a factor of two unless
the work hardened layer (~1 mm) is removed by water polishing.
Mechanically finished surfaces of somewhat harder ZnS crystals can
be 30% harder than as-grown or etched surfaces, though more commonly
this effect is 10% or less.[62] Since grinding and mechanical polishing
are more common methods of preparing specimens for hardness measurements than are cleavage and chemical polishing, values shown in
Table 2 will generally be too high, though the effect probably
decreases with increasing hardness.

Most hardness testing of ceramics is done at low loads to avoid
cracking. However, hardness values generally rise substantially
with decreasing load, attributed to two factors; (1) increasing
errors due to difficulties in determining smaller indent dimensions
and (2) greater proportional amounts of elastic recovery (due to
the absence of cracking) as the indent size decreases with load.
This results in hardness values being substantially high (e.g.,
20-40% at 100 g).[63,64] Recently Haglund[65] has shown that optical
measurements give shorter values of indent diagonals, and hence
higher hardness values, than do measurements by either scanning or
replication electron microscopy.

A recent survey shows that the average Knoop hardness of most
ceramics is about 20% lower than the average Vickers hardness.[49]
Since it is not known whether the local deformation (and hence
workhardening) of a hardness indent is substantially greater than
expected in a corresponding volume of material undergoing microplastic deformation from compression, and since the greater surface
area of the Knoop indenter should lead to less local work hardening
(which is believed to be an important factor in its tendency to
lower values), it is difficult to determine which should be used.
Later discussions on ballistic effects indicate that Knoop values
are too low, but could be due to strain rate-yield point effects.

Hardness is not an isotropic property. For cubic materials
it can vary by 20% or more between different crystal planes.[58,66]
Many moderately anisotropic materials will show similar variations,
but some (e.g., those with würtzite structure) can vary by
40-50%[58,62,66] Highly anisotropic materials such as graphite may
show as much as an order of magnitude difference in hardness with
orientation.[67] Part of this anisotropy is attributable to the lack
of constraint on deformation above the indented surface, in contrast

to deformation occurring internally in a solid. However, much of
this anisotropy may result from a greater portion of the necessary
deformation occurring by means of easily activated slip systems.
Since grains that are favorably oriented for such easier slip may
be most important, a value of hardness closer to the lowest value
may be more nearly representative of compressive behavior.
Certainly, when a polycrystalline body is textured, the anisotropy
in hardness should correlate with texture-related anisotropy in
compressive behavior. Robertson[67] has measured an approximate
fourfold hardness anisotropy (12.5 : 44 kg/mm^2) in natural graphite,
which corresponds with compressive strength ratios (14,000 :
50,000-70,000 psi) in pyrolytic graphite.[68,69] The lower strengths
(6,000 : 21,000 psi) predicted by these hardness values for natural
graphite are reasonable in view of its weaker nature (due in part
to large grain size). The net effect of all of these variations
is that hardness values will generally be somewhat high.

STRENGTHS BELOW MACROSCOPIC YIELD STRESSES

While corrections for measurements of hardness and compressive
strength will generally bring the yield and compressive strengths
substantially closer, the latter will still often be lower as a
consequence of other general factors: (1) imperfections, (2) twinning,
and (3) internal stresses.

Body Imperfections

Imperfections of the body include cracks, voids, and impurities.
Favorably oriented large cracks will clearly cause premature failure
and smaller cracks may also cause failure at significantly lower
stresses if sufficiently dense and favorably oriented. However,
as will be discussed later, compressive stresses tend to prevent
crack opening, so that microplastic effects need not be precluded
by the presence of some cracks. In fact, stress concentration from
crack tips, jogs, or areas where they do not close evenly, may act
as sources of slip and lead to slip-induced failure. This can
occur either because slip provides an easier means of failure than
the original crack, or because it has made the crack more difficult
to propagate (e.g., by blunting). However, local stress concentrations would allow such failures to occur below the macroscopic
yield stress.

Porosity will also obviously lower compressive strengths.
Extremes of pore size or pore volume might induce a truly brittle
failure. However, the local stress concentration effects of pores
should again enhance the occurrence of slip. Evidence of pores
acting as sources of slip has been discussed elsewhere.[70] This
includes the specific observation by the author that corners or

steps in pore walls in MgO crystals are the preferential sources
of slip during compressive straining. Experimental data indicate
that compressive strengths are somewhat more sensitive to porosity
than are tensile strengths. Most of the earlier data on compressive
strengths are representative of specimens containing appreciable
porosity.

Impurities can have two effects: (1) those which inhibit slip
or twinning (e.g., by solid solution or fairly homogeneous precipitation) will encourage flaw-related brittle failure; (2) those
which segregate substantially (e.g., at grain boundaries) may also
lead to brittle behavior due to cracks, weakened bonding, and
enhanced stress concentrations. However, such concentrations may
also cause slip which may be the ultimate cause of failure.

Twinning

Crystallographic deformation twinning has been observed in
many materials, including ceramics and related materials.[71,72]
A few others not listed in these sources are B_4C[73], SiC[74,75],
WC[76], $CdTe$[77], and InS[78], as well as those listed in Table 1.
A few, such as PbS[79] (and probably NaCl), twin only under special
conditions. On the other hand, twinning occurs very readily even
at room temperature and below in many, especially in noncubic
materials.

Yield, as from a hardness indent, is a generally homogeneous
process, while twinning is not. Thus microhardness and the
resulting yield stress do not represent the stress conditions for
twinning. When twinning is commonly observed, it will normally
occur at a stress below the yield stress. Since it can lead to
crack nucleation in the same fashion as slip, it will therefore
cause failure below the yield stress. These considerations are
especially important to compressive strengths since twinning is
usually highly favored in compression, but not in tension. Stofel
and Conrad[80] observed twinning (and resultant crack nucleation) in
sapphire at elevated temperatures only in compression, not in
tension.

Bridgman[81] observed compressive strengths of 300,000 and
600,000 psi in sapphire crystals oriented with the optic axis at
84° and at 7° to the compressive axis (length), respectively. Use
of hydrostatic supporting pressures of 350,000 and 390,000 psi
produced compressive strengths respectively of 1,300,000 psi in a
specimen with the crystal axis at 0° to the length, and 750,000 psi
in one oriented at 87°. Since these specimens were carefully
selected to be free of internal bubbles, the pronounced orientation
dependence appears to be a material characteristic (See also
Table 4). While there is some question about details of Bridgman's

discussion of slip and twinning,[82] it is clear that twinning was
an important mechanism of failure. The upper strengths
(1.0 - 1.3 x 10^6 psi, observed in other tests[81]) generally agree
with those predicted by hardness (neglecting orientation effects).
These findings indicate that twinning may cause failure below the
predicted yield stress in Al_2O_3. Similarly, Griggs[83] has observed
profuse twinning in calcite single crystals in compression with and
without supporting pressure. The failure stresses (without
supporting pressure) of ~2,000 psi are well below the yield stress,
~64,000 psi. Metal-bonded WC is stronger than self-bonded
WC (Table 2). While other factors probably also contribute to
this difference, the metal grain boundary phase may be important
as a "buffer" to minimize effects of twinning, since the metal could
deform slightly in compression (but probably much less in tension)
to accommodate the shear-offset due to twinning.

Another factor illustrative of the importance of preferred
deformation not indicated by hardness is the high degree of
anisotropy in slip in many crystals. For example, ratios of compressive yield stresses along the <111> and <100> axes of NaCl
structured materials (i.e., directions which activate different
slip systems) range between about 5 and 20 at room temperature for
materials such as LiF[30], MgO[29], and CaO[85]. Similar differences in
activation of different slip systems are indicated in CaF_2[86],
PbS under pressure[79], and possibly ZrC[87] and are probably typical
of many other materials. Substantially easier activation of one
set of slip systems (insufficient for the homogeneous deformation
required for a hardness indent) can lead to single crystal compressive
strengths being very much lower than predicted (e.g.,greater than
ten for MgO[29]). Bodies with very large grains are sensitive to
this effect, probably accounting for the relatively low compressive
strengths reported for LiF specimens[84] having grain sizes in the
millimeter range.

Elastic and Thermal Expansion Anisotropy

Thermal expansion anisotropy can result in substantial residual
stresses despite some microplastic relief during cooling. For
example, Smith and Weissman[88] report residual stresses of 28,000
and 14,000 psi, respectively, for randomly oriented and textured
BeO. They noted that these stresses resulted in microplastic
deformation, predominantly near the grain boundaries. The author
has recently shown that such stress may reach the order of 10%
of the compressive strength.[32]

Chung and Buessem[89] have shown that most materials, including
most cubic structured ones, are elastically anisotropic. Hasselman[90]
has shown that such elastic anisotropy can lead to local increases

in stress of 5-15% for equiaxial grains, with greater increases
as the shape of the grains also becomes anisotropic. For most
materials, grain length-to-diameter ratios of 2 and 5 lead to
stress increases of the order of 10-40% and 15-80%, respectively.

EVALUATION OF MARSH'S HARDNESS THEORY

In considering the deformation of glass caused by indents,
and relating this to possible fracture mechanisms, Marsh[92,93] was
lead to derive a new theory of hardness. For low values of the
ratio of hardness (H) to elastic modulus (E), this reduces to the
established relation for metals that the yield stress (Y) is one-
third of the hardness. However, for higher H/E ratios, the Y/H
ratio increases. For many of the materials considered, this
alternate theory gives similar results (e.g., only about 10%
different for Al_2O_3).

For harder substances yield stress differences are substantial,
e.g., Marsh's theory would predict a yield stress of about 5×10^6
psi for B_4C, about twice that predicted by using H/Y = 3. Marsh's
theory does not appear to be truly applicable to these hard materials.
First, his theory should become applicable when a material can
undergo a substantial amount of elastic deformation. Glasses,
which have low elastic moduli (thus giving high H/E ratios), can
do this. However, very hard materials such as B_4C have very high
elastic moduli, and hence undergo much less elastic deformation;
high H/E ratios for such materials result from the high hardness.
Second, Marsh's theory may not hold for all glasses, e.g., SiO_2[92].
The general correspondence between Y = H/3 and compressive data
for glassy SiO_2 and C with that for other crystalline ceramics
(Table 2) indicates that his theory may not hold for either of
these. Materials which are in accord with the Marsh theory appear
to have a common feature: they all have "something to drag along"
during deformation. All of this group of glasses are very likely
to be phase-segregated, and thus have a "microstructure somewhat
similar to highly hardened metals", which incidentally, also fit
his theory. A high density of substantially harder objects (with
spacings of the order of 0.1 µm) are present in these materials
and must be carried along even in the limited deformation required
for indention. Plastics, the third class of materials to which his
theory applies, are somewhat similar in that they contain long
chain molecules with much higher in-chain than chain-to-chain
strengths. thus, their flow must involve dragging along parts of
these chains.

It appears that Marsh's theory, if correct, is most applicable
to materials of relatively low elastic moduli which contain a
finely dispersed second phase or a chain structure, a group which
probably includes most glasses with two or more cations. Thus, the

Marsh relationship should be considered in evaluating compressive
yield stresses of such materials. He reports correlations between
his calculated glass yield stresses and ultimate tensile strengths
of fibers. Since there is no basic reason to expect different
behavior in compression, his results lend support to the concept of
microplasticity-induced failure.

INFLUENCE OF MICROSTRUCTURE

Compressive Strength

Compressive strength-microstructure data is limited, but enough
does exist to show a clear trend. Bridgman's data[81] for sapphire
(300,000 and 600,000 psi for the c axes oriented approximately 90°
and 0°, respectively, with the compression axis), along with polycrystalline data are shown in Fig. 1. [Strengths in this and subsequent figures are plotted versus the inverse square root of
grain size following N. J. Petch's original analysis of brittle
metals   (J. Iron Steel Inst. 174 25, 1953) since it allows a
clearer evaluation of microplastic effects.[32]]   Many alumina

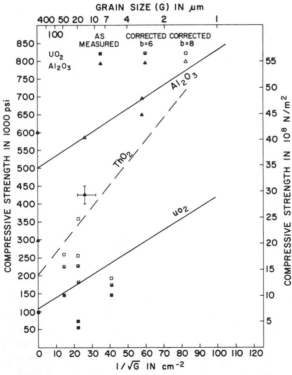

Fig. 1.   Compressive strength-grain size relationships for Al_2O_3,
          ThO_2, and UO_2.

manufacturers simply list the strength of their denser, purer Al_2O_3 bodies as being in excess of 300,000 while several give strengths of 400,000-450,000 psi in agreement with values obtained by Ryshkewitch.[94] Such bodies typically have grain sizes in the range of 10-20 μm and porosities of the order of 5%, which can be conservatively corrected to zero porosity[95] using b = 6. More recently, a finer grain size, ~2 μm, denser, 1-2% porosity, commercial sintered Al_2O_3[AD-999, a product of Coors Porcelain Co., Golden, Colo.] has been reported to have a compressive strength of 650,000 psi[41] (Fig. 1). These data clearly indicate a grain size dependence as predicted by the Petch relationship. The intercept at infinite grain size (equivalent to single crystal size) falls between observed single crystal values and closer to the higher value, which is consistent with a microplastic process.

The author has shown elsewhere that the data of Curtis and Johnson[45] and of Knudsen[97] for ThO_2 together clearly show that the compressive strength of ThO_2 decreases with increasing grain size to a substantial non-zero intercept. The estimated average curve is shown in Fig. 1.

Igata, et al.,[96] have reported room temperature compressive strengths of 99,000 psi for UO_2 crystals and 141,000 psi (average) for sintered polycrystalline urania bodies having ~6 μm grain size and 3-4% porosity; they noted evidences of microplasticity. Figure 1 combines their data with that of Knudsen, et al.[98] The latter indicates an average porosity correction factor, b, of between 6 and 7 (the two more porous specimens would be consistent with b~3 similar to that for flexural data[57]). As with alumina, compressive strength of UO_2 is shown increasing with decreasing grain size. Since porosity corrections tend to decrease with increasing grain size, these strengths were corrected with b values of 8, 6, 6, and 3 respectively, from smallest to largest grain size; this procedure minimized scatter of data.

Compressive data of Alliegro[35] for hot pressed ZrB_2 (corrected for 1-8% porosity using a conservative b = 3) is shown in Fig. 2. Porosity correction at b = 5, as Mandorf and Hartwig[99] observed for TiB_2, would approximately double the slope of the curve. The large grained TiB_2 body, substantially purer than the others in Fig. 2, displayed lower relative strength; however, it showed no relative difference in flexural strength, where flaws are presumed to be controlling. Alliegro's compressive data for finer grain TiB_2, which was pore-free, is in good agreement with the data of Mandorf and Hartwig extrapolated to zero porosity (Fig. 2). Alliegro's specimen of largest grain size, which has been corrected for 7% porosity using b = 5, is substantially lower, indicating a fairly steep grain size dependence. However, this specimen was also of greater purity, hence as a conservative estimate, a slope similar to that indicated

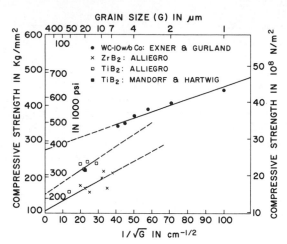

Fig. 2.   Compressive strength-grain size relationships for
          WC-10 w/o C, ZrB₂, and TiB₂.

for ZrB$_2$ has been sketched in Fig. 2.   In contrast to this indicated
lower strength of more pure TiB$_2$, Hansen[40] notes that addition of
20% SiC to ZrB$_2$ results in doubling of its strength, to 500,000 psi.
Only a limited portion can be explained by the reduction in grain
size from 10–20 to 5–10 μm.   This effect of additives or impurities
was not observed in tension, where brittle failure was indicated.[32]

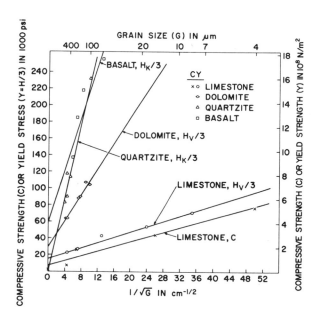

Fig. 3.   Yield stress and compressive strength versus grain size
          of various rocks.

The resultant increase in compressive strength is clearly consistent with a microplastic effect.

Data for WC-10 w/o Co from Exner and Gurland[100] shows a definite grain size dependence (Fig. 2). Since the use of metal-bonded carbide data might be questioned, it should be noted that the author has observed good agreement between the tensile strength data of bonded and unbonded WC in the same range of grain sizes and metal contents.[32] However, as noted earlier, the metal phase may inhibit premature crack opening, e.g., due to twinning, which could lead to lower strengths in unbonded bodies.

Griggs' single crystal calcite data[83] and Handin's data on marble and limestone[44] indicate a definite grain size dependence for $CaCO_3$ (Fig. 3). Values for these are undoubtedly depressed by impurities and porosity, which may account for the near-zero intercept shown for Quartzite.

One of the important predictions of brittle fracture-flaw theories for compressive behavior concerns the ratio of compressive to tensile strengths, e.g., 8 for the Griffith theory and about 3-5 for the maximum tensile strain theory. As shown in Table 3, the ratio depends on the material and its grain size, and can be substantially greater than the above predictions. Since several of

Table 3.   Room Temperature Compressive and Tensile (Bending) Strength Ratios

Material	Grain size range, μm	C/T Ratio*
TiB_2[†]	20-50	4-6
ZrB_2[†]	20-50	4-6
B_4C	15	7
WC	2-6	4-6
Al_2O_3	1-100	4-30
$MgAl_2O_4$	1	7
ThO_2[†]	4-60	13-17
UO_2[†]	20-50	5-18

* C/T ratios are given in the same order as grain size ranges, i.e., smaller ratios are for smaller grain sizes.
† C/T ratios based upon related specimens from same investigator.

these ratios are taken from systematic studies that measured both
tensile and compressive behavior, the observed low ratios are not
attributable to fabrication differences. The grain size dependence
and the range of ratios can be accounted for by microplastic
mechanisms but not by these brittle failure theories.

Hardness

While the effect of grain size on the hardness of ceramics has
been almost totally neglected (in fact the grain size of specimens
is usually not even mentioned), such a dependence would be expected
since it involves microplasticity. A grain size dependence might
also be expected from the increase in length of slip bands associated
with hardness indents of decreasing hardness.[101-103] For example,
in Fig. 4a the slip bands impinging on the grain boundary from
Vickers indents (25g) are shorter than those not impinging on the
boundary.[104] Similarly in Fig. 4b 100 g indents on comparable
fracture faces of a coarse grained MgO body have rosettes impinging
on all boundaries. Such rosettes usually are due only to the
easiest operating slip system, and thus constraining them will
affect hardness only to the degree that it must overcome this
constraint. No effect was seen in coarse grained material but if
the grain size became much smaller than typical single crystal
rosette size, measurable effects would be expected.

There are limited experimental indications of the effect of
grain size on hardness, e.g., Lendvay and Fock[62] reported Vickers
hardness of 149-213 for ZnS crystals depending on orientation,
and Carnall[105] showed Knoop hardnesses of 192 and 292 for polycrystalline ZnS with grain sizes of 2-4 and <1 μm, respectively.

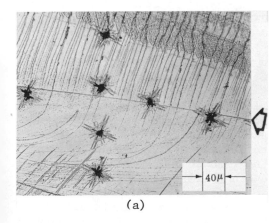
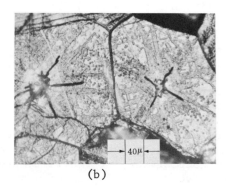

(a)   (b)

Fig. 4.   Vickers indentation rosettes in MgO (a) Cleavage {100}
          surface, bicrystal (arrow indicates boundary), 25 g load;
          (b) fractured, indented, and etched polycrystal, 100 g
          load.

Similarly, Carnall reported a Knoop hardness of 637 for dense MgF_2 with 0.5–1 μm grain size, while Buckner, *et al.*[106] gave a value of 585 for bodies with grains of 3 μm. Also, higher hardness values are frequently reported for fine grain samples in comparison to single crystals. Tashiro[107] has clearly shown that hardness increases with decreasing grain size in Li_2O-SiO_2 crystallized glasses. It is interesting to note that a Petch plot of his hardness data shows an intercept of about 400 kg/mm^2 for bodies with $Li_2O-2SiO_2$ crystallites, and about 800 kg/mm^2 for bodies with Li_2O-SiO_2 crystallites, corresponding to yield stresses of 190,000 and 380,000 psi, respectively. Brace[61] reported data showing similar effects in calcite and subsequently showed a clear grain size dependence for hardness at high loads[108] (Fig. 3). Thus, to be accurate, hardness values for the same grain size should be used, or a value calculated from single crystal data coupled with grain size parameters using a Petch-type equation. Unfortunately, data for the normalizing procedures are generally not available. For present purposes the uncertainties are not too severe, but such errors should not be ignored.

Porosity will normally not affect hardness as much as strength. In many bodies, with much of the porosity at grain boundaries, little effect on hardness is seen. However, as the porosity distribution becomes finer, or the indent larger, then significant effects on hardness can be observed.[109]

DISCUSSION OF COMPRESSIVE BEHAVIOR

Compressive Strengths at Room Temperature

Three general questions must be answered in considering microplasticity as a mechanism of compressive failure: (1) why can microplasticity occur in compression and not necessarily in tension? (2) why does microplastic flow not lead to macroplastic behavior? (3) how can microplasticity lead to the observed differences in tensile and compressive strengths of brittle materials?

Tensile failures of many ceramics of low and intermediate hardness (e.g., halides and most oxides) frequently may be due to microplasticity, while those of hard materials (e.g., borides, nitrides, and carbides) are more likely due to brittle Griffith-type behavior.[32] Cracks are far less stable in tension than in compression. Surface cracks that can cause failure in tension will generally not propagate in compression. Further, compressive failure generally appears to involve the linking up of many small cracks to cause failure.[110,111] Thus cracks must generally be available internally, which probably is not the case in many well fabricated ceramic bodies. In compression, flaws are more stable and the loaded body reaches high stress levels, especially locally in the presence of stress concentrators. Such stress levels should be adequate to induce microplasticity in many bodies.

The high values of yield stress derived from hardness measurements usually are sufficient to activate all slip systems necessary for general ductility. However, operation of all the necessary slip systems is alone not sufficient to achieve generalized plastic flow. As Stokes[112] has discussed, they must also operate homogeneously (i.e., be capable of interpenetration) in the body. From earlier sections, one concludes that much of the microplasticity in normal compressive behavior probably occurs due to local stress concentrations (e.g., near grain boundaries), and that homogeniety of local plastic flow may be hindered by the high resistance to interpenetration of different slip systems except at quite high temperatures.[113] Another limiting factor (partly related to the problem of interpenetrability of slip bands) is the lack of cross-slip in typical ceramics until high temperatures are reached, normally resulting in narrow slip bands and high stress concentrations wherever the bands are blocked.[114] Thus, localization of microplasticity by stress concentrations, poor interpenetrability of slip bands, and limited cross slip will generally preclude macroscopic ductility, except possibly at high confining pressures (discussed later). At high temperatures, interpenetrability and cross-slip become easier, permitting generalized ductility. On the other hand, the lack of surface constraints and the very small volume affected beneath an indenter are probably key factors permitting microplastic events even at low temperatures, allowing small hardness indents to be made without cracking.

Microplastic compressive failure is also consistent with high ratios of compressive to tensile (C/T) strengths. This is true for the same general reason as with flaw mechanisms of failure: in compression, high stresses can be reached because cracks (in this case, nucleated by slip) are not immediately unstable. However, in contrast to brittle failure theories, microplastic failure would result in a broad range of C/T ratios: (1) for materials which fail due to flaws in tension, microplastic failure in compression would not lead to a fixed C/T ratio since the two processes (flaw propagation and microplasticity) are not directly related and will depend differently on grain size; (2) for materials which fail due to microplasticity in tension, C/T ratios will depend on the ratio of stresses required to operate the different slip (or twin) systems involved in the respective tensile or compressive failure modes. As discussed earlier, the ratios of stress required to operate different slip systems can be high, consistent with some high C/T ratios. However, such stress ratios can be expected to depend on the type of microplasticity (e.g., slip, twinning, or a mixture), and on the type of bonding, which in turn will lead to variable C/T ratios. Differences in stress concentration effects on local microplastic events might also affect C/T ratios.

Compressive Failure Modes at Room Temperature

Most cracks can be expected to form at grain boundaries since
boundaries are: (a) effective in blocking slip, (b) frequently
associated with and altered by impurities and voids, and (c) subject
to concentrated anisotropic stresses. Crack formation at grain
boundaries is observed in MgO bicrystals.[114-116] As expected, when
formed in compression, such cracks did not propagate catastrophically.
While transgranular cracks can be formed, they usually do so only
at lower boundary misorientations, and hence intergranular cracks
are expected to be the predominant type of crack formed in a polycrystalline body.[116] In some cases, cracks are not formed unless
slip bands intersect the grain boundary within 1-2 μm of each other.[115]
Since these cracks are relatively stable, many will have to form in
order to cause failure. Since they lie mostly along the grain
boundaries, much of the resulting fracture should be intergranular.
Where fracture modes have been studied (in PbS and MgO), intergranular
cracking has predominated.[79,117,118] As temperatures are raised, some
of the propensity for intergranular fracturing may be relaxed (at
least until grain boundary sliding becomes effective) as indicated
in studies of CaF_2.[119]

Corroboration of Microplastic Mechanisms of Failure

The microplastic mechanisms of failure proposed here draw cor-
roborative support from four general areas of study: (1) dynamic
(shock) behavior, (2) high pressure testing, (3) high temperature
testing, and (4) microhardness testing.

Dynamic (Shock) Behavior. Gilman[120] showed that the Hugoniot
elastic limit or dynamic yield stress [See also W. H. Gust and
E. B. Royce, "Dynamic yield strengths of B_4C, BeO, and Al_2O_3 ceramics",
J. Appl. Phys. 42 (1) 276-95 (1971).] correlated well with hardness,
citing Cline's data for $TiBe_{12}$, Be_2B, B, BeO, Al_2O_3, SiC and B_4C.
The results showed a nearly constant ratio (Y/H) of ~0.45. Two
factors may make this ratio higher than expected: (1) strain rate
effects, known to be important in ceramics, and (2) differences in
yielding and elastic recovery between Knoop and Vickers hardness
measurements. In the first the sensitivity of the dynamic yield
stress to upper yield points and of hardness to lower yield points
also should be considered. The second effect may be a factor in
Gilman's use[120] of Cline's Knoop values (See Ref. 66) measured at
high magnifications. Improved agreement between measured dynamic
yield stresses and those calculated from hardness measurements seems
reasonable since the various stress concentration effects noted under
static conditions would be substantially diminished or eliminated
under shock wave conditions because of the small amount of material
included across the shock front at any given time.

Table 4. Strengths of Ceramics as a Function of Hydrostatic Restraint

Material	Yield Stress (Y=H/3) psix10³	Compression without pressure psix10³	Compression with pressure psix10³	Tension with pressure psix10³
B₂O₃ Glass	140[1] (400)[1,2]			145
Pyrex Glass	284 (550)[2]		700	
SiO₂, Glass	260 (~600)[2]	190	560	
SiO₂ Quartz	516	360 C axis ‖ length	230–2130	
Sapphire	1100–1270	300 C axis ⊥ length	750	> 280
	1250–1470	600 C axis ⊥ length	1300	> 290
MgO	312	200	140–220	
Spinel (crystal?)	780		880	
Diamond (crystal)	4700		1800	
TaC+2% VC+1% MoC	> 850?		510	
TaC + 0.3% Co	~ 850?		440	
Carboloy 905 WC+4% TaC+4% Co	> 850?	620	1570	
Carboloy 999 WC+3% Co	~ 850	620		775
PbS	36	40	30 (yield)	
NaCl	10	6	9	> 1.3 (single crystal yielded)

[1] Estimates based on an assumed $H_V \sim 300$ kg/mm² since other soft (e.g., PbO) glasses have $H \sim 350$–400.
[2] Yield stresses based on hardness theory of Marsh.[92,93]

High Pressure Testing. Materials tested while under hydrostatic
pressure also provide insight into strength behavior since the
superimposed pressure further inhibits cracking and associated
premature failure. Because of inhibited cracking and work hardening,
effects attributable to twinning and local stress concentrations
should be reduced. Further, and possibly more important, the superimposed hydrostatic pressure should compensate for many of the
parasitic stresses resulting from imperfect compressive loading.
These effects are indicated by the data of Table 4 where strengths
under pressure (taken from Handin's compilation[44] except for data
on PbS[79] and MgO[118]) more closely approach the hardness-derived
yield stress than do static strengths. Some tensile strength data
also follow this trend, especially when it is recalled that single
crystals (e.g.,diamond and NaCl) may exhibit lower strengths due
to extensive yielding on easy slip systems (e.g., the NaCl was
unbroken after 22% strain). The high compressive values for the
glasses can probably be accounted for by densification and by Marsh's
theory. Because SiO_2 glass does densify substantially, this
phenomenon might account for the marked difference between estimated
and observed strengths. The very high values of compressive
strength of quartz under hydrostatic pressure have not been repeated
consistently; most values are in the range of 330,000 - 480,000 psi,
i.e., approaching the yield stress. Since high values are obtained
only in some tests at higher confining pressures, a phase transformation to a denser and hence possibly stronger form of SiO_2 might
be inferred.

It is not surprising that the yield stress should generally be
the limit of strengths even under hydrostatic restraint since
pressure *per se* has little effect on yield of ductile materials
unless the pressure is high enough to cause phase transformation
or significantly affect lattice spacings.[79,121,122] Even in ceramics,
it appears to hold true, since confining pressure has little effect
on the strengths of MgO[121] or PbS[79] crystals oriented for easy slip
at room temperature or on polycrystalline MgO at elevated temperatures.[118] The increases in strength generally observed upon application of confining pressure should stop when general yielding
begins (unless significant work hardening occurs).

It should be noted that pressure testing also corroborates
several other proposed aspects of compressive behavior. Patterson
and Weaver[118] observed Petch-type behavior for MgO under pressure.
They found that wavy slip occurred only in some grains at 400°C,
and complete ductility did not occur below temperatures of the order
of 750°C. Microfracturing probably occurred at lower temperatures,
and brittle failure was primarily intergranular (also observed by
Auten[117]). As noted earlier, Tyall and Patersen's[79] results for
PbS indicated increasing strength with decreasing grain size and
porosity, and exhibited primarily intergranular fracture. Recent

theoretical work by Franoise and Wilshaw[123] indicated the stability
of many cracks under pressure, predicted a high incidence of grain
boundary fracturing, and dependence of strength upon the inverse
square root of grain size.

High Temperature Testing. If the yield stress is the limit of
room temperature compressive strengths, it also should be true at
elevated temperatures; indeed, under these conditions, the two values
should approach more closely. While measurements of hardness at
elevated temperatures are limited, Atkins and Tabor[124-127] showed
that hardness is clearly related to plastic deformation. Koester
and Moak's[87] observation that semilogarithmic plots of yield stress
and of hardness of Al_2O_3 and TiC as functions of temperature are
both linear (as found for metals) also supports this concept.
Comparisons of their hardness data with compressive strengths of
Al_2O_3 and ZrO_2 (Table 5) indicates that yield stresses and compressive
strengths have their nearest approach at intermediate temperatures
(1000-1200°C), where most of the room temperature effects (elastic
recovery, stress concentrations, etc.) will have been considerably
relaxed, yet grain boundary sliding due to porosity and impurities
will not have become pronounced. The data for spinel (Table 5) does
not agree; however, it is probabl that Westbrook[130] worked with
nonstoichiometric spinel crystals which would undergo precipitation
hardening and would not be comparable to the more stoichiometric
polycrystalline specimens used in compression studies. However, it
is again observed that the hardness-derived yield stress is the
upper limit of compressive strengths. The data of Evans,[131] showing
that elevated temperature compressive strengths of MgO and Al_2O_3
obey the Petch equation with intercepts (σ_o's) decreasing with
increasing temperature, also support the concept of yield phenomena
limiting compressive strengths.

When ductility in tension becomes feasible as temperature is
increased, it is likely that the tensile and compressive strengths
should move closer together. They will not necessarily become equal,
since effects due to pores and the relative roles of slip and twinning
are not necessarily the same for both stress states. However, σ_o's
of the Petch equation for the tensile and compressive states should
become equal or nearly so as shown for Al_2O_3 at 1600°C (Fig. 5).
Tensile strengths are somewhat lower than compressive strengths,
but not much, if just Lucalox bodies are considered. The tensile
and compressive intercept values are not only similar but are also
in good agreement with single crystal compressive data. The small,
medium, and large grain sizes for the Lucalox data[132] were given
numerical values by referring to Charles' work,[133] while other Al_2O_3
data is from extrapolated curves compiled by Rice.[134] Such behavior
is exhibited by tungsten, where compressive strengths are higher than
tensile strengths well below the ductile-brittle transition temperature, but fall to the level of the tensile strengths as the transition
is approached.[122]

Table 5. Effect of Temperature on Compressive Strength and Hardness-Derived Yield Stress*

| Temp | Alumina | | | | Spinel | | | Zirconia | | |
| | Strength | | Yield Stress (Y=H/3) | | Strength | Yield Stress (Y=H/3) | | Strength | Yield Stress (Y=H/3) | |
°C	single crystals psix10^{-3}	poly-crystals psix10^{-3}	kg/mm^2	psix10^{-3}	poly-crystals psix10^{-3}	kg/mm^2	psix10^{-3}	poly-crystals psix10^{-3}	kg/mm^2	psix10^{-3}
RT		427 – R	786	1120	405 – P	407	578	300 – R	350	497
RT	300 – L	650						303 – Z		
400		214 – R	500	710	199 – R	367	521	228 – R	300	426
500										
600		199 – R	267	379						
800		128 – R	167	237	171 – R	333	473	171 – R	133	189
1000										
1100		85 – R	100	142	85.5 – R	267	379	100 – Z	100	142
1200		71 – R	83	118	71.0 – R					
1300	21.3 – C		67	95						
1328					52 – P					
1400	9.9 – C	35 – R	50	71	21.4 – R			18.5 – R	83	118
1428					40 – P					
1500	6.4 – C	14 – R	37	52				2.8 – R	57	81
1550					6.2 – P					
1600	3.5 – C	7 – R	33	47	8.5 – R					
1600(Lucalox)		22 – 40								
1700	2.1 – C		27	38						
1800					2.2 – P					

*Sources of compressive strength data, as indicated by letters, are as follows:

L: Linde Division of Union Carbide Corp.
R: Ryshkewitch[94]
P: H. Palmour III[129]
Z: Zircoa Corp.
C: Data of Conrad after Koester and Moak[128]

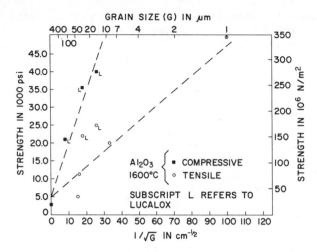

Fig. 5.   Short term tensile and compressive strength of Al_2O_3 at
1600°C.

Indentation Testing of Rocks. Brace[61] has reported that hardness (Vickers indenter) correlated well with compressive strengths
of homogeneous rocks (Table 6). Compressive strengths were determined at a confining pressure of one-sixth of the hardness, to a
strain of 7 ± 5%. (Compressive strengths on anhydrite and limestone
from the same source and for larger grain NaCl without confining

Table 6.   Compressive Strength (as Function of Hydrostatic Restraint)
and Hardness of Rocks.

Material	Grain size (μm)	Strengths (S_c) psix10^3 Without pressure[b]	With pressure[a]	Hardness (Vickers)[a,d] kg/mm² psix10^{-3}		H_v/S_c[a]
Rock salt (NaCl)	400	6[c]	7.7	17	24	3.2
Anhydrite (CaSO₄)	50	19	64	132	187	2.0
Marble (CaCO₃)	200		26	47	67	2.6
Limestone (Solenhofen) (CaCO₃)	4	40-50	77	147	209	2.8

a = Data of Brace[61]
b = Data from Handin[44]
c = Natural rock salt, of larger grain size than Brace's artificial
    rock salt
d = Microhardness values are typically at 21 kg.

pressure are shown for comparison.) Indenting loads were 7 to 75 kg
in order to make impressions at least 10 grains wide. Indentation
deformation was predominantly by slip in rock salt, while limestone,
marble, and anhydrite exhibited both twinning and slip within grains
along with intergranular and transgranular cracking, especially near
the center of the indent. Considerable work hardening was observed
with rock salt, some with marble.

Only for NaCl was the ratio of hardness to compressive strengths
greater than three, probably due to its more extensive work hardening
during indentation. The other materials showed ratios somewhat less
than three, attributed mainly to the fairly extensive cracking that
occurred under the heavy indenting loads. Under such conditions,
hardness corrections (e.g., due to elastic recovery) are negligible,
and cracking reduces the apparent hardness below the true hardness.
Thus, the approximate agreement of yield and compressive stresses,
as well as their deviations from a 1:1 ratio, both corroborate the
previous discussion of hardness and compressive strength.

Since Brace's work with rocks showed encouraging results, an
extension to ceramics was undertaken with lighter loads to reduce
but not eliminate cracking, on the basis that some crack generation
normally accompanies room temperature compressive failure. Commercial ceramic materials (for which manufacturers' literature cites
typical compressive strengths) were selected; several of them were
anisotropic. Knoop hardness was used because it was the most
readily available, and is the most sensitive to anisotropy. The
results of these cursory trials (Table 7) indicate that the loads
employed were generally too high since the hardness-derived yield
stress values are too low. In most materials, considerable cracking
occurred, consistent with low hardness values. The data for Al_2O_3
and $BN-SiO_2$ were quite scattered (as shown for Al_2O_3), indicating
that more thorough work at lower loads might be fruitful.

Applicability of Microplastic Failure Mechanisms

It is clear that bodies severely weakened by flaws and impurities
will be more prone to follow flaw mechanisms of failure, especially
when stresses needed to activate slip or twinning are high. The
presence of high local tensile stresses due to porosity, other
phases, anisotropy, or nonuniform loading, will also favor flaw-
induced failure. In addition to certain ceramics, this category
includes many rocks, for which flaw concepts have been more extensively developed. On the other hand, purer, less imperfect bodies
will be strong and more prone to microplastic failure mechanisms.
This category includes most of the modern single-phase crystalline
ceramic materials. There will also be a variety of "in-between"
materials wherein competition and/or cooperation of both failure
mechanisms may occur. Slip contributions to growth of cracks, and

Table 7.   Hardness-Derived Yield Stress and Compressive Strength of
                        Commercial Ceramics

Material	Designation**	Hardness K_{1000} kg/mm²	Yield Stress $(Y=H_K/3)$ kg/mm²	psix10^{-3}	Compressive* strength psix10^{-3}
BN	H. P.[a]	15	5.0	7	15
		19	6.3	9	16
BN	A[a]	37	12	17	34
		38	13	18	45
BN-SiO$_2$	M[a]	60	20	28	42
		90	30	43	46
Graphite Poco[b]		28	9.3	13	18
Graphite,[+]Pyrolytic[c]		33	11	16	14
		103	34	48	54-82
Glassy Carbon	LSMC 2000[d]	231	77	110	128
Al$_2$O$_3$	AD-94[e]	860$^{+250}_{-170}$	370	300	
			290	210	300
			230	160	

* The first value listed loaded parallel to "c" axis, and the
   second normal to it.
** Suppliers: a = The Carborundum Co., b = Pure Oil Co., c = General
   Electric Co., d = Lockheed Missile & Space Co., e = Coors
   Porcelain Co.
+ Origin of pyrolytic graphite somewhat uncertain.

crack initiation or enhancement of slip not only at tips but at jogs
or other high friction points, may lead to cooperative or competitive
failure modes. Glassy bodies have been discussed in part in considering Marsh's theory. While glasses have not been subjected to
extensive scrutiny in this paper, it appears from the evidence cited
that some sort of microdeformation failure mechanism should be
considered for vitreous materials, with Marsh's theory possibly being
applicable, especially for phase-segregated glasses.

                        SUMMARY AND CONCLUSIONS

     An extensive review of the literature has shown that yield
stress (which is defined as approximately one-third of the indentation
microhardness for most ceramics, as in metals) is generally the
upper limit for compressive strengths of ceramics. A similar concept,
though necessarily involving different mechanisms of "plastic" flow,

may apply to glassy bodies, with Marsh's theory probably being applicable for phase-segregated glasses, and possibly for crystalline ceramics having a dense population of second phase particles. The general tendency is for measured compressive strengths to be too low, and correspondingly, for hardness values to be somewhat high.

When interpreted in terms of stress concentration effects due to flaws, impurities, porosity, and elastic and thermal anisotropies, these findings indicate that microplastic effects can occur, at least locally, in many ceramics. Microplasticity does not, however, necessarily lead to macroscopic ductility. Evidences drawn from ballistic, high pressure, high temperature, and higher load indentation testing all corroborate this concept. Therefore, microplastic mechanisms of failure should be considered as competitive or cooperative with the classic brittle flaw mechanisms, with flaw predominating in weaker bodies and microplasticity in stronger bodies.

ACKNOWLEDGMENT

The aid of Dr. W. McDonough in making some hardness measurements is gratefully acknowledged.

REFERENCES

1. A. Rudnick, C. W. Marschall, W. H. Duckworth, and B. R. Emrich, "The Evaluation and Interpretation of Mechanical Properties of Brittle Materials", Defense Ceramic Information Center Report DCIC 68-3, 1968.
2. R. W. Davidge and R. W. Whitworth, Phil. Mag. 62, 217-224 (1964).
3. J. S. Dryden, S. Morimoto, and J. S. Cook, Phil. Mag. 12 [116] 379-91 (1965).
4. R. I. Garbe, I. I. Soleshenko, and O. A. Khaldie, Soviet Phy.-Solid State 7 [9] 2147-52 (1966).
5. A. S. Keh, J. Appl. Phys. 31 [9] 1538-45 (1960).
6. A. A. Urusovskaya and V. G. Govorkov, Soviet Phy-Solid State 10 [4] 437-441 (1967).
7. D. J. Barber, J. Appl. Phys. 36 [10] 3342-49 (1965).
8. R. M. Katz and R. L. Coble, J. Appl. Phys. 41 [4] 1871-73 (1970).
9. T. S. Liu and C. H. Li, J. Appl. Phys. 35 [11] 3325-30 (1964).
10. P. G. Riewald and L. H. vanVlack, J. Am. Ceram. Soc. 53 [4] 219 (1970).
11. F. W. Vahldiek, S. A. Mersol, C. T. Lynch, Science 149 748 (13 Aug. 1965).
12. D. J. Rowcliffe and W. J. Warren, J. Mat. Sci. 5 345 (1970).
13. G. G. Bentle, F. E. Ekstrom, and R. Chang, "Plastic Deformation of Uranium Carbide Crystals", Atomics International Report NAA-SR-8108, 1963.
14. S. B. Luyckx, Acta Met. 18 233 (1970).

15. T. Takahashi and E. J. Freise, Phil. Mag. 12 1 (1965).

16. L. Pons, p. 393 in Anisotropy in Single Crystal Refractory
    Compounds, Vol. 2 Ed. by F. Vahldiek and S. Mersol. Plenum
    Press, New York, 1968.

17. H. Palmour III; pp.320 in Mechanical Properties of Engineering
    Ceramics, Ed. by W. Kriegel and H. Palmour III. Interscience,
    New York, 1961.

18. B. J. Hockey, Am. Ceram. Soc. Bull. 48 [4] 393 (1969).

19. G. F. Hurley, Met. Trans. 1 2029 (1970).

20. R. R. Vandervoort and W. L. Barmore, J. Appl. Phys. 37 [12] 4483
    (1966).

21. G. G. Bentle and K. J. Miller, J. Appl. Phys. 38 [11] 4248 (1967).

22. W. F. Brace, Bull. Geol. Soc. Am. 69 [12-2] 1539 (1958).

23. O. W. Johnson, J. Appl. Phys. 37 [7] 2521 (1966).

24. J. L. Daniel and S. Takahashi, "Fracture of Uranium Dioxide",
    Proc. 1st Mat. Conf. on Fracture V3, 1967, Japanese Soc. for
    Strength and Fracture of Materials, 1966.

25. W. S. Rothwell, J. Am. Ceram. Soc. 47 [8] 409 (1964).

26. E. Votava, S. Amelinckx, and W. Dekeyser, Acta Met. 3 89 (1955).

27. R. W. Sauer and E. J. Freise; p. 459 in Anisotropy in Single-
    Crystal Refractory Compounds, Vol. 2 Ed. by F. Vahldiek and
    S. Mersol. Plenum Press, New York, 1966.

28. S. A. Mersol, F. W. Vahldiek, and C. T. Lynch, Trans. Met. Soc.
    AIME, 233 1658 (1965).

29. C. O. Hulse, S. M. Copley, and J. A. Pask, J. Am. Ceram. Soc.
    46 [7] 317-23 (1963).

30. D. W. Budworth and J. A. Pask, J. Am. Ceram. Soc. 46 [11] 560-61
    (1963).

31. J. H. Westbrook, "Flow in Rock Salt Structures", General Electric
    Research Laboratory Report No. 58-RL-2033, 1958.

32. R. W. Rice, "Strength-Grain Size Effects in Ceramics", presented
    at the British Ceramic Society Conference on "Textural Studies
    of Ceramics", London, Dec., 1970. Proceedings to be published.

33. A. H. Heuer and J. P. Roberts, Proc. Brit. Ceram. Soc. No. 6
    17-27 (1966).

34. R. A. Alliegro, pp. 1125-30 in The Encyclopedia of Electrochemistry
    Ed. by C. A. Hampel. Reinhold Pub. Co., 1964.

35. G. R. Finlay, Chem. in Canada, 41-43 (Mar. 1952).

36. P. T. B. Shaffer, High-Temperature Materials. Plenum Press, New
    York, 1964.

37. W. S. Williams and R. D. Schaal, J. Appl. Phys. 33 [3] 955-62
    (1962).

38. E. Ryschkewitsch, "Properties and Physical Constants Data of High
    Refractory Materials", Air Force Technical Report No. 6330,
    (Aug. 1950.

39. K. M. Taylor and Camille Lenie, J. Electrochem. Soc. 107 [4]
    308-14 (1960).

40. J. V. E. Hansen, The Norton Co., Private communication, 1970.

41. Anonymous, Materials Engineering, 22 (June 1967).
42. J. Elston and C. Labbe, J. Nuc. Mat. 4 [2] 143-64 (1961).
43. D. M. Chay, J. Palmour III, and W. W. Kriegel, J. Am. Ceram. Soc. 51 [1] 10-16 (1968).
44. J. Handin, p. 223-89 in Handbook of Physical Constants. Ed by S. Clark, Jr. Geological Soc. Am. Memoir 97, New York, 1966.
45. C. E. Curtis and J. R. Johnson, J. Am. Ceram. Soc. 40 [2] 63-68 (1957).
46. M. D. Burdick and H. S. Parker, J. Am. Ceram. Soc. 39 [5] 181-87 (1956).
47. H. T. Hall, Science 169, 868-69 (Aug. 1970).
48. W. G. Bradshaw, Lockheed Palo Alto Research Laboratory, private communication, 1970.
49. R. W. Rice, "Comparison of Knoop and Vickers Microhardness", to be published.
50. P. J. Soltis, "Anisotropic Mechanical Behavior in Sapphire (Al_2O_3) Whiskers", Naval Air Engineering Center, Aeronautical Material Lab. Rpt. No. NAE-AML-1831, April 1964.
51. P. D. Bayer and R. E. Cooper, J. Mat. Sci. 2 347-53 (1967).
52. F. P. Mallinder and B. A. Proctor, Phil. Mag. 13 [121] 197-207 (Jan. 1966).
53. F. P. Mallinder and B. A. Proctor, Proc. Brit. Ceram. Soc. No. 6, 9-16 (1966).
54. G. F. Hurley, "Progress Report for the VIII Refractory Commposites Working Group", Tyco Lab. Inc., Waltham, Mass., 1970.
55. L. Mordfin and M. J. Kerper; pp.243-61 in Mechanical and Thermal Properties of Ceramics. Ed. by J. B. Wachtman, Jr. U. S. Natl. Bur. of Stds., Special Pub. 303 May 1969.
56. J. D. Jenkins, pp.102-103 in Mechanical Properties of Non-Metallic Brittle Materials. Ed. by W. Walton. Interscience Publ., London, 1958.
57. W. F. Brace, in Proceedings of a Conference on the State of Stress in the Earth's Crust, 1963. See also J. C. Jaeger, p. 268-283 in Fracture 1st. Tewksbury Symposium, C. J. Osborn, Ed., Univ. of Melbourne, 1965.
58. J. H. Westbrook and P. J. Jorgensen, Am. Minerol. 53 1899-1909 (1968).
59. R. W. Rice, "Machining and Surface Workhardening of MgO", submitted for publication.
60. B. J. Hockey, "Observations of Plastic Deformation in Alumina due to Mechanical Abrasion", Presented at the 72nd Annual Meeting of the Am. Ceram. Soc. [Abstr. Am. Ceram. Soc. Bull. 49 (4) 498 (1970)].
61. W. F. Brace, J. Geophy. Res. 65 [6] 1773-88 (1960).
62. F. Lendvay and M. V. Fock, J. Mat. Sci. [4] 747-52 (1969).
63. D. R. Tate, Trans. ASM 35 374 (1945).
64. L. P. Tarasov and N. W. Thibault, Trans. ASM 38 331 (1947).
65. B. O. Haglund, Prakt. Metallog. 7, 173-82 (April 1970).
66. C. F. Cline and J. S. Kahn, J. Electrochem. Soc. 110 [7] 773-75 (1963).

67. F. Robertson, Geolog. Soc. Am. Bull. 72 621-37 (1961).
68. Pyrolytic Graphite Engineering Handbook, General Electric Co.
    Schenectady, N. Y., 1963.
69. W. C. Riley, pp.14-75 in Ceramics for Advanced Technologies.
    Ed. by J. E. Hove and W. C. Riley. John Wiley & Sons, Inc.,
    New York, 1965.
70. R. W. Rice, "The Effect of Porosity on the Mechanical Properties
    of Ceramics", to be published.
71. M. V. Klassen-Neklyudova, Mechanical Twinning of Crystals;
    pp.167-176. Consultants Bureau, New York. 1964.
72. A. S. Tetelman and A. J. McEvily, Jr., Fracture of Structural
    Materials; p. 159. John Wiley & Sons, New York, 1967.
73. D. Kalish, E. V. Clougherty, J. Ryan, "Fabrication of Dense
    Fine Grain Ceramic Materials", Final report for Contract
    DA-19-066-AMC-283(X), 1966.
74. T. D. Gulden, J. Am. Ceram. Soc. 52 [11] 585 (1969).
75. R. W. Bartlett and G. W. Martin, J. Appl. Phys. 39 [5] 2324
    (Apr. 1969).
76. J. Corteville, Compt. Rend. 260 [7] 4477-80 (Apr. 1965).
77. T. Simecek, "Twinning of CdTe Crystals", Ceskoslov casopisfys
    10 180-1 (1960). [Chemical Abs. 54:23573h)].
78. A. G. Fitzgerald and G. Thomas, Phy. Stat. Sol. 25 263 (1968).
79. K. D. Lyall and M. S. Paterson Acta Met. 14 371-83 (Mar. 1966).
80. E. Stofel and H. Conrad, Trans. Met. Soc. AIME 227 1053 (1963).
81. P. W. Bridgman,Studies in Large Plastic Flow and Fracture; p. 120
    McGraw-Hill, New York, 1952.
82. M. L. Kronberg, p. 329 in Mechanical Properties of Engineering
    Ceramic. Ed. by W. W. Kriegel and H. Palmour III. Inter-
    science Pub., New YOrk, 1961.
83. D. Griggs, Am. Min. 23 28-33 (1938).
84. D. W. Budworth and J. A. Pask, Trans. Brit. Ceram. Soc. 62 763
    (1963).
85. C. O. Hulse, "Mechanical Properties of CaO Single Crystals",
    to be publsihed.
86. A. G. Evans, C. Roy and P. L. Pratt, Proc. Brit. Ceram. Soc.
    No. 6 173-188 (1966).
87. D. W. Lee and J. S. Haggerty, J. Am. Ceram. Soc. 52 [12] 641-47
    (1969).
88. D. K. Smith, Jr. and S. Weissman, J. Am. Ceram. Soc. 51 [6]
    33-36 (1968).
89. D. H. Chung and W. R. Buessen, pp.217-45 in Anisotropy in
    Single-Crystal Refractory Compounds, Vol. 2. Ed. by F.
    Vahldiek and S. Mersol. Plenum Press, New York, 1968.
90. D. P. H. Hasselman, p. 247-65, ibid.
91. H. P. Kirchner and R. M. Gruver, J. Am. Ceram. Soc. 53 [5]
    232-36 (1970).
92. D. M. Marsh, "Plastic Flow in Glass", Proc. Roy. Soc. (London)
    279-A [1378] 420-35 (1964).
93. D. M. Marsh, pp.143-55 in Fracture of Solids. Ed. by D. C. Drucker
    and J. J. Gilman. Interscience Publishers, New York, 1963.

94. E. Ryschkewitsch, Ber. Deutsch. Keram. Gesel 22 [2] 54-65 (1941).
95. E. Ryschkewitsh, J. Am. Ceram. Soc. 36 [2] 65-68 (1953).
96. N. Igata, R. R. Hasiguti, and K. Domoto; pp.883-98 in Proc. 1st Int. Conf. on Fracture, Vol. 2, Japanese Society for Strength and Fracture of Materials, 1966.
97. F. P. Knudsen, J. Am. Ceram. Soc. 42 [8] 376-390 (1959).
98. F. P. Knudsen, H. S. Parker, M. O. Burdick, J. Am. Ceram. Soc. 43 [12] 641-47 (1960).
99. V. Mandorf and J. Hartwig; pp.455-66 in High Temperature Materials II. Ed. by G. Ault, W. Varclay, and H. Munger, Interscience Publishers, New York, 1963.
100. H. E. Exner and J. Gurland, Pwd. Met. 13 [75] 13-31 (1970).
101. D. I. Matkin and J. E. Caffyn, Trans. Brit. Ceram. Soc. 62 753-61 (1963).
102. N. S. Stoloff and T. L. Johnston, "Formation and Structure of Alloy Layers in the MgO-Mn Systems" presented at the 66th Annual American Ceramic Society meeting, April 1964 [Abstr., Am. Ceram. Soc. Bul. 43 [3] 339 (1964)].
103. G. W. Groves and M. E. Fine, J. Appl. Phys. 35 [12], 3587-93 (1964).
104. R. W. Rice, J. G. Hunt, G. I. Friedman and J. L. Sliney, "Identifying Optimum Parameters of Hot Extrusions", Final report for NASA Contract NAS7-276, Aug. 1968.
105. E. Carnall, Eastman Kodak Company, private communication, 1970.
106. D. A. Buckner, H. C. Hafner, and N. J. Kreidl, J. Am. Ceram. Soc. 45 [9] 435-38 (1962).
107. M. Tashiro, "Chemical composition of glass - ceramics, part two", The Glass Ind. p. 428, Aug. 1966.
108. W. F. Brace, Penn. State Univ. Mineral Expt. Sta. Bul. 76 99-103 (1961).
109. I. Soroka and P. J. Sereda, J. Am. Ceram. Soc. 51 [6] 337-41 (1968).
110. E. Hoek and Z. T. Bieniawski, Intl. J. Fract. Mechan. 1 [3] 137-55 (Sept. 1965).
111. J. S. Rinehart, "Fracture of Rocks", Intl. J. Fract. Mechan. 2 [3] 534-51 (Sept. 1966).
112. R. J. Stokes; pp.379-405 in Ceramic Microstructures, Their Analysis, Significance and Production. Ed. by R. Fulrath and J. Pask. John Wiley & Sons, New York, 1968.
113. R. B. Day and R. J. Stokes; pp.355-86 in Materials Science Research, Vol. 3. Ed. by W. W. Kriegel and H. Palmour III. Plenum Press, New York, 1966.
114. R. J. Stokes and C. H. Li; pp. 133-57 in Materials Science Research, Vol. 1. Ed. by H. Stadelmaier and W. Austin. Plenum Press, New York, 1963.
115. A. R. C. Westwood, Phil. Mag. 6 [62] 195-200 (Feb. 1961).
116. T. L. Johnston, R. J. Stokes, and C. H. Li, Phil. Mag. 7 [73] 23-34 (Jan. 1962).
117. T. Auten, Case Western Reserve U., private communication, 1970.

118. M. S. Paterson and C. W. Weaver, J. Am. Ceram. Soc. 53 [8]
        463-71 (1970).
119. A. G. Evans, C. Roy, P. L. Pratt, Proc. Brit. Ceram. Soc. 6,
        173 (June 1966).
120. J. J. Gilman, J. Appl. Phys.41 [4] 1664-66 (1970).
121. C. W. Weaver and M. S. Paterson, J. Am. Ceram. Soc. 52 [6]
        293-302 (1969).
122. A. S. Wronski and A. C. Chilton, Scripta Metall. 3 394-400
        (1969).
123. D. Francois and T. R. Wilshaw, J. Appl. Phys. 39 [9] 4170-77
        (1968).
124. A. G. Atkins and D. Tabor, J. Inst. Metals 94 107-115 (1966).
125. A. G. Atkins, A. Silverio, and D. Tabor, J. Inst. Metals
        94, 369-78 (1966).
126. A. G. Atkins and D. Tabor, Proc. Roy. Soc. (London) A292,
        441-459 (1966).
127. A. G. Atkins and D. Tabor, J. Am. Ceram. Soc. 50 [4] 195-98
        (1967).
128. R. D. Koester and D. P. Moak, J. Am. Ceram. Soc. 50 [6]
        290-96 (1967).
129. H. Palmour, III, "Flow and Fracture in Spinel Structured
        Ceramics", Final Report, Contract DA31-124-ARD-D-207,
        Jan. 1970.
130. J. H. Westbrook, Rev. Hautes Temper. Refract. 3 47-57, (1966).
131. P. R. V. Evans, "Effect of Microstructure"; pp.164-202 in
        Studies of The Brittle Behavior of Ceramic Materials, Ed. by
        N. A. Weil, Report ASD-TR-61-628, Part II, Contract
        AF 33(616)-7465, April 1963.
132. R. R. Matheson, Ceramic Age, p. 54-58, June 1963.
133. R. J. Charles, "Static Fatigue; Delayed Fracture", p. 468-519
        in Ref. 131.
134. R.Rice, "Tensile Strength and Fracture of Al_2O_3," to be
        published.

DISCUSSION

D. D. Briggs (Coors Porcelain Co.): How can you expect to correlate
single crystal microhardness data with polycrystalline compressive
strengths?
Author: Most microhardness data are from polycrystalline materials.
Differences between mono-and polycrystalline values are generally
not extreme, the basic deformation processes are related, and hence,
general correlations are feasable. There are many complexities
which enter into establishing more detailed correlations. The prime
points I would make are (a) the hardness-derived yield stress is
the upper limit to compressive strength, and (b) microplastic processes
appear to be very important in compressive failure mechanisms. These
points have important implications for improvements in compressive
behavior of practical ceramics.

D. D. Briggs: One probable explanation for the low quoted strengths
(Table 2, Fig. 1) relates to stress corrosion effects in room
temperature tests.
Author: Stress corrosion might have an effect, though probably quite
a small one, since it is basically a tensile (rather than compressive)
phenomenon.

R. N. Katz (U. S. Army Materials and Mechanics Research Center):
There is no essential difference in what is measured in a microhardness test for a ceramic or for a metal. The indenter imposes an
arbitrary shape change on the material; therefore, the von Mises
criterion [having five available (operative) slip systems] must be
fulfilled if fracturing is to be avoided. Apparent differences in
the yield stress-hardness relationship observed between metals and
ceramics are consequences of ceramics having much greater spreads
between the stress levels required to activate primary and secondary
slip systems. In a normal metal at room temperature, five slip
systems are operative, so the measured yield stress *is* the stress
which activates the most difficult system. In ceramics, yield
stress measurements usually involve slip on the most easily activated
system; additional stress results in fracture since the tensile
fracture stress is commonly less than that required to activate
secondary slip systems. However, in microhardness testing, a hydrostatic stress field confines the material under the indenter, thereby
suppressing tensile failure. This allows all the stress to build up,
activating all five slip systems for general yielding. Thus, $H \approx 3 Y$
for ceramics, as for metals. The dislocation rosette observed upon
etching relates only to slip on primary systems; the so-called wing
spacing is related to yield stress on primary slip planes.   Hence,
properly conducted microhardness measurements on ceramics can yield
much information about relative stresses required to operate primary
and secondary slip systems.
Author: We are in basic agreement. However, the additional requirement that the five slip systems must be able to interpenetrate may
account for the fact that small indentations are crack-free, whereas
large (high load) ones are not.

L. L. Fehrenbacher (U. S. Air Force Aerospace Research Laboratory):
The hardness - compressive strength relationship of refractory ceramics
is to be expected at room temperature. However, deviations might
become marked at temperatures where slip on secondary systems occurs
readily. What correlations are there between hardness and compressive
strengths at elevated temperature, and are there any interesting
discontinuities?
Author: Microhardness indentation requires the operation of five
(or nearly five) independent systems; good hardness - compressive
strength correlations imply similar operative slip systems (at
least locally) during compression. Thus I do not expect any significant deviations at elevated temperatures; indeed closer correlations

are expected, at least for cases where porosity and impurity effects
are limited. This view is corroborated by the available high
temperature data.

G. Mayer (U. S. Army Research Office - Durham): Emphasis upon
requiring activation of at least five independent slip systems to
satisfy a specific shape change is perhaps misplaced when one
refers to hardness tests. First, this quite restrictive criterion
has been realistically modified to account for combinations of slip
and twinning (Chin and colleagues, Bell Telephone Laboratories, Inc.).
Second, the slip-line field (and associated strain) under an indenter
is quite nonuniform. Third, ridging around the indenter, which
varies with the load, negates the assumption of *specific* shape change.
Author: Generally I agree, having stated that for indentations
with access to free surfaces slip must approach but not necessarily
attain operation on five systems, thus allowing for possible slip-
twin combinations.

Editors: Our early findings on microindentation of sapphire
[Brittleness in Ceramics: Engineering Study Report, Contract
DA-36-034-ORD-2645, Part I, Jan. 1966; Part II, June, 1961] are in
basic accord with Dr. Katz's comments on the role of hydrostatic
restraint at and near indentations. Ridging is not a major factor
in Al_2O_3, but it is not uncommon to find cracks initiating at
distances >20 μm away from an otherwise crack-free K_{100} indentation
(i.e., outside the elastically coupled hydrostatic stress field).
Furthermore, such indentations on (000$\bar{1}$) are sensitive to minor
axis orientation with either slip <11$\bar{2}$0> or twin <10$\bar{1}$0> vectors;
in the latter case, one side deforms, the other cracks, alternating
from side to side as orientations are varied to successive twin
directions, consonant with unidirectional twinning vectors. Von
Mises' original treatment of generalized plasticity chronologically
preceded the theory of dislocations; in it he established the
criterion for five independent (but undesignated) *shear modes*. As
pointed out in the paper and the discussion, slip, twinning, or
combinations thereof may qualify as shear modes, so long as they
are capable of operating independently, which ultimately implies
interpenetratability. At least in part, interpenetration should
depend upon stress levels; at the extreme stress states generated
in microindentation, there is much greater likelihood of forced
interpenetration. Finally, increased tendencies toward cracking
at larger loads may well be related to greater workhardening
(i.e., yield stresses increasing with total strain) for larger
indent sizes. For them, greater total shear displacements (and
commonly, shear rates) must be accomodated; however, probably
without corresponding increases in the linear distances into
adjoining areas which gain protection from cracking through coupled
hydrostatic restraining forces.

OXIDE CERAMICS IN METAL CUTTING APPLICATIONS

J. L. Pentecost and N. Levy, Jr.

W. R. Grace and Company

Clarksville, Maryland

and

Hayne Palmour III

North Carolina State University

Raleigh, North Carolina

ABSTRACT

The chemical, mechanical, and thermal environment which exists at the cutting edge of an oxide tool in contact with a metal is summarized. Constitution and microstructure of the tool are related to its interactions with this dynamic environment. Performance tests, analyses, and scanning electron microscopic studies are presented, including examinations of wear mechanisms operative on tool surfaces.

BACKGROUND

Oxide cutting tools stem from many developments in the fabrication of dense alumina ceramics. During World War II, initial efforts on oxide tools were carried out in Germany.[1] After the war, further research was pioneered in this country by Watertown Arsenal.[2,3] By the mid-1950's oxide cutting tools were available which began to find unique applications in metal cutting operations, replacing conventional steel and carbide tools.

231

Table 1[4] compares high speed steel, carbide and oxide cutting
tool materials. It can be seen that in hardness and melting point,
oxides are unique, and that tensile (or transverse rupture) strength
is their major limitation. For satisfactory cutting, oxide tools
must have transverse rupture strengths above 90,000 psi and compressive strengths well in excess of 200,000 psi. Since transverse
rupture strength for an oxide is far less than that for a competitive
steel or carbide tool material, one must seek other reasons to
account for outstanding cutting performance on the part of oxides.
The environment in which the cutting tool operates must be described
in order to understand its cutting behavior and unusual durability.

Table 1. Physical Properties of Tool Materials (After King and
          Wheildon)[4]

Measurements	High speed steel	C-2 Carbide	C-6 Carbide	TiC	Ceramic
Transverse rupture (psi)	50×10^4	23×10^4	25×10^4	17.5×10^4	9×10^4
Compressive (psi)	60×10^4	65×10^4	61×10^4	45×10^4	50×10^4
Modulus of elasticity (psi)	32×10^6	100×10^6	80×10^6	60×10^6	60×10^6
Microhardness (R_A)	85	92	91	93	93
Microhardness[a] (Knoop 100)	740	1800	1410	----	1780
Melting Point[a] (°C)	1300	1400	1400	1400	2000
Density (g/cm³)	8.6	14.9	12.7	6.0	3.98
Grain size (µm)	10	2	3	3	3

[a] Variable depending on composition

Temperature

Anyone witnessing a hard steel cylinder being turned at 600-800
sfpm with an oxide tool has noted the bright red color of the chip-
tool contact. Temperatures of 800°C or higher are apparent, and
it is obvious that the temperature gradient is large. Only a small
volume element of the tool near the cutting edge experiences any
significant temperature rise. Additionally, entry of the cutting
edge into the metal is rapid during start-up; temperatures of ~800°C

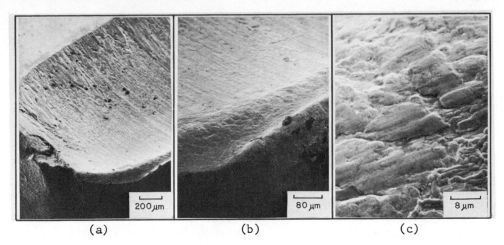

(a)   (b)   (c)

Fig. 1. Typical wear surface on metal bonded WC tool after machining
         steel (a) nose, cutting edge, and crater, (b) detail of
         worn edge and crater, (c) adherent metal at beginning of
         main crater.

are achieved in a fraction of a second.   From careful examination
of the chip and tool surface, even higher surface temperatures are
achieved.[4] The temperature of the metal chips being removed
approaches the melting point (~1500°C) which is verified by metallurgical examination.

    With the average temperatures experienced by the cutting edge
in excess of 800°C, including highly localized temperatures far above
that level, while subject to intense mechanical loading, intimate
metal-tool contact at these temperatures results in a severe chemical
environment. For most carbide tools (Fig. 1), the basic failure
mechanism is dissolution of the carbide material by the hot metal
chip. For oxide tools, this mechanism also exists, but poor wetting,
low thermal conductivity and the limited chemical reactivity of
ceramics combine to minimize its effect.

Mechanical Stress

    Previous investigators have shown evidences of plasticity
(involving slip processes) in sapphire under high mechanical
stresses.[5,6] More recently, direct evidences of plasticity in polycrystalline alumina within the temperature range of interest have
been reported.[7] However, brittle failure mechanisms are frequently
observed with poor quality oxide cutting tool materials. Thus one
expects the types of flow or brittle failure processes which are
possible in oxide cutting tool applications to be environment-
sensitive, depending strongly on interfacial chemical affinities,
mechanical stresses and stress distributions, as well as on temperatures and temperature gradients.

Table 2.   Estimates of Stress Levels in Tool Wear Areas

Material Cut	Rc50 4340 Steel	1045 Steel	Class 30 Gray Cast Iron
Machining Speed, sfpm	600	1500	1500
Vertical Load, kbars	45	28	10-16
Horizontal Thrust, kbars	29	13.5	7.5-11

Table 2 provides estimates of stress levels achieved in this investigation, as calculated from the studies of Krabacher and Haggerty[8] and various oxide tool application data,[9,10] together with the appropriate geometrical factors. Cutting loads of 400 - 650 lbs and thrust loads of 200 - 550 lbs were recorded on soft steel.[8] For typical depths of cut and feed rates employed in this investigation, the cross-section of the chip (and hence the crater area) is in the range 0.5-2.0 x 10^{-3} in^2.

Fatigue

Some cyclic loading is superimposed on the average mechanical stresses shown in Table 2, resulting from the natural resonances in machine, tool holder or work piece, as well as from interrupted cuts, repeated cuts and material inhomogeneities. The evidence for fatigue damage and failures in oxide tools has been presented by many investigators.[4,11,13]

Material Machined

The type of alloy being cut is most important in determining the environment present at the working edge of the tool. Cast iron, which is commonly machined with oxide tools, contains 2-4% C, producing a soft metal matrix containing hard carbide particles. Hard steel (Rc 45-60) has a hard, tough matrix, with a tensile strength typically >200,000 psi. Titanium, having a great affinity for oxygen, reacts rapidly with most oxides. Thus the type of material machined strongly influences the physicochemical environment at the tool edge.

Machining Speed and Geometry

The speed with which the chip moves past the cutting edge was used by King[4] to explain the type of reaction zone present on the basis of limited oxygen diffusion and availability. The machining speed also greatly affects the coefficient of friction between the chip and tool.[8] The rake angle, side relief, edge cutting angle, depth of cut and feed all obviously affect the total load, as well as the area of the tool which is actually loaded.[14]

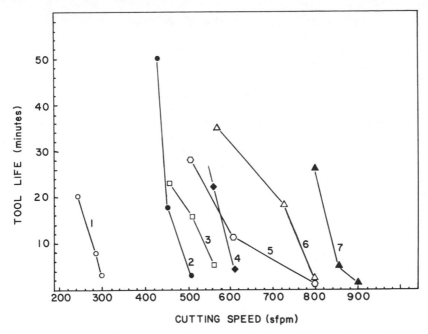

Fig. 2. Cutting performance of commercial oxide tools on 4340 steel (Rc50). Both sintered and hot pressed tools are represented. Numbers correspond to those of Fig. 3.

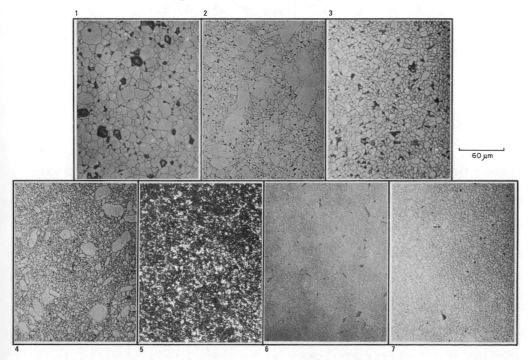

Fig. 3. Microstructures of commercial oxide tools.

Microstructure of the Oxide Tool

Early in this study of the cutting behavior of oxide tools, microstructures were examined and compared with cutting performance for seven commercially available ceramic tools, representative of both hot press and cold press-and-sinter fabrication technologies. The material machined was Rc 50 4340 steel, providing a severe cutting environment which tends to magnify differences in cutting behavior. Comparative cutting performances and microstructures for the seven tool types (identified by numbers on the figures) are shown in Figs. 2 and 3, respectively.

From the data of Figs. 2 and 3, it is evident that the fine, uniform grain size desired in an oxide tool can be achieved by hot pressing or by cold pressing and sintering. The prime requirements for outstanding performance in this difficult application appear to be fine grain size (below 3 μm average), uniformity of grain size, and minimum porosity.

EXPERIMENTAL WORK

To determine the mechanism of failure and the type of environment causing failure, a typical commercial oxide tool type was subjected to machining tests. Machining conditions were standardized with respect to tool geometry (type SN-433; -5° rake, 15° edge, and 5° side relief angles), depth of cut (0.050 in.), feed rate (0.005 in./rev) and duration (16 min, or failure, if sooner). Cutting speeds were selected to achieve representative tool wear with each of the metals tested: (1) 4340 Steel, Rc 50 (600, 900 sfpm); (2) 1045 Steel, 187 BHN (1200, 1500, 1700 sfpm); and (3) Grey Cast Iron, 187 BHN (1250, 1500 sfpm). After testing, individual tool edges representative of the principal experimental variables were examined by scanning electron microscopy. All had survived 16 min cuts.

The tool material selected for this study was an aluminum oxide tool which is 99.9% Al_2O_3 and over 99% of theoretical density, with average grain size 2.5 - 4 μm (comparable to Fig. 3, No. 7). Fabrication was by pressing and sintering reactive alumina. Specimens had been subjected to post-firing abrasive finishing to meet current industry standards for initial smoothness of tool surfaces and edges.

RESULTS

Steel

Figure 4 shows the wear surface on tools subjected to hardened (4340, Rc 50) and mild (1045, 187 BHN) steels. Although generally similar wear patterns are common to both on flanks, edges and in

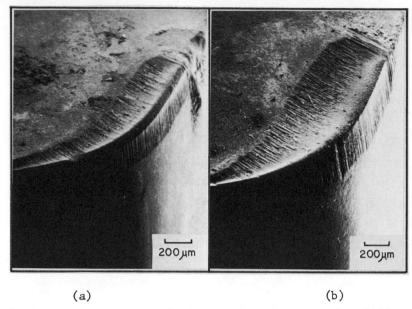

(a)   (b)

Fig. 4. Wear surfaces on alumina tools after cutting different
         steels, (a) 4340, 600 sfpm; (b) 1045, 1500 sfpm.

crater areas, several significant differences in the respective
wear mechanisms are obvious even at low magnification.

    With 4340 steel, the crater is narrow, the cutting edge
rounded and some minor chipping of the edge has occurred. The
static yield stress of the steel is ~200,000 kpsi (13.3 kb), leading
to very high dynamic stresses in the crater area, probably well
over 30 kb. The chip velocity approached 600 ft/min. Shear
markings indicative of plastic flow of the alumina tool material
under this severe stress-strain environment are clearly evident in
both crater and flank areas.

    For 1045 steel, the crater is wider, and the cutting edge is
sharper. The higher velocity of the softer steel chip
(~1700 ft/min) both broadens the dynamically stressed (constrained)
areas and apparently tends to sharpen the cutting edge by flow
processes dividing between crater and flank. Plastic flow of tool
material in the crater and flank areas is generally similar to that
for the harder 4340 steel. A subsequent section treats these
evidences for plasticity in greater detail.

    In contrast to that for steel, the crater area of a tool sub-
jected to cast iron machining is narrow, well defined, and polished

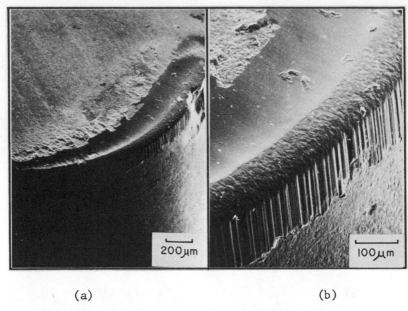

(a)   (b)

Fig. 5.   Wear surfaces on alumina tool after cutting cast iron at
          1500 sfpm (a) overall view of cutting edge, flank, and
          crater; (b) detail at higher magnification.

smooth (Fig. 5).   There is little direct evidence of plastic flow
in the crater area.   The edge is rounded and a distinct crater
exists at the depth of cut line.   A marked transition from smooth
wear to irregular chipping occurred at the perimeter of the crater.
In this case, shear and impact stresses were high but were not
accompanied by constraint from the moving metal.

     Further detail of the rounded cutting edge is shown in Fig. 6.
Only a portion of the grains present are load bearing at any given
time; individual grains apparently fail due to fatigue or some other
time-dependent process.   Some evidences of randomly directed fine
scratches across the alumina grains have been observed at higher
magnification.   This suggests that hard carbide particles in the
cast iron damage these alumina grains by abrasive wear, presumably
accompanied by localized plastic deformation of the sort described
by Hockey.[15]

     For 4340 and 1045 steels tool edge wear (Fig. 6) also involves
breakdown of individual grains.   Small scratches were not observed.
Rather, plastic flow away from the edge occurs in both crater and
flank directions, contributing to a continuing "sharpening" effect.
Local chipping and subsequent "healing" by plastic flow and wear
is frequently observed with the harder steel.

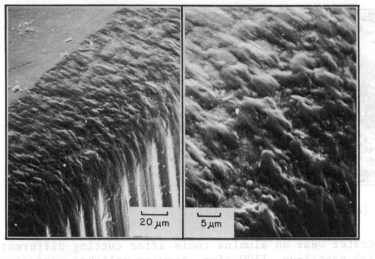

(a)   (b)

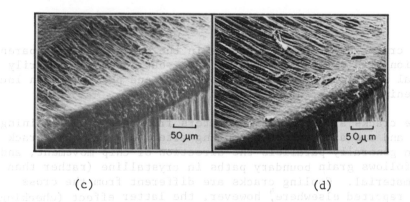

(c)   (d)

Fig. 6.   Edge wear on alumina tools after cutting different metals
          (a) cast iron, 1500 sfpm; (b) detail of (a) at higher
          magnification; (c) 4340 steel, 600 sfpm; (d) 1045 steel,
          1500 sfpm.

Figure 7 compares crater areas of tools subjected to wear at
optimum cutting conditions for various metals. Examination of a
transverse fracture across a tool used to machine cast iron shows
that the microstructure of the tool is unchanged beneath the polished
surface and that the polished layer is less than one grain deep;
no evidence of melting in the crater was observed.

For tools used to machine 4340 steel and 1045 steel, extensive
plastic flow of the alumina surface is evidenced within the heavily

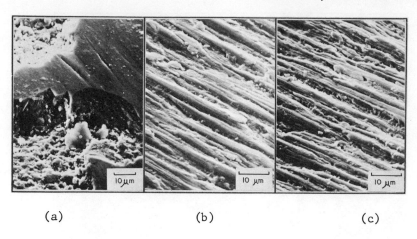

(a)   (b)   (c)

Fig. 7. Crater wear on alumina tools after cutting different metals
(a) cast iron, 1500 sfpm, showing polished surface and
subsurface microstructure exposed by post-machining fracture;
(b) 4340 steel, 600 sfpm; (c) 1045 steel, 1500 sfpm.

stressed crater area. Flow patterns in the crater had no apparent
correlation with original microstructure, but depended primarily
upon metal shear directions and velocities, and possibly upon local
workhardening effects in the alumina.

    Some cracks are observed in craters from cast iron machining
(Fig. 8) and are attributed to cooling after testing. The crack
direction generally parallels the direction of chip movement, and
clearly follows grain boundary paths in crystalline (rather than
glassy) material. Cooling cracks are different from the cross
checking reported elsewhere[4] however, the latter effect (checking
perpendicular to direction of chip movement) was observed on
some tools used at higher machining speeds with cast iron. The
appearance of cross checking suggests that limited slip or plastic
flow had occurred in the wear area under the prevailing stress
state.

    Plastic flow occurring in the crater area of tools used on
steel is shown at higher magnification in Fig. 9. Flow is extensive, characterized by folding, transverse shear directions, and
what may be either inclusions or workhardened grains. At these
velocities (1045 steel ~25 ft/sec, ~400,000 psi) it is difficult
to picture this pattern as due to oxygen diffusion into the hot
chip and reaction of the resulting metal oxide with the alumina
surface to create an intermediate ductile spinel phase.[4] The

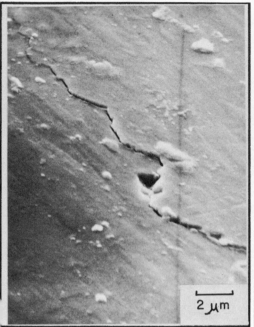 

2 μm   2 μm

Fig. 8. Crack paralleling metal   Fig. 9. Plastic flow in crater
        flow direction in polished   area of alumina tool
        crater area of alumina   after machining 1045
        tool after machining cast   steel at 1700 sfpm.
        iron, 1250 sfpm. (Vertical
        line is an artifact intro-
        duced during microscopic
        examination.)

depth of cut crater resulting from 1045 steel machining is shown
in Fig. 10. A sharp transition from plastic flow to brittle
fracture as the hydrostatic constraint and temperature decrease is
clearly evidenced. From these observations, it is apparent that
plastic flow within crater and flank areas involves the alumina
tool material itself. Furthermore, since high levels of hydrostatic stress together with some localized frictional heating are
present in the flow areas but are not in the exterior regions
marked by chipping, it seems reasonable to associate the observed
ductility with the achievement of multiple slip in polycrystalline
alumina, in general accord with the findings of Becher[7] and
Heuer *et al.*[16]   The demonstrated dependence of such processes
in alumina cutting tools upon local pressures, temperatures, and
strain rates lends additional support to the concepts presented
by Petch[17] and Rice.[18]

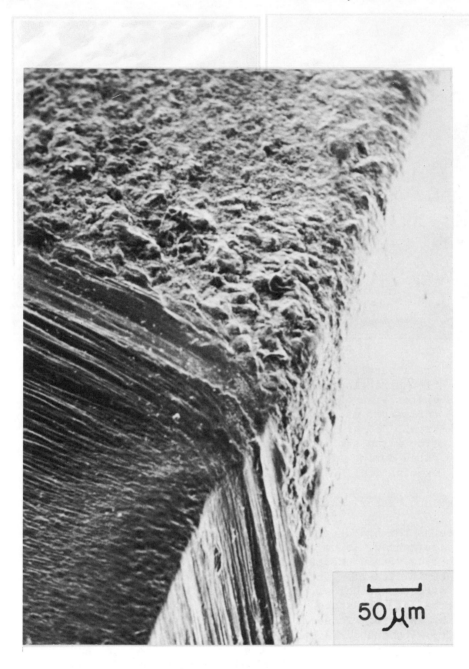

Fig. 10.   Transition from plastic flow to chipping (brittle
           fracture) at depth-of-cut line on alumina tool after
           machining 1045 steel, 1500 sfpm.

CONCLUSIONS

1. Plastic flow is usually observed in the heavily loaded crater areas of an oxide tool used to machine steel. The ability to survive the severe load-temperature environment depends on the physicochemical characteristics of the tool, and especially the mechanical strength of the cutting edge *in this environment*, and upon the operative wear mechanisms.

2. Fine grain structure, high purity, and absence of porosity in alumina cutting tools enhance resistance to fatigue and impact damage, and promote uniform wear.

3. For a given grade of metal being machined, little difference in tool wear mechanism was noted over the range of speeds considered.

4. With cast iron, the crater area of an oxide tool is polished with little or no plastic flow. The greater pressures encountered in machining steel are responsible for extensive plastic flow of tool material in the crater area.

5. Edge wear occurs as individual grains fail by fatigue or other time dependent processes. Plastic flow occurring during machining of steels may contribute to sharpening of edges.

6. Oxide tools available today perform well, but an understanding of wear and failure mechanisms offers possible avenues of improvement such as dispersion strengthening with second phase particles.

REFERENCES

1. E. Ryschkewitch, Ceramic cutting tool. U. S. Patent 2,270,607 (1942).
2. Watertown Arsenal Rept. No. RPL 23/2, Minutes of symposium on ceramic cutting tools, (1955).
3. W. M. Wheildon, Notes on the development and performance of ceramic cutting tools. Presented by J. K. Sjogren, at 58th Annual Meeting, Am. Ceram. Soc., 1956.
4. A. C. King and W. M. Wheildon, Ceramics in Machining Processes, Academic Press, Inc., New York, 1966.
5. R. P. Steijn, J. Appl. Phys. 32, 1951-1958 (1961).
6. E. J. Duwell, J. Appl. Phys. 33, 2691-2698 (1962).
7. P. F. Becher, p. 315, this volume.

8. E. J. Krabacher and W. A. Haggerty, Performance characteristics of ceramic tools in turning and milling. ASTE Tech. Paper 145, Book 2 (1958).

9. The Carborundum Company. Ceramic cutting tool design and application data.

10. Greenleaf Corporation. Cutting tool calculation slide rule.

11. H. D. Moore and D. R. Kibbey, Development of accelerated testing procedures for ceramic cutting tools. Ohio State University, Eng. Expt. Sta. Report 169 (1962).

12. L. S. Williams, pp. 245-302 in Mechanical Properties of Engineering Ceramics , W. W. Kriegel and H. Palmour, III, Eds., Interscience Publishers, Inc., New York, 1961.

13. E. B. Shand, Am. Ceram. Soc. Bull. 38, 653-660 (1959).

14. H. J. Siekmann and L. A. Sowinski, Am. Machinist pp. 113-122 (1957).

15. B. J. Hockey, Am. Ceram. Soc. Bull. 49 (4) 498 (1970).

16. A. H. Heuer, R. F. Firestone, J. D. Snow, and J. Tullis, p. 331 , this volume.

17. N. J. Petch, p. 185, this volume.

18. R. W. Rice, p. 195, this volume.

DISCUSSION

J. E. Burke (General Electric Research and Development Center):
How can you distinguish between plastic flow and selective wear as a cause of grooving in the crater zone? For example, have there been microscopic studies of the immediate subsurface region?
Authors: We have not made transmission electron micrographs capable of revealing the individual dislocations involved in plastic flow in these specimens, though it would be both possible and very interesting to do so (e.g., see Refs. 7, 15 and 16). However, scanning electron micrographs at higher magnifications indicate shallow worked zones (typically < 1 grain deep)-substantially equivalent to polishing - in areas subjected to moderate normal stresses and shear velocities (e.g., cast iron, Fig. 7), but more extensively worked, folded and extended flow textures at high working pressures (e.g., 1045 steel, 1700 sfpm, Fig. 9). We consider plastic deformation to be involved in selective wear (polishing) as well as in ductile flow (grooving) on oxide tools, i.e., it is likely a matter of degree rather than kind. For a given ceramic, the wear observed seems to be strongly dependent upon confining pressures and local shear rates, and these in turn are related to the dynamic yield stress of the metal being cut.

THE FRACTURE BEHAVIOUR OF CARBON FIBRE REINFORCED COMPOSITES

M. G. Bader, J. E. Bailey and I. Bell

University of Surrey

Guildford, Surrey, England

ABSTRACT

Carbon fiber-epoxy resin composites containing up to 0.80 V_f of unidirectionally disposed fibers have been developed with flexural moduli and strengths very close to the theoretical values. However, the fiber-matrix interface is relatively weak resulting in low interlaminar shear strength. Attempts to raise the shear strength have led to brittleness problems. The fracture behavior of a variety of fiber-resin systems has been studied in the flexural and shear modes under static and repeated loading (fatigue), and under sustained loading (creep). Mechanical test data is correlated with fracture mode and morphology.

INTRODUCTION

Carbon fibres having very high strength and stiffness have become available.[1] Their present high cost, however, means that they can only be economically exploited when their superior properties are fully utilized. In practice this demands composites containing high volume fractions of unidirectionally oriented fibre. Current viable systems use polymeric matrices of which those employing epoxy resins are the most popular. These systems are being developed for use in aerospace and other applications demanding high specific mechanical properties. Two basic fabrication techniques are used, both of which are based on the well established procedures used with glass fibres. They are the so-called "wet lay up" and "prepreg" methods. These have been adequately described by Phillips.[2] The "prepreg" method is generally the best suited to aerospace production techniques since it avoids handling both the raw fibre and the messy liquid resins.

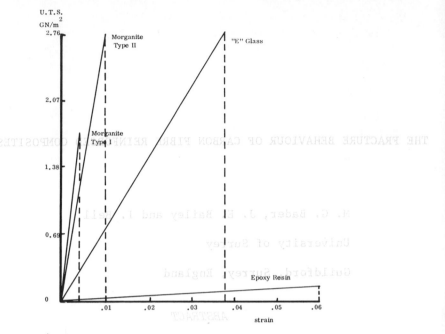

Fig. 1. Stress/strain curves for fibres and matrix materials.

In comparison with glass fibre based composites the use of carbon fibre introduces a number of problems. Firstly, very high volume fractions, (V_f), of fibre must be used – at leat 0.6 is usually demanded. This introduces difficulties in ensuring adequate wetting and impregnation of the fibre especially as it is supplied as "tows" of 10^4 fibres each of ~7μm dia. Secondly, the carbon fibres have much higher stiffness than glass and this accentuates the modulus mismatch at the resin-fibre interface as is shown in Fig. 1.

Properties sensitive to the interface condition are the interlaminar shear strength, compressive strength parallel to the fibre, and properties involving crack initiation and propagation in any direction. Our present studies are aimed at exploring those regions of the stress-strain-time-temperature envelope where fracture considerations are dominant.

Owing to the high degree of anisotropy in the mechanical properties of these materials, conventional testing techniques are seldom entirely satisfactory. The ratio of the tensile strength parallel to the fibre axis to that in the transverse direction may be as high as 40:1, and there is also a considerable difference between the axial flexure and the interlaminar shear strengths. For this reason there has been some difficulty in arriving at a satisfactory form of tensile test piece. The conventionally waisted coupon test piece would require an excessively large radius in the shoulder

to ensure tensile failure in the gauge length rather than shear
failure in the shoulders, due to the shear gradient in that region.
Ewins[3] has proposed an alternative form which consists of a strip
waisted in the thickness rather than width. This allows a reduction
in the shear gradient and ensures satisfactory performance.

It has been found convenient to use the three point bend test
for much composite testing; this uses a very simple form of test
piece, and by virtue of the large differences between flexural and
shear strength, a short beam test piece can be used for the determination of shear strength, whilst a more conventionally dimensioned
test piece will allow the flexural strength and modulus to be
determined.

We have measured the flexure and shear strengths of several
carbon fibre reinforced composite (CFRP) materials using these test
pieces and have also conducted fatigue and creep tests on the same
basis. In addition we have measured impact and fracture strength
using both conventional pendulum impact tests and other forms of
fracture test. We have been particularly concerned with the effects
of varying the interfacial bond strength. The mechanical results
are presented and an attempt has been made to correlate the fracture
morphology with the form of test.

MATERIALS

We have tested composites manufactured by both the "wet lay up"
and the "prepreg" processes and utilising the two principal fibres,
i.e., the high strength and high modulus types. The fibres used
were the Morganite "Modmor" Type I which is a high modulus fibre,
Modmor Type II which is the high strength type and Courtaulds
"Grafil" HT-S which is again the high strength type. The Modmor
fibres were long staple of 1 m length whilst the Grafil was supplied
as 25 mm wide "prepreg" tape 0.25 mm thick. This was manufactured
from continuous fibre. Details of the normal fibre properties are
given in Table 1.

Table 1. Graphite Fibre Properties

Fibre	Code	Max. Tensile Strength GN/m^2	Youngs Modulus GN/m^2	Fibre dia. μm	Density g/cm^3
Modmor Type I	MI	1.00–2.30	300–420	8.0	1.95
Modmor Type II	MII	1.50–3.40	180–250	7.7	1.78
Grafil HT-S	HT-S*	2.10–2.80	225–280	8.9	1.80

* S refers to surface treatment.

The Modmor fibres were supplied in both the untreated condition and after a chemical surface treatment designed to improve the resin-fibre bond. The Grafil fibre had been similarly treated.

The wet lay up composites were made from the Modmor fibre using a standard leaky mould technique. The mould size was 100 x 12.5 mm and mouldings of 2 - 2.5 mm thickness were made for the shear and flexure tests and up to 6 mm thick for the impact test pieces. The resin system was "Shell Epikote" 828 (DGEBA) 100 parts by weight, hexahydrophthallic anhydride (HPA) 90 parts and benzyldimethylamine (BDMA) 1 part. The cure cycle was 3 hr at 80°C followed by 4 hr at 150°C, which gave a fully cured resin with reproducible properties. Four series of composites were prepared from MI and MII fibre each in the treated and untreated conditions and having a V_f between 0.3 and 0.8.

The Grafil prepreg tape was laminated into strips 25 x 2.5 x 100 mm (approx) in a steam heated autoclave, additional pressure being applied by an evacuated rubber capsule. Composites containing 0.60 V_f were made. The prepreg resin was an epoxy of CIBA Ltd. manufacture designated "DLS-60". No further details of this are available.

Some of the early impact and fracture tests reported here used composites made from MI-S fibre in the DLS 60 and a Shell 828/DDM/BF$_3$400 resin system.

EXPERIMENTAL

Physical Testing

<u>Flexure and Shear Tests</u>. These tests were performed on an Instron testing machine using a simple 3-point bending jig. The loading rollers were 6 mm dia. and the crosshead speed 0.5 mm/min. Figure 2 shows details of the two test pieces used.

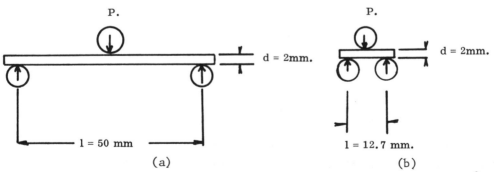

Fig. 2.   Test configurations (a) conventional flexure, $\sigma=3P\ell/2bd^2$; (b) short beam shear, $\tau=3P/4bd$.

Fatigue Tests. The same basic arrangement was used but with the machine set to a load cycling mode and the crosshead speed increased to 5 mm/min so that a cycle speed of up to 10/min could be achieved. The load cycle was unidirectional from a small positive load (~10N) to the selected maximum. This avoided the zero load situation, prevented the test piece from wandering and thus avoided the need for elaborate location fixtures.

Creep Tests. These were conducted on a modified 1 ton capacity lever type creep machine. All tests to date have been at room temperature and only the short beam interlaminar shear strength, (ILSS), specimen has been used. The deflection at the center of the beam was measured with a dial test gauge.

Impact Tests. A specially constructed miniature Charpy pendulum type machine was used. The impact energy was 5 joules and the impact velocity 3 m/sec.

Controlled Fracture Tests. A number of tests using the controlled fracture 3-point beam test piece described by Tattersall and Tappin[4] have been conducted. The test piece is shown in Fig. 3c. Its principal advantage is that the fracture is initiated at a low load and propagates in a controlled manner across the test piece. The Instron testing machine is used with a high value load cell to reduce the external energy in the system. A limited number of double cantilever beam cleavage tests have been carried out using the test method described by Berry[5] but a technique which gives consistent results is yet to be established. The main difficulties arise from the fact that the fibre is inevitably imperfectly aligned and this precludes

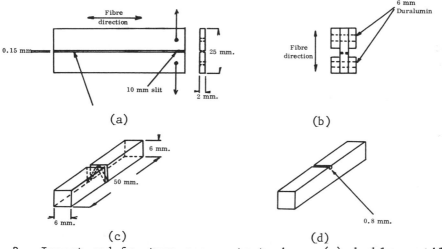

Fig. 3. Impact and fracture energy test pieces (a) double cantilever beam[5] work-of-fracture; (b) end view of modified d-c-b; (c) "Harwell-type"[4] slow bend work-of-fracture; (d) Keyhole-notch Charpy impact.

a plane crack path. We are, however, continuing to evaluate this type of test for both axial and transverse cracking directions.

Fractography and Microscopy

Fracture surfaces and sections cut through broken and partly broken test peices have been thoroughly investigated using both optical reflected light microscopy and the "Stereoscan" scanning electron microscope (SEM), this latter technique proving to be the most rewarding.

RESULTS AND DISCUSSION

Modulus and Flexural Strength

The flexural modulus and strengths of the various composites are shown in Table 2.

Table 2. Flexural Modulus and Strength of Composites.

Composite System		V_f	Flexural Modulus		Flexural Strength	
Fibre	Resin		E	E_t*	σ	σ_t*
			(GN/m^2)		(GN/m^2)	
MI+	828/HPA	0.35	97.0	121	0.42	0.53
		0.60	155.0	208	0.60	0.92
		0.80	158.3	259	>0.32	1.22
MI-S+	"	0.35	98.2	121	0.45	0.53
		0.60	193.6	208	0.77	0.92
		0.80	200.0	259	0.82	1.22
MII+	"	0.35	57.5	77	0.74	0.86
		0.60	114.1	132	1.16	1.48
		0.80	122.4	176	1.17	1.84
MII-S+	"	0.35	75.9	77.0	0.78	0.86
		0.60	123.6	132	1.37	1.48
		0.80	132.1	176	1.65	1.84
HT-S†	DLS 60	0.60	-		1.22	1.48
MI-S†	DLS 60	0.60	-		0.58	0.92

*E_t and σ_t are values predicted from law of mixtures rule
+Specimens made by the "wet lay-up" method
†Specimens made by the "prepreg" method

The values for the "wet lay-up" specimens deviate from a simple law of mixtures rule and the fall off in both modulus and strength increases at the high fibre loadings. This effect is most marked with the MI fibre and is attributed to the difficulty in ensuring good penetration of the tows by the resin adequate wetting of the individual fibres, and also to their misalignment. Microscopic examination has shown that these highly filled composites are more susceptible to dry fibre contact regions and also to voids. Our results indicate an optimum practical fibre fraction of about 0.6. The values for the "prepreg" composites show that similar strengths and stiffnesses are possible in what may be regarded as a normal production composite.

Interlaminar Shear Strength (ILSS)

The ILSS is calculated from the results of the short beam tests and the results are shown in Fig. 4. The general trend is to increased ILSS up to about 0.6 V_f above which some fall off is again apparent. The surface treated fibres give much better ILSS values and the high strength fibres (MII and HT-S) are better than the high modulus ones. It should be emphasised that the points on Fig. 4 are the average of at least five separate tests involving several different mouldings and the apparently anomalous curve for the MI-S is quite reproducible.

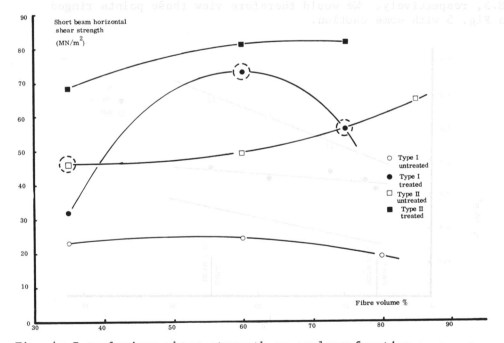

Fig. 4. Interlaminar shear strength vs. volume fraction.

These shear test values must, however, be examined with some caution since their validity depends on the critical shear stress being reached before the critical flexural stress during the test. To some extent this can be determined by a visual examination of the failed test piece for signs of interlaminar failure (Fig. 9), but since many failures show mixed shear and flexural modes this test is not always sufficient. In many cases there are indications that the shear event is often consequential to the flexural failure.

From elementary elastic beam theory it can be shown that the relationship between the maximum flexural stress and the maximum shear stress in a 3-point loaded beam is

$$\sigma/\tau = 2\ell/d.$$

In the case of the short beam used in this investigation the value of $2 \ell/d$ is 13, so that in order for the shear strength result to be valid it must not be greater than 1/13 of the flexural strength of the material. We have conducted a series of tests in beams of various ℓ/d ratios and shown that the recorded flexural strength decreases as the ℓ/d is reduced (Fig. 5). This has also been reported by other workers.[6] The reduction factors for: short beam strength/long beam strength amount to 0.9 for the MI composites and 0.7 for the MII which increases the effective σ/τ ratio to 14.5 and 18.5, respectively. We would therefore view those points ringed in Fig. 5 with some caution.

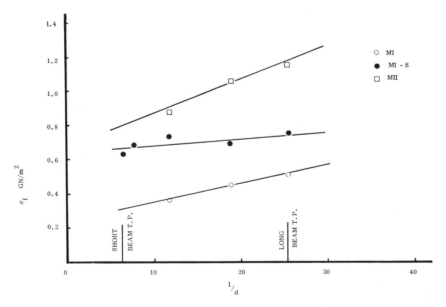

Fig. 5. Variation of flexural strength with ℓ/d, 3-point bending.

Fatigue

Figure 6 shows the S-log N curves for the tests on long and short beam specimens of the HT-S/DLS 60 composites. The stress scales have been arranged so that the short beam results may be read on a basis of either shear or flexural stress. The fatigue strengths of the short specimens are very nearly 0.7 of those of the long specimens (broken line, Fig. 6) at the higher endurances but are lower in the shorter life region (i.e., below N = 100). This is substantiated by an observation of the fracture mode which was interlaminar shear at low endurances and flexural at high endurances. This observation strengthens the arguments of the previous section.

The curves are very shallow indicating a low fatigue sensitivity as reported by Owen[7] in earlier work.

Creep

The creep tests were on the short beam specimens only and the HT-S/DLS 60 composite was used. To date only short term tests at high stresses have been completed. These show a conventional strain/time relationship (Fig. 7). An intermittent test has also been carried out by loading for 100 hr followed by a 100 hr rest period. There was no significant recovery and the net creep curve (not shown) was similar to that for uninterrupted creep. Creep

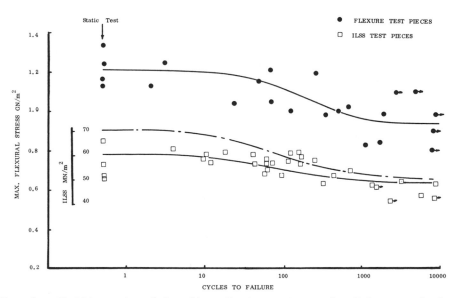

Fig. 6.   Unidirectional bending fatigue strength of long and short test pieces.

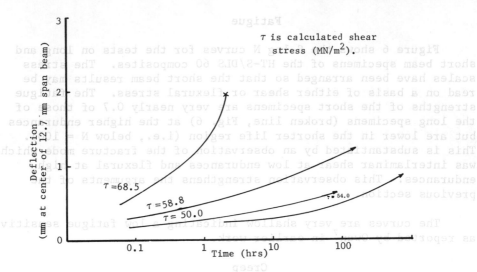

Fig. 7. Creep deflection curves for short beam test pieces of
         HT-S/DLS 60 composite (see Table 2).

failures in these and other creep/rupture tests, not reported here,
have invariably been interlaminar and microscopic examination has
revealed strong evidence of resin flow as mentioned in a later
section.

Table 3. Pendulum Impact Energies

Composite System		V_f	Pendulum Impact Energy	
			Notched	Un-Notched
			(kJ/M^2)	
MI	828/HPA	0.35	20.3	47.2
		0.60	55.5	50.5
		0.80	56.2	47.4
MI-S	828/HPA	0.35	8.3	12.1
		0.60	10.8	35.5
		0.80	14.2	25.3
MII	828/HPA	0.35	41.6	61.7
		0.60	88.0	115.7
		0.80	114.4	>128.6
MII-S	828/HPA	0.35	19.4	41.3
		0.60	58.5	104.0
		0.80	102.0	112.2
*MI-S	DLS 60	0.60	6.0 - 10.2	–
*MI-S	828/DDM	0.60	4.5 - 8.0	–

*Prepreg composites: part of earlier programme of work.

Impact Tests

Table 3 summarises the impact test results for both the wet
lay up and prepreg systems. These results are to some extent
confusing. There is a clear impact energy/V_f relationship in all
cases (Fig. 8) and a general superiority of the high strength
fibres. The untreated fibre composites have higher impact energies
but this cannot be related directly to the interlaminar shear
strength. This is possibly due to the questionable validity of
the high ILSS values discussed previously. On a qualitative basis
Fig. 9 demonstrates that high impact energy is associated with
multiple and complex failures and that delamination is a feature
of the "toughest" composites.

The results of Harris et al.[8] suggest a simple relationship
between impact energy and ILSS. However, their results were on
composites of only one V_f. It is apparent from the present work
that the relationship is more complex. The V_f and fibre strength
(or perhaps strain at fracture) clearly influence the impact energy
most strongly whilst the interface bond condition is also important
as evidenced by the superiority of the untreated fibre composites.

Controlled Fracture Energy Tests

Owing to a shortage of material in the form of 6 x 6 mm bar,
we have completed only a few tests on the main composites discussed
in this paper. However, during previous work[9] two comparable
materials were studied. They were both laminated from prepregged
tows of the MI-S fibre to a V_f of 0.6 using the DLS 60 and a Shell
828/DDM/BF$_3$400 resin system respectively.

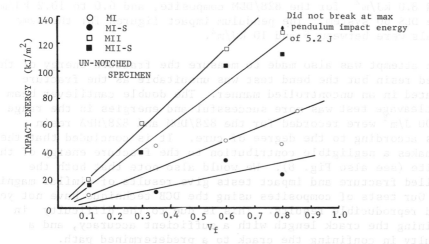

Fig. 8. Impact energy vs. volume fraction for unnotched composites.

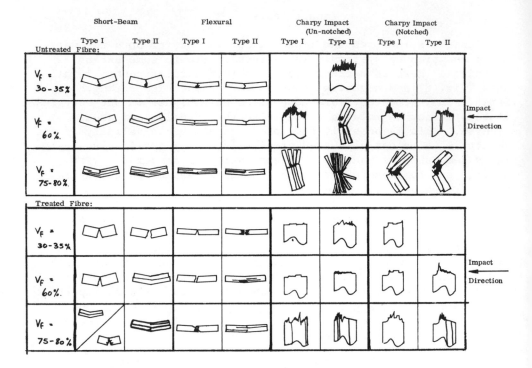

Fig. 9.   Fracture modes of carbon fibre reinforced composites.

A slow bend fracture test using the triangular notched test piece shown in Fig. 4 was used, the fracture energy being computed from the area under the force/deflexion trace. Over a wide range of resin cure conditions the fracture energies fell between 4.5 and 8.0 kJ/m^2 for the 828/DDM composite, and 6.0 to 10.2 kJ/m^2 for the DLS 60 system. The pendulum impact figures for the same materials were between 5 and 10 kJ/m^2.

An attempt was also made to measure the fracture energy of the unfilled resin but the bend test was unsuitable as the fracture propagated in an uncontrolled manner. The double cantilever beam (DCB) cleavage test was more successful and energies in the range 20 - 100 J/m^3 were recorded for the 828/DDM and 828/HPA resin systems according to the degree of cure. It is concluded that the resin makes a negligible contribution to the fracture energy of the composite (see also Fig. 8). We would also note that both the controlled fracture and impact tests give results of similar magnitude. Our tests of composites using the DCB technique have not yet yielded reproducible results. This is due to the difficulty in determining the crack length with a sufficient accuracy, and a difficulty in confining the crack to a predetermined path.

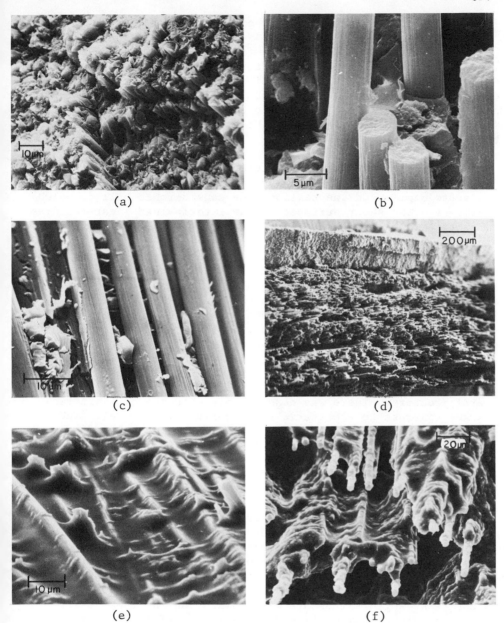

Fig. 10. Scanning electron micrographs of fracture surfaces (a)
HT-S/DLS 60 flexural failure, compressive zone; (b)
HTS/DLS 60 flexural failure, tensile zone; (c) MI-S/828
0.6V$_f$, ILSS failure; (d) HT-S/DLS 60, fatigue failure;
(e) HT-S/DLS 60 creep, ILSS surface; (f) HT-S/DLS 60
creep, flexure failure.

Fracture Morphology

Two techniques have been used: fracture surfaces have been
examined in the scanning electron microscope and also sections have
been cut and polished for optical microscopy. For electron microscopy the samples were vacuum coated with a gold/palladium alloy to
render the surface conductive but no further preparation was necessary.
In optical microscopy polarised light was used to improve the resin/
fibre contrast.

The general features of the flexural fracture were a compressive
failure zone (Fig. 10a) showing much fibre debris and fibres fractured
at an acute angle. On the tension side considerable pull out (10b)
was evident. The shear surfaces of the interlaminar shear failures
were smooth and free from debris. There appeared to be a clean
fibre/resin debonding (10c).

The longer life fatigue fractures especially those on the short
beam specimens showed two distinct failure zones (Fig. 10d). The
compressive zone was more extensive than in the static flexural
failure and was often bounded by some interlaminar cracking as shown.
The rougher zone showed extensive pull out and was similar to the
static flexural fracture. Note that all the fibres break normal
to their axes in tension and at an angle in the compressive zone.

The creep fractures all showed interlaminar shear combined with
flexural fracture. There was considerable evidence of resin flow
in marked contrast to the static shear failure (Fig. 10e). The
appearance of some of the pull out regions was also quite distinctive
showing bundles of fibres coated with resin (10f). It seems quite
clear that extensive resin flow was associated with the failure, and
that the fibres pulled out slowly following fibre fracture. There
was no fibre/matrix debonding.

Sections cut through both flexural and fatigue fractures show
a fibre buckling effect in the compression zone (Fig. 11a). This
appears to be the first failure event and precedes the tension
failure. It is believed that in general, compressive failure
precedes tensile failure in both static flexure and flexural fatigue.
This view is supported by Harris et al.[8] Their published composite
compressive strength values are much lower than the tensile strength
values. True compressive strength is however difficult to measure
and in spite of considerable development work[10] reliable figures
are not generally available.

Figures 11 b, c, and d are of impact fracture surfaces. Figure 1
is from a 0.8 V_f MII-S composite. The smooth zone on the compressive
side and the clumped pull out regions on the tension side are typical
of the moderately high impact strength materials. Figures 11 c and d

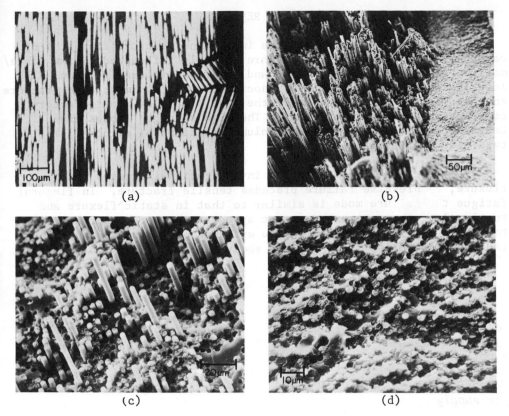

(a)   (b)

(c)   (d)

Fig. 11.   Fractographic details (a) HT-S/DLS 60, longitudinal
cross-section showing flexural failure compressive zone,
optical micrograph; (b) MII-S 0.8 V_f, impact failure,
SEM; (c) MI-S 0.35 V_f, impact failure, SEM; (d) MI-S,
0.6 V_f, impact failure, SEM.

are from MI-S composites of 0.35 and 0.6 V_f, respectively. They
show the reduction in pull out length which is observed with
increasing V_f and ILSS. The 0.6 V_f material has an impact energy
of only 35 kJ/m^2 in the unnotched impact test. This is the nearest
approach to completely brittle behaviour.

As illustrated in Fig. 9, the fractures in the low V_f MI-S
composites are almost completely brittle but as V_f is increased the
mode changes to one of clumped pull out with axial splitting and
high impact energies are recorded. The MII composites show more
pull out than the MI under all conditions but the greatest amount
of debonding is apparent at high V_f and with the untreated fibres.

CONCLUDING REMARKS

Our results indicate that the impact and fracture strengths of carbon fibre reinforced plastics are strongly influenced by the fibre/matrix bond, the fibre strength, and the fibre volume fraction. High interlaminar shear strength is associated with low impact and fracture energies but the inadequacies of the test precludes a satisfactory correlation at the present time. The high strength fibres give tougher composites. The higher volume fractions of either fibre type also increase the toughness.

The fracture morphology investigation indicates that in flexure, compressive failure precedes tensile fracture. In flexural fatigue the failure mode is similar to that in static flexure and the level of shear stress does not appear to be very important. On the other hand creep failures show evidence of resin flow which indicates the predominance of the shear mode.

ACKNOWLEDGMENTS

This work has been carried out partly with support from the Science Research Council and partly under a co-operative research agreement with the British Aircraft Corporation Ltd. who have supplied most of the prepreg composites tested. We would also gratefully acknowledge the assistance of both the Shell Chemical Company Ltd. for supply of resin materials and Morganite Modmor Ltd. for supply of fibre.

REFERENCES

1. W. Watt, L. N. Phillips and W. Johnson, The Engineer 221, 815-816, (May 27, 1966).
2. L. N. Phillips, R.A.E. Technical Report 67088, Royal Aircraft Establishment, Farnborough, England, 1967.
3. P. D. Ewins, Royal Aircraft Establishment, Farnborough, England. Private communication.
4. H. G. Tattersall and G. Tappin, J. Mat. Sci. 1 296-301 (1966).
5. J. P. Berry, J. Appl. Phys. 34 (1) 62-68 (1963).
6. H. Wells, W. J. Colclough and P. R. Goggin, Paper 2C, Proc. 24th Annual Technical Conference, S.P.I., R.P./Composites Div., Feb. 4-7, 1969.
7. M. J. Owen and S. Morris, Modern Plastics 47 (4) 158 (1970).
8. B. Harris, P.W.R. Beaumont, E. Moncunill de Ferran and M.A, McGuire Technical Report No. 69/Mat/10, 1969. University of Sussex, Brighton, England.
9. P. L. Brazenor, I. Bell and M. G. Bader, Dept. of Metallurgy and Materials Technology, University of Surrey, Guildford, England, Unpublished work.
10. P. D. Ewins, R.A.E. Technical Report 70007. Royal Aircraft Establishment, Farnborough, England, 1970.

EFFECT OF MICROSTRUCTURE ON THE BALLISTIC PERFORMANCE OF ALUMINA

W. J. Ferguson and R. W. Rice

Naval Research Laboratory

Washington, D. C.

ABSTRACT

Results of a study of various dense Al_2O_3 bodies using a 22-caliber fragment simulator are presented showing that there is no significant correlation of ballistic performance with static tensile strength, surface finishes or single crystal orientations. In polycrystalline alumina a limited reduction in performance occurred with increasing grain size and a larger reduction was associated with certain impurities. The effect of impurities depends on their state and thus on methods of addition and thermal histories. Results are discussed in terms of possible microplasticity.

INTRODUCTION

In 1962, armor consisting of a ceramic facing backed with glass reinforced plastic (GRP) was introduced to provide protection from attack by armor piercing projectiles such as 30- and 50-caliber service rounds. Subsequent development has led to increased understanding of the characteristics of materials which influence their performance as an armor. For example, harder materials generally perform better than softer ones. Performance usually decreases with increasing porosity but does not appear to depend significantly on static tensile strength or grain size. However, the effect of porosity, grain size and other microstructural variables such as the type and distribution of impurities have not been fully explored. This is due in part to the large (6 x 6 in.) size and number of tiles normally required for ballistic testing which sets a practical limit on the range of variables that can be examined.

Ceramic faced armor has been considered also for protection from fragments and other non-armor piercing projectiles. Research on this application has been limited by early indications that such ceramic armor is superior to other materials only with targets which are thick relative to the fragment size. Earlier data suggests a bimodal distribution with different lots of the same alumina bodies falling on either an upper or lower curve.

The present work was initiated to explain the effect of microstructure and strength determining factors (e.g., surface finish) on the ballistic performance of alumina. It was hoped that such basic information would provide insights into the causes of the differences in supposedly identical materials as suggested by earlier data. The fragment simulator was selected for the evaluations rather than armor piercing projectiles for several reasons. The penetration mechanisms for armor piercing projectiles are different from those for fragment simulators; however, the general trends should be similar. Furthermore effective scaling laws are available to normalize various sizes of fragment simulating projectiles for the class of target materials considered, and of course, fragment armor data is useful in itself. The use of the much smaller fragment simulating projectile makes it possible to obtain more samples with a given quantity of material and a wider range of microstructural variables.

Alumina was chosen for study since fabrication technology and effects of microstructure are better established for it than for other ceramics. Alumina is generally less expensive than other ceramics used for armor and is useful as a component of layered armor.

MATERIALS AND EXPERIMENTAL PROCEDURES

The method of fabrication, composition, strength and density of the various materials tested are shown in Table 1. Tests were performed with 6 x 6 and 4 x 4 plates and 1.5 to 2.5 in. dia disks. A few rectangular 1.35 x 2.0 in. fused cast alumina bodies were used. The Lucalox and single crystal samples were 1.375 in. dia disks. Surfaces were tested in as-fired, diamond cut, conventionally ground, or polished conditions.

Ballistic limit velocities were obtained with a 22 caliber fragment simulating missile having a truncated chisel front. The missile weigh 17 grains and are made of a moderately hard steel (Rockwell C-2

Alumina densities were determined by the Archemides principle. Static tensile strengths were taken as the modulus of rupture measure in three point bending on a span of 0.5 in. with bars of width at

Table 1. Characterization of Alumina Samples

Description and Source*	Condition**	Min. Al_2O_3 Content %	Density g/cm^2	Grain Size, μm	$\sigma \times 10^3$, psi
HOT PRESSED					
Linde A (1)	h.p.	99.9	3.92	2	65
	h.p.,anneal.			10	40
+2% LiF (1)	h.p.	99.0	3.89	2	27
	h.p.,anneal.	99.5	3.84	4	52
+2% LiF,					
+2% TiO_2 (1)	h.p.	97.0	3.91	2	48
+2% TiO_2 (1)	h.p.	97.9	3.81	4	24
+2% La_2O_3 (1)	h.p.	97.9	3.87	1	55
+2% Cr_2O_3 (1)	h.p.	97.9	3.86	1	68
SINTERED					
High purity body (2)	sintered	99.9	3.96	3	70
G.McB-352 (3)	sintered	99.3	3.81	15	35
Lucalox (4)	sintered	99.9	3.98	60	35
AD-94 (5)	sintered	94.0	3.60	20	51
AD-85 (5)	sintered	85.0	3.40	10	46
FUSED					
Single Crystal (6) (Czochralski)	as grown	99.9	3.99		60
Casting Monofrax A (7)	as cast	99.3	3.76 †	300-1000	15
Monofrax M (7)	as cast	94.5	3.41 †	250	15
Al_2O_3 (7) +.4% MgO	as cast	99.1	3.72 †	20	20
Al_2O_3 (7) +Ti_2O_3	as cast	96.5	3.94	1000	15

*   Sources: (1) Naval Research Lab., (2) American Lava Corp., (3) Interpace, (4) General Electric Lamp Division, (5) Coors Porcelain Co., (6) Union Carbide, Crystal Products, (7) Carborundum Co. (Courtesy Dr. R. C. LaBar).

** Conditions: h.p. = as hot pressed; anneal = annealed after hot pressing; sintered = as sintered by manufacturer.

†   Low densities attributed to inhomogeneously distributed small cavities. Test specimens selected from areas of maximum density.

least twice the thickness (e.g., 0.2 x 0.1 in.or 0.15 x 0.070 in.).
Grain sizes were taken as the average linear intercept lengths
measured on fracture surfaces.

RESULTS

Alumina Armor Data and Development of Small Target Testing

Figure 1a shows the trend in ballistic limit velocity versus
areal density for alumina-GRP targets tested with the 17 grain fragment simulating missile. A bimodal distribution of data is apparent.
All of the data for targets prepared from the commercial aluminas
which are normally considered for armor fall very close to either
one of the two lines. The separation of these two lines is in
excess of 400 fps near the right hand edge, representing a separation
of about fifteen percent in the ballistic limit velocity. A
statistical analysis showed that it was highly unlikely that the
data for the two lines were from the same population. The alumina
specimens were obtained from two different commercial sources,
tested in three different laboratories, and the purities ranged
from 85 to 99.3%. About half of the alumina specimens tested had
as-fired surfaces and the remainder had ground surfaces. All of the
alumina tiles were at least 4 x 4 in. with the majority about 6 x 6 in
surely an adequate size. The bimodal difference in performance is
an unexplained example of the variability mentioned earlier, and
was one of the factors stimulating the present study. Unfortunately,
samples of aluminas from various ballistic test lots represented
in Fig. 1a were no longer available; therefore supplemental information was sought from a general microstructural study and other special
tests. Since smaller size ceramic specimens would greatly facilitate
such work, disks 1.5 in. dia were cut from plates of two different

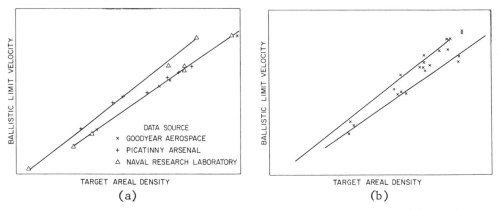

Fig. 1.   Ballistic performance of commercial aluminas; (a) earlier
          tests by various laboratories; (b) fragment simulator tests
          by NRL.

thicknesses of a commercial alumina previously tested as 4 x 4 and
6 x 6 in. plates. The total spread in these limit velocities for a
given ceramic thickness was less than 1.5%, i.e., about one tenth
the separation of the two lines shown in Fig. 1a. Additional
1.5 in. dia disks cut from the commercial plates were ground on
an iron lap with 220 grit SiC, and polished with fine alumina. This
polishing had no discernable effect upon ballistic limit.

Effect of Microstructure and other Strength Variables

Most of the targets used for the following evaluations consisted
of 1.5 in. dia disks of alumina bonded to 12 x 12 in. GRP panels
with a 0.020 in. thick layer of Proseal 890 (a polysulfide resin).
These data are shown in Fig. 1b, on which the lines from Fig. 1a
have been superposed, with points generally falling on or between
them. These ballistic evaluations are based upon limited numbers
of specimens (typically 4-5) of the hot pressed and fusion-cast
bodies. Testing of additional targets would undoubtedly result in
moderate revisions of the ballistic limit velocities; possibly with
an occasional significant change, particularly for highly non-
homogeneous samples. Modification of techniques and/or numbers
counted for grain size determinations would also cause some revisions.
However, the range covered is about 1000:1, so such experimental
refinements would be of little consequence for present purposes.

In order to more explicitly evaluate the effect of grain size,
data were normalized by dividing the ballistic limit by the areal
density of the target. Most of the ceramics used in this study were
of nearly the same areal density, thus this modest normalization
was considered appropriate, facilitating delineation of those factors
significantly influencing ballistic performance. In Fig. 2, normalized
ballistic data (BL/AD) are plotted against the inverse square root

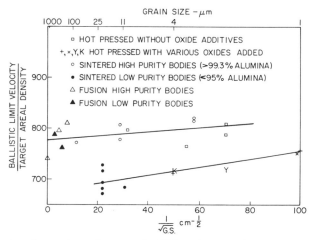

Fig. 2. Effect of grain size on ballistic performance.

of grain size, consistent with other recent studies of ceramics.[1]
Values of about 700 and 800 correspond approximately to the upper
and lower lines of Figs. 1a and 1b. The total spread in performance
of the commercial sintered aluminas was almost the same as shown
for all of the experimental materials used in this study.

    Two sets of values were obtained for hot pressed Linde A alumina
(1) as-hot pressed and (2) the same material after annealing.
Annealing increased the grain size by a factor of about five with
a corresponding decrease in flexural strength of ~40% yet there
was not a significant change in ballistic performance. The use of
2 w/o LiF in hot pressing Linde A powder did not result in a significant change in average grain size, but reduced flexural strength
by more than 50%. However, no change was noted in the ballistic
performance. Subsequent annealing of this material doubled both
the grain size and flexural strength, due to reductions in impurity
and additive content. The ballistic limit decreased by only ~6%.

    The single data point for Lucalox (Fig. 2) and some of the
data for AD-85 and 99.3% alumina used for Fig. 1 shows the ballistic
results to be near maximum for those commercial sintered products.
There are four data points for specimens diamond cut from fusion
cast bricks. Three of these fusion cast materials, having the
lowest flexural strength and the largest grain size of any of the
bodies, were among the best as judged by this ballistic test, even
though they were tested in as-cut surface condition. The fused
Al_2O_3 - 3% Ti_2O_3 mixture behaved similarly. Single crystal sapphire
slabs oriented for impact at angles of 0°, 30°, 60°, and 90° with
the C axis all showed the same level of performance, falling somewhat below that of the fused cast materials.

    Poor ballistic performance resulted from all of the oxide
additions to alumina, whether tested in as-hot pressed or annealed
conditions. Hot pressed, annealed specimens containing added TiO_2,
and as-hot pressed specimens containing both LiF and TiO_2 showed
the poorest performance, although their flexure strengths were higher
and the grain sizes smaller than for other bodies. These findings
suggest that specific impurities present in some commercial materials
may be responsible for their inferior ballistic performance. All
annealed samples containing oxide additives were tested with as-
annealed surfaces. For other alumina bodies, the ballistic limit
velocity was essentially independent of surface finishing as noted

    Impurities appeared to have a much greater effect on
ballistic performance than microstructure. Impurities are
present in commercial alumina bodies, and their distribution can
be affected by rates of cooling; therefore, specimens (AD-94
alumina) were heated in air to 1500°C for ~16 hrs, lowered as fast

as possible (~5-10 sec.) from a commercial bottom loading furnace
and moved rapidly from their lowered position (~3 ft below the
furnace). The specimens were 2 x 2 in. squares, supported by one
corner, inserted ~0.5 in. into a slotted fire brick ~2 in. apart,
with their faces at a substantial angle to one another in order to
maximize the rate of cooling. Fragment simulator tests showed
these fast-cooled bodies to have a ballistic limit velocity ~5%
higher than the as-fired commercial production lot from which the
quenching experiment specimens were taken.

DISCUSSION

The Effect of Impurities and Grain Size

The ballistic data show only moderate sensitivity to the various
material parameters, including microstructure and strength. The
most pronounced effect is the presence of certain impurities
(Fig. 2). However, similarities in performance between fused alumina
(a) with and (b) without titanium additives suggests that the state
and/or distribution of the impurity is also important. As indicated by
cooling   rate   experiments, quenching may change impurity
distributions.

In a more selective set of data (from those sintered ceramic
bodies made by commercial producers, and of ballistic tests performed
at NRL on them) a more definitive relationship may be considered
between purity and fragment armor performance (Fig. 3). As noted
previously, purity had the principal effect; further inspection of
purity groups suggested a limited decrease in ballistic performance
with increasing grain size. This trend is even clearer if one
limits the population to the hot pressed doped specimens and the
lower purity commercial aluminas. Decreasing ballistic performance

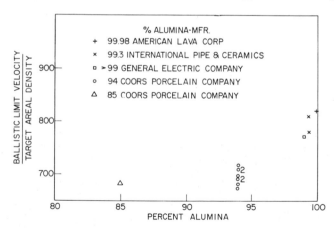

Fig. 3. Effect of purity on ballistic performance.

with increasing grain size is considered to be consistent with
(a) hardness increases with decreasing grain sizes[2] (b) probable
failure mechanisms and (c) possible correlations with compressive
strengths (discussed later).

The Mechanism of Failure

Since the forces generated in stopping a projectile are proportional to the change in momentum, the ordinate of Fig. 2 is
proportional to the stress. This means that the ballistic limit,
and hence the ballistic force, obey an equation of the form of:

$$F = F_o + KG^{-\frac{1}{2}}$$
(1)

where F_o and K are constants and G is the grain size. Clearly F_o
and K are dependent on parameters relating either to the system
(e.g., missile and backing) or material (e.g., hardness) or both.
While a final system/material separation cannot be made, the above
equation may indicate that microplastic processes are important
in ceramic armor. F_o is clearly the major factor in determining F.
Since F is significantly different for materials of different hardness or bodies of different porosity, material parameters are very
important. The system parameters were fixed in this study and since
material variables are important in F and hence F_o, the material
may be obeying the Petch equation:

$$\sigma_f = \sigma_o + K\ G^{-\frac{1}{2}}$$
(2)

where σ_f is related to a dynamic fracture stress, σ_o = the stress to
activate microplastic process, k = a constant, and G = the grain
size, and each of these, especially σ_f and σ_o, are factors in the
corresponding terms in Eq. 1.

This suggestion of microplastic behavior is consistent with
other results. Recently Palmour $et\ al$[3] have shown direct evidence
of microplasticity in ballistically damaged Al_2O_3. Further, Gilman[4]
has shown that hardnesses and Hugoniot elastic limits of hard
ceramics tested for armor purposes are directly correlated. Since
Rice[2] has shown that hardness is essentially a measure of microplastic
yielding of ceramics, Gilman's results also imply that microplastic
effects are important in ballistic behavior, influencing F_o in
particular.

Rice[2] has shown that the yield stress (Y = H/3) of ceramics is
the upper limit of static compressive strengths of ceramics, thus

suggesting a general correlation of compressive strength and ballistic strength and ballistic behavior. However, the correlation will often not be very good since a variety of factors change from static to shock wave conditions, where the stresses are respectively long range and short range (across the shock front). The short range of the ballistic stress at any one time would be expected to reduce the sensitivity to grain size since the shock front will be narrower than many grains. However, some effect is still expected, consistent with the trend in Fig. 2.

The effect of impurities (and therefore also of quenching) is consistent with a microplastic mechanism of ballistic behavior. Some inhibition of slip and twinning by additives or impurities could increase both hardness and dynamic yield stress. This may be the case in some of the fusion cast materials. However, too much hardening could reduce microplasticity to the point where little or no dynamic yielding occurs, which means less energy is absorbed, penetration probably proceeds faster, and more brittle fracture occurs (as appears to be the case here for doped and lower purity commercial bodies). Because of the high stresses involved, multiple shear modes (slip and twinning) can be expected, minimizing crystal orientation and surface finish effects.

If one proposes a microplastic mechanism of failure, then the question of why macroscopic ductility is not observed must be answered. Three reasons can be given: (1) deformation may not occur on enough different systems to produce a general plastic deformation without cracking; (2) even if enough slip systems are activated, they may not be able to interpenetrate to produce the homogeneity required for ductility; (3) very high stresses are required for extensive slip in these hard materials. Thus, microplasticity is probably confined to localized regions under the area of impact and (at any one time) around the shock front.

SUMMARY

Ballistic performance appears to follow a Petch type equation in which the grain size term has only a limited effect. Since tensile (or flexural) strength depends substantially on grain size in contrast to its limited role in ballistics, it is not surprising that the correlation between tensile strength and ballistic limit is poor. Impurities can have a substantially greater effect on ballistic performance, depending on the type and state of the additive. The influence of impurities is consistent with cooling rate effects, and hence may account for variability between different lots of materials. Different mechanical finishes and different single crystal orientations were found to have no effect on ballistic performance. All of these findings appear to be consistent with a microplastic process.

ACKNOWLEDGMENTS

The assistance of J. R. Spann in hot pressing the doped bodies and D. P. McGogney for ballistic limit testing is gratefully acknowledged.

Data from an earlier paper, "Comparison of Static Strength, Ballistic Behavior, and Microstructure of Al_2O_3," presented at the 72nd Annual Meeting of the American Ceramic Society in Philadelphia, Pa. [Abstract: Am. Ceram. Soc. Bull., 49 (4) 487 (1970)] is included here with the permission of the American Ceramic Society.

REFERENCES

1. R. W. Rice, "Strength-Grain Size Effects in Ceramics" submitted for publication in the Proceedings of the British Ceramic Society.
2. R. W. Rice, p. 195, this volume.
3. H. Palmour III, C. H. Kim, D. R. Johnson, C. E. Zimmer, "Fractographic and Thermal Analysis of Shocked Alumina", Tech. Rept. No. 69-5 for Office of Naval Research, Contract No. N00014-68-A-0187, NR 064-504, April 1969.
4. J. J. Gilman, J. Appl. Phys. 41 (4) 1664-66 (1970).

DISCUSSION

S. K. Dutta (Army Materials and Mechanics Research Center): Did you observe any grain size effect on ballistic performance of boron carbide? Also, did you notice any effect of grain orientation on ballistic properties of alumina, i.e., grains oriented perpendicular compared with parallel to the direction of projectiles?
Authors: We have not yet studied B_4C, only Al_2O_3. The Al_2O_3 specimen generally had equiaxial grains of nearly random orientation, thus we had no opportunity to observe grain orientation effects. However, no orientation effect of single crystals was observed indicating that there would be no effect due to a crystallographic orientation of grains. Since there is some effect of grain size, orientation of columnar grains could have some effect, but I would not expect this to be very large.

Hayne Palmour III (North Carolina State University): In recent work [C.H. Kim and H. Palmour III, Tech. Rept. 70-8, Contract N00014-68-A-0187, December, 1970] we have examined ion beam thinned foils from ballistically damaged AD-94 alumina, finding conclusive evidence of intensive but localized plasticity. Dislocations of $<11\bar{2}0>$ and $<0111>$ types are observed at densities $>1 \times 10^{11}/cm^2$ within the shocked bulk and at distances up to 20 μm inward from fracture surfaces. In our view, plasticity in shocked alumina is real at the microscopic level, hence it is gratifying to have your macroscopic ballistic findings arrive at the same conclusion.

FRACTOGRAPHY OF HIGH BORON CERAMICS SUBJECTED TO BALLISTIC LOADING

R. Nathan Katz and William A. Brantley*

Army Materials and Mechanics Research Center

Watertown, Massachusetts

ABSTRACT

Fractographic studies of B_4C and AlB_{12} ceramics subjected to ballistic impact are presented. Results suggest that macroscopic textures of fracture-exposed surfaces are indicative of stress states occurring during the fracture event, whereas microscopic topography is strongly influenced as well by microstructural features. Variation observed as a function of distance from the impact axis allows some differentiation between the strain rate sensitivity of the fracture processes of the two materials.

INTRODUCTION

Within recent years ceramics have been employed for lightweight armor applications by the United States armed forces, and considerable research has been conducted to elucidate the mechanics of penetration, fracture sequence, and the role of material properties during ballistic impact of ceramics.[1-6] Many material properties significant for ballistic response such as density, elastic modulus, hardness, and wave propagation velocity are essentially structure insensitive. Much less understood concerning ballistic response are effects of ceramic microstructures and concomitant structure-sensitive properties such as initiation stress, propagation stress, and propagation velocity for fractures. Because of limitations of ceramic processing technologies and prohibitive expense, even if technology were available, it would not be possible to evaluate

---

* Present address: Bell Telephone Laboratories, Inc. Murray Hill, New Jersey.

271

ballistically ceramic armor specimens in which single microstructural
features had been independently varied. Alternatively, considerable
insight may be obtained from microstructural and fractographic
studies of present ceramic armor materials after ballistic impact.
Microstructural alterations following shock or ballistic loading
of MgO[7] and Al_2O_3[8] have previously been described. Fractographic
studies of a variety[3-5,8] of ballistically tested ceramics have
provided considerable information about the fracture sequence
induced by ballistic impact and suggest that the fine scale of this
fracture may be strongly dependent on ceramic microstructure.

In this paper will be presented the principal results of our
fractographic analysis of ballistically loaded boron carbide (B_4C)
and aluminum dodecaboride (AlB_{12}). Boron carbide is currently
utilized for ceramic personnel armor[1] and AlB_{12} is a relevant
experimental material because of ballistic performance comparable
to that of B_4C.[3]

MATERIALS

The materials used in this investigation were hot pressed B_4C
and AlB_{12}. The B_4C has a rather complex polyphasic microstructure.[5]
Based on electron beam microprobe analysis, compositions of phases
appearing in Fig. 1a can be elucidated. The primary phase (phase 1)
is B_4C, and the average grain size of the matrix is approximately
10μm. Phase 2, typically appearing as a group of discrete particles

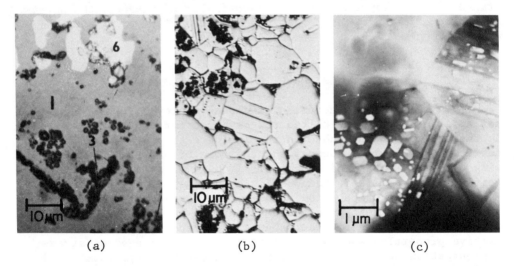

(a)   (b)   (c)

Fig. 1. Microstructure of hot pressed B_4C (a) phase identification
         (see text), polished, polarized reflected light; (b) after
         electrolytic etching in HC1; (c) microtwinning and
         associated porosity, transmission electron microscopy.[17]

Table 1. Nonballistic Properties of Hot Pressed High Boron Ceramics

Property	B$_4$C	AlB$_{12}$
Flexural strength (4-point bend, psix10^{-3})	47.8	31.1
Young's modulus (psix10^6)		
Static	66.5	57
Dynamic	64.7	–
Poisson's ratio (dynamic)	0.17	–
Density (measured, g/cm^3)	2.51	2.52
Apparent porosity (%)	< 0.1	$\sim$ 1-2
Microhardness (K$_{100}$)	2990-3200	2600

is rich in Al, Ca and Si and the whitish phase (6) is Ti-rich. Due
to the small size of the particles comprising phase 2, their compositions are indeterminate. Phase 3 is a graphitic phase with
varying composition. Twins produced during hot pressing and
probably associated with impurities[9] are shown in Fig. 1b. The
apparent porosity tends to be faceted in shape and nonuniformly distributed in the microstructure. These features are more clearly
displayed in a transmission electron micrograph of B$_4$C shown in
Fig. 1c. This material is characterized by the mechanical properties
presented in Table 1.

The hot pressed AlB$_{12}$ also possesses a complex polyphasic microstructure.[10] Figure 2a shows the AlB$_{12}$ matrix and the two principal
secondary phases; the dark gray phase has been identified as Al$_2$O$_3$

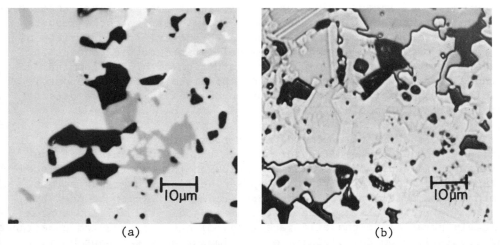

(a)   (b)

Fig. 2. Microstructure of AlB$_{12}$ (a) phase identification (see text),
polished, reflected light; (b) specimen showing twins,
electrolytically etched in HCl.

by use of electron beam microprobe analysis, and the white phase
(not present in all AlB_{12}) has been identified as Ta_2O_5 by the same
technique. The average grain size of the matrix is approximately
6µm. Twins are apparent in the etched microstructure shown in
Fig. 2b; as for B_4C, it is premature to speculate on the genesis of
these twins. The apparent porosity of the AlB_{12} (1 to 2%) is not
as low as in the boron carbide. The porosity is closed, nonfaceted,
uniformly distributed, and approximately the size of the grains;
some of the apparent porosity may be due to pullouts. The AlB_{12} used
in this study is characterized by the mechanical properties presented
in Table 1.

Composite armor specimens were fabricated from the materials
described above and subjected to ballistic impact by armor-piercing
(hardened steel) projectiles. The term composite armor conventionally
refers to the fact that the ceramic face plate is bonded with an
adhesive to a ductile back-up plate, in this case woven roving.
Fracture-exposed surfaces were examined and the fracture path related
to the stress conditions and/or microstructure, wherever possible.
The fracture characteristics of the ceramic only have been considered.

GROSS TOPOGRAPHY AND FRACTURE SEQUENCE

The typical fracture appearance of both B_4C and AlB_{12} composite
armor subsequent to impact with hardened steel projectiles is shown
in Fig. 3. The macroscopic appearances of these fracture surfaces
are indicative of both the axially symmetric and dynamic natures of
ballistic loading. From examination of macrophotographs such as
Fig. 3 many features of the ballistic fracture sequence can be
deduced. The fracture sequence and prominent topographical characteristics for these high boron compounds are similar to those reported
by Frechette and Cline[4] and by Long and Brantley[11] for a variety of
ceramics. Formation of the central ballistic fracture conoid has
been described previously.[1-4] Reconstruction of highly comminuted
fragments from conoid regions was not possible, and this report is
limited to a description of fracture processes in the area outside
of the fracture conoid.

After the fracture conoid, major radial cracks (MRC in Fig. 3)
having included angles of about 30° are the first elements of the
fracture patterns to form. This conclusion follows from the discon-
tinuity (D, Fig. 3) in fracture-exposed surface topography across
each major radial crack. Within each resulting sector bounded by
two major radial cracks, the fracture-exposed surface lying approximately parallel to the tile face (Fig. 3) has a corrugated appearance
particularly from the vicinity of the impact axis to about halfway
toward the tile periphery. It is plausible that within each sector
a fracture front initiated near the impact axis (at or within the

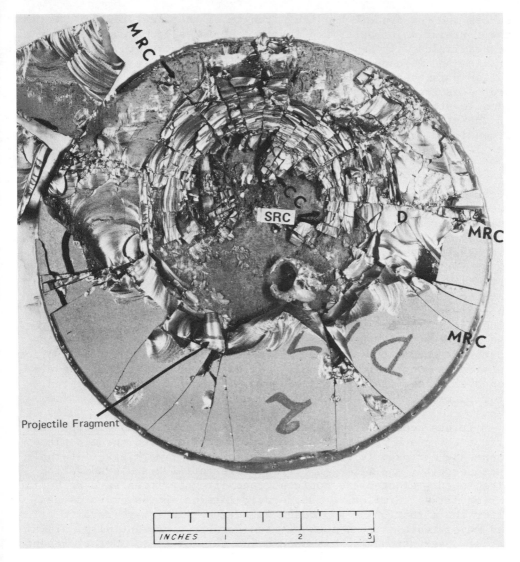

Fig. 3. Typical macroscopic fracture-exposed surfaces, composite
armor specimen impacted by hardened steel projectile.

the initial fracture conoid) sweeps radially outward. The corrugated
topography is attributed to modulation of the axial tensile stress
by radial and circumferential tensile stresses, with the corrugation
wave-length indicative of stress wave reverberation. In fact, within
each major sector there are a number of approximately parallel
fracture surfaces having such corrugated appearances, particularly
near the impact axis. In principle, it should be possible to relate
the distance between these fracture surfaces to the axial impact

stress and the dynamic tensile strength of these materials. However, the correlation of ballistic fracture surfaces from reverberating stress fields with impact stress and material strength is a formidable problem and is undoubtedly best attempted by computer studies; detailed parameters for the fracture process, presently unavailable, are necessary to obtain meaningful information. Indeed, the dynamic tensile strength should be a function of stress state and details of the stress pulse;[12] some aspects of the much simpler one-dimensional case are available elsewhere.[13]

Secondary radial cracks, (SRC, Fig. 3) having included angles of about 10° are also prominent within each major fracture section. Because of the continuity of corrugated surface topography across secondary radial cracks, these latter features occur after the corrugated surfaces are formed; it is presumed that the secondary radial cracks also are initiated near the impact axis. Less prominent are circumferential cracks (CC, Fig. 3) which occur subsequent to major radial cracks but typically precede secondary radial cracks. Late-time, spall-like fractures frequently are observed toward the periphery of the front face; a number of these features are apparent in Fig. 3, the fractures having been initiated at various points along radial cracks separating the large fragments most remote from the impact site. Other late-time fractures are conspicuous at the northwest quadrant of Fig. 3 (where mating fragments have been placed outside the sector). Generally, curved topographical features are concave in the direction of the initiation point for each region of fracture.

It has been observed elsewhere[11] that three gross factors can affect the crack pattern on a ballistically impacted ceramic plate. These are macroscopic surface irregularities or markings, internal macroporosity, and tile geometry. Preferential crack propagation often occurs along macroscopic surface irregularities such as surface scribe marks. Similarly, macroscopic internal pores or voids can alter the direction of a propagating crack. In square ceramic tiles the major radial cracks seek out the corners preferentially. This latter observation also has been borne out for AlB_{12} in the present investigation.

Based on the axial symmetry of the fracture patterns and the similarity of those patterns for widely differing ceramic materials, one can conclude[5] that the macroscopic fracture patterns and topography are principally determined by the local stress states existing during the fracture event rather than the microstructure of the given material.

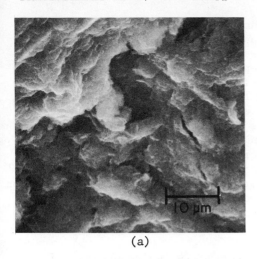
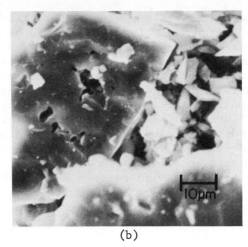

(a)   (b)

Fig. 4. Microtopography of impact fragments (a) B_4C, fragment A;
(b) AlB_{12}, fragment C.

MICROTOPOGRAPHY

In contrast to the gross fracture appearance described above, one expects that details of the fracture appearance on a fine scale will be influenced by such factors as grain boundaries, presence or absence of cleavage planes, porosity, second phases and impurities. To elucidate the role of such features, both scanning electron microscopy (SEM)[14] and replication electron microscopy (REM)[15] were utilized. Surfaces of fracture fragments from various distances and orientations with respect to the impact axis were examined.

From the shattered B_4C plate shown in Fig. 3, two fragments were selected for SEM examination: fragment A, located from ~0.5 - 1.0 in. outside the impact axis at the rear of the ceramic, and fragment B, located at ~1.0 - 1.5 in. from the impact axis and having the previously discussed corrugated fracture surface. From a similar ballistically impacted AlB_{12} plate, two samples were selected for SEM examination: fragment C, located ~0.5 in from the impact axis at the rear of the ceramic, and fragment D, located ~1.5 in. from the impact axis and also displaying a corrugated fracture-exposed surface.

The roughest microtopography is exhibited by surfaces nearest to the impact axis and concentric with it. Figure 4a illustrates the microtopography in such an area of fragment A (B_4C). The scale of features in this figure is less than the grain size and has not yet been related to any subgrain structural feature. Figure 4b shows the face of fragment C (AlB_{12}) also concentric to the impact axis.

Again extreme surface roughness is noted, but in this case the micro-topography consists of a rubble of angular shards. Consideration of size and porosity distributions as well as shapes of these shards leads to the conclusion that they represent a region of decohered grains (and therefore intergranular fractures). The smaller shards most probably are AlB_{12} grains; the larger ones, Al_2O_3. Thus, in contrast to B_4C the microtopography of AlB_{12} in this area can be related to the microstructure. This extreme surface roughness is indicative of substantial conversion from ballistic impact energy to fracture surface energy. The microstructure of AlB_{12} in this region shows evidence of deformation twinning, which is also indi-cative of a high rate of energy absorption.[10] Evidence of deformation twinning upon ballistic impact was not noted in B_4C, although such twinning may have occured within the central fracture conoid regions not examined in this investigation.

At farther distances from the impact axis the dependence of the fracture path of both materials on microstructural constituents is much more evident. Fragment B (B_4C) includes an area of the corrugated fracture parallel to the plate face; a portion of this area is shown in Fig. 5 at two magnifications. Corrugated topo-graphical features lying in approximately a horizontal direction are visible in Fig. 5a, along with parallel vertical features lying along the path of fracture propagation. At higher magnification, it is found that the fracture mode is predominantly intergranular though some evidence of transgranular fracture is observed. The influence of porosity and secondary phases on fracture behavior is observable in Fig. 5b as fracture fronts often are induced to inter-sect these defects. The macroscopic topography of the corrugated

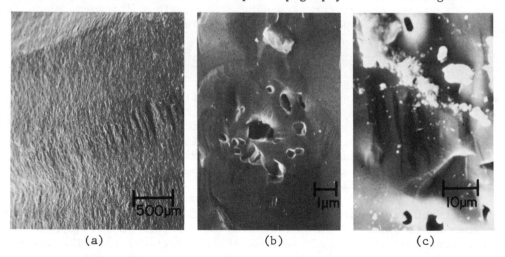

(a)	(b)	(c)

Fig. 5.   Wedge shaped impact fragments (a) B_4C, fragment B; (b) detail of (a) at higher magnification; (c) AlB_{12}, fragment D.

surface of sample D (AlB_{12}) is similar to that of the B_4C fragment
shown in Fig. 5a. The microtopography moreover, here apparently
remains corrugated to magnifications as high as 1000X (Fig. 5c).   A
unique form of this undulating fracture surface is the terraced
fracture pattern.   It is difficult to assess the predominant mode
of fracture in this case, but it appears to be mixed inter- and transgranular.   Again there is some evidence of the fracture front interacting with porosity.

A region from the lateral surface of fragment B, representative
of radial fracture-exposed surfaces, is illustrated in Fig. 6a.
Although present results are preliminary, the roughness of topo-
graphical features appears greater here than for Fig. 5.   Again
fracture is generally intergranular, but some indications of transgranular fracture (presumably river patterns) are visible in Fig. 6a.
The lateral surface of fragment D is shown in Fig. 6b.   This surface
is considerably smoother than the other AlB_{12} surface examined, but
as in Fig. 5c the fracture appears to be of mixed inter-intragranular
modes.   The central portions of Figs. 5c and 6b may provide evidence
of crystallographic fracture.

Further illustrations of the interaction of pores (or secondary
phases) and grain boundaries with the propagating fracture of the
corrugated surfaces have been obtained from replication electron
microscopy of B_4C and are presented in Fig. 7.   Figure 7a is from
a corrugated surface near impact axis, and Fig. 7b is from a similar
surface at the periphery of the tile.   In both fractographs, river
patterns characteristic of transgranular fracture are evident. The
local fracture propagation directions (arrows) Fig. 7 a are parallel
to the river patterns and in the direction of converging tributaries
for a given river marking.[14,16]   The transgranular fracture swept
approximately south to north, and changed to an intergranular mode
at the grain boundary.   Secondary crack initiation is observed at
the almost spherical second phase particle at the left of this
figure.   In Fig. 7b the cracks interact with the shallow features
(most likely pores, but possibly fractured second phase particles),
in several cases, either initiating or reinitiating at them.   No
REM work has been carried out on AlB_{12}.

The microtopographic studies of boron carbide indicate that
although some transgranular fracture occurs upon ballistic impact,
there are apparently no preferential fracture planes giving rise to
cleavage as is the case for alumina.[8]   This observation has also
been made for statically loaded boron carbide.[6]   One cannot be
certain whether the observed fracture behavior is intrinsic to B_4C,
or the result of impure material with possible variations in local
B:C stoichiometry.   The microtopographic studies of AlB_{12} indicate
that the fracture mode changes from extremely intergranular (to the
point of gross decohesion) near the impact axis to a mixed mode

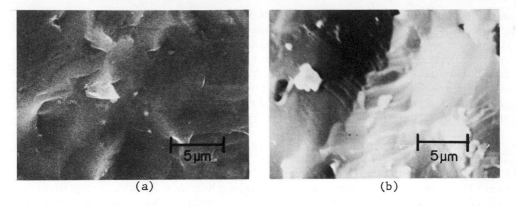

(a)   (b)

Fig. 6.   Lateral surfaces of fragments from Fig. 5 (a) B_4C, fragment
          B; (b) AlB_{12}, fragment D.

farther out from the axis.   On the basis of the above evidence, it
appears that the microscopic fracture processes in AlB_{12} may be
more strain rate sensitive than B_4C, which exhibits a less variable
microtopography.

     It must again be emphasized that due to our inability to re-
construct the conoid area, it is nearly impossible to adequately
study the nature of the fracture processes in the portion of the

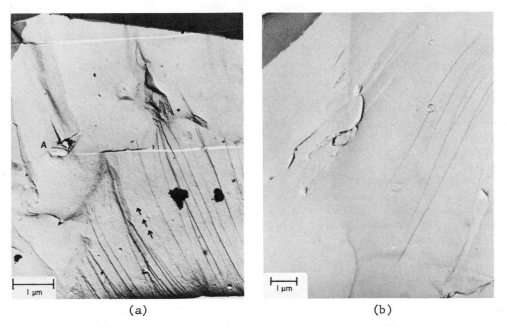

(a)   (b)

Fig. 7. B_4C top surface fragments (a) near impact axis; (b) near
          periphery of plate.

ceramic most significant to the defeat of the projectile. It has generally been observed [5,8] that the microtopography of ballistically impacted ceramics at distances ~1.5 in. from the impact axis is virtually identical with that from statically induced failures, and in this investigation changes in microtopography in ballistically impacted ceramics are evident at distances of ~0.5 in. from the impact axis. These facts strongly suggest that the unique aspects of ballistically induced fractures are to be found in the conoid.

SUMMARY

It has been shown that by utilizing macroscopic fractography it is possible to deduce much of the fracture sequence in ballistically impacted ceramic tiles. Through the use of scanning electron microscopy and replication electron microscopy, the microtopography of ballistically induced failure can be elucidated. The results of such examinations of ballistically impacted high boron ceramics suggest that *macroscopic* topography and crack propagation behavior is indicative of the stress states occurring during the ballistic impact, whereas *microscopic* topography and crack propagation behavior is strongly influenced by microstructural features. Examination of microtopography as a function of distance from the impact axis allows some differentiation between the strain rate sensitivities of the fracture processes of the two materials examined.

ACKNOWLEDGMENTS

The authors would like to thank G. Bruno of Advanced Metals Research Corp. for conducting the electron microprobe analysis and for his help with the SEM work. We would also like to acknowledge the valuable assistance on SEM studies of AlB₁₂ rendered by Prof. Jacques Sultan of M.I.T. Particular thanks must go to K. H. Abbott and F. Hannon of AMMRC, through whom the ballistically impacted plates were obtained.

REFERENCES

1. D. M. Martin, p. 5 in Ceramic - Armor Technology, Proceedings of Symposium. DCIC Report 69-1 Part I, 1969.
2. C. F. Cline and M. L. Wilkins, p. 13 ibid.
3. M. L. Wilkins, C. F. Cline, and C. A. Honodel, "Fourth Progress Report of Light Armor Program" Lawrence Radiation Lab. Report - UCRL 50694 June 4, 1969.
4. V. D. Frechette and C. F.Cline,"Fractography of Ballistically Tested Ceramics", Bull. Amer. Ceram. Soc. 49 (11) 994-997 (1970).

5.W. A. Brantley, "Microstructural and Fractographic Studies of Boron Carbide Sujected to Ballistic Impact and Static Flexural Loading" Army Materials and Mechanics Research Center, Watertown, Mass. AMMRC TR 70-18, August 1970.

6.E. G. Bodine, J. G. Dunleavy, and R. F. Rolsten; p. 21 in Ceramic-Armor Technology Proceedings of Symposium, DCIC Report 69-1 Part I, 1969.

7.M. J. Klein; p. 572 in Ceramic Microstructures. Ed. by R. M. Fulrath and J. A. Pask. John Wiley and Sons, New York, 1968.

8.H. Palmour III, C. H. Kim, D. R. Johnson, and C. E. Zimmer, Fractographic and Thermal Analyses of Shocked Alumina. North Carolina State University Technical Report 69-5 (AD 687415), Raleigh, N. C. 1969.

9.L. J. Beaudin (Norton Research Corp. [Canada] Ltd.) and C. Q. Weaverr (Norton Co., Worcester, Mass.) private communication.

10.R. N. Katz and J. McCauley "Microstructural and Mechanical Characterization of AlB_{12}" AMMRC TR in press.

11.W. D. Long and W. A. Brantley, Fractographic Analyses of Densified Ceramics and Glass Ceramics Ballistically Impacted by Caliber .30 M2 Projectiles, Army Materials and Mechanics Research Center, AMMRC TR 70-17, July 1970.

12.F. R. Tuler and B. M. Butcher, Int. J. of Fract. Mech. 4 p. 431 (1968).

13.S. Kumar and N. Davids, J. Frank. Inst., 263 (4) 295 (1957).

14.O. Johari, J. Metals, 20 26 (1968).

15.A. Phillips, V. Kerlins, and B. V. Whiteson, Electron Fractography Handbook. Douglas Aircraft Company, Inc., Contract No. AF33-(657) - 11127, Air Force Materials Lab., Wright-Patterson Air Force Base, Technical Report ML-TDR-64-416, January 1965. (AD 612912).

16.V. D. Frechette, "Characteristics of Fracture-Exposed Surfaces". Proceedings of the Brit. Ceram. Soc., no. 5, 97-106 (Dec. 1966).

17.R. N. Katz and A. E. King, "Transmission Microscopy of Ion-Bombardment Thinned Boron Carbide", Metallography, 4 87-89 (1971).

FACTORS INFLUENCING THE STRESS-STRAIN BEHAVIOR OF CERAMIC MATERIALS

Terence G. Langdon[*] and Joseph A. Pask

University of California

Berkeley, California

ABSTRACT

The stress-strain behavior of ceramic materials is greatly influenced by microstructural features ranging from the presence of point defects in single crystals to the size and location of pores and nature of grain boundaries in polycrystals. Several factors may affect the behavior at any one time, and the analysis of experimental data, particularly for polycrystals, is thus extremely difficult. This review examines the interpretation of mechanical behavior in materials having the rock salt structure, with particular emphasis on the role of impurities, the significance of grain boundary and/or intragranular porosity, and the problems associated with the intersection of slip bands.

INTRODUCTION

Although ceramic materials are of considerable potential interest, because of their ability to withstand high temperatures and severe corrosion environments, their use has so far been limited by brittleness and poor resistance to thermal shock. A large volume of work over the last decade has shown that several factors influence the stress-strain behavior observed in ceramic systems, ranging from the presence of point defects in single crystals to the size and location of pores in polycrystals. Furthermore, the complexity of this influence may be illustrated by noting that the behavior of nominally

---

[*] Now at University of British Columbia, Vancouver, B. C. Canada.

identical single crystals is markedly dependent on whether impurities
are present as isolated point defects, as aggregates, or in the form
of pairs of impurity atoms and charge-compensating vacancies.

This paper reviews the primary factors influencing the stress-
strain behavior of single crystals, and discusses some recent results
obtained on polycrystalline magnesium oxide. For simplicity, emphasis
is placed on systems having the rock salt structure. The effect of
changes in stoichiometry, important in materials such as UO_2 and
ThO_2, is not included.

FACTORS INFLUENCING THE BEHAVIOR OF SINGLE CRYSTALS

Under a constant rate of strain or loading, a single crystal
tested in tension or compression deforms elastically up to the
critical resolved shear stress (CRSS) at which plastic flow begins.
An analysis of the temperature dependence of the CRSS (or the yield
stress) is most conveniently carried out by considering the schematic
curves in Fig. 1, in which the behavior for a given strain rate, $\dot{\varepsilon}_1$,
is divided into three distinct regions. At low temperatures (region I)
typically $\lesssim 0.2-0.3\ T_m$ where T_m is the absolute melting temperature,
the CRSS decreases rapidly with increasing temperature, and the
behavior is controlled by thermal fluctuations which permit the dislocations to overcome short-range obstacles in the glide plane, e.g.,
solute atoms or the Peierls barrier. In region II, the CRSS drops
only slightly with increasing temperature, and the controlling
mechanism is then athermal (e.g., overcoming of long-range stress

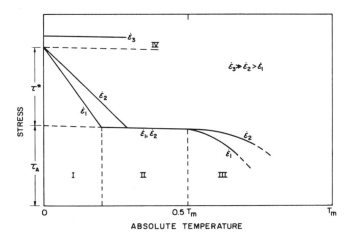

Fig. 1   Temperature dependence of the CRSS up to the absolute melting
temperature, T_m, for strain rates of $\dot{\varepsilon}_1$, $\dot{\varepsilon}_2$ and $\dot{\varepsilon}_3$. Region I
is thermally-activated, region II is athermal, region III is
diffusion controlled, and region IV represents viscous damping.

fields); the slight temperature dependence in this region arises
from the decrease in shear modulus with increasing temperature. At
high temperatures (region III), typically $\gtrsim 0.5\ T_m$, diffusion plays
a significant role, and the CRSS again drops. These three regions
are not delineated by fixed temperature boundaries; a faster strain
rate, such as $\dot{\varepsilon}_2$ ($>\dot{\varepsilon}_1$), moves the transitions to higher temperatures.
Finally, region IV corresponds to very rapid strain rates ($\dot{\varepsilon}_3 \gg \dot{\varepsilon}_2$),
as in impact tests, where the motion of dislocations is limited by
viscous damping.

Factors influencing the behavior in regions I and II are now
examined; the deformation mechanisms of importance in region III,
corresponding essentially to the "high-temperature creep" range,
are not considered here, but are dealt with in detail in the following
paper.[1]

Effect of Surface Condition

The surface condition of ionic crystals is important in determining the shape of the stress-strain curve; a detailed review of
the extensive work on magnesium oxide is given elsewhere.[2] Stokes[3]
showed that MgO crystals containing "fresh" dislocation sources,
due to cleavage or mechanical contact, yielded smoothly in tension
at ~6-8000 psi. However, if the "fresh" sources were eliminated
by chemical polishing, the crystals deformed elastically up to
stresses of ~30-50,000 psi before yielding with a sharp drop in
stress down to the level at which the "fresh" dislocations moved.
These experiments emphasize the significance of the surface condition,
and serve to show that the "grown in" dislocations in MgO are immobile.

Effect of Strain Rate

As indicated in Fig. 1, an increase in strain rate at any given
temperature results in a higher CRSS in the thermally-activated
region I, but there is no change in region II. An experimental
example is shown in Fig. 2 for NaCl crystals having a <100> axis
tested at two different strain rates;[4] the CRSS, τ_c, is divided by
the shear modulus, G, so that the results in region II are independent
of temperature. At temperatures less than ~250°K, the experimental
points for the two strain rates are different, and, within the error
bars indicated, the results are reasonably approximated by the two
straight lines converging to τ_c/G ~7.8 x 10^5 at T = 0°K.

Although Fig. 2 refers to only a modest increase in $\dot{\varepsilon}$, tests
have also been carried out on MgO crystals at very rapid strain rates.[5]
Results are given in Fig. 3 for values of $\dot{\varepsilon}$ ranging from 10^{-4} sec^{-1}
(which is comparable to the usual stress-strain data) to 3 x 10^3 sec^{-1},
thereby showing the transition to region IV indicated in Fig. 1.

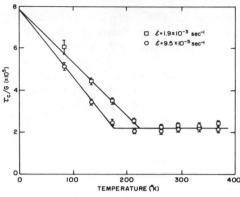

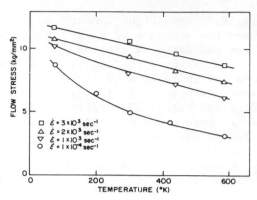

Fig. 2. Normalized CRSS (τ_c/G) as
a function of temperature
for NaCl single crystals
tested at two different
strain rates.[4]

Fig. 3. Flow stress as a function
of temperature for MgO
single crystals tested at
strain rates differing by
over seven orders of
magnitude.[5]

Effect of Orientation

The orientation of a crystal determines the magnitude of the
resolved shear stress acting on the slip planes. For crystals of
the rock salt structure, slip takes place preferentially on the
{110} <1$\bar{1}$0> slip systems, although flow is also possible on the
{001} <1$\bar{1}$0> systems at higher temperatures or stresses. For a
crystal tested in compression with a <100> longitudinal axis, which
represents the usual cleaved condition, there is a resolved shear
stress equal to one half of the applied stress acting on four of
the six {110} <1$\bar{1}$0> slip systems; the other two systems experience
no stress. For a <111> loading axis, however, there is no resolved
shear stress acting on any of the {110} <1$\bar{1}$0> slip systems, so that
crystals then deform at the higher temperatures by slip on the
{001} <1$\bar{1}$0> slip systems; at low temperatures the crystals fracture
without any plastic deformation. The importance of this orientation
difference is illustrated in Fig. 4 for MgO crystals tested at a
constant rate of loading of 20 psi/sec;[6-8] for the <100> axis there
is only a slight decrease for temperatures above ~1000°C, but the
values obtained for crystals with a <111> axis are substantially
higher (143,400 psi at 360°C; 71,000 psi at 560°C, not plotted) and
are still decreasing sharply even at the highest indicated test
temperature.

Effect of Impurity Concentration

Impurities play a major role in determining stress-strain
behavior, even when the concentration is relatively small. For
example, Gorum *et al.*[9] showed that the yield stress for an MgO
crystal containing 30 ppm of iron was about four times higher than
for a crystal containing only 10 ppm. By contrast, a further

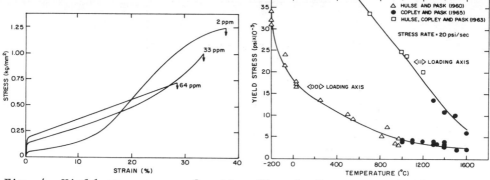

Fig. 4. Yield stress as a function   Fig. 5. Stress-strain curves at
        of temperature for MgO   294°K for NaCl single
        single crystals having   crystals having Ca^{2+} con-
        <100> and <111> loading   centrations of 2, 33 and
        axes.[6-8]   64 ppm, respectively.[11]

increase in iron content to 3000 ppm increased the yield stress only
slightly.

The CRSS in alkali halide crystals is extremely sensitive to
the presence of divalent cations. Using NaCl crystals, Hesse[10]
showed that an increase in Ca^{2+} concentration from 2 to 20 ppm
increased the athermal plateau (region II) and, because of the larger
number of obstacles, introduced a greater temperature-sensitivity
in the thermally-activated region I.   There is also a marked difference
in the shape of the stress-strain curves for different impurity
concentrations, as shown in Fig. 5 for NaCl crystals tested at 294°K
with Ca^{2+} concentrations of 2, 33 and 64 ppm, respectively.[11]   In
particular, the slope and length of gradual workhardening (stage I)
increases with increasing impurity content.

Effect of Nature of Impurities

Whilst small variations in the concentration of impurities
significantly affect the stress-strain behavior, the magnitude of
this influence is dependent on both the state of dispersion and the
valency of the impurities in question.   This is illustrated in Fig. 6.
For "pure" LiF crystals, containing 3 ppm Mg^{2+}, the CRSS increases
only slightly at very low temperatures, and there is no significant
difference between crystals slowly cooled (~0.002°C/min; solid line)
and air-cooled (~50°C/min; dashed line) from an annealing treatment
at 300°C. By contrast, the behavior of "impure" crystals, containing
75 ppm Mg^{2+}, depends critically on thermal history: for air-cooled
crystals, the athermal stress level (region II) is only slightly
higher than for the "pure" crystals, but there is a pronounced increase in CRSS in the thermally-activated region I; for slowly
cooled crystals, the CRSS in region II is very much increased, but

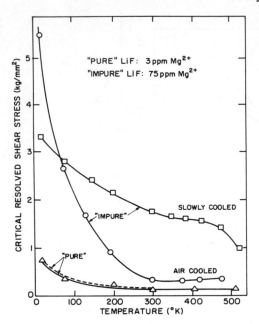

Fig. 6.   Critical resolved shear stress as a function of temperature
          for "pure" and "impure" LiF single crystals showing effect
          of thermal history.[12]

the temperature dependence in region I is less pronounced.[12]   This
suggests that in the slowly cooled crystals the impurities are primarily present in clusters or precipitates. However, further complexity may arise, e.g., these results for Mg^{2+} in LiF show a
quench softening, so that the CRSS at room temperature is less for
air-cooled crystals than for slow-cooled, whereas NaCl crystals
doped with large concentrations of Mn^{2+} and Cd^{2+} show a *quench
hardening*.[13]

   In MgO, the presence of Fe^{3+} leads to considerable hardening,
but Ni^{2+} in the absence of iron has only a small effect. This is
shown in Fig. 7, using room temperature data reported by Srinivasan
and Stoebe[14] and Moon and Pratt[15] (note that the point for 130 ppm
Fe^{3+} was obtained by extrapolation of the published data[15]). This
plot suggests that, over the limited range examined, τ_c is approximately linearly proportional to c, where c is the defect concentration,
but results by Davidge[16] indicate τ_c is proportional to $c^{1/2}$ over a
range of 100-1400 ppm of Fe^{3+} . The differing hardening effects
of Ni^{2+} and Fe^{3+} in MgO are thus analogous to the effects of mono-
and divalent impurities in the alkali halides (e.g., compare Fig. 7
with Fig. 23 of the review by Gilman[17]). The significance of valency
is also shown by noting that the CRSS for MgO crystals heat treated
in air to give ~6 ± 1 ppm of iron in the Fe^{3+} state was 1.1 ± 0.07
kg/mm^2 (1560 ± 100 psi), whereas identical crystals given a reducing

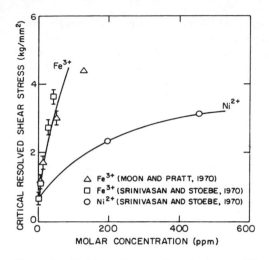

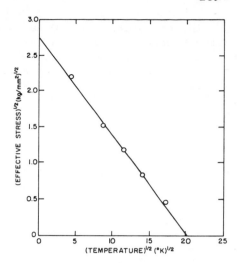

Fig. 7. Critical resolved shear
stress for MgO single
crystals as a function
of the molar concentra-
tion of Fe^{3+} and Ni^{2+}
(tested at room
temperature).

Fig. 8. Temperature variation of the
effective stress for LiF
single crystals containing
75 ppm of magnesium in solution (experimental data of
Johnston[12]).[18]

treatment at 2000°C, so that the iron was present as Fe^{2+}, gave a
CRSS of only 0.73 $\pm$ 0.07 kg/mm^2 (1040 $\pm$ 100 psi).[14]

It may be initially anticipated that plastic flow in ionic
crystals at the low temperatures is governed by a form of Peierls
mechanism, because of the necessity for the dislocations to overcome
Coulombic interactions. However, the experimental data presented
in Fig. 6 show that the behavior is dependent on the presence of
impurities. To explain the large temperature sensitivity of the
air-cooled "impure" LiF crystals, a theory was developed in which
the hardening was attributed to the tetragonal lattice distortions
produced by an impurity (Mg^{2+})-vacancy dipole.[18] This theory predicts
a linear relationship between $\tau^{*\frac{1}{2}}$ and $T^{\frac{1}{2}}$, where τ^* is the effective
stress (equal to the measured CRSS minus the athermal stress, τ_A;
see Fig. 1) and T is the absolute temperature, in agreement with the
data shown in Fig. 8. The theory also predicts a linearity between
τ^* and $c^{\frac{1}{2}}$; this agrees with experiments at low temperatures but
some results at higher temperatures (e.g., in NaCl at room temperature
for Ca^{2+} concentrations of up to 900 mole ppm[13]) show τ^* is linearly
proportional to c. These latter results have been interpreted by
assuming that the impurity-vacancy dipoles are then able to rotate
in the strain field of the dislocations, thereby lowering the elastic
energy of the system.[19]

The situation for MgO is somewhat different since, although one cation vacancy is required for charge compensation for every two Fe^{3+} ions incorporated in the lattice, electron spin resonance indicates that these are not contained in the form of ion-vacancy pairs;[16] in this case the hardening may arise because the excess vacancies, or the presence of Fe^{3+} ions on interstitial sites, increase the rate of jog formation on the screw dislocations.

Effect of Pressure

Experiments show that the CRSS is independent of pressure for NaCl crystals of <100> orientation up to 10 kbars[20] and MgO crystals of <111> orientation up to 5 kbars,[21] respectively. But it should be noted that the situation is more complex for polycrystalline materials, and in MgO there is a brittle-ductile transition with increasing pressure, so that large strains ($\gtrsim$ 10%) may be achieved at room temperature without fracture.[21]

Effect of Irradiation

Irradiation produces a hardening in alkali halides, the magnitude of which depends on the time, temperature, intensity, and nature of the radiation, and the impurity level of the crystal. The theory of asymmetrical defects[18] has been used to interpret the magnitude and temperature dependence of radiation hardening in Ag-doped KCl;[22] but the theory predicts that defect alignment will lead to anisotropic hardening, and this is not observed experimentally.[23]

FACTORS INFLUENCING THE BEHAVIOR OF POLYCRYSTALS

Role of Grain Boundaries

Whilst the preceding section shows that several factors significantly influence the stress-strain behavior of single crystals, the situation for polycrystalline materials is more complex. In particular four additional features should be considered.

Impurity Segregation. Impurities may further influence the deformation behavior in polycrystals due to preferential segregation at the grain boundaries, either as a second phase or in solid solution. For example, it has been shown that some common impurities in polycrystalline MgO, such as Al, Ca and Si, segregate to the boundaries even when only present in amounts as small as 30 atomic ppm, although other impurities, such as Fe, tend to be more uniformly distributed.[24] More complex ceramic systems may have a silicate phase at the boundary.

Residual Porosity. The density of polycrystalline ceramics is invariably less than the theoretical value for single crystals. As

discussed in the following section, the influence of this residual
porosity depends on such features as the primary pore location
(i.e., whether along the grain boundaries or within the grains), the
size and distribution of the intragranular pores, and the average
size of the boundary pores with respect to the grain size.

Shear Requirements for Generalized Deformation. A polycrystalline
material is only capable of deforming plastically, without the
nucleation of internal voids, when it possesses five independent slip
systems. This requirement is fulfilled in the rock salt structure
when slip occurs on both the {110} <1$\bar{1}$0> and {001} <1$\bar{1}$0> slip systems;
in MgO, this necessitates either high temperatures or the ability
for large stress concentrations to build up at the boundaries and
thereby nucleate slip on the {100} planes.

Interpenetrating Slip. It is necessary that these various slip
systems are capable of interpenetration. A detailed investigation
of MgO single crystals, tested in tension at an initial strain rate
of ~6.7 x 10^{-4} sec^{-1}, showed that the 90° intersections due to slip
on conjugate planes (which form neutral —i.e., uncharged- jogs) are
only possible at temperatures of 1300°C and above, and the 60° and
120° intersections due to slip on oblique planes (which form electro-
statically charged jogs) only occur above 1700°C;[25] although these
temperatures may be lowered by a reduction in strain rate. In con-
formity with these results, tensile tests on recrystallized polycrystalline MgO show a brittle-ductile transition at ~1700°C.[26]
In compression tests, however, catastrophic failure is more difficult,
and large stresses can then build up to aid the nucleation and interpenetration of the different slip systems at lower temperatures; e.g.,
the 90° intersections become possible at temperatures as low as 1200°C.

Stress-Strain Behavior of Polycrystalline MgO

The considerable significance of small changes in microstructural
detail is demonstrated by results obtained from compression tests on
six different types of polycrystalline MgO.[27] All of these materials
deformed plastically at 1200°C and above, but two types (5 and 6)
exhibited plastic flow at temperatures as low as 800°C and a third
(type 4) at 1000°C.

Examples of the microstructures of types 1-4 are shown in Fig. 9.
Type 1 was produced by hot pressing MgO powder with a 3 w/o LiF
additive, giving a transparent material which was nominally fully
dense but with a residual Li content of 75 ppm. Although electron
microscopy showed that no second phase existed at the grain boundaries,[28]
the occurrence of much intergranular fracture at low temperatures in
comparison with specimens without an additive suggested that LiF was
probably preferentially segregated at the boundaries in solid solution.
Types 2, 3 and 4 were produced by isostatic pressing and sintering,

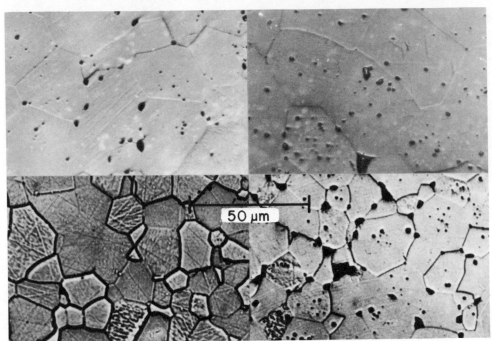

Fig. 9. Microstructures of four types of polycrystalline MgO: types 1 (lower left), 2 (lower right), 3 (upper left) and 4 (upper right).

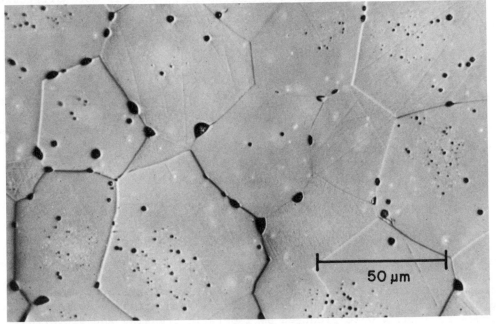

Fig. 10.   Microstructure of type 5 polycrystalline MgO.

and were typically ∿1.5% porous but with variations in the average
pore size and distribution. The microstructure of type 5, which
was one of the two materials deforming plastically at the lower
temperatures, is shown in greater detail in Fig. 10. This material
was also ∿1.5% porous, and was obtained by annealing the type 6
material at 1800°C. The pore-free regions adjacent to many of the
grain boundaries probably represent the areas swept out during grain
growth from the initial average grain size of 25 μm to the final size
of 80 μm.

Although the microstructures of these materials appear fairly
similar, the stress-strain curves reveal significant differences;
this may be seen by comparing the results at 800°C and 1000°C, as
shown in Figs. 11 and 12, respectively. At 800°C, types 1-4 were
not ductile, but types 5 and 6 deformed to strains >0.02. Type 1
fractured at ∿40,000 psi, represented by the vertical arrow, but
the tests on types 2-4 were discontinued prior to fracture at
∿45,000 psi. At 1000°C, types 1-3 fractured with little significant
plastic deformation at stresses in the range ∿29-45,000 psi, but
types 4-6 deformed to strains >0.02. All specimens were ductile to
strains in excess of 0.02 at 1200°C.

Figure 13 shows the yield stresses for the six types calculated
from a strain offset of 5×10^{-4} as a function of temperature, together
with the data described earlier for single crystals having <100> and
<111> loading axes; the closed symbol for type 1 at 1000°C represents
a fracture stress when no yielding occurred. The values recorded
for types 5 and 6 are lower than for the <111> loaded single crystals,
and this at first suggests that no slip occurred on the {001} <1$\bar{1}$0>
slip systems. However, wavy slip lines resulting from slip on both
the {110} and {00} systems were visible on the surfaces of these
specimens after testing at temperatures as low as 800°C, although no

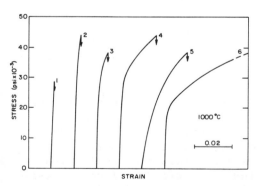

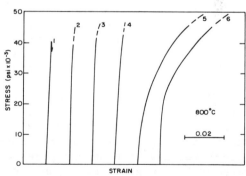

Fig. 11. Stress-strain curves for
         six types of polycrystal-
         line MgO at 800°C.

Fig. 12. Stress-strain curves for
         six types of polycrystal-
         line MgO at 1000°C.

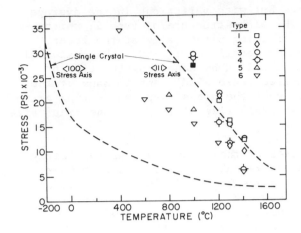
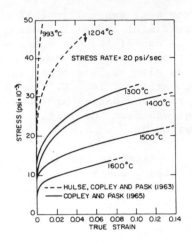

Fig. 13. Yield stress as a function of temperature for <100> and <111> MgO single crystals and polycrystalline specimens, types 1-6. Type 1 (closed symbol, 1000°C) fractured without yielding.

Fig. 14. Stress-strain curves for MgO single crystals having a <111> loading axis.[7][8]

wavy slip was observed in types 1-3 at temperatures below 1200°C. This indicates that stress levels of sufficient magnitude were able to build up at the grain boundaries in types 5 and 6 to realize flow on the secondary slip systems.

The marked differences in behavior for materials of similar density (∿98.5%) show that the microstructural features influencing the stress-strain relations in polycrystals are fairly complex, particularly since the realization of general ductility at 1200°C corresponds with the temperature at which such behavior is obtained for single crystals with <100> and <111> orientations,[7,8] as indicated for the latter orientation in Fig. 14. An analysis of the experimental observations in terms of the material characteristics of the various specimen types, described in detail elsewhere,[27] suggests two factors which aid plasticity at temperatures below ∿1200°C.

Boundary Strength. The grain boundaries should be sufficiently strong to allow the build-up of stress concentrations and nucleation of slip on the {100} system and to permit extension of slip bands across the boundaries. This requires (a) that they are relatively pore-free, and if boundary pores exist they should be small with respect to the grain size, and (b) that they are free from excessive amounts of impurities in solid solution. Condition (a) is not fulfilled in type 2, where the ratio of boundary pores to grain size is ∿0.2; condition (b) is not fulfilled in type 1 due to the residual LiF which appears to weaken the grain boundaries, or in type 3 due to a higher SiO_2 and CaO content which appear to interfere with dislocation motion.

Intragranular Porosity. Very fine pores distributed within the grains appear to be beneficial, probably because they permit some mass accommodation required due to the limited slip activity. Types 5 and 6 are examples of fairly pure materials containing fine intragranular porosity.

SUMMARY AND CONCLUSIONS

The factors influencing the stress–strain behavior of single crystals are briefly reviewed. It is shown that the results obtained under a given set of experimental conditions depend on features ranging from crystal orientation, which is readily determined, to less controllable parameters such as the presence, and in particular the state of dispersion, of certain impurities. Further complexities arise in polycrystals because of impurity segregation at the boundaries, the presence and nature of residual porosity, and problems associated with the availability, and interpenetration, of the various slip systems.

An examination of experimental results obtained on several different types of polycrystalline MgO suggests that ductility is aided at temperatures less than $\sim 1200°C$ by the presence of strong grain boundaries and by very fine intragranular porosity.

ACKNOWLEDGMENT

This work was done under the auspices of the United States Atomic Energy Commission.

REFERENCES

1. T. G. Langdon, D. R. Cropper and J. A. Pask, p.297 this volume.
2. T. G. Langdon and J. A. Pask; p. 53–127 in High Temperature Oxides– Volume III. Ed. by A. Alper. Academic Press, New York, 1970.
3. R. J. Stokes, Trans. AIME, 224 (12) 1227–37 (1962).
4. J. Hesse, Phys. Stat. Sol. 9 (1) 209–30 (1965).
5. A. Kumar, p. 67–96 in High Speed Testing – Volume VII. Interscience, New York, 1969.
6. C. O. Hulse and J. A. Pask, J. Am. Ceram. Soc., 43 (7) 373–78 (1960).
7. C. O. Hulse, S. M. Copley and J. A. Pask, J. Am. Ceram. Soc., 46 (7) 317–23 (1963).

8. S. M. Copley and J. A. Pask, J. Am. Ceram. Soc., 48 (3) 139-46 (1965).

9. A. E. Gorum, W. J. Luhman and J. A. Pask, J. Am. Ceram. Soc., 43 (5) 241-45 (1960).

10. J. Hesse, p. 413-20 in Reinststoffprobleme - Volume III. Ed.by E. Rexer. Akademie-Verlag, Berlin, 1967.

11. J. Hesse, Phys. Stat. Sol., 21 (2) 495-505 (1967).

12. W. G. Johnston, J. Appl. Phys., 33 (6) 2050-58 (1962).

13. P. L. Pratt, R. P. Harrison and C. W. A. Newey, Discs. Faraday Soc., 38 211-17 (1964).

14. M. Srinivasan and T. G. Stoebe J. Appl. Phys., 41 (9) 3726-30 (1970).

15. R. L. Moon and P. L. Pratt, Proc. Brit. Ceram. Soc., 15 203-14 (1970)

16. R. W. Davidge, J. Mater. Sci., 2 (4) 339-46 (1967).

17. J. J. Gilman, Progress in Ceramic Science, 1 146-99 (1961).

18. R. L. Fleischer, J. Appl. Phys., 33 (12) 3504-08 (1962).

19. P. L. Pratt, R. Chang and C. W. A. Newey, Appl. Phys. Letters, 3 (5) 83-85 (1963).

20. E. Aladag, L. A. Davis and R. B. Gordon, Phil. Mag. 21 (171) 469-78 (1970).

21. M. S. Paterson and C. W. Weaver, J. Am. Ceram. Soc. 53 (8) 463-71 (1970).

22. J. S. Nadeau, J. Appl. Phys. 37 (4) 1602-08 (1966).

23. J. S. Nadeau, p. 149-68 in Proc. Conf. on Nuclear Applications of Nonfissionable Ceramics. A. Boltax and J. H. Handwerk, Eds. American Nuclear Society, Hinsdale, Illinois, 1966.

24. M. H. Leipold, J. Am. Ceram. Soc., 49 (9) 498-502 (1966).

25. R. B. Day and R. J. Stokes, J. Am. Ceram. Soc., 47 (10) 493-503 (1964).

26. R. B. Day and R. J. Stokes, J. Am. Ceram. Soc. 49 (7) 345-54 (1966)

27. T. G. Langdon and J. A. Pask, "Effect of Microstructure on Deformation of Polycrystalline Magnesium Oxide", J. Am. Ceram. Soc., (in press).

28. T. G. Langdon and J. A. Pask, p. 594-602 in Ceramic Microstructures Ed. by R. M. Fulrath and J. A. Pask, John Wiley and Sons, New York, 1968.

CREEP MECHANISMS IN CERAMIC MATERIALS AT ELEVATED TEMPERATURES

Terence G. Langdon,[*] Donald R. Cropper and Joseph A. Pask

University of California

Berkeley, California.

ABSTRACT

The creep of ceramic materials at elevated temperatures may take place by the movement of dislocations within the lattice, by grain boundary sliding, and/or by stress-directed diffusion either through the lattice or along the grain boundaries. Other accommodation mechanisms, such as grain boundary separations, may also occur. Some indication of the significant creep mechanism may be obtained by determining the dependence of steady-state creep rate on stress, grain size, and temperature. A comparison is made between the predictions arising from the theoretical models and recent experimental data obtained on several materials in both single crystal and polycrystalline forms.

INTRODUCTION

An attraction of ceramic materials is the possibility of their use as structural components at high temperatures, but some knowledge is then required of the plastic deformation, or creep, which occurs when the materials are subjected to a constant load or stress. Creep tests in the laboratory are primarily concerned with a determination of the rate-controlling mechanism under a given set of experimental conditions; but it is also necessary to know whether these results may be extrapolated to predict the behavior of the material under the much slower creep rates usually required in practice. Although it is not normally possible at the present time to meaningfully

---

[*] Now at the University of British Columbia, Vancouver, B. C., Canada.

extrapolate existing data over wide ranges of stress or temperature, recent creep studies on ceramics have added substantially to our knowledge of creep mechanisms and are providing a foundation on which further work may be based.

This report is divided into two parts: (1) possible deformation mechanisms occurring during high temperature steady-state creep are reviewed and (2) these mechanisms are examined with particular reference to the experimental results obtained from the research program on the compressive creep of ceramic materials conducted at this University.

THE MECHANISMS OF CREEP

In reviewing the mechanisms operating during high temperature creep, it is convenient to make a division according to whether they are controlled by the movement of point or line defects. Furthermore, a direct comparison is made easier by using one basic creep equation for all mechanism types.

Creep is a thermally-activated process, and the secondary or steady-state creep rate, $\dot\varepsilon$, may be generally formalized under a given set of experimental conditions by an equation of the form

$$\dot\varepsilon = \frac{A}{kT} \left(\frac{\ell}{d}\right)^m \frac{\sigma^n}{G^{n-1}} D \tag{1}$$

where A is a constant independent of temperature and having dimensions of length, k is Boltzmann's constant, T is the absolute temperature, ℓ is a length, d is the average grain size, σ is the applied stress, G is the shear modulus, D is the coefficient of diffusion ($=D_0 \exp(-\Delta H/kT)$, where D_0 is a frequency factor and ΔH is the activation energy for creep), and m ($= -(\partial\ln\dot\varepsilon/\partial\ln d)_{\sigma,T}$) and n ($= (\partial\ln\dot\varepsilon/\partial\ln\sigma)_{d,T}$) are constants. The significance of the length ℓ depends on the particular model under consideration, but it may be equated with $\Omega^{1/3}$, where Ω is the atomic volume and is equivalent to ~$0.7b^3$ where b is the Burgers vector.

In general, each creep mechanism predicts a specific value for the constants m and n, but the constant A is usually not too well defined and depends on such unknown factors as the height of grain boundary ledges or the density of mobile dislocations. The mechanisms are therefore compared in Table 1 by presenting the predicted values for m and n and also noting the relevant value for D, where D_ℓ is the coefficient for lattice self-diffusion, D_{gb} is the coefficient for grain boundary diffusion, D_p is the coefficient for pipe diffusion along the dislocation cores, and D^* is a diffusion coefficient related to the second phase rather than to the host lattice.

Table 1.   Creep Mechanisms at Elevated Temperatures

Mechanism (Point Defect Mechanisms)	m	n	D	Ref.
(I) Vacancy diffusion through the lattice	2	1	D_ℓ	1,2
(II) Vacancy diffusion along the grain boundaries	3	1	D_{gb}	3
Vacancy diffusion around grain boundary ledges:				
(III) (a) through the lattice	1	1	D_ℓ	4
(IV) (b) along the grain boundaries	1	1	D_{gb}	4
(V) Grain rearrangement by viscous flow due to second phase at boundaries	1	1	$D*$	–
Mechanism (Line Defect Mechanisms)				
(VI) Dislocation glide/climb controlled by climb	0	4.5	D_ℓ	5
(VII) Dislocation glide/climb controlled by glide	0	3	D_ℓ	6
(VIII) Dissolution of dislocation dipoles	0	5	D_p	7
(IX) Dissolution of dislocation loops	0	4	D_ℓ	7
Dislocation climb without glide:				
(X) (a) dislocations are sources and sinks for vacancies	0	3	D_ℓ	8
(XI) (b) pipe diffusion occurs along dislocation cores	0	5	D_p	8
Grain boundary sliding by dislocation glide/climb:				
(XII) (a) in zone near boundary	1	2	D_ℓ	9
(XIII) (b) along boundary	1	2	D_{gb}	9

    With the exception of mechanisms V, VIII and IX, these processes were developed for the creep of metals, and, whilst the major deformation modes are probably included, the list is not exhaustive. For example, Table 1 does not include the possibility of lattice diffusional creep between subgrain boundaries, due to the lack of experimental evidence for such a process; this would give a creep rate faster than that predicted by the Nabarro–Herring equation,[1,2] a value of m = 0 since the controlling subgrain size is independent of grain size, and, if the size of the subgrains decreases with increasing applied stress, n > 1. Similarly, the theory of creep due to the motion of

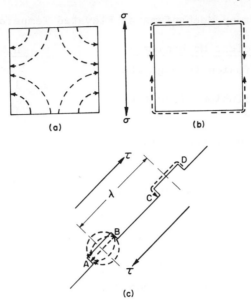

Fig. 1. Types of vacancy diffusional creep in tension: (a) through
        lattice; (b) along grain boundaries; (c) at grain boundary
        ledges either through the lattice (AB) or along the
        boundary (CD).

jogged screw dislocations is not included because of the problems
associated with the interpretation of this model,[10] although a
recent analysis provides some evidence in favor of jog-limited
behavior.[11]

It is convenient to briefly examine these mechanisms before
considering in more detail the methods of determining the rate
controlling process.

Point Defect Mechanisms

Mechanisms I-IV are illustrated schematically in Fig. 1 for
tests performed under an applied stress, σ. In Fig. 1a, the grain
boundaries perpendicular to the stress axis experience a tensile
stress and thus have an excess vacancy concentration, whereas
boundaries parallel to the stress axis have a depletion of vacancies.
This gives rise to a vacancy flow, as indicated by the arrows, and
a consequent elongation of the grain. In the original theory[1,2] the
flow of vacancies was assumed to occur through the lattice (I), but
a later modification considered flow along the grain boundaries (II)[3],
as indicated in Fig. 1b. Under a compressive stress, the direction
for vacancy flow is reversed and the grains become flattened.
Mechanism II predicts a higher value for m and a diffusion coefficient
of D_{gb} rather than $D_{\ell j}$ since the activation energy for diffusion along

the grain boundaries is lower than that for diffusion through the lattice, it follows that this mechanism becomes dominant under conditions of lower temperatures (typically $\lesssim 0.6\ T_m$, where T_m is the absolute melting point) and smaller grain sizes. This diffusional flow is associated with some relative grain movement as an accommodation process to maintain coherency across the grain boundaries;[12] this movement gives rise to offsets in surface marker lines which are similar in appearance to those due to high temperature grain boundary sliding[13] (c.f. mechanisms XII and XIII).

Ashby *et al*[4] have developed a model for diffusion-controlled sliding in a grain boundary containing a periodic array of ledge-pairs, such as AB and CD in Fig. 1c, where sliding occurs due to the vacancy flow arising from the shear stress, τ, acting on the ledges. In this theory, which represents an improvement on the earlier model of Gifkins and Snowden,[14] the rate of sliding on any boundary is controlled by the ledge-pair having the greatest height and length, and vacancy flow may again take place either through the lattice (AB, III) or along the grain boundaries (CD, IV). It has been shown that mechanism III is dominant at the higher temperatures (*vide* mechanisms I and II) and when the ledge-pairs are widely separated [λ large in Fig. 1c].

Mechanism V refers specifically to a phenomenon associated with the presence of a liquid phase along the grain boundaries at the creep temperature, which is of greater importance in ceramic materials rather than metals. It is probably particularly significant in hot-pressed polycrystalline ceramics where an additive is used to aid densification; e.g., this may account for the results obtained at high temperatures on hot-pressed BeO when there was an MgF_2 phase at the boundaries.[15] In addition, tests on ceramics have shown that impurities may segregate at the grain boundaries, even when only present in amounts as small as ~30 ppm,[16] but their effect on the nature of the creep mechanism is not clear.

Line Defect Mechanisms

Mechanisms VI to XIII refer specifically to models involving the movement of dislocations. In the original theory of climb-controlled creep (VI), it was assumed that the dislocations glide on their slip planes from Frank-Read sources, until the motion of the leading edge dislocations was blocked by dislocations of opposite sign on the adjacent planes.[5] A pile-up then occurred, and the back stress was relieved by the climbing together, and consequent annihilation, of the leading edge dislocations in each pair. A recent modification of this theory, involving groups of dislocation dipoles, yields the same values for m, n and D.[17] If some form of interaction occurs on the glide plane, such as solute drag, the rate of glide may be slower than the rate of recovery by climb, and the value of n is

reduced from 4.5 to 3 (VII)[6]; an example of the latter is provided
by the experimental results obtained on NaCl-KCl solid-solution
alloys.[18]

Whilst mechanism VI predicts a stress dependence in reasonable
agreement with that often observed experimentally in metals (n ~4.5),
a disadvantage of the original theory was that it depended on the
formation of piled-up arrays of edge dislocations. However, these
arrays are not usually observed with the electron microscope, either
in metals or in ceramics such as MgO, and it has also been shown
that they are unstable. Since dislocation dipoles and loops are
readily observed in deformed crystals, including MgO, an alternative
procedure was suggested in which creep is controlled by dipole
(VIII) or loop (IX) dissolution and n is 5 or 4, respectively.[7]

A modified form of diffusional creep is represented by mechanisms
X and XI, for the situation where vacancies flow between dislocations
within the lattice and the resultant climb occurs without glide. A
constant internal stress was assumed in the original theory, and this
gives values for n of 3 or 5 depending on whether the dislocations
are sources and sinks for vacancies or, at lower temperatures, pipe
diffusion occurs along the dislocation cores.[8] It has been pointed
out that the introduction of an effective stress for climb and a
periodic internal stress field gives values for n which can be very
large.[19]

Grain boundary sliding is often overlooked as a possible deformation mechanism, although there is good evidence that it can be of
considerable importance in polycrystalline ceramics with small grain
sizes.[20] A possible mechanism for sliding is by the glide/climb
movement of dislocations either in a zone adjacent to the boundary
(XII) or along the boundary (XIII).[9] An alternative possibility
is that sliding occurs by the pure glide of grain boundary dislocations
having their Burgers vectors in the plane of the boundary, but this
seems unable to account for the experimental observations of sliding
on a macroscopic scale.[21]

Identifying the Rate-Controlling Mechanism

Creep investigations are usually concerned with an identification
of the principal deformation mechanism, although this determination
may be difficult if the tests are conducted in a stress or temperature range which is a transition region where more than one process
is significant. In some instances, such as for diffusional creep
(I or II) and dislocation climb (VI), the mechanisms may contribute
strain rates which are essentially additive; but in other cases, as
when grain boundary sliding occurs, the interaction between the
individual processes is less clear.

The rate-controlling mechanism is usually investigated by determining the values of ΔH, m and n. However, a measure of ΔH may be of only limited use in ceramic systems at the present time, because the diffusion coefficients (whether intrinsic, extrinsic or grain boundary) are usually not known with a high degree of accuracy. Furthermore, the atomistic nature of the deformation mechanism is poorly understood so that it is not clear which of the two atomic species controls the behavior; whilst the slower moving ion should be rate-controlling, such as the anion in stoichiometric oxides, there is some contrary evidence in materials of very fine grain size.

More information is obtained from a measure of the constants m and n, since some generalizations are then possible. Firstly, mechanisms involving point defects give Newtonian viscous behavior with n = 1, whereas line defect mechanisms give n > 1. However, it should be noted that the Nabarro-Herring model of diffusional creep assumes the grain boundaries to act as perfect sinks and sources for vacancies;[1,2] if this is not the case, so that the absorption or emission of vacancies influences the process rate, there may exist a threshold stress to give Bingham flow.[22] Secondly, line defect mechanisms occurring within the lattice show no dependence on grain size (i.e., m = 0, mechanisms VI to XI). The constant m is only non-zero when the rate-controlling mechanism involves either (i) movement of vacancies along a diffusional path governed by the separation between the boundaries (as in I), or (ii) a deformation process occurring in the boundary zone (as in II-V, XII and XIII).

The diffusional path, for mechanisms I and II, whether through the lattice (m = 2) or along the grain boundaries (m = 3), is identified at low stress levels by testing specimens covering a range of grain sizes; although even here the result may appear ambiguous, as shown by the value of m = 2.5 reported for polycrystalline MgO in bending.[23] In some ceramic materials there is a change in the value of n from unity at low stresses to ~4.5 at high stress levels, suggesting a change in mechanism from diffusional creep to a dislocation process within the lattice. In these cases, an indication of the diffusional path may be obtained, even when tests are conducted on specimens of only one grain size, from the temperature dependence of the transition stress between the two mechanisms.[24]

Some information on the rate-controlling mechanism is also provided by the shape of the creep curves: two examples are shown in Fig. 2. In A, there is an instantaneous strain, ε_o, a primary stage I in which the creep rate is decreasing, a secondary, or steady-state, stage II in which the creep rate is constant with increasing strain, and a tertiary stage III (observed in tensile creep) in which the creep rate increases to fracture. Stage I is associated with the formation of an intragranular substructure which then remains

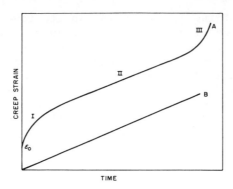

Fig. 2.   Typical creep curves for mechanisms involving dislocation
          movement (type A) and vacancy diffusion (type B).

essentially constant during stage II; this behavior typically relates to mechanisms controlled by dislocation movement (with the
exception of mechanism VII). In contrast, creep processes dependent
on diffusion-controlled mechanisms exhibit no primary stage, and
the deformation proceeds immediately in "steady-state" (curve B);
although this statement should be qualified, since, as the average
path length in mechanisms such as I is changing with increasing
strain, the concept of a continuous steady-state is not strictly
valid. For polygonal grains tested in axial tension, the strain
rate tends to diminish with increasing strain; for axial compression,
a quasi-steady-state may be maintained to a shortening of ~50%.[25]

COMPARISON WITH EXPERIMENTAL RESULTS

All of the experimental results discussed here were obtained
from tests conducted in compression, thereby avoiding the problems
associated with the correlation and interpretation of data from
bending tests. For brevity, details of the various experimental
procedures are not given, but full particulars are included in the
appropriate references.

Lithium Fluoride

Lithium fluoride single crystals of <100> and <111> orientations
were creep tested at temperatures from 650° to 750°C (~0.8-0.9 T_m).[26]
Figure 3 shows the steady-state creep rates obtained from isothermal
tests plotted as a function of applied stress for <100> crystals;
the points marked with an asterisk represent cube specimens. Values
of n calculated from the three principal sets of data (open symbols)
average 3.9. The effect of crystal orientation, asperity ratio,
and purity appears to be minimal, at least in the high temperature
range of this investigation; crystals of lower and higher purities

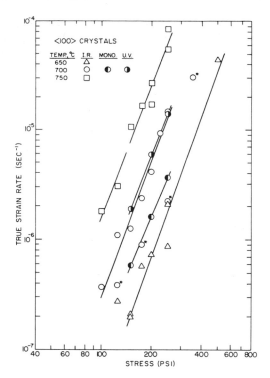

Fig. 3.   True strain rate versus stress for LiF single crystals
          of three different purities.[26]

(half shaded symbols) as well as crystals in <111> orientation exhibited similar behavior with n ranging from 3.4 to 3.9.   The
activation energy in all cases was ∿55 kcal/mole, which is similar
to the value for intrinsic diffusion of F-ions.

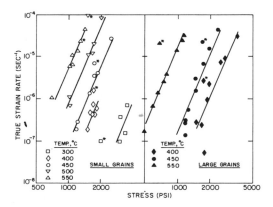

Fig. 4.   Creep rate versus stress for polycrystalline LiF of two
          different grain sizes (∿160 μm and ∿3000 μm, respectively).[27]
          Asterisks indicate initial points which may not relate to
          steady-state.

Tests were also carried out on polycrystalline LiF of two
different grain sizes which differed by more than an order of magnitude ($\sim$160 μm and 3000 μm, respectively), and at temperatures in
the range 300-550°C ($\sim$0.50-0.72 T_m).[27] Steady state creep rates,
determined by change in stress experiments under isothermal conditions, are plotted as a function of the applied stress in Fig. 4.
The results show no dependence on grain size (m = 0) so that the
possibility of a point defect mechanism is excluded (Table 1).
This is further proven by a plot of the ratios of the steady-state
creep rates ($\dot{\varepsilon}_1$ and $\dot{\varepsilon}_2$) and the stresses (σ_1 and σ_2) before and
after a change in stress, as shown in Fig. 5. In this plot, a
least squares analysis yields n=7.6, and the experimental points
are clearly far removed from the expected trend for mechanisms in
which n = 1, as indicated by the line for Nabarro-Herring creep
through the point $\dot{\varepsilon}_2/\dot{\varepsilon}_1=1=\sigma_2/\sigma_1$. Since the creep curves also
exhibited both primary and secondary stages (type A of Fig. 2), it
appears that creep is controlled by dislocation motion.

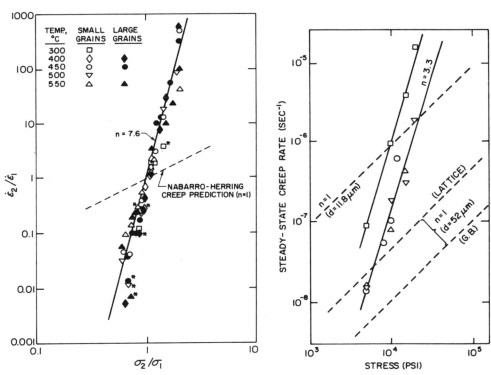

Fig. 5. Creep rate versus stress
ratios for polycrystal-
line LiF obtained by in-
stantaneous changes in
the stress at constant
temperature.[27]

Fig. 6. Steady-state creep rate
versus stress for polycrystalline MgO having grain
sizes of 11.8 μm (◻:ℓ/w=4;
∇:ℓ/w=1.52), 33 μm (△:ℓ/w=
1.52) and 52 μm (O:ℓ/w=1.52),
respectively.[28]

Magnesium Oxide

Transparent polycrystalline MgO, produced with the aid of a processing addition of LiF, was tested at 1200°C ($\sim$0.5 T_m), and the steady-state creep rates were found to be independent of grain size (m $\doteq$ 0) in the range 11.8–52 µm.[28] The results are shown in Fig. 6 for specimens having length: width (ℓ/w) ratios of 4.0 and 1.52, respectively. The value obtained for n was 3.3 and the deviation from theories of diffusional creep may be appreciated by a comparison of the line of slope n = 1 drawn through the point for d = 11.8 µm at the highest stress and the predicted lines for d = 52 µm for lattice (I, m = 2) and grain boundary (II, m = 3) mechanisms, respectively.

Figure 7 shows the data for the specimens having ℓ/w=1.52 replotted as $\dot{\varepsilon}$/D versus σ, where D is calculated from the extrinsic lattice diffusivity of 0^{2-}. Also shown are the theoretical

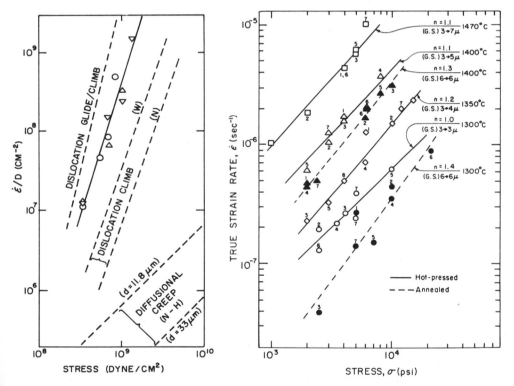

Fig. 7. Diffusion compensated creep rate ($\dot{\varepsilon}$/D) versus stress for polycrystalline MgO, taking D for extrinsic lattice diffusion of 0^{2-}.[28]

Fig. 8. Creep rate versus stress for alumina specimens of two different heat treatments.[29]

predictions arising from Nabarro-Herring diffusional creep (I) for
d = 11.8 μm and 33 μm, respectively, the dislocation glide/climb
mechanism controlled by climb (VI), and dislocation climb (X) as
formulated by Nabarro (N)[8] and reanalyzed by Weertman (W)[17]. The
experimental values for $\dot{\epsilon}/D$ are intermediate between those predicted
by mechanisms VI and X, but the stress dependence is in better agreement with the latter.

Doped Polycrystalline Aluminum Oxide

Polycrystalline Al_2O_3 doped with 0.23 w/o MgO and 0.22 w/o NiO
was creep tested at temperatures in the range 1300-1470°C (∿0.68-
0.75 T_m) both after hot-pressing at 1450°C ("hot-pressed") and after
a subsequent long anneal at 1430°C ("annealed").[29] The results are
shown in Fig. 8 for experiments where the stress was changed periodically during the test: the numbers represent the order of changing.
Due to accompanying grain growth, the notation indicates both the
initial and final average grain sizes. The results show n ∿1.1
for "hot-pressed" specimens and ∿1.3 for "annealed", with no
apparent dependence on grain size (compare points 1 and 6 for the
"hot-pressed" material at 1470°C). The activation energy for
creep of the former specimens was ∿95 kcal/mole, and for the latter
∿125 kcal/mole.

The observation of m = 0 suggests that deformation does not
proceed by the usual mechanisms of diffusional creep, and the low
value of n precludes the normal dislocation models. The results
suggest instead the presence of localized plastic deformation within
the grains at stress concentration points (such as triple edges)
with accommodation by grain boundary sliding and then separation
resulting from the low grain boundary strength.

Polycrystalline Magnesium Oxide With Lithium Fluoride

An unannealed specimen of polycrystalline MgO containing
∿0.95 w/o LiF, and having ∿99.4% theoretical density relative to
MgO, was placed under a compressive stress of 750 psi and heated at
a rate of ∿250°C/hr.[30] Under these conditions, the specimen started
straining at ∿630°C, and showed a rapid increase in strain rate at
∿830°C (Fig. 9). This corresponds closely to the melting point
of LiF (848°C) and the lower temperature activity may be due to the
presence of hydroxyl and carbonate anion impurities. True strains
in excess of 55% were achieved without fracture, thus demonstrating
a behavior similar to superplasticity. In contrast, a specimen
annealed at 1300°C, which reduces the Li content, showed only
thermal expansion under equivalent conditions. Values for n of
1.2 and 1.1 were obtained for the unannealed specimen from changes
in stress tests at 770°C and 850°C suggesting that primarily viscous
flow occurs in a liquid phase at the grain boundaries, possibly as
in mechanism V. The dependence on grain size was not evaluated.

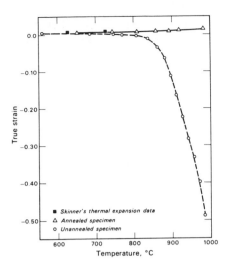

Fig. 9. True strain on heating annealed and unannealed MgO_{30} specimens at a rate of ~250°C/hr under a stress of 750 psi. Two points based on thermal expansion data are also shown [B. J. Skinner, Am. Mineral, 42 (1-2) 39-55 (1957)].

DISCUSSION AND CONCLUSIONS

The creep mechanisms summarized in Table 1 were developed on the basis of the existence of a theoretically dense continuum which maintains coherent grain boundaries during deformation. Some of these relationships may need to be modified for use with ceramic materials due to problems associated with such features as preferential impurity segregation at the grain boundaries (either in solution or as a separate phase), the presence of significant porosity, and the limited number of active slip systems in some crystal structures.

The effect of porosity on creep is not covered here, but the need for some modification in existing theories is seen by the experimental observation that the steady-state creep rate is markedly affected by the degree of porosity within the system, even for theoretical densities exceeding ~95% as reported for BeO.[31] This sensitivity could arise if dislocations were absorbed or emitted at pores within the lattice, although the size and location of the pores may be of more importance than the total volume of porosity *per se*.[32]

An indication of the rate-controlling mechanism is at present most readily obtained from a measure of the grain size sensitivity, -m, or the stress sensitivity, n. Intragranular deformation mechanisms, such as the motion of lattice dislocations, are

independent of grain size (m = 0); those processes directly controlled by the presence of a grain boundary or free surface are
dependent on grain size (m $\geq$ 1). Similarly, the mechanisms controlled
by the motion of line defects have n > 1, whereas those controlled
by diffusion of point defects have n = 1.

It is also apparent that the observed behavior may be dependent
on the method of testing, e.g., an examination of the published
creep data on polycrystalline MgO indicates some values of m greater
than O and large variations in n. A direct comparison of data
obtained in bending and compression is therefore difficult, especially
if grain boundary separation is important. Furthermore, it is expected that problems may arise with the occurrence of barrelling at
high strains in compressive creep tests; this factor has not as yet
been completely evaluated.

In conclusion, it appears that the complexity of the creep
behavior of ceramic systems has not always been fully appreciated.
However, the large volume of creep data now available is providing
a background of knowledge and some understanding upon which future
meaningful studies can be designed and performed.

ACKNOWLEDGMENT

*This work was done under the auspices of the United States
Atomic Energy Commission.*

REFERENCES

1. F. R. N. Nabarro; pp. 75-90 in Rpt. of a Conf. on Strength of
      Solids, (University of Bristol, July 1947). The Physical
      Society, London, 1948.
2. C. Herring, J. Appl. Phys. 21 (5) 437-45 (1950).
3. R. L. Coble, J. Appl. Phys. 34 (6) 1679-82 (1963).
4. M. F. Ashby, R. Raj and R. C. Gifkins, Scripta Met. 4 (9) 737-41
      (1970).
5. J. Weertman, J. Appl. Phys. 28 (3) 362-64 (1957).
6. J. Weertman, J. Appl. Phys. 28 (10) 1185-89 (1957).
7. R. Chang; pp. 275-85 in Physics and Chemistry of Ceramics. Ed.
      by C. Klingsberg. Gordon and Breach, New York, 1963.
8. F. R. N. Nabarro, Phil. Mag. 16 (140) 231-37 (1967).
9. T. G. Langdon, Phil. Mag. 22 (178) 689-700 (1970).
10. A. K. Mukherjee, J. E. Bird, and J. E. Dorn, Trans. Amer. Soc.
      Metals 62 (1) 155-79 (1969).
11. N. Balasubramanian and J. C. M. Li, J. Mat. Sci. 5 (5) 434-44
      (1970).
12. I. M. Lifshitz, Soviet Physics JETP 17 (4) 909-20 (1963).

13. R. C. Gifkins and T. G. Langdon, Scripta Met. 4 (8) 563-66 (1970).
14. R. C. Gifkins and K. U. Snowden, Nature (London) 212 (5065)
    916-17 (1966).
15. G. G. Bentle and R. M. Kniefel, J. Amer. Ceram. Soc. 48 (11)
    570-77 (1965).
16. M. H. Leipold, J. Amer. Ceram. Soc. 49 (9) 498-502 (1966).
17. J. Weertman, Trans. Amer. Soc. Metals 61 (4) 681-94 (1968).
18. W. R. Cannon and O. D. Sherby, J. Amer. Ceram. Soc. 53 (6)
    346-49 (1970).
19. J. M. Dupouy, Phil. Mag. 22 (175) 205-7 (1970).
20. R. L. Bell and T. G. Langdon; pp. 115-37 in Interfaces Conference.
    Ed. by R. C. Gifkins. Butterworths, Sydney, 1969.
21. T. G. Langdon, Mat. Sci. Eng. 7 (2) 117-18 (1971).
22. M. F. Ashby, Scripta Met. 3 (11) 837-42 (1969).
23. E. M. Passmore, R. H. Duff and T. Vasilos, J. Amer. Ceram. Soc.
    49 (11) 594-600 (1966).
24. T. G. Langdon, J. Nucl. Mat. 38 (1) 88-92 (1971).
25. H. W. Green, J. Appl. Phys. 41 (9) 3899-3902 (1970).
26. D. R. Cropper, "Creep in LiF Single Crystals at Elevated
    Temperatures", UCRL-20350, University of California at
    Berkeley, (Ph.D. Thesis), 1970.
27. D. R. Cropper and T. G. Langdon, Phil. Mag. 18 (156) 1181-92 (1968).
28. T. G. Langdon and J. A. Pask, Acta Met. 18 (5) 505-10 (1970).
29. T. Sugita and J. A. Pask, J. Amer. Ceram. Soc. 53 (11) 609-13
    (1970).
30. P. E. Hart and J. A. Pask, J. Amer. Ceram. Soc., (in press).
    (UCRL-19699, July 1970).
31. N. V. Shishkov, P. P. Budnikov and P. L. Volodin; pp. 225-43 in
    New Nuclear Materials Including Nonmetallic Fuels, Vol. 1.
    International Atomic Energy Agency, Vienna, 1963.
32. T. G. Langdon and J. A. Pask, p. 283 this volume.

DISCUSSION

B. A. Wilcox (Battelle Memorial Institute): In your LiF single
crystal creep tests, what creep rate was plotted, steady-state creep
rate? Taylor and Nix at Stanford found sigmoidal creep plots for LiF
crystals. Only in region III was there true steady state behavior,
and often it was impossible to reach region III before marked
barrelling occurs.
Authors: Our rates were steady-state, which were reached after ~20%
strain when loading in the <100> direction, and thus refer to the
so-called "region III" in the Stanford work.

R. W. Rice (U. S. Naval Research Laboratory): I think it is meaningful that greater ductility is observed in specimens with some porosity,
much of which is within the grains. There are a number of reasons
why one might expect pores to be sources of dislocations. Further,
I have directly observed corners and steps in pore walls acting as
preferred sources of slip in room temperature compressive stressing

of MgO crystals. This has two important effects: (a) it provides
more homogeneous sources of slip which, because of the shorter
source - grain boundary distance when pores are in the grain (in
comparison to sources at the grain boundary), are less likely to
initiate fracture, and (b) pores provide a means of initiating slip
in grains that are unfavorably oriented for slip due to the applied
stress alone. Thus I feel that pore location is quite important.
However, I feel it is very important to emphasize the state of stress.
It is generally not possible to get all pores within the grains. One
can tolerate some porosity at grain boundaries in compression much
more than in tension. Thus it is quite likely that this effect of
enhanced ductility due to location of some pores within grains is
effective primarily, if not exclusively, in compression.

It is also interesting to note that locating pores within grains
also appears to significantly increase fracture surface energy.
Authors: We are in complete agreement. We have concluded that intragranular pores are beneficial because they allow some mass accommodation
by acting as sources and sinks for dislocations. It is important,
however, that the pores are very small in size with respect to the
grain size, and that they are distributed throughout most, if not
all, of the grains. These experimental observations refer to tests
in compression, and the situation could well be different when testing
in tension.

A. H. Heuer (Case Western Reserve University): We have also studied
deformation in alumina as a function of grain size (1-10 μm) in the
same temperature range as you used but in bending. We find similar
n values (1.1-1.5) but a very large grain size exponent (m of 2-3).
Could you comment on why there is so little grain size effect in
your work?
Authors: We do not believe that it is valid to make such comparisons
with the present state of our knowledge. There are basic differences:
the alumina compositions, and testing method, are not the same. Also,
the characters of the specimens probably are not the same both before
and after creeping. The m and n values would be expected to be
quite sensitive to such differences. The same material should be
tested by both techniques for comparison.

J. E. Burke (General Electric Company): I should like to extend
Dr. Heuer's question about a lack of grain size effect on the creep
rate in polycrystalline Al_2O_3. Folweiler, in my laboratory, also
carried out extensive creep measurements (in bending I grant) on
polycrystalline, theoretically dense alumina and observed quite
closely that the strain rate at a given stress varied inversely
with the square of the grain size as predicted by the Nabarro-Herring
relationship. The only deviations were at higher strain rates in
the coarsest grained samples, where the stress exponent appeared to
increase above 1, but microscopic examination showed that in these

deviating cases, porosity had developed at the tension grain boundaries, so of course the presumed vacancy source-vacancy sink distance was drastically lowered. In bend tests, half of the specimen deforms in tension, and half in compression, but that should make no difference in the grain size effect if deformation occurs either by dislocation movement or vacancy transport.

I should like to ask what range of grain sizes were studied in these experiments, and whether you have a rationale for the observed difference in behavior?

Authors: The starting grain size was the same for all hot-pressed specimens, and also for the annealed; grain growth occurred during the experiments with no effect on the creep rate as indicated by the method of experiment. No experiments were made on specimens of similar character but with varying grain size; also the compositions of our specimens are not the same as those of **Folweiler.** It would be highly informative to study similar specimens by both methods of testing.

It is possible that the stress range used by **Folweiler** was low enough so that the deformation for most specimens was essentially in the diffusion-controlled region. Also, the description of the appearance of the specimen after creeping indicates that the grain boundary coherency was not maintained which may require a modification of the analysis.

R. W. Rice (U. S. Naval Research Laboratory): Another factor that would have to be considered in comparing bend and compressive creep of MgO doped with LiF are the differences in resulting LiF contents. Compressive specimens are generally bulkier and so can retain more LiF internally where stresses are high. On the other hand flexure specimens are normally smaller so it is easier to diffuse more of the LiF out. Further, even if the cross-sections of the two types of specimens are the same, the surfaces should have lower residual Li and F contents. Thus, since the load is carried primarily by the surface layer in bending the bend test would show less effect of residual Li and F.

Authors: Both compressive and bend tests were not made on MgO with LiF, only in compression. The last traces of LiF are difficult to remove, and only traces are necessary to modify the grain boundary strength. On that basis it is quite possible that different results would have been obtained by the two methods of testing.

DEFORMATION BEHAVIOR OF ALUMINA AT ELEVATED TEMPERATURES

Paul F. Becher

U. S. Naval Research Laboratory

Washington, D. C.

ABSTRACT

The thermomechanical behavior and microstructural observations of several polycrystalline alumina bodies varying from 2-80 μm in grain size with typical purities of ≥99% tested in compression at T ≥1210°C are presented. The mechanical behavior of alumina is considered in terms of these results and similar investigations of the behavior of sapphire bicrystals. Plastic deformation is observed to be related to dislocations, mechanical twinning and/or grain boundary shear. In general, decreasing the grain size gives rise to a transition in deformation mechanisms and leads to increased mechanical strength. Impurities, however, strongly modify this relationship and are instrumental in enhancing grain boundary shear processes in alumina.

INTRODUCTION

The response of ceramics subjected to a variety of thermal and mechanical conditions has been receiving considerable attention. Studies of ceramics subjected to dynamic stress (constant strain rate) conditions have been concerned primarily with single crystals and cubic polycrystalline oxides. These and other findings[1-7] have shown that dislocation and diffusional processes contribute to the elevated temperature deformation of ceramics. In addition, grain boundary sliding can also play an important role in the behavior of polycrystalline bodies.[8] Furthermore, the mechanical behavior of polycrystalline ceramics is found to be sensitive to a variety of microstructural (and compositional) characteristics, e.g., grain size,

315

porosity, and impurities. The present paper reports observations
on the mechanical behavior and deformation microstructures of
several polycrystalline alumina materials which were studied in
conjunction with sapphire double bicrystals. These results are
discussed together with those observed in double bicrystal
experiments.

EXPERIMENTAL TECHNIQUES

The techniques used for fabricating and testing of α-Al_2O_3
double bicrystals have been previously described;[9] thus the following
procedures apply primarily to the polycrystalline alumina bodies.
Test specimens suitable for compressive loading were prepared by
conventional cutting and surface grinding (100-mesh diamond metal
bonded abrasives) to obtain rectangular parallelepipeds with 2:1
aspect ratios. Mechanical polishing with diamond pastes (30 through
1/4 µm grit sizes) followed by a vacuum anneal at 1750°C for two
hours, was utilized for subsequent microstructural observations.
Specimens were tested in a Brew furnace mounted on an Instron physical
testing machine. Testing was accomplished *in vacuo* ($<10^{-4}$ torr) at
a strain rate of $5x10^{-3}$ min^{-1} over a temperature range of 1210°C–1820°C.
A graphite ball joint was incorporated into the lower loading column
to prevent skew loading; polycrystalline and single crystal alumina
disks were employed as loading surfaces.

Samples were examined before and after testing, using standard
optical microscopy techniques, and selected deformed specimens were
chemically etched (H_3PO_4 at ~300°C) to delineate microstructural
features. In addition, thin sections were prepared by the polishing
techniques described above, providing ~0.002 in. thick samples for
subsequent ion beam thinning. This technique was utilized to produce
thin foils (<2000Å) which were subjected to transmission electron
microscopy analysis employing the analytical techniques described
by Hirsch *et al*.[11]

RESULTS

Characterization of Polycrystalline Alumina Bodies

As the polycrystalline ceramic had been selected to obtain
variations in grain sizes and purities, materials characterization
served to illustrate some of the material parameters involved
(Table 1). These results show these materials to be generally
comparable in purity, with the notable exception of Body III and its
high silica content. This trend was substantiated microstructurally
in that second phases along boundaries and at triple points were
observed solely in Body III. In addition to these obvious second

Table 1. Analysis of Polycrystalline Alumina

	Body				
	I[a]	II[a]	III[a]	IV[b]	V[b]
Annealed Grain Size μm	2-4	20	20	20	80
Density (25°C), g/cm^3	3.956 (98.9%)	3.986 (99.9%)	3.901 (97.8%)	3.972 (99.5%)	3.972 (99.5%)
Major Impurity Content, ppm	SiO_2-100	SiO_2-230	$SiO_2 \sim 5000$	$SiO_2 < 500$	
	Y_2O_3-500[c]	CaO-110	CaO<500	CaO<100	Same
	MgO-500[c]	MgO-360	MgO<5000	MgO<100	as
	Fe_2O_3-25	Fe_2O_3-42	Fe_2O_3<500	Fe_2O_3<10	IV
	PbO-15	ZrO_2-230	TiO_2 B_2O_3 <100 MnO	B_2O_3<10 $MoO_3 \sim 5000$ $WO_3 < 1000$	

[a] Sintered bodies, supplied by Coors Porcelain Co., Golden, Colorado. Body I is an experimental material; Bodies II and III are in commercial production.

[b] Press forged disk, supplied by Systems Division, AVCO Corp., Lowell, Mass.

[c] As added to raw material; not significantly altered after firing.

phases exhibiting amorphous diffraction behavior, localized strain contrast configurations suggested some impurity precipitation at boundaries.

The grain sizes determined after annealing were stable during the short time of testing (<30 min.) even at the highest test temperatures. The major portion of the observed porosity was intra-granular, with the remainder primarily at triple points, with no difference in pore location being noted in these bodies. Dislocations were frequently associated with intragranular pores, as previously observed.[12,13] A somewhat higher dislocation content observed in Body III was felt to be related to the higher pore content of this body. Dislocations (singularities and networks) and twins and basal twin boundary dislocations as observed previously[14,15] were typical of all bodies, although twin populations were quite sparse. These typical structural features are illustrated in Fig. 1.

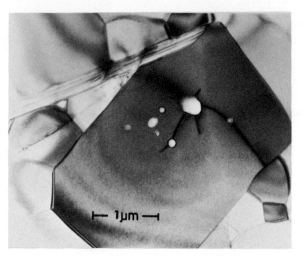

Fig. 1.   Characteristic substructure in the polycrystalline,
          sintered alumina bodies prior to deformation.

        Finally, x-ray analysis (using peak area comparison) revealed
that Bodies I, II and III contained randomly oriented grains, while
Bodies IV and V exhibited some preferred crystallographic texture.
Some basal plane texturing perpendicular to the forging direction

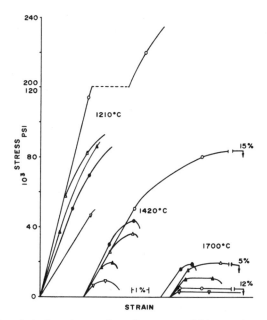

Fig. 2. Mechanical behavior of polycrystalline alumina bodies com-
        pressively deformed at a constant strain rate $(5 \times 10^{-3} \text{min}^{-1})$
        Body I - 0; II - Δ; III - ∇; IV - $\bullet$; V - $\blacktriangle$.

(i.e., at 0° and 45° to the compressive axes of the two orientations employed for testing) and a slight prismatic texture parallel to the forging direction was detected. The crystallographic texture resulted from the press forging process utilized in fabrication as reported by Rhodes *et al.*[16]

Mechanical Behavior of Polycrystalline Bodies

The stress-strain responses of these alumina bodies at T⩾1210°C (Fig. 2) were illustrative of variations in mechanical behavior as related to material parameters. The decrease in yield stress with both increasing temperature and grain size was consistent with plastic flow by dislocation (possibly twinning) processes; notable exceptions to this yield stress behavior were observed for Body I at 1700°C and Body III above 1210°C. The rate of strain hardening decreased while the amount of plastic flow generally increased over the range 1210–1700°C, suggestive of more extensive activity of slip and diffusion processes (i.e., dislocation climb and poly-gonization) as the test temperature was increased. Limited testing at 1820°C revealed additional decreases in the yield stresses and corresponding increases in permanent strains.

Polycrystalline Deformation Microstructures

Optical microscopy of all deformed samples, in conjunction with transmission electron microscopy of those having 1.5-2% permanent strains, were utilized in delineating grain deformation and grain boundary behavior.

Twinning. Deformation twinning on the rhombohedral and (to a limited extent) basal systems was prevalent at 1210°C, as was crack initiation by twin interactions, particularly for 80 μm grain size (Body V, Fig. 3a). Above 1210°C, mechanical twins played an important role in Body V, while decreasing activity of twins with temperature was observed in the remaining bodies. Further, twins detected in the ⩽20 μm grain size materials were quite thin (⩽1 μm at T>1210°C, and at these temperatures the shear induced at grain boundary twin barriers was often accommodated by slip (Fig. 3b). As a result of the low incidence of twinning and the fineness of twins in the moderate to fine grained materials, crack initiation was seldom observed in these bodies at T>1210°C. Above 1420°C, mechanical twinning is substantially reduced in Body V, and other (slip) mechanisms predominated.

Dislocation Mechanisms. In association with deformation twinning, dislocation mechanisms were found to contribute to plastic flow (Fig. 4). Evidence for basal slip ($<11\bar{2}0>\{0001\}$) was found in all bodies as was $<\bar{1}011>\{10\bar{1}2\}$ (rhombohedral) slip. The activity of dislocations involved in slip on these systems was observed to

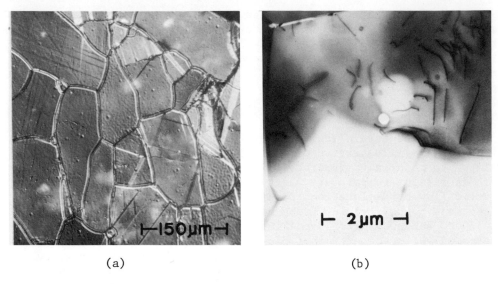

(a)   (b)

Fig. 3.   Deformation twins which have been initiated during compres-
          sive loading of alumina at (a) 1210°C and (b) 1420°C.

increase with temperature and with reduction in grain size (<80 μm).
The degree of macroscopic plastic flow obtained in Body I at
T≥1420°C and in Body II at 1700°C correlated well with an increased
activity of dislocations (especially $\bar{b} = 1/3 <\bar{1}011>$) detected
microscopically in these alumina materials.   Impurities apparently
had a strong repressive influence on dislocation mechanisms, as
Body III exhibited only limited evidence for slip, even at 1700°C.

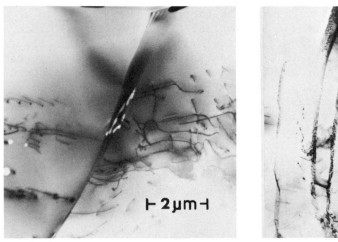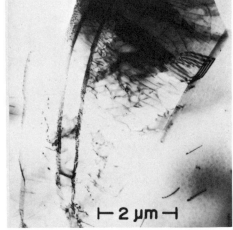

Fig. 4.   Dislocation structures developed in alumina during high
          temperature plastic deformation; (a) 1420°C and (b) 1700°C.

Observations of dislocation debris (i.e., dipoles and loops) were
suggestive of the interaction of the two (basal, rhombohedral) slip
systems, but were generally limited to those specimens tested at
T⩾1420°C. Although grain boundaries restricted dislocation motion,
crack initiation was not detected at dislocation pile-ups at 1.5-2%
permanent strain levels. At the higher temperatures (e.g., 1420°C
and 1700°C) dislocation climb occurred, relieving internal stresses
and, at the higher temperatures, resulting in rapid polygonization
within extensively deformed grains (Fig. 4b). At 1420°C, rearrangement of dislocations by diffusional processes was more limited, as
evidenced by the small degree of sub-boundary formation in heavily
deformed grains.

   In addition to intracrystalline deformation, grain boundary
offsets were optically discerned in coarser grained bodies, especially
body III. Voids (detected as light scattering centers) were also
observed along grain boundaries in body V at T⩾1420°C. This behavior
was confirmed with the aid of transmission electron microscopy: interfacial porosity appeared to propagate from one triple point to the
next (Fig. 5). This interfacial porosity was also associated with
triple point offsets accompanied by some accommodation by plastic
flow in opposing grains.[7] In addition, serrations in these boundaries
contained dislocation segments, suggesting their plastic deformation
during boundary shear. The above features were indicative of grain
boundary sliding and were especially prevalent after testing at 1420°C.
Grain boundary shear was quite evident in Body III at T⩾1420°C, and
was often related to triple point and boundary second phases (Fig. 6).

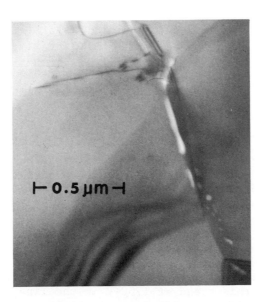

Fig. 5.   Interfacial porosity present after 1-2% permanent strain
          which is associated with grain boundary sliding.

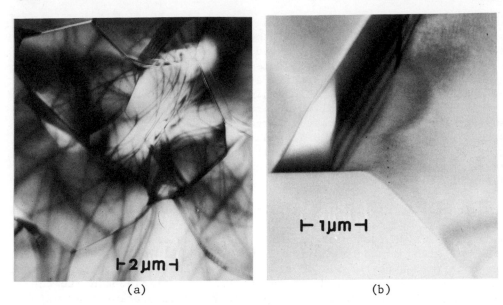

(a)   (b)

Fig. 6.   Grain boundary shear detected in a polycrystalline alumina
          containing glassy, intergranular phases (a) second phase
          boundaries with local sliding; (b) shear in triple point.

Electron diffraction analysis of these regions showed them to be
amorphous in nature, comparable to their condition prior to deformation

        In conjunction with the above boundary shear charactertistics,
boundary shear dislocations were found in the 2 μm and 20 μm grained
bodies (Fig. 7).   These boundary sliding dislocations were observed
in interfaces devoid of second phase or porosity, and were distinguishe
by image contrast behavior as described by Gleider et al [17] and
Ishida and Henderson-Brown. [18]   Electron diffraction analyses obtained
for these dislocations have shown them to have <112̄0> Burgers vectors.

                          DISCUSSION

        The pronounced grain size and temperature dependence of yield
stress (i.e., inverse proportionality) is consistent with plastic
flow involving dislocation mechanisms.   As discussed by Heuer et al, [19]
when diffusional processes (Nabarro-Herring or Coble creep) and
boundary shear are invoked during deformation, the yield stress
would be expected to decrease with decreasing grain size, having
grain size exponents of 1, 2 or 3.   Analysis of the present mechanical
test data has shown the yield stress to be inversely proportional to
the grain size, with grain size exponents slightly less than 0.5.
The alteration of this trend at 1700°C in the fine grained alumina
and the generally different behavior of the lower purity 20 μm grain
size body reflect the importance of various other mechanisms of
deformation.

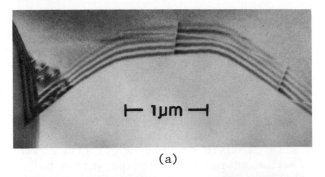

(a)

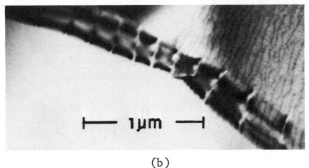

(b)

Fig. 7.   Dislocations involved in grain boundary sliding which
          occurred at T>1210°C (a) 99.9% Al_2O_3 body; (b) 99% Al_2O_3.

Deformation twinning has been observed to contribute quite
significantly to the thermomechanical behavior of single crystals[2,21]
and bicrystals[9,22] of α-Al_2O_3. Decreasing temperatures and increasing strain rates promote twinning - as opposed to basal slip-
in sapphire.[21] A transition from rhombohedral twinning to rhombohedral slip in α-Al_2O_3 double bicrystals (where [0001] coincided
with the compressive stress axis) was noted earlier,[9] supported by
more recent transmission electron microscopy.[10] Evidence for the
rhombohedral slip system at elevated temperatures was initially
reported for sapphire whiskers by Bayer and Cooper,[23] while Hockey[14]
has observed this slip system in the region of room temperature
microhardness indentations in sapphire. The behavior of the polycrystalline bodies shows a dependence on grain size, as well as
temperature, for the activation of deformation twinning. As noted,
twinning was prominent in $\geqslant$20 μm grain sized bodies at 1210°C, while
at 1420°C it was significant only in the 80 μm body. This suggests
a Petch relationship for twinning as observed in metals[22]

$$\sigma_y = \sigma_i + Kd^{-\frac{1}{2}}$$

where σ_y is the yield stress, σ_i is the friction stress opposing twin

growth, K is the stress concentration at a twin tip or stress for
local yielding, and d is the grain diameter. A similar relationship
is found for yield stresses involving plastic flow by dislocation
processes. However, in metals, the grain size coefficients and
stress intercepts indicate that at larger grain size, twinning
prevails, while at lower grain size, slip is dominant. The lower
stress required for twinning versus slip in sapphire double bicrystals
at the same temperature and strain rates[9] is suggestive of similar
behavior in alumina. Increases in temperature cause the transition
from twinning to slip to shift towards larger grain sizes, and indicate a lower activation energy for twinning versus slip. This may
be indicated by the activation energy value for rhombohedral twinning
for alumina which is less than that for basal slip[22] and is also
expected to be less than that for rhombohedral slip. On the basis
of the findings presented here, this type of behavior (first observed
in metals) is considered to occur during compressive deformation of
polycrystalline alumina at elevated temperatures.

The influence of temperature on twinning is seen to be related
to the activation of other modes of deformation (slip,etc.) and
possibly to removal of sites of twin initiation. Cahn[25] has proposed that dislocation pileups could act as source of twins as has
been noted in zinc.[26] Further, deformation twins were not observed
in tin with increasing test temperature,[27] and the results of that
study indicated that dislocation climb occurred which relieved
dislocation pile-up stresses and led to the absence of twinning. This
is felt to explain the twinning behavior of the polycrystalline
alumina at 1420°C and above, where dislocation climb processes were
found to become important. Similar behavior might explain the
sapphire double bicrystal twinning-slip transition; however, twins
in these samples appeared to initiate at free surfaces rather than
at the grain boundaries, which did act as dislocation barriers and
thus as possible twin sources. It has been recently shown that
surface machining results in considerable surface damage (dislocations and twins).[14,15] Thus, although elevated temperatures would
be conducive to dislocation climb, it may also result in the
removal of existing surface twins at the higher temperatures,
favoring a transition from twinning to slip in the double bicrystals.

The role of deformation twinning in crack initiation by various
twin interactions [as observed in (a) single crystals,[20] (b) double
bicrystals[9,22] and (c) the 80 μm grain size alumina described here]
is detrimental to the achievement of appreciable plasticity. The
observation of microtwins having ≤1 μm thicknesses in the finer
grain sized bodies, and of twin accommodations at barriers (grain
boundaries), suggests a relationship between twin thickness and
grain size, influencing the ability to initiate twin fractures. In
addition, Heuer *et al*[19] have proposed that microtwins induced by
grain boundary shear can play an important role in the plasticity
of fine grained (≤10 μm) alumina tested in bending at elevated

temperatures. Deformation twinning can also result in permanent
strains of up to 30% in polycrystalline metals (e.g., zirconium)[28]
in the absence of significant fracture. Although this cannot be
directly transposed to the behavior of alumina, it does indicate
in conjunction with the previous observations that deformation
twinning (especially microtwins) can be accommodated, permitting
them to contribute to the plastic deformation of alumina.

Previously alumina has been considered as a semi-brittle ceramic
at elevated temperatures as a result of the activation of basal
slip alone. However, rhombohedral ($<\bar{1}011>\{10\bar{1}2\}$) slip can also
contribute to plastic flow as noted herein and in the results from
double bicrystals.[9,10] As a result of the $\alpha-Al_2O_3$ structure, only
three $<1011>$ translations are allowed, resulting in three independent
rhombohedral slip systems. These, coupled with the two independent
basal slip systems, are sufficient to satisfy the Taylor-von Mises
criteron for ductile behavior in a polycrystalline body. The
ability of these slip systems to interpenetrate is also a factor in
achieving truly ductile behavior as discussed for MgO,[1] a factor
which has not been explicitly determined for $\alpha-Al_2O_3$. The ductile
macroscopic behavior of Body I supported by direct observations of
both $<\bar{1}011>$ and $<11\bar{2}0>$ glide dislocations are illustrative of truly
ductile behavior in alumina, suggesting that these slip systems
are able to interpenetrate.* The ability to achieve ductile behavior
is also strongly affected by various microstructural parameters
(porosity, impurities) as observed in MgO[1] and is reflected by the
behavior of the different polycrystalline aluminas presented here.
In addition, polygonization of dislocations at 1420°C and above can
bring about a reduction in strain energy and strain hardening effects
and thus contribute to the plastic deformation process. Polygonization has also been observed during basal kinking in sapphire[9]
over a similar temperature range. Kinetic analysis indicated dislocation climb to be the rate-controlling process[29] limiting plastic
strain.

The influence of impurities (especially silica) on the behavior
of polycrystalline alumina is particularly evident in Body III
(99% Al_2O_3). Impurities enhanced grain boundary sliding over other
deformation modes, as indicated by the decreased inactivity of
dislocations in this body coupled with accentuated grain boundary
sliding. The similar behavior of mullite (3 Al_2O_3 · 2 SiO_2)-doped
double bicrystals, which exhibited rapid boundary shear occurring
at quite low (near zero) stress levels, was seen to be related to
the presence of a viscous amorphous boundary phase. The ease of
sliding in Body III also resulted from a reduced grain boundary
viscosity associated with intergranular, amorphous second phase.
* Compressive deformation of polycrystalline alumina under hydrostatic constraint at elevated temperatures has substantiated
  ductile behavior by dislocation mechanisms (i.e., five independent
  slip systems).[30]

It was previously shown that nondisruptive (i.e., noncata-
strophic) grain boundary sliding results when $MgAl_2O_4$ spinel
impurities are present at grain boundaries.[9] More recent studies
of spinel-doped bicrystals have shown that grain boundary sliding
is not associated with second phases and that dislocation mechanisms
are not responsible for the sliding behavior.[10] The lack of grain
boundary sliding[9] and the absence of enhanced grain boundary
diffusion of ^{18}O[31] in similar nondoped bicrystal grain boundaries,
together with the preceding findings, have been cited as evidence
in support of an impurity-enhanced diffusional grain boundary shear
process.[10] The occurrence of impurity-enhanced grain boundary
diffusion in various oxides[32-34] gives further support to this
concept. In addition, boundary sliding rates (calculated on the
basis of a relationship for sliding employing boundary diffusion in
nonplanar (sinusoidual) boundaries together with volume diffusion,
presented by Ashby and Raj[35] are in good agreement with those
observed in spinel-doped boundaries. These observations indicate
that solute impurities (especially Mg^{+2}) can result in boundary
shear by enhancing boundary diffusion. Further, this type of
boundary sliding results in generation of interfacial porosity in
the spinel-doped bicrystals; similar behavior in certain of the
polycrystalline bodies associated with grain boundary sliding points
to similar impurity-enhanced diffusional sliding phenomena in polycrystalline bodies.

Although impurities appear to strongly influence grain boundary
sliding in alumina, boundary shear by dislocation mechanisms was
also observed. The observation of grain boundary sliding dislocations
with <$11\bar{2}0$> type Burgers vectors, and the earlier finding that
boundary sliding occurs only in nondoped bicrystals having (0001)
interfaces,[20] indicate that nonviscous boundary sliding occurs
either by glide (and climb) of matrix dislocations in the interface[36]
or by zonal shear in the adjacent grain regions.[37] The present
observations of grain boundary sliding dislocations give support to
a glide and climb process of dislocation-related grain boundary
sliding. The behavior of the 2 μm grain sized body at 1700°C indicates a possible increase in this nonviscous boundary shear behavior
with decreasing grain size as suggested previously.[19] The limited
data obtained in this region suggests that for the finer grain sizes
a transition in kinetics occurred above 1420°C, where the yield
stress is proportional to the grain size, having an exponent which
approaches 1 (boundary shear in metals is seen to have similar
behavior). Although these data do not explicitly identify grain
boundary sliding, when evaluated in conjunction with the observed
boundary shear in alumina, they are suggestive of an increasing
role of boundary shear at finer grain sizes, as noted previously,[19]
and similar to the behavior of metals.

SUMMARY

The mechanical deformation of polycrystalline alumina at elevated temperatures is a complex process related to microstructural
and compositional parameters, stress states and strain rates, etc.
It has been shown that dislocation mechanisms (glide and climb)
can strongly contribute to plastic deformation in alumina.
Basal and rhombohedral slip systems result in a sufficient number
of independent shear modes to contribute to ductile behavior in
alumina under compressive stress at elevated temperatures. Although
these slip systems are observed over a grain size range of 2-80 µm
(especially at 1700°C), there is only limited plastic flow in the
larger grain sizes. The large plastic strains obtained at T⩾1420°C
in the 2 µm grain sized material, together with the slip systems
observed in transmission electron microscopy, are indicative of
ductile behavior in alumina. To a limited extent, similar behavior
occurred at 1700°C in a 20 µm body. In addition, deformation twinning
is also induced and is prominent in large grained bodies and/or at
lower temperatures; under these conditions twinning results in
crack initiation.

Grain boundary sliding is particularly evident in the behavior
of spinel- and mullite-doped sapphire double bicrystals. Impurity-
enhanced grain boundary diffusion is considered responsible for
boundary shear obtained in the spinel-doped boundaries. However,
viscous flow occurs when siliceous interfacial phases are present.
The observations in polycrystalline materials indicate that these
impurity-related processes are significant in the deformation
behavior of most commercial aluminas. Interfacial shearing by grain
boundary dislocation processes also contributes to the deformation
of alumina, probably being more prominent in fine grained materials.
In general, the roles of these several different mechanisms (and
competetive selections among them) during the deformation of alumina
at elevated temperatures are strongly related to microstructure and
composition together with the thermomechanical environment.

ACKNOWLEDGMENTS

*The author extends his sincere appreciation to colleagues
Drs. Hayne Palmour III, W. W. Kriegel and Ray B. Benson of North
Carolina State University, and Roy W. Rice of the Naval Research
Laboratory for their contributions. This paper is based on
research undertaken by the author while at North Carolina State
University. The support of the Department of Engineering Research,
NCSU, and that of the U. S. Atomic Energy Commission under
Contract At-(40-1)-3328, is gratefully acknowledged.*

REFERENCES

1. T. G. Langdon and J. A. Pask; pp. 53-127 in High Temperature Oxides, Part III, Ed. by A. M. Alper, Academic Press, New York, 1970.
2. R. M. Fulrath and J. A. Pask, Ceramic Microstructures, John Wiley & Sons, Inc. New York, 1966.
3. J. J. Burke, N. L. Reed, and V. Weiss eds. Ultrafine Grain Ceramics. Syracuse University Press, Syracuse, N. Y. 1970.
4. F. M. Vahldiek and S. A. Mersol, Eds. Anistropy in Single-Crystal Refractory Compounds. Plenum Press, New York, 1968.
5. T. Vasilos and E. M. Passmore, pp. 406-430 in Ref. 2.
6. A. H. Heuer, Proc. Brit. Ceram. Soc. 15 173-184 (1970).
7. P. F. Becher, "Deformation Substructure in Polycrystalline Alumina" to be published in J. Mater. Sci.
8. R. L. Bell and T. G. Langdon; pp. 115-137 in Interfaces Conference, Ed. R. C. Gifkins, Butterworths, Sidney (Australia), 1969.
9. P. F. Becher and H. Palmour III, J. Am. Ceram. Soc. 53 (3) 119-123 (1970).
10. P. F. Becher and H. Palmour III, "Grain Boundary Shear in α-Al$_2$O$_3$", submitted for publication.
11. P. B. Hirsch, A. Howie, R. B. Nicholson, D. W. Pashley and M. J. Whelan, Electron Microscopy of Thin Crsytals. Butterworths, London, 1965.
12. N. J. Tighe and A. Hyman, pp. 121-136 in Ref. 4.
13. N. J. Tighe, pp. 109-133 in Ref. 3.
14. B. J. Hockey, "Plastic Deformation of Alumina by Indentation and Abrasion", submitted for publication.
15. B. J. Hockey, "Observation on Mechanically Abraded Aluminum Oxide Crystals", in Proceedings of Symposium on the Science of Ceramic Machining & Surface Finishing, National Bureau of Standards, Gaithersburg, Md., Nov. 1970. (in press).
16. W. H. Rhodes and R. M. Cannon, "Microstructure Studies of Refractory Polycrystalline Oxides", Summary Report AVSD-0038-70-RR, Contract N00019-69-C-0198, Dec. 1969.
17. H. Gleiter, E. Hornbogen and G. Baro, Acta. Met. 16, 1053-1067 (1968)
18. Y. Ishida and M. Henderson-Brown, Acta Met. 15, 857-860 (1967).
19. A. H. Heuer, R. M. Cannon and N. J. Tighe pp. 339-360 in Ref. 3.
20. A. H. Heuer, Phil Mag. 13, 379-393 (1966).
21. H. Conrad, K. Janowski and E. Stofel, Trans. AIME 233, 255-256 (1965)
22. R. L. Bertolloti, "Creep of Aluminum Oxide Single Crystals and Bicrystals", Ph.D. Thesis, Department of Ceramic Engineering, University of Washington, Seattle, 1970.
23. P. D. Bayer and R. E. Cooper, J. Mater. Sci. 2 (3) 301-302 (1967).
24. A. S. Teleman and A. J. McEvily, Jr., Fracture of Structural Materials. John Wiley and Sons, Inc., New York, 1967.
25. R. W. Cahn; pp. 1-28 in Deformation Twinning. Ed. by R. E. Reed-Hill, J. P. Hirth and H. C. Rogers, Gordon and Breach Scientific Publishers, New York, 1964.

26. P. B. Price; pp. 41-130 in Electron Microscopy and Strength of
    Crystals. Ed. by G. Thomas and J. Washburn, Interscience
    Publishers, Inc., New York, 1963.
27. S. Maruyama, J. Phys. Soc. Japan 15, 1248-1251 (1960).
28. R. E. Reed-Hill, pp. 295-320 in Ref. 23.
29. H. Palmour III, W. W. Kriegel, P. F. Becher and M. L. Huckabee,
    "Grain Boundary Sliding in Alumina Bicrystals", Final Report
    ORD-3328-17, Contract AT-(40-1)-3328, July 1970.
30. A. H. Heuer, R. F. Firestone, and J. D. Snow, "Non-basal Slip in
    Aluminum Oxide (Al₂O₃)", Tech. Rept. 2, Contract N00014-67-A-
    0404-0003, NR 032-058, Sept. 1970.
31. J. B. Holt; pp. 169-190 in Sintering and Related Phenomena, Ed.
    by G. C. Kuczynski, N. A. Hooton and C. F. Gibbon. Gordon and
    Breach Scientific Publishers, New York, 1967.
32. B. J. Wuensch and T. Vasilos, J. Am. Ceram. Soc. 48 (8), 433-436
    (1966).
33. J. B. Holt and R. H. Condit; pp. 13-30 in Materials Science
    Research Vol. 3. Ed. by W. W. Kriegel and H. Palmour III,
    Plenum Press, New York, 1966.
34. K. R. Riggs and M. Wuttig, J. Appl. Phys. 40 (11) 4682-4683 (1969).
35. M. F. Ashby and R. Raj, "On Continum Aspects of Grain Boundary
    Sliding and Diffusional Creep", Tech. Rept. 2, Contract
    N00014-67-A-0298-0020, NR-031-732, June, 1970.
36. M. F. Ashby and R. Raj, "The Use of a Bubble Model to Study Stress-
    induced Migration and Sliding of Grain Boundaries", Tech.
    Report 556, Contract N00014-67-A-0298-0010, Nr-031-503, 1968.
37. H. Conrad; pp. 218-269 in Mechanical Behavior of Materials at
    Elevated Temperatures. Ed. by J. E. Dorn, McGraw-Hill Book
    Co., Inc., New York, 1961.

NONBASAL SLIP IN ALUMINA AT HIGH TEMPERATURES AND PRESSURES

A. H. Heuer,[*] R. F. Firestone,[*] J. D. Snow[**] and
J. Tullis [***]

[*]Case Western Reserve University, Cleveland, Ohio
[**]Ferro Corporation, Cleveland, Ohio
[***]University of California at Los Angeles, Los Angeles,
   California. Presently with Brown University,
   Providence, R. I.

ABSTRACT

*Czochralski (0°) sapphire has been deformed in tensile creep
at 1600-1800°C., and fine-grained alumina polycrystals have been
deformed under hydrostatic confining pressure at 1000-1200°C., in
order to activate nonbasal deformation modes. The tensile creep
data is in agreement with Nabarro-climb, although evidence for
pyramidal slip on the {1$\bar{1}$02} planes was also observed. Either of
these deformation modes, in conjunction with basal and prismatic
plane slip, permit general deformation of polycrystalline specimens.
In addition, "water weakening" was observed to occur in the deformation of polycrystalline alumina under hydrostatic confining pressures.*

INTRODUCTION

Although basal slip (0001) [11$\bar{2}$0] has the lowest critical
resolved shear stress of any slip system in Al_2O_3[1-3], there is considerable scientific as well as technological interest in ascertaining
and characterizing the important nonbasal slip modes. For example,
it is well known that basal slip provides only two of the five independent slip systems[4] required to satisfy the von Mises criterion
for homogeneous plastic deformation of a polycrystalline body without
loss of integrity. Table 1 summarizes all the slip systems that have
been reported to date for Al_2O_3; the most important of the nonbasal
systems will be briefly discussed here.

331

Table 1. Slip Systems in Sapphire

(SP)[*]	(SD)[**]	Comments	Min. Temp. Operation	Ref. Nos.
		BASAL		
a (0001)	1/3 [11$\bar{2}$0]	SP and SD confirmed lowest CRSS†	>900°C	1,2,3,12
b (0001)	[1$\bar{1}$00]	SD inferred from x-ray asterism	>1700°C	29
		PRISMATIC		
c {11$\bar{2}$0}	[1$\bar{1}$00]	SP confirmed SD suggested by crystallography	∿1800°C	5,6,7,30
d {11$\bar{2}$0}	1/3 [1$\bar{1}$01]	Only nonbasal SD identified by TEM††	?	8
e {1$\bar{1}$00}	1/3 [11$\bar{2}$0]	Inferred (dislocation decoration and x-ray asterism)	>1830°C	31
		PYRAMIDAL		
f {$\bar{1}$102}	1/3 [1$\bar{1}$01]	SP from slip traces SD by TEM	>1200°C	14
g {$\bar{1}$101}		SP from TEM	?	32
h {$\bar{1}$102}	1/3 [11$\bar{2}$0]	Inferred (dislocation decoration and x-ray asterism)	1870°C	30
i {$\bar{1}\bar{1}$22}	[1$\bar{1}$00]	Inferred (dislocation decoration)	1870°C	31
j {$\bar{1}\bar{1}$23}	[1$\bar{1}$00]	Inferred (dislocation decoration)	1870°C	30

    * SP   = Slip plane
    ** SD   = Slip direction
    †   CRSS = Critical resolved shear stress
    †† TEM = Transmission electron microscopy

    Klassen-Neklyudova first reported slip on the {11$\bar{2}$0} prism plane[5] and subsequently[6] suggested, based on crystallographic reasoning, that [10$\bar{1}$0] is the slip direction. Although it is now clear that other nonbasal slip modes operate in Al_2O_3 (Table 1), the {11$\bar{2}$0} slip plane has been confirmed.[7] However, there are two difficulties with the   [10$\bar{1}$0]   direction: (i)   dislocations with $\bar{b}$ equal to

[10$\bar{1}$0] should be unstable; decomposition to two   1/3 [11$\bar{2}$0] dislocations is energetically favorable, as was first pointed out by
Scheuplein and Gibbs[7] and later emphasized by Gulden[8]; and (ii) all
nonbasal dislocations that have been unambiguously identified by
transmission electron microscopy[8,9] have had b equal to 1/3 [10$\bar{1}$1],
and it has been proposed[8] that [10$\bar{1}$1] may be the slip direction for
{11$\bar{2}$0} slip.

Operation of basal and prismatic plane slip will not, however,
allow a general deformation in alumina polycrystals, inasmuch as no
extension along <0001> is possible. Groves and Kelly[10] have argued
that Nabarro-type climb[11] may be an important deformation mechanism
in alumina, and suggested that the tensile creep observed by Wachtman
and Maxwell[12] in 0° and 90° single crystals occurred by this mechanism.
On the other hand, Table 1 indicates that several pyramidal slip
modes are possible in alumina. Of these, "rhombohedral" slip* (f in
Table 1, first reported by Bayer and Cooper[14]) is probably the most
important. The operation of any of the pyramidal systems, in conjunction with basal and prismatic plane slip, will satisfy the von
Mises criterion for randomly oriented polycrystals.

The work reported in the present paper (some of which has been
reported elsewhere[16-18]) has been concerned with the tensile deformation of 0° sapphire crystals (an orientation chosen to suppress
basal and prismatic slip) at elevated temperatures (1600-1800°C)
and with the deformation of fine grained alumina polycrystals under
hydrostatic confining pressures at relatively low temperatures
(1000-1200°C). This deformation condition was chosen in order to
suppress diffusional creep and grain boundary sliding, important
deformation mechanisms in fine grained alumina at atmospheric
pressures.[15]

TENSILE DEFORMATION OF 0° SINGLE CRYSTALS

The most straightforward way to activate pyramidal slip in
sapphire is by tensile deformation of 0° crystals - the resolved
shear stress is then zero on both basal and prismatic planes. The

---

* The {$\bar{1}$012} plane (structural unit cell, c/a = 2.73) is equivalent
to the {100} morphological rhombohedral plane (morphological unit
cell, c/a = 1.365). To be consistent with previous literature on
the mechanical properties of alumina, the {$\bar{1}$012} plane will be
referred to as "rhombohedral". From a crystallographic point of
view, this is incorrect - the {10$\bar{1}$1} plane is by definition the
rhombohedral plane, being indexed as {100} in the rhombohedral unit
cell. For certain mechanical properties, e.g., rhombohedral deformation twinning[13], the morphological unit cell is more convenient to
use.[2] For studying dislocation phenomena by transmission electron
microscopy, however, use of the structural unit cell is mandatory
and is the basis for notation throughout this paper.

samples used for this study were centerless ground rods (3.2 mm dia, 305 mm long) of 0° Czochralski-grown sapphire, having an initial dislocation density less than $10^2/mm^2$. An internally wound, platinum-40% rhodium furnace provided temperatures up to 1800°C in air; cold grips of the Diehl type were employed. Initially, it was hoped to use an Instron tensile testing machine and a specimen with a 25 mm "thermal" gauge section. However, even at temperatures up to 1800°C, brittle behavior was observed at the lowest crosshead extension rate available, $3.3 \times 10^{-4}/sec$. Under these conditions, the fracture stress, $16 \times 10^7 N/m^2$, was undoubtedly lower than the stress needed to activate nonbasal deformation. Accordingly, tensile creep tests have been used. It was necessary to use specimens with a 25 mm long machined gauge section of 2.3 mm dia because delayed failure was observed to occur at that point in the rod where the temperature was between 300 and 800°C, due to the strength minimum which has been reported in sapphire in this temperature range.[19]

Creep could readily be induced at temperatures above 1600°C for stresses $>86 \times 10^6$ N/m^2. One surprising feature of the creep tests (not yet fully understood) is the lack of an induction period (equivalent to a yield point in a constant strain rate test) or of any primary or tertiary creep; steady state (secondary) creep was established immediately (within 5 min) for all tests.

Table 2. Tensile Creep in 0° Czochralski Sapphire

Temp. (°C)	Environment	Tensile Stress (MN/m^2)	Creep Strain Rate (nm/m/sec)	
			Observed	Calculated
1600	Ambient Air	86	14	9–44
1600	Ambient Air	114	56	20–106
1650	Ambient Air	86	28	22–116
1650	Ambient Air	114	83	48–134
1700	Ambient Air	84	44	60
1700	Ambient Air	112	140	162
1700	Vacuum, 10^5	220	170*	1100
1800	Ambient Air	38	< 5	4
1800	Ambient Air	114	330	1040
1900	Vacuum, 10^{-5}	172	4750*	20000

*Data from P. Shahinian[23]

Both stress changes and temperatures changes have been employed to characterize the tensile deformation; typical results are given in Table 2. The creep strengths of 0° sapphire are high, being possibly the best for any materials yet reported in this temperature

range. These tests indicate an activation energy of 375 to 490 kJ/mole
consistent with other values reported for deformation of alumina, and
a stress exponent (n in the relation $\dot{\varepsilon} = \sigma^n$) of ~ 3.

This value for the stress exponent is consistent with deformation
due to Nabarro climb[11] and thus may be confirmation of Grove's and
Kelly's[10] assertation of the importance of this deformation mode in
alumina. The strain rates calculated using Weertman's modification
of Nabarro's equation[20] and extrapolated literature values for oxygen
self-diffusion[21] and shear modulus[22] are compared with the observed
values in Table 2. Calculated values are also shown for Shahinian's
data on 0° sapphire filaments.[23] The agreement between calculated
and experimentally determined creep rates is quite good.

Preliminary observations of slip dislocations suggest that there
may be a concurrent slip mechanism. In no case have plastic strains
in excess of 10% been achieved during tensile creep and the fracture
surfaces have been characterized by large pyramidal facets. This
limited ductility may have resulted from the intersection of several
crystallographically equivalent but noninterpenetrating slip systems.
Furthermore, etch-pit studies have shown that the dislocation density
on the basal plane increases from $<10^2$ to $>10^5$/mm^2 during creep. The
observation of dislocations intersecting the basal plane is clear and
unambiguous evidence for the generation of nonbasal dislocations
during creep.

A complete transmission electron microscopy study of deformed
specimens is not yet available. However, long straight dislocations
have been observed lying on $\{\bar{1}012\}$ "rhombohedral" planes, suggesting
that "rhombohedral" slip had been activated during tensile creep.

DEFORMATION OF ALUMINA POLYCRYSTALS
UNDER HYDROSTATIC CONFINING PRESSURES

Although deformation under high hydrostatic confining pressures
has been investigated in a number of metals and some minerals (the
literature in this field has been reviewed recently[24]), with the
exception of some work on MgO,[25] little research of this type has
been attempted for refractory oxides. In the context of the present
work, the deformation of polycrystalline alumina at elevated temperatures under hydrostatic confining pressures was both novel and
valuable in diminishing possible contributions of grain boundary sliding
and Nabarro-Herring creep.

The specimens used for the present study were cylinders, cut from
a single hot-pressed Al_2O_3 + 1/4% MgO disc of 2 μm average grain size.
They were deformed in a solid-medium pressure apparatus (described
by Griggs[26]) using both "Alsimag" and talc as the pressure transmitting

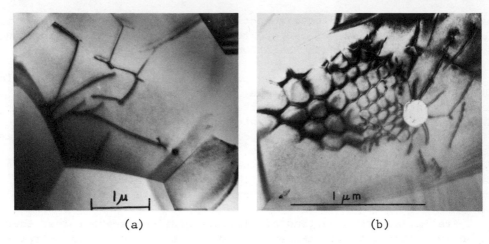

(a)   (b)

Fig. 1.   Dislocation networks in annealed, hydrostatically deformed
          polycrystal. (a) The plane of the grain containing the
          dislocations is {10$\bar{1}$1}.   (b) The plane of the grain con-
          taining this net is {63$\bar{3}$7}. (Bright field).

media. Preliminary results for the "Alsimag" tests have been reported
elsewhere.[16]   Briefly, it was found that fine-grained (2 μm) polycrystals were stronger than coarser grained material (10 μm) at
1000-1200°C,[*]   and that deformed specimens had substructures similar
to that of cold-worked metals when examined by transmission electron
microscopy.   Furthermore, diffraction contrast experiments on specimens
annealed after deformation suggested that some nonbasal dislocations
had been present during the deformation. Evidence indicating that
pyramidal slip had been activated in the high pressure samples is
available from the dislocation networks shown in Fig. 1. The length
of the dislocation segments, combined with the thickness of the foil
(~3 x 10^{-7}m) and the orientation of the two grains, indicates unequivocally that the dislocation segments could not be lying in either a
basal or a prism plane, but must be on pyromidal planes.

     The strengths of the samples deformed in talc were much lower
than those deformed in "Alsimag" (Fig. 2), and this can be correlated
with the fact that talc dehydrates under pressure at 950°C. The
water liberated within the pressure chamber is apparently able to
diffuse into the specimen and lowers its strength; polycrystalline

---

     [*]Although it is well known[27] that fine grained metals are usually
stronger than those with a coarser grain size, the opposite effect is
commonly observed in polycrystal ceramics at elevated temperatures at
atmospheric pressure. This softening with decreasing grain size in
ceramics can be correlated with the importance of grain boundary
sliding and diffusional creep in fine-grained polycrystals.[15] It is
of interest to note that polycrystalline Al_2O_3 appears to be the
strongest material tested to date at elevated temperatures under
pressure.

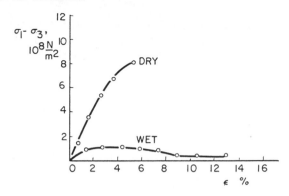

Fig. 2.   Stress-strain curves for 2 μm grain size alumina poly-
          crystals tested in compression at 1200°C at a strain rate
          of 10^6/sec under a hydrostatic confining pressure of
          14×10^8 N/m^2. The specimen marked "wet" was deformed in
          talc, while that marked "dry" was in "Alsimag".

alumina thus appears to be as susceptible to "water weakening" as
are quartz and other silicates.[26]   In the case of quartz, where substantial quantities of "water" can be incorporated into the lattice
as OH$^-$ during crystal growth, "wet" crystals (high OH$^-$ content) are
stronger than "dry" crystals (low OH$^-$ content).[28] However, dry

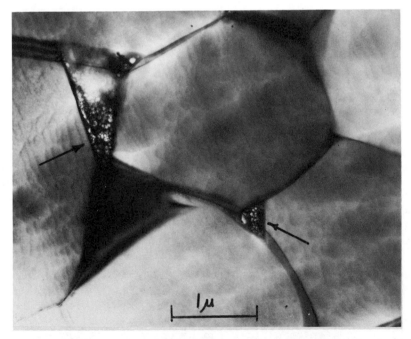

Fig. 3.   "Foam" structure (arrowed) in "water-weakened" alumina
          polycrystal. (Bright field).

crystals of quartz can be readily weakened by deforming in talc
above 950°C.[26] Griggs has hypothesized that "hydrolytic" weakening
in silicates occurs when Si-O-Si bridges adjacent to a dislocation
are hydrolyzed by the migration of water, thus allowing easy slip
to occur by hydrogen bond exchange (Frank-Griggs hydrolytic weakening).
Although the solubility of "water" in Al_2O_3 is apparently very much
lower than in quartz,[29] a similar phenomena may be occurring.
However, grain-boundary sliding type phenomena may be even more
important in causing water weakening in fine grained alumina polycrystals. Figure 3 shows a "foam" structure visible at triple
points in water-weakened polycrystals, and also shows an example
of a grain having a rounded appearance; in addition, there is an
absence of sharp contrast at grain boundaries. None of these effects
have been seen heretofore in deformed alumina polycrystals. It is
suggested that the liberated water diffused along the grain boundaries
of the polycrystal and reacted to form a grain boundary "lubricating"
phase, which enhanced grain boundary sliding and thus caused a
marked weakening. It is not suggested that dislocation activity was
absent in water-weakened polycrystals. Indeed, Hockey[18] has studied
the substructure of coarser-grained water-weakened alumina polycrystals
by transmission electron microscopy, and found the dislocation
structure to be much more marked than in the fine-grained sample
shown in Fig. 3. High pressure deformation studies of single crystals
in both talc and "Alsimag" are needed to assess the magnitude of
Frank-Griggs hydrolytic weakening in alumina.

SUMMARY AND CONCLUSIONS

The tensile creep data on 0° alumina single crystals are in
agreement with the operation of Nabarro-climb deformation, although
evidence for pyramidal slip was also obtained. Both modes of
deformation provide a means of obtaining elongations parallel to
<0001> and of satisfying the von Mises criterion in polycrystals.

Water weakening was observed during the deformation of alumina
polycrystals under hydrostatic confining pressures. In fine grained
samples, an unusual grain-boundary sliding phenomena appeared to be
involved, although dissolved "water" may also cause easy slip to
occur by hydrolytic weakening. Because of the low solubility for
water in alumina, this latter mechanism may be less important than
in silicates.

ACKNOWLEDGMENT

*The work performed at Case Western Reserve University was sup-
ported by the Office of Naval Research under Contract N00014-67-A-
0404-0003. We are grateful to W. Roger Cannon, Stanford University,
for pointing out to us the good correlation of our creep data with
Nabarro-climb deformation.*

REFERENCES

1. J. B. Wachtman, Jr. and L. H. Maxwell, J. Amer. Ceram. Soc. 37
   291 (1954).
2. M. L. Kronberg, Acta Met. 5 507 (1957); ibid, J. Amer. Ceram.
   Soc. 45 274 (1962).
3. H. Conrad, J. Amer. Ceram. Soc., 48 195 (1965).
4. G. W. Groves and A. Kelly, Phil. Mag., 8 877 (1963).
5. M. V. Klassen-Neklyudova, J. Tech. Phys. (U.S.S.R.), 12 519,
   535 (1942).
6. M. V. Klassen-Neklyudova and G. E. Tomilevskii, Trudy Inst. Krist.
   Akad. Nauk., U.S.S.R., 8 237 (1953).
7. R. Scheuplein and P. Gibbs, J. Amer. Ceram. Soc., 43 450 (1960);
   ibid, 45 439 (1962).
8. T. D. Gulden, J. Amer. Ceram. Soc., 50 472 (1967).
9. D. J. Barber and N. J. Tighe, Phil. Mag., 11 495 (1965); ibid,
   14,531 (1966).
10. G. W. Groves and A. Kelly, Phil. Mag., 19 977 (1969).
11. F. R. N. Nabarro, Phil. Mag., 16 231 (1967).
12. J. B. Wachtman, Jr. and L. H. Maxwell, J. Amer. Ceram. Soc.,
    40 377 (1957).
13. A. H. Heuer, Phil. Mag., 13 379 (1966).
14. P. D. Bayer and R. E. Cooper, J. Mater. Sci., 2 301 (1967).
15. A. H. Heuer, R. M. Cannon, and N. J. Tighe; in Ultrafine-Grain
    Ceramics. Eds. J. J. Burke, N. L. Reed, and V. Weiss, Syracuse
    University Press, Syracuse, N. Y. 1970.
16. A. H. Heuer, R. F. Firestone, J. D. Snow and J. Tullis, to be
    published in the Proc. 2nd Int. Conf. on The Strength of
    Metals and Alloys, Amer. Soc. Metals.
17. R. F. Firestone and A. H. Heuer, to be published.
18. A. H. Heuer, J. D. Snow, B. J. Hockey, and J. Tullis, to be
    published.
19. A. H. Heuer and J. P. Roberts, Proc. Brit. Ceram. Soc., 6 17 (1966).
20. W. J. Weertman, Trans. ASM, 61 6811 (1968).
21. Y. Oishi and W. D. Kingery, J. Chem. Phys., 33 480 (1960.
22. N. Soga and O. L. Anderson, J. Amer. Ceram. Soc. 49 355 (1966).
23. P. Shahinian, J. Amer. Ceram. Soc., 54 67 (1971).
24. Effect of Pressure on Mechanical Behavior of Materials. Ed.
    H. L. L. D. Pugh, Elsevier Press (The Netherlands), 1970.
25. (a) C. W. Weaver and M. S. Patterson, J. Amer. Ceram. Soc., 52
    293 (1969); (b) M. S. Patterson, C. W. Weaver, Ibid, 53
    463 (1970); (c) T. A. Auten, Ph.D. thesis, Case Western
    Reserve University, Cleveland, Ohio. 1970.
26. D. T. Griggs, Geophys. J. Roy. Ast. Soc., 14 19 (1967).
27. R. W. Armstrong, Metal Trans., 1 1169 (1970).
28. D. T. Griggs and J. D. Blacic, Science, 147 292 (1965).
29. R. A. Waugh, Ph.D. Thesis, University of Leeds, Leeds, U.K. 1969.

30. M. V. Klassen-Neklyudova, V. G. Govorkov, A. A. Urusovskaya, M. N.
       Voinova and E. P. Kozlovskaya, Phys. Stat. Solidi, 39 679 (1970)
31. V. N. Rozhanskii, M. P. Nazarova, I. L. Svetlov and L. K.
       Kalashnidova, in Proceedings of a Conference on Deformation
       Hardening of Single Crystals, Charkov, 1969.
32. B.J. Hockey, submitted for publication, J. Amer. Ceram. Soc.

DISCUSSION

L. L. Hench (University of Florida): What annealing conditions were
necessary to obtain polygonization in the worked sample?
Authors: Several hours at 1500°C were required.

B. A. Wilcox (Battelle Memorial Institute): Regarding the polycrystalline Al_2O_3 deformation studies under hydrostatic pressure,
after testing did you find by metallography that grain boundary
cracking (or void formation) was suppressed by the hydrostatic
pressure, i.e., similar to what Hall and Rimmer found for polycrystalline Cu?
Authors: After pressure quenching the specimens are so highly
strained (internally) that metallographic preparation is extremely
difficult; furthermore, the small grain size prevents meaningful
optical microscopic examination. Therefore we directed our efforts
to electron microscopy. Nevertheless, the point you make is correct,
i.e., the effect of the hydrostatic pressure was to suppress both
grain boundary separation and void formation at triple points.

CERAMIC FIBRES FOR THE REINFORCEMENT OF GAS TURBINE BLADES

J. E. Bailey and H. A. Barker

University of Surrey

Guildford, Surrey, England

ABSTRACT

Continuous filament reinforcement of nickel base alloy turbine blades with SiC, W, or Al₂O₃ has been considered as a means of improving creep resistance at high temperatures (up to 1200°C). SiC is readily available in fibre form, but reacts chemically with the matrix; tungsten wires are excessively dense; and alumina fibres have not been available until recently. Extrusion or pulling of molten alumina was tried but abandoned at an early stage. Extrusion and sintering of a very concentrated aqueous dispersion of hydrated alumina has been moderately successful. Bend strengths of about 80×10^3 psi at 85% theoretical density are better than most commercially available sintered alumina rods of larger diameter and are capable of withstanding stresses $>10 \times 10^3$ psi for 100 hr at 1100°C. Fabrication of composite blades using polycrystalline alumina (or continuous sapphire single crystal) fibres has not proved successful by liquid infiltration since severe damage is caused to the fibres.

INTRODUCTION

It is well known that the efficiency of gas turbines could be increased by raising the operating temperature. This necessitates new materials with improved creep resistance while still retaining oxidation, thermal shock and impact resistance at least equal to the present nickel-base alloys. The brittleness and poor corrosion resistance of alloys based on niobium or chromium prevents their use as turbine blade materials. A possible solution to the problem

341

may lie in the use of composite materials. With this in mind, the
reinforcement of nickel alloy turbine blades with fibres has been
studied fairly extensively by a number of workers. However, the
present discussion will be limited to reinforcement with continuous
filaments aligned along the direction of the blade axis, that is
parallel to radius of the turbine rotor. Such fibres must carry
the major part of the creep loading and also be compatible with the
matrix for at least 100 hours under service conditions.

Tungsten or molybdenum alloys, silicon carbide, and alumina
were thought to have the greatest potential as fibre materials for
this application. A programme of work has been carried out at the
National Gas Turbine Establishment (NGTE) to investigate the use
of these materials. Tungsten wires clearly have the advantage of
ready availability in a variety of diameters. Tests showed[1,2] that
tungsten, or preferably W-5% Rh, has as a very high creep-rupture
strength and is in fact superior to either Al_2O_3 or SiC in this
respect. However, W-alloys are comparatively dense (19.4 g/cm^3)
and since the creep loading of a turbine blade is due largely to
centrifugal force it is the ratio of the ultimate tensile strength
(UTS) to the density, or specific strength, that must be considered.
The specific creep-rupture strength of W-5% Rh wires for 100 hr at
1000°C is 4.3 km (172 x 10^3 in*), approximately twice that of pure
tungsten wires. Certain Mo-alloy wires were found to have specific
creep-rupture strengths somewhat higher than W-5 Re, but their
stability in a nickel matrix was poor. Nickel-base alloy matrix-
tungsten wire composite blades have in fact been fabricated at
NGTE by conventional casting techniques.[3,4]

The SiC filaments used in the investigation were supplied by
the United Aircraft Corporation. They had a 0.07 mm (0.0028 in.)
dia and a density of 3.18 g/cm^3 (0.115 lb/in^3). Their room
temperature UTS, 2.41 GN/m^2 (350 x 10^3 psi) was very close to that
of W-5% Re wire. The specific strength of the SiC was therefore
much greater. This is reflected in the value of 6.75 km
(270 x 10^3 in.) obtained for specific creep-rupture strength for
100 hr at 1000°C. However, by 1100°C the specific creep-rupture
strength ahs dropped to about 2.38 km (95 x 10^3 in.) compared to
about 3.08km (123 x 10^3 in.) for the W-alloy under the same conditions.
Silicon carbide has the further disadvantage of undesireable chemical
interaction with Ni-alloys, although protective coatings of W, which
would be feasible for filaments of 0.5 mm (0.020 in.) dia might
overcome this problem.

Continuous high strength Al_2O_3 filaments have not been available
until recently. - In view of the potentially high specific strength
of alumina reflected in the value 160 km (6.4 x 10^6 in.) for

---

*   Units of length are commonly used but more correctly these figures
    should be multiplied by the acceleration due to gravity.

sapphire whiskers compared to about 65 km (2.6×10^6 in.) for
continuous SiC filament and 13 km (0.5×10^6 in.) for W–5% Re wire,
it was thought worthwhile to study possible methods for the production of strong Al_2O_3 filaments. Extrusion techniques have been
found to be unsuitable for molten Al_2O_3 due to the lack of a viscous
region near the melting point. Alumina degraded with 40% silica
could be fiberized in this way but would not be expected to have
any value as a high temperature reinforcement. However, a fairly
conventional ceramic procedure involving the extrusion and firing
of a plastic mix of hydrated alumina was moderately successful.
The method (described in detail later) is capable of producing
straight continuous filaments, perhaps more appropriately referred
to as rods, of 0.12 – 0.75 mm (0.005 to 0.030 in.) dia with negligible variation in diameter over lengths of many metres. They have
mean room remperature three-point-bend strengths of about 0.55 to
0.70 GN/m^2 ($80 - 100 \times 10^3$ psi) with values for individual specimens
sometimes reaching 0.90 GN/m^2 (130×10^3 psi). Other attempts to
produce strong alumina fibres have been fairly limited and there is
often some difficulty in obtaining information in this area of
research. A brief review of reported work is given below.

REVIEW OF FABRICATION AND PROPERTIES OF ALUMINA FILAMENTS

Kliman's work[5] in 1962 was apparently the first attempt to
produce Al_2O_3 filaments by an extrusion and firing method. Suspensions
of fused, calcined or colloidal alumina powders in ammonium alginate
solution were extruded, dried and sintered in batches to produce
fibres about 1 m long and 0.05 to 3.50 mm (0.002 to 0.140 in.) in
dia. The predominant range was, however, 0.05 to 0.50 mm. The bend
filaments had transverse bend strengths of about 0.70 GN/m^2
(100×10^3 psi) after firing for 3–4 hr at 1650°C. After this
treatment there was little difference netween calcined and colloidal
alumina as raw materials, but prolonged firing had a weakening
effect on fibres prepared from the latter, apparently due to
excessive grain growth. Firing to temperatures as high as 1900°C
did not improve the strength, but sintering times could be reduced
to 1 hr at this temperature.

Some years later Parratt[6] at the Explosives Research and Development Establishment used essentially the same technique, with a submicron particle size partly calcined alumina, to produce rather
thicker, 0.1 to 1.0 mm (0.004 to 0.040 in) dia. fibres. The transverse rupture strength of these was about 0.35 GN/m^2 (50×10^3 psi).
The process was slightly different in that it made fuller use of
the properties of the ammonium alginate. By extruding directly into
dilute HCl the alginate was gelled to yield strong green fibres
which could be handled easily. Excess acid was subsequently removed
by washing in water, and magnesium ions were introduced into the
fibre by soaking in $MgSO_4$ solution.

Further work of this kind was carried out at Illinois Institute
of Technology Research Institute.[7,8] Suspensions of a fibrillar
colloidal boehmite, with no added plasticiser but with a small
quantity of glycerine or oleic acid as lubricant, were extruded
through an orifice of about 0.125 mm (0.005 in.) and flame sintered
to yield fibres of about 0.063 mm (0.0025 in) dia. Their room
temperature U.T.S. was 0.213 GN/m^2 (31 x 10^3 psi) with values for
individual specimens of up to 0.48 GN/m^2. Electron microscopy revealed regions of undefined structure and areas of well developed
1-2 μm grains within the fibres.

More recently Nazarenko *et al*[9] has extruded 1 μm particle size
α-Al_2O_3 plus sintering aids such as 1% MgO plasticised with a variety
of organic gels, glues, cellulose, rubbers and waxes. The 0.2 to
0.5 mm (0.008 to 0.030 in) dia fibres were heated to 1650°C over
a period of 5 hr, soaked for 1 hr and cooled slowly. Despite a
density of only about 70% of theoretical and a correspondingly poor
elastic modulus of 200-250 GN/m^2 (29 - 36 x 10^6 psi), tensile
strengths as high as 2.20 to 3.00 GN/m^2 (320 - 430 x 10^3 psi) were
claimed. Stabilised ZrO_2 fibres were also produced by this process,
for which the tensile strength was reported to be 1.0 - 2.0 Gn/m^2
(145 - 290 x 10^3 psi). Ceramic and metal matrix composites were
said to have been fabricated from these oxide fibres.

It is worthwhile mentioning several other approaches to alumina
fibres production before returning to a more detailed account of
the present work which was very similar to the methods reported above.

Thermal decomposition of salts has been used by Blaze[10] to
manufacture oxide fibres, including alumina, of a semicontinuous
nature. The composition of solutions containing very large concentrations of aluminium ions was adjusted so that highly viscous
liquids were formed. This was achieved partly by the addition of
sufficient colloidal silica to yield 15 - 25% SiO_2 in the final
product. Other additions included B_2O_3 and Cr_2O_3. The liquid can
be fiberised by centrifugal spinning or preferably by a specially
developed extrusion attenuation method. The fibres were dried by a
variety of methods such as hot air currents or infra-red lamps. They
were then calcined at relatively low temperatures. The diameters
of the fibres were about 0.2 to 10 μm and their average U.T.S. was
in the region of 1.4 GN/m^2 (200 x 10^3 psi). The extrusion-attenuation
technique presumably overcomes the disadvantages of diameter variation
and the discontinuous nature of fibres associated with centrifugal
spinning processes, but with such high silica content good creep
resistance at elevated temperatures would not be expected.

Two processes employing organic precursor fibres from which
ceramic relic fibres are obtained by pyrolysis have been developed.
The first of these due to Hamling *et al*[11,12] involved the impregnation

of cellulosic fibre with an aluminium salt. This was dried and
pyrolysed to remove organic material, oxidised to yield alumina,
and finally sintered. Clearly it is potentially a continuous process. The grain size in these fibres was said to be only about 1000Å.
The fibre diameter was typically about 5 μm and their U.T.S. was
0.61 - 0.90 GN/m² (85 - 130 x 10³ psi). The other method was
developed by Wizon and Robertson.[13,14] Inorganic polymers were poly-
merised in a cellulosic matrix to form the precursor fibre material.
Filaments formed from this were similarly pyrolysed to yield pure
or mixed oxide filaments with high tensile strengths. The microstructure was again extremely fine and there was a tendency for the
10 μm filaments to fibrillate into 0.2 - 0.3 μm fibrils. Firing
under tension markedly increased their tensile strength.

Perhaps the most interesting pure alumina filaments are those
currently produced by Tyco Laboratories.[16] These are single crystals
of unlimited length (in the c-axis direction) and about 150 - 500 μm
(0.006 to 0.020 in.) dia. They are produced by a carefully controlled
crystal growth process in which the filament is drawn upwards through
an orifice in a refractory metal boat floating on molten alumina.
Their U.T.S. is ~2.1 GN/m² (300 x 10³ psi) and their modulus of
elasticity if 460 GN/m² (67 x 10⁶ psi). However, their present very
high cost is prohibitive. Even so this material has been considered
for turbine blade reinforcement in the expectation that the cost will
eventually be greatly reduced.[4]

SOME RECENT STUDIES OF POLYCRYSTALLINE ALUMINA FIBRES AND RODS

The work described below was initiated at Morganite Research and
Development and continued by the present authors at the University
of Surrey. A full account of the process has already been published
elsewhere[17] and only an outline will be given here.

The raw material which consistently gives the best results is
"Balgel", a non-fibrillar colloidal boehmite produced on a moderate
scale at the Atomic Energy Research Establishment.[18] The only additive apart from the water in which the powder is dispersed, is a
sintering aid (0.25% MgO) added by coprecipitation with the boehmite.
A colloidal boehmite, supplied by Cawood Wharton Ltd. under the
name G. C. Alumina hydrate, with magnesium added as an aqueous
solution of a salt has given almost as good results. A very stiff
paste containing about 60% alumina hydrate and 40% water is extruded
through a nozzle which is commonly between 0.75 and 1.50 mm dia,
although nozzles down to 0.50 mm have been used successfully. A
loop of the wet filament is allowed to hang freely before passing
over a uide wheel and into a preheat furnace maintained typically
at about 1350°C. This serves to dry and partly sinter the filament
before passing through the final sintering stage. The latter consists
of either an oxypropane flame, the temperature of which can be varied

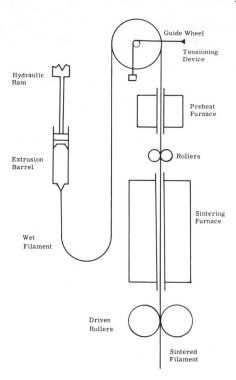

Fig. 1. Diagram of the production of alumina filaments and thin
         rods by continuous sintering.

to well above 2050°C but is normally about 1850°C, or an air atmosphere tube furnace at 1700-1850°C. A roller tensioning device keeps
the filament straight during sintering (Fig. 1.) Throughput rates
which are about 10 cm/min are determined largely by the need to
provide the correct properties such as flexibility, dryness and
strength at the various process stages. The filaments shrink to
about 60% of their extruded diameter during firing and attain
densities of 75 - 90% of theoretical. The addition of 0.3 µm
particle size α-alumina (Linde A) increases their density but seriously weakens them.

    The strength of the thin alumina rods or filaments fabricated
in this way was evaluated initially by 3-point and 4-point flexural
loading. The loading spans were arranged in each type of test to
give the same maximum moment at a given load. The ratio of the
measured 3- and 4 point strengths was found to be 1.47 ± 0.17. In
view of the fairly small standard deviation from the average ratio,
only 3-point loading was subsequently used. Mean flexural strengths
(3-point) of about 0.55 - 0.70 GN/m^2 (80 - 100 x 10^3 psi) can be
achieved quite consistently by oxy-propane flame firing. Furnace
sintering in air at 1850°C generally yields slightly higher strengths

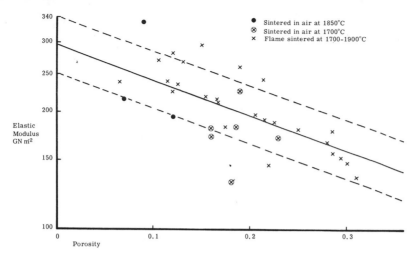

Fig. 2.   The variation of elastic modulus with porosity for thin
          alumina rods prepared from Balge 1/0.25 MgO.

with mean values occasionally in excess of 0.70 GN/m^2 (100 x 10^3 psi).
Values as high as 0.90 GN/m^2 (130 x 10^3 psi) have been recorded for
individual specimens.

The elastic modulus of the rods, also determined by flexural
loading, is between 120 - 350 GN/m^2 (20 - 50 x 10^6 psi) depending
on the density of the specimen.   The observed increase in modulus
with increasing density is roughly as expected for a porous ceramic
(Fig. 2)[19] and the actual values are in agreement with others reported
for sintered alumina.[20]

The strength (Fig. 3) even at 20% porosity, is considerably above
the range 0.20 - 0.35 GN/m^2 (30 - 50 x 10^3 psi) which is typical for
most commercially available sintered alumina.   Although there is a
wide scatter in the strength/porosity data it is clear that strength
increases with decreasing porosity in the present material within
the range 25% to 10% porosity.   A number of factors could contribute
to the unexpectedly high strength, but a definite evaluation of these
is difficult since the relationship between strength and microstructur
of ceramics is complex and only partially understood.[19]

A better understanding of the properties of the filaments has
been sought through scanning electron microscopy of external surfaces
and fracture surfaces.   Surface micrographs (Fig. 4) of the specimens
prepared from Balge1/0.25% MgO show irregularities of various kinds
including longitudinal ridges, presumably caused by defects in the
extrusion nozzle, and small hills and depressions up to about 30 μm
in size.   These are probably not serious stress concentrators.   It

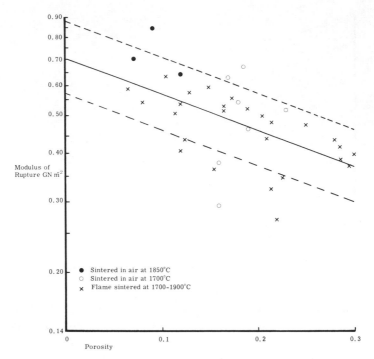

Fig. 3.   The effect of porosity on the strength in 3-point flexure
           of thin alumina rods prepared from Balgel/0.25 Mgo.

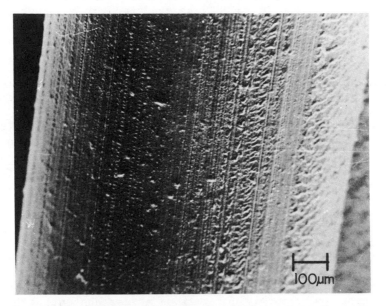

Fig. 4.   Scanning electron micrograph of the external surface of
           a furnace sintered thin alumina rod.

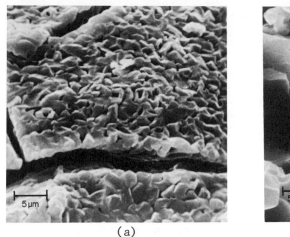 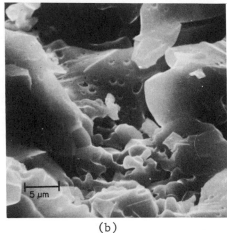

(a)   (b)

Fig. 5.   Scanning electron micrographs of flame sintered thin
          alumina rods (a) severe cracking of external rod surface
          (Balgel/0.25 MgO-Linde A); (b) irregular grain structure
          on fracture surface (G. C. Hydrate-Linde A).

is more likely that small sharp cracks extending along grain
boundaries from the surface are the cause of crack initiation. In
contrast, large surface cracks of the order of 50 μm can be seen in
specimens containing Linde A in both the external surface and the
fracture surfaces (Fig. 5). The grain structure of such filaments

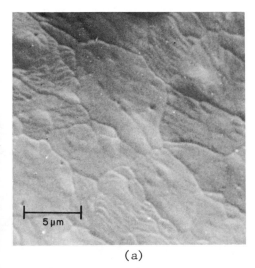 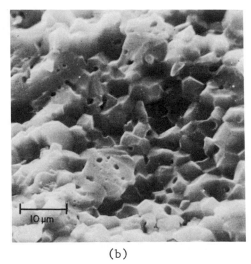

(a)   (b)

Fig. 6.   Scanning electron micrographs of furnace sintered thin
          alumina rods (Balgel/0.25% MgO) (a) as fired surface texture;
          (b) fine regular grain texture, fracture surface.

is coarser and more irregular than those prepared from Balgel/0.25%
MgO along (Fig. 6). These observations must certainly be connected
with the lower strength observed for specimens containing Linde A.

A dependence of strength on a statistical distribution of flaws
should be reflected in a difference between the rupture strength in
3- and 4-point loading since a greater volume of material is subjected to the maximum stress in the latter. As mentioned above a
ratio of about 1.5 between these strengths was in fact observed
for the filaments.

As observed for many other thin filaments or fibres there appears
to be a slight tendency towards higher strengths for smaller diameters.
It is difficult to define precisely the factors which may lead to
this behaviour, but they may include the competing effects of surface
to volume ratio and a limitation on the inherent crack size imposed
by the fibre diameter. In particular the effect of thermal shock
during manufacture may be less severe at smaller diameters.

Both surface and fracture surface micrographs revealed a grain
size which was particularly fine and regular in specimens prepared
from Balgel/0.25% MgO (Fig. 6). The grain diameter as observed by
scanning electron microscopy is of the order of 1 - 5 μm. The grain
sizes for thin rods prepared from various hydrated aluminas by
Blakelock et al[17] were plotted against strength (Fig. 7). For rods
of similar density and diameter a relationship of the kind

$$\sigma = \sigma_o + k_1 \, G^{-\frac{1}{2}} \tag{1}$$

was obtained, where σ_o and k_1 are arbitrary constants and G is the
grain diameter. When the grain size becomes comparable to the

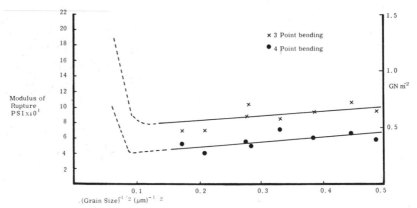

Fig. 7. Modulus of rupture versus grain size of thin alumina rods
prepared from various alumina hydrates.[17]

diameter of the rod or filament the strength observed appears to be
appreciably greater than predicted by Eq. 1. Howvever, they
obtained insufficient data to determine the full extent of this
change of behaviour.

A relation of the form of Eq. 1 has been suggested by
Carniglia[21] to be applicable to fine grain size ceramics, but more
usually[19] the dependence of strength on grain size can be described
by

$$\sigma = K_2 \, G^{-\frac{1}{2}}$$ (2)

A direct relationship between grain size and the size of inherent
flaws may be inferred from Eq. 2. Following Griffith[22] the minimum
size, c, of cracks energetically capable of propagation at a given
stress, σ is given by

$$\sigma = (\frac{2 \, \gamma \, E}{\pi \, c})^{\frac{1}{2}}$$ (3)

Taking the value $\sigma = 0.48$ GN/m^2 which is approximately the flexural
strength in 4-point loading of the better filaments, putting Young's
modulus, E, equal to 400 GN/m^2 and taking the value 10 J/m^2 for the
effective surface energy, γ, as measured in double cantilever beam
experiments by Gross and Gutshall,[23] we obtain c $\simeq$ 10 μm. Gross
and Gutshall have recently shown that while the effective surface
energy increases with increasing grain size, it is the over-riding
effect of surface flaws which causes the measured flexural strength
of polycrystalline alumina to decrease with increasing grain size.
Davidge and Tappin[24] have obtained results for two polycrystalline
aluminas which indicate that the Griffith flaw size is equal to the
maximum grain size plus pore size. As seen from the micrograph
(Fig. 8) the pores in the present material are only of the order
of 0.2 - 0.4 μm dia and are spaced at intervals which are certainly
less than 1 μm. Thus they probably do not contribute to the crack
length in this instance. The calculations above indicate a crack
size somewhat larger than the observed grain diameter but the value
of γ used was derived from experiments on hot-pressed alumina and
is not necessarily applicable to the present material. Passmore
et al[25] have found that the ratio of the critical crack size calculated from Eq. 3 to the mean grain diameter is independent of
grain size and equal to about one for hot pressed alumina (porosity 3%)
of 2 - 100 μm grain size; however, for similar material with 6%
porosity this ratio falls from 1.5 at 2 μm to 0.7 at 100 μm. The
actual values of the ratio are suspect since the intrinsic surface
energy, 1.0 J/m^2, was used in the calculation of the crack length.
Hoever, it is clear that one must take account of porosity when
attempting to relate the crack size to the grain size. Moreover,

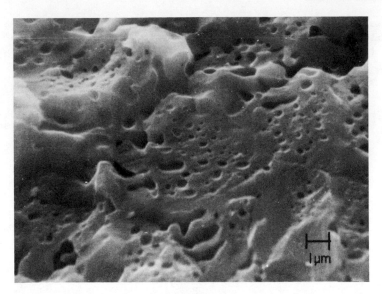

Fig. 8.   Scanning electron micrograph showing a large amount of fine
           porosity on a fracture surface of a flame sintered thin
           alumina rod (Balgel/0.25% MgO).

the pore size and distribution must be considered as well as the
total porosity.   This is clear from the results for alumina filaments having at least 7% porosity yet developing strength as high as
that of fully dense hot-pressed alumina of 1 - 2 μm grain size.[26]

The unusually high strength for alumina with this level of
porosity may be due to the structure which develops as a result of
the abnormal sintering behaviour of colloidal boehmite.[27] The
phenomenon is not fully understood but is apparently associated with
the phase change which occurs at 1200°C.   It is known that shrinkage
of boehmite compacts is very rapid near this temperature and that
very fast rates of heating lead to higher sintered densities.[28]
The generally low density seems to be due to the trapping of spherical
pores within grains at an early stage of the sintering process.   At
80 - 90% of theoretical density the porosity in conventional sintered
alumina is mainly open in character and situated at grain boundaries.
Subsequent grain growth leads to the trapping of a few pores but
necessarily produces larger grain sizes.   In the present instance,
however, a fairly perfect grain boundary structure may develop between
grains which already contain trapped pores, while still retaining a
fine grain size.   As already mentioned scanning electron microscopy
of fracture surfaces has provided evidence for the existence of a
large number of pores of the order of 0.5 μm or less within the
grains (Fig. 8).

Whereas large grain boundary pores may contribute to the inherent crack length and so reduce the measured strength as suggested by Davidge *et al*,[24] small internal pores are not likely to contribute in this way. Indeed one could imagine that they might even enhance the strength by acting as obstacles to the propagation of cracks through grains. The critical length of a crack blunted by a pore of radius 0.1 μm would be about 100 μm according to the relation

$$\sigma_m = \sigma \left(1 + \frac{2c}{r}\right)^{\frac{1}{2}} \tag{4}$$

where $\sigma_m = (2\,E/a)^{\frac{1}{2}} = 35$ GN/m^2 is the theoretical strength of alumina, r is crack tip radius and c is its length. The limitation of the effectiveness of Griffith flaws caused by a distribution of spherical particles or pores has been discussed by Hasselman and Fulrath.[29] Unfortunately the theory, originally applied to dispersion hardened glass, is not directly applicable to a polycrystalline ceramic.

THE CREEP-RUPTURE BEHAVIOUR OF ALUMINA FILAMENTS IN RELATION TO TURBINE BLADE REINFORCEMENT

The creep-rupture of the alumina filaments discussed above has been investigated at NGTE over the temperature range 1000 to 1300°C and has been reported in detail by Walles.[2] The filaments used in these experiments were from several different batches produced during the early stages of development of t e process, but the results showed reasonable consistency with the exception that Balgel/0.25% MgO dilaments were decidedly superior with regard to creep-rupture behaviour (Fig. 9).

Tilly and Walles[30] have previously shown that creep and creep-rupture of a very wide range of metals can be described by an empirical equation of the form

$$\varepsilon = C_1\sigma^{\beta_1}\phi_1^{K_1} + C_2\sigma^{\beta_2}\phi_2^{K_2} + \ldots C_r\sigma^{\beta r}\phi_r^{Kr} \tag{5}$$

where ε is strain, σ stress and ϕ a time, t, and temperature, T, dependent variable, the constants C_r are arbitrary, and the exponents K and β are taken from the series

$$K = \ldots \tfrac{1}{3}, \quad 1, \quad 3 \ldots \tag{6}$$

$$\beta/K = 1, 2, 4, \quad \ldots \tag{7}$$

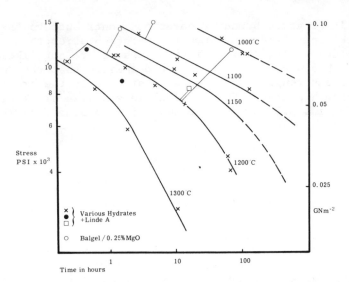

Fig. 9.   Creep-rupture data for thin alumina rods prepared from
          various alumina hydrates.[2]

When applying this to creep-rupture two assumptions are made: only
terms with a single value of K contribute to strain at rupture;
and variations in strain to rupture, ε_c, are small.   Equation 5
then reduces to

$$\varepsilon_c = \sum_r C_r \sigma^{\beta r} \tau^K (T_r^{-1} - T)^{\pm 20K} \tag{8}$$

which predicts a series of straight lines with slopes -1, $-\frac{1}{2}$, $-\frac{1}{4}$,
etc. for a creep-rupture graph of log stress v. log time.   Although
the equations apparently fit experimental data from a wide range of
experiments they have little physical basis and several arbitrary
variables.   However, they do have a distinct advantage over other
formulae in that they provide an apparently reliable extrapolation
to longer times or higher temperatures.   Thus long term creep
resistance can be predicted from the results of short creep-rupture
experiments at temperatures higher than would be encountered in
normal service.

     In the present experiments only the times to rupture at constant
stress and temperature were measured, partly because of the experimental difficulty of strain measurement, but mainly since strains
of the order of 1% are not likely to affect the serviceability of
turbine blades.   The data from these and similar tests on SiC
filaments were found to be consistent with Eq. 8 and the values of
K/β in the range studied were $-\frac{1}{8}$ and $-\frac{1}{2}$.   As seen in Fig. 9 the

100 hr creep-rupture strength was about 0.086 GN/m^2 (12 x 10^3 psi)
at 1000°C. The limited data for MgO doped filaments suggests
similar strength at 1200°C. Using the extrapolation referred to
above, this infers more than 0.14 GN/m^2 (20 x 10^3 psi) for 100 hr
at 1000°C.

    The volume fraction of reinforcing fibres which could in practice
be incorporated in a turbine blade is not likely to exceed 30% except
possibly in localised regions of exceptionally high stress or temperature. Using the relationship

$$\sigma = \sigma_F V_F + (1-V_F) \sigma_m \qquad\qquad (10)$$

where σ_F is the creep-rupture strength of the fibres, σ_m that of the
Ni-alloy matrix, and V_F is the fibre volume fraction, Walles[31]
calculated σ, the creep-rupture strength of various composites. The
acceptable level for specific creep-rupturing strength is indicated
in Fig. 10 at about 1.6 km (65 x 10^3 in). On this basis both SiC
wires would be useful, even at low concentrations at 1000°C, but
at 1100°C W-5% Re would be acceptable only above 20% V_F. The best

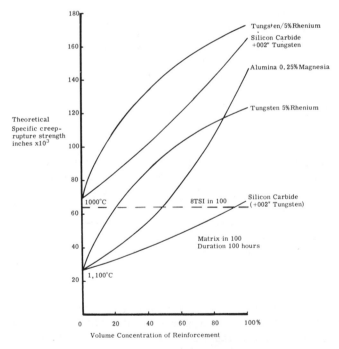

Fig. 10.   The specific creep-rupture strength of various composites
           calculated by the mixture law.[31]

Al_2O_3 filaments would provide useful reinforcement only at concentrations higher than 50% at 1100°C. Thus the creep-rupture strength of these must be raised by a factor of at least two before they are likely to be of any real value for turbine blade reinforcement.

Some recent work at NGTE[4] has shown that the creep-rupture resistence of the single crystal sapphire filament manufactured by Tyco Laboratories is very much better than the present polycrystalline filaments. The 100 hr strength at 1100°C was found to be 0.41 GN/m^2 (60 x 10^3 psi). This means that $V_F \simeq$ 30% would certainly give useful reinforcement. Electron beam microprobe analysis also showed that it is very stable for periods up to 300 hr in a Ni-matrix at 1100°C. However, when they were leached following this heat treatment the filaments were found to be very severely internally cracked. Two factors could contribute to this damage: thermal shock during the casting process, and mechanical stress due to the difference (5 x 10^6 °C^{-1}) in thermal expansion coefficients of the Al_2O_3 fibres and Ni-matrix. Obviously these are serious obstacles to the successful manufacture and use of composite turbine blades.

CONCLUSION

Of the various materials considered for continuous fibre reinforcement of gas turbine blades at high temperatures, W-Re alloys have shown the greatest promise with regard to creep-rupture behaviour, stability in Ni-base alloys and incorporation into the matrix. Indeed it has been demonstrated that such composites can be fabricated both by conventional casting techniques and by extrusion, and that they possess enhanced creep-resistance.

Ceramic fibres have considerable potential which is not yet realised, but neither SiC nor Al_2O_3 filaments are really developed to the point where they may be seriously considered for high temperature reinforcement. Silicon carbide, although showing reasonable creep performance at 1000°C does not have adequate strength at 1100°C, and in addition interacts seriously with the Ni-matrix. The continuous single crystal Al_2O_3 filament developed by Tyco Laboratories also shows an attractive specific creep-rupture strength several times greater than that of SiC at 1100°C. This filament also appears to have reasonable resistance to chemical attack by the Ni-matrix.

Several processes have been devised for producing polycrystalline alumina filaments which deserve further attention. In particular the polycrystalline filaments or thin rods produced by extrusion and sintering of hydrated aluminas, discussed in some detail above, show certain unusual and interesting deviations from the normal properties of sintered alumina. The elastic modulus and 3-point bend strength of the best samples of this material are respectively about 350 GN/m^2

and 0.8 GN/m^2. The 100 hour creep rupture strength at 1200°C is about 0.09 GN/m^2. A complete explanation of these properties in terms of the microstructure requires more data than is at present available but it appears that the unusual porosity distribution may play a significant role.

However, should a sufficiently creep-resistant filament be achieved there would remain the thermo-mechanical stress problems associated with composites of alumina and nickel alloys, which may prove very difficult to solve.

ACKNOWLEDGMENTS

The authors gratefully acknowledge the financial support received from the Department of Trade and Industry (formerly Ministry of Technology) for this work.

REFERENCES

1. A. Burwood-Smith, Fibre Science and Technology, 3 (2) 105-17 (1970).
2. K. F. A. Walles, Proc. Brit. Ceramic Soc. Vol. 15, 157-171 (1970).
3. A. W. H. Morris and A. Burwood-Smith, Fibre Science and Technology, 3, (1) 53-78 (1970).
4. A. W. H. Morris and A. Burwood-Smith, Fibre Strengthened Nickel-base Alloy. Paper presented at AGARD 36th Meeting P.E.P. on High Temperature Turbines, Florence, September 1970.
5. M. I. Kliman, Watertown Arsenal Laboratories. Tech. Report WAL TR 371/50, 1962.
6. N. J. Parratt, Explosives Research and Development Establishment, Waltham Abbey, England. Private Communication.
7. S. A. Bortz (IITRI, Chicago) "Refractory fiber composite systems". Paper presented at 21st Annual Meeting of the Reinforced Plastics Div. of the Soc. of Plastics Industry, Chicago.
8. S. A. Bortz and P. C. Li (IITRI, Chicago) "Continuous Filament Refractory Fibers". Paper presented at 22nd Annual Meeting of the Reinforced Plastics Division of the Soc. of Plastics Industry, Washington, D. C.
9. N. D. Nazarenko, V. F. Nechitailo and N. I. Vlasko. Soviet Powder Metallurgy 4 265-67 (1969).
10. Babcock and Wilcox Company (U.S.A.), British Patent No. 1,141,207. (1969).
11. B. H. Hamling, A. W. Nauman and W. H. Dresher. Applied Polymer Symposia, 9 387-94 (1969).
12. B. H. Hamling, Union Carbide Corporation (U.S.A.) British Patent No. 1,144,033 (1969).
13. I. Wizon and J. A. Robertson, J. Polymer Sci. C. (19) 267-81 (1967).
14. I. Wizon, Applied Polymer Symposia 9 395-409 (1969).
15. Rolls Royce Ltd. British Patent No. 1,069,472 (1967).
16. Tyco Laboratories, Massachussetts Bulletin 102 (1969).

17. H. D. Blakelock, N. A. Hill, S. A. Lee and C. Goatcher. Proc.
    Brit. Ceram. Soc. No. 15 69-83 (1970).
18. J. M. Fletcher and C. J. Hardy. Chemistry and Industry 48-51,
    13 Jan 1968.
19. J. E. Bailey and N. A. Hill. Proc. Brit. Ceramic. Soc. No. 15
    15-35 (1970).
20. R. L. Coble and W. D. Kingery. J. Amer. Ceram. Soc. 39 (1)
    377-85 (1956).
21. S. C. Carniglia. J. Amer. Ceram. Soc. 48 (11) 580-83. (1965).
22. A. A. Griffith, Phil Trans. Roy. Soc. (London) Vol. A.221,
    163 (1920).
23. P. L. Gutshall and G. E. Gross, Eng. Fracture Mech. Vol. 1 463-71
    (1969).
24. R. W. Davidge and G. Tappin, Proc. Brit. Ceramic Soc. No. 15
    p. 47-60 (1970).
25. E. M. Passmore, R. M. Spriggs and T. Vasilos, J. Amer. Ceram.
    Soc. 48 (1) 1-7 (1965).
26. R. M. Spriggs, T. Vasilos and L. A. Brissette; pp. 313-44 in
    Materials Science Research, Vol. 3. W. W. Kriegel and Hayne
    Palmour III, Eds., Plenum Press, New York, 1967.
27. B. J. Baggaley, A. S. Malin and E. R. McCartney. J. Australian
    Ceram. Soc. 4 (2) 46-50 (1967).
28. P. Vergnon, F. Juillet and S. J. Teichner. Rev. Int. Hautes
    Tempér et Refract., 3. 409-19 (1966).
29. D. P. H. Hasselman and R. M. Fulrath. J. Amer. Ceram. Soc. 50
    (8) 399-404 (1967).
30. G. P. Tilly and K. F. A. Walles, The Engineer p. 551, 27 Oct. 1967.
31. K. F. A. Walles, National Gas Turbine Establishment, Private
    Communication.

DISCUSSION

R. W. Rice (U. S. Naval Research Laboratory): Your fiber strengths
are not necessarily higher than might be expected. I have found
that the room temperature ambient flexural strengths of hot-pressed
Al_2O_3 typically are about one-half of such strengths of sintered
Al_2O_3 of the same microstructure (i.e., grain size, porosity and
size distribution). Further, finer pores generally are less
detrimental to strength than are larger pores; differences between
your measured strengths and what they would be at zero porosity
is likely to be less than in larger Al_2O_3 bodies having the same
V_f porosity but generally much larger pore sizes.
Authors: We believe, as we say in our paper, that the comparatively
high strengths of our alumina filaments are exceptional in view of
their high porosity (~10%); examples of sintered aluminas with
comparable strengths exist in the literature but these have much
lower porosities (~1-2%) and comparable grain sizes.
    Concerning the first point we are not aware that a comparison

of hot pressed and normal sintered aluminas of similar structure (i.e., porosity, pore size distribution and grain size) showed that the strengths of the hot pressed samples are about one half those of the sintered samples.

As discussed in our paper, we agree with the second comment.

S. A. Bortz (I.I.T. Research Institute): IITRI has been working in the same area, making both extruded and spun Al_2O_3 fibres. We find a diameter-strength relationship as well as a grain size relationship. For a 35 to 40 µm diameter and a 2-5 µm grain size and a gage section of 3/4 in., we obtained strengths as high as 200,000 psi. By reducing the diameter to 12-15 µm we anticipate strengths of 400,000 to 500,000 psi.

THE EFFECT OF NONSTOICHIOMETRY ON CREEP OF OXIDES

A. H. Clauer, M. S. Seltzer and B. A. Wilcox

Battelle Memorial Institute, Columbus Laboratories

Columbus, Ohio   43201

ABSTRACT

High temperature creep in oxides is often quite sensitive to stoichiometry, primarily because creep is often a diffusion-controlled process, and diffusion occurs by the movement of point defects. The point defect concentration depends on both the type and amount of impurities present, and on the ambient oxygen pressure. Therefore, oxide stoichiometry is particularly dependent upon the environment at high temperatures, where diffusion is rapid. The relation between the creep behavior and stoichiometry in oxides is reviewed in detail.

INTRODUCTION

All oxides exist as a single phase over a range of compositions. In some such as Al_2O_3 and MgO the homogeneity range is very narrow, amounting to only a few parts per million deviation from the stoichiometric composition. However, other oxides such as FeO and UO_2 remain single phase over a wide range of compositions encompassing tens of atomic percent. The variation in composition is accommodated in several ways, and these have been extensively reviewed by Kroger.[1] Small deviations from stoichiometry are achieved by introducing anion and cation vacancies or interstitials into the crystal lattice. Larger deviations from stoichiometry can result in the formation of a high density of vacant lattice sites, superlattice ordering of point defects on the sublattices, or the formation of Magneli phases.[2]

361

The introduction of lattice defects usually strengthens crystals at low temperatures, but at high temperatures they decrease the strength. The influence of small deviations from stoichiometry on creep has been studied in detail in only a few oxides. The effects of the various types of accommodation of nonstoichiometry which occur at large deviations have not been investigated in detail.

In this paper we will review the studies which have been made of the effect of nonstoichiometry on creep in selected oxides. The review is limited to those oxides in which this has been studied most thoroughly: CoO, FeO, CeO_2, TiO_2, and UO_2. Since the number is not large they will be presented in detail. To properly understand and interpret the high temperature creep behavior of oxides the self-diffusion coefficients and defect structure must also be known. Table 1 contains selected diffusion data for the oxides discussed here. Where diffusion of a given species has been studied by several investigators, such as for UO_2, the data chosen for inclusion in Table 1 were those considered most representative, or which best show the influence of nonstoichiometry. The defect structures of these oxides are still controversial, and only the more recent suggestions regarding the defect structures are presented here. Table 2 summarizes creep results for the various oxides, and these are discussed in some detail in the following sections.

REVIEW OF STOICHIOMETRY EFFECTS IN SOME OXIDES

CoO

Cobalt monoxide is an oxygen deficient oxide having the sodium chloride structure, with the cobalt ions and oxygen ions located on interpenetrating face centered cubic lattice sites. The primary slip system is $\{110\}<1\bar{1}0>$[12] although $\{100\}<0\bar{1}1>$ has been observed in other NaCl-structure materials. Fisher and Tannhauser[20] suggested that the predominant defects are singly charged cobalt vacancies at high oxygen pressures and doubly charged cobalt vacancies at low oxygen pressures. Creep experiments have been performed on both single crystals[12] and polycrystalline[13] CoO by Clauer, Seltzer, and Wilcox.

The single crystals were tested in compression parallel to a $<100>$ direction. Slip occurred primarily on two symmetrical orthogonal $\{110\}<\bar{1}10>$ slip systems with the development of a subboundary structure during creep. The creep curves had a sigmoidal shape which extended to such high strains that barreling and cracking of the specimen prohibited study of steady state creep beyond the sigmoidal region. Therefore, the inflection creep rate, $\dot{\varepsilon}_2$, was used to characterize the creep behavior and typical creep curves are shown in Fig. 1 with $\dot{\varepsilon}_2$ indicated. The activation energy, Q_c, for $\dot{\varepsilon}_2$ was measured at $P_{O_2} = 1$ and 0.01 atm, giving $Q_c = 100 \pm 16$

Table 1. Self-Diffusion in Selected Oxides

Oxide	Ref.	Element	Temperature Range, °C	Atmosphere	D_o, $\frac{cm^2}{sec}$	$Q, \frac{cal}{mol}$	Comments
CeO	3	O	767–1262	O_2	–	74	Diffusion measured by color change
$Co_{1-x}O$[a]	4	Co	800–1350	1 atm O_2	2.15×10^{-3}	34.5	$D \propto P_{O_2}^{0.35}$ at 1150°C
CoO	5	O	1175–1560	0.21 atm O_2	50	95	Li doping increases D; Al doping decreases D
FeO	6	Fe	700–1000	H_2/H_2O	0.118	29.7	$Fe_{0.907}O$
TiO_2	7	Ti	900–1300	Air	6.4×10^{-2}	61.4	Extrinsic
TiO_{2-x}	8	O	710–1300	10^{-3}–760 torr O_2	2.0×10^{-3}	60	D independent of P_{O_2}
UO_{2+x}[a,b]	9	U	1350–1650	CO/CO_2	1.1–2.0×10^{-3}	81–105	$D \propto x^2$
UO_2[a,b]	10	O	780–1250	H_2,A	0.26	59.3	O/U ~2.000

(a) More diffusion data have been obtained but these values were considered most representative.

(b) See Ref. 52 for a review of diffusion in UO_2.

Table 2. Creep of Oxides Where the Relation $\dot{\varepsilon} = A\sigma^n \exp-(Q_c/kT)$ is Assumed

Oxide	Ref.	Temperature Range, °C	Stress Range, psi	Atmosphere	Q_c, $\frac{kcal}{mol}$	n	Type of Test[c]	Comments[a]
CeO_{2-x}	11	1350-1450	91-363	Air, He	92.5-39	2.2-1.2	B	P.C., $\sim 0 < x < 0.14$
$Co_{1-x}O$	12	1000-1200	850-1700	A/O_2, to give 10^4 -1 atm O_2	87-103	7.1	C	S.C., $8 \times 10^4 < x < 1 \times 10^2$ at 1100°C, $\dot{\varepsilon} \propto P_{O_2}^{0.45}$ at 1100°C
$Co_{1-x}O$	13	925-1250	400-5500	A/O_2, to give 10^5 -1 atm O_2	79-60 / 51	6.5 / 2	C / C	P.C., $\dot{\varepsilon} \propto P_{O_2}^m$, $0 < m < 1/2$ at 1100°C / P.C., $m \simeq 0$
$Co_{1-x}O$	30	925-1050	4000-5600	10^2 -1 atm O_2	46	2.3	B	P.C., $\dot{\varepsilon} \propto P_{O_2}^{0.2}$ at 1000°C
$Fe_{1-x}O$	14,15	1000-1300	320-2670	CO/CO_2	78	4.2	C	P.C., S.C., $0.054 < x < 0.11$, $\dot{\varepsilon} \propto x$
TiO_{2-x}	16	830-1030 / 815-1030	3500-12,000 / 2000-8500	Air / 10^5 torr vac.	67 / 40	1.9 / 1.5	C / C	S.C., Q_c independent of σ, $\dot{\varepsilon}$ increases with x
TiO_{2-x}	17	1103-1225 / 1103-1225	5500-7250 / 5500-7250	O_2 / N_2	115-162 / 135-179	8 / -	B / B	S.C., $\dot{\varepsilon} \propto \dot{\varepsilon} \exp(B\sigma)$, $B = 1 \times 10^3$ psi^{-1}, $B = 1.22-1.56 \times 10^{-3}$ psi^{-1}, $\dot{\varepsilon}$ increases with x
UO_{2+x}	18	1100-1400	2500-14,000	CO/CO_2	57.5-135	17-7	C	S.C., $\dot{\varepsilon} \propto x^2$
UO_{2+x} (b)	18	1100 / 1100-1400	6000-25,000 / 2000-13,000	CO/CO_2 / CO/CO_2	90,for $x \simeq 0$	4 / 1	C / C	P.C., $\dot{\varepsilon} \propto x^{1.75}$ / P.C., $\dot{\varepsilon} \propto x$
UO_{2+x}	19	975-1400	1000-10,000	A/O_2	55.7	1	B	P.C., $\dot{\varepsilon} \propto x$

(a) P.C. = polycrystalline, S.C. = single crystal
(b) More data on stoichiometric UO_2 are available. These are taken as representative.
(c) B = bending, C = compression.

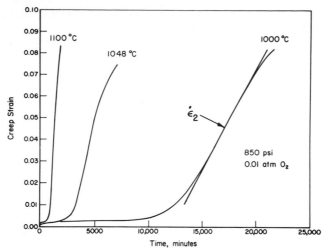

Fig. 1.   Sigmoidal creep curves of <100> oriented $Co_{1-x}O$ single
          crystals in compression.[12]

and 87 ± 6 kcal/mol, respectively. The stress dependence, when
fitted to a power law relation, $\dot{\varepsilon}_2 \propto \sigma^n$, gave n = 7.1.   The oxygen
pressure dependence of the creep rate is shown in Fig. 2 to be
$\dot{\varepsilon}_2 \propto p_{O_2}^{0.45}$, or $\dot{\varepsilon}_2 \propto x^{1.7}$.

     If the creep rate is diffusion controlled, then that of the
slower diffusing species should be rate controlling. Table 1 shows
that at 1100°C oxygen diffuses about 10^{-4} times as fast as cobalt.
Furthermore, the values of Q_c are nearly equal to the activation
evergy for oxygen diffussion. Thus it is   concluded   that the

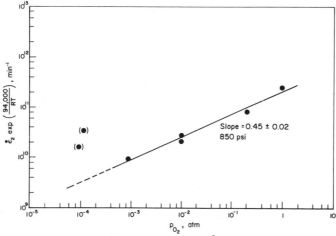

Fig. 2.   Oxygen pressure dependence of $\dot{\varepsilon}_2$ in $Co_{1-x}O$ single crystals.[12]
          The reason for the relatively high creep rates of the specimens
          at 10^{-4} atm O_2 (the points in parentheses) is not understood.

creep of single crystal CoO is controlled by oxygen diffusion has
not been determined. If creep were controlled by oxygen diffusion,
then $\dot{\epsilon}_2$ should decrease with increasing p_{O_2} if oxygen diffused via
vacancies in the oxygen sublattice. However, as seen in Fig. 2,
the creep rate increases with increasing p_{O_2} and thus it was suggested
that oxygen diffuses either as a neutral interstitial, which gives
$\dot{\epsilon}_2 \propto$ oxygen diffusion $\propto p_{O_2}^{1/2}$, or that oxygen diffusion involves
complex defects.[12]

The significance of the stress dependence of the creep rate is
not clear. The dislocation dynamics models describing sigmoidal
creep curves[21-24] relate $\dot{\epsilon}_2$ to the dislocation glide velocity and
to a hardening rate due to accumulation of dislocations. This
description gives excellent results for the diamond structure
materials,[21,25] where the stress dependence and activation energy
of the dislocation glide velocity are known, and the dislocation
densities can be measured quantitatively. In CoO, precipitation
of Co_3O_4 upon cooling from the creep temperature prohibited a
quantitative study of the dislocation substructure formed during
creep, and the stress dependence and activation energy of the dislocation glide velocity were unknown. However, if dislocations
moved by diffusion controlled viscous glide, such as by motion of
jogged screws[26,27] or by drag of atmospheres of charged defects,[28,29]
and recovery processes were not dominant, then $\epsilon \propto \sigma^{\sim 3}$ should have
been observed.[21,22] Therefore, the dislocations probably do not move
by a viscous glide mechanism and some other stress-dislocation
velocity relation is appropriate. Alternatively, recovery effects
could complicate the models, and these effects are difficult to
assess quantitatively.[23]

Creep of polycrystalline CoO has also been studied by the
authors.[13] The material was hot pressed to ~0.94 theoretical density
with a grain size of 25 - 30 μm. The specimens were tested in compression in O_2 and Ar-O_2 atmospheres over the ranges p_{O_2} = 1 to 10^5 atm
925 to 1250°C, and 400 to 5500 psi. For polycrystalline CoO,
sigmoidal creep was not observed and the data presented are steady
state creep rates. The stress dependence of the steady state creep
rate at 1100°C could be separated into two regions: at high stresses
n ~6.5, while at low stresses n ~2. The stress exponent was independen
of oxygen pressure. Strafford and Gartside[30] investigated bending
creep between 925° and 1050°C, under stresses of 4000 to 5600 psi
in polycrystalline CoO made by oxidizing cobalt coupons. They found
n ~2.3 at 1000°C and 1 atm O_2.

At high stresses Q_c varied from 59.6 kcal/mol at 1 atm O_2 to
about 80 kcal/mol over the p_{O_2} range 10^2 to 10^5 atm, while at low
stresses Q_c = 51 kcal/mol, independent of p_{O_2} (Fig. 3). Included
in the low stress region in this figure are the data of Strafford
and Gartside, who observed Q_c = 46 $\pm$ 5 kcal/mol at 0.01 atm O_2.

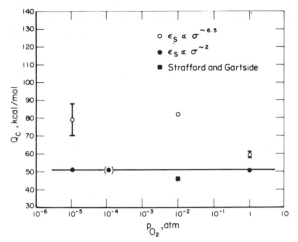

Fig. 3.   Oxygen pressure dependence of the creep activation energy
          for polycrystalline $Co_{1-x}O$. [13,30]

For high stresses at 1100°C, where n ~6.5, the creep rate was nearly
independent of p_{O_2} at 10^{-5} atm O₂ but increased with increasing p_{O_2}
and approached the dependence $\dot{\varepsilon} \propto p_{O_2}^{0.5}$ at 1 atm O₂. At low stresses,
where n ~ 2, the creep rate was essentially independent of p_{O_2} over
the whole range. However, Strafford and Gartside[30] found that the
creep rate increased slightly with **increasing** p_{O_2} at 1000°C. Their
data are not sufficient to differentiate between a linear or power
law dependence of the creep rate on x, and give an *approximate*
oxygen pressure dependence of the creep rate $\dot{\varepsilon} \propto p_{O_2}^{0.2}$.

     As for the case of CoO single crystals, if the creep rate of
polycrystalline CoO is diffusion controlled, then the diffusion of
oxygen as the slower moving species would be expected to be rate
controlling. However, there is no agreement between the values of
Q_c in the polycrystalline CoO and the activation energies for
diffusion of either oxygen or cobalt as given in Table 1. In both
the high and low stress regions, Q_c is greater than the activation
energy for cobalt diffusion, ruling out cobalt diffusion as the
primary rate controlling mechanism. In the high stress region, Q_c
approaches the activation energy for oxygen diffusion only at low
p_{O_2}. Near one atm O₂, the oxygen pressure dependence of the creep
rates of the single crystal and polycrystalline CoO were similar,
but Q_c for the polycrystalline CoO was significantly lower.

     The high stress region in metals is usually characterized by
Q_c being equal to the activation energy for diffusion and by creep
rates which are independent of grain size. Therefore, the creep
mechanism in the high stress region for polycrystalline CoO is
either not predominantly diffusion controlled, or else short

circuiting processes such as diffusion along dislocations or grain boundaries are making significant contributions to diffusion controlled creep.

Creep rates in the low stress region for polycrystalline oxides are usually grain size dependent as in the UO_2 results discussed later. The increasing contribution of Nabarro-Herring creep and grain boundary related creep mechanisms at lower stress levels causes an accompanying decrease in the stress dependence of the creep rate. The limit is n = 1 for various diffusion and grain boundary sliding models.[31-35] The low stress region for CoO is probably dominated by a grain boundary creep mechanism. The p_{O_2}-independent Q_c and creep rates probably reflect a grain boundary diffusion controlled mechanism wherein, over the p_{O_2} range studied, changes in bulk composition do not influence grain boundary diffusion. At low stresses, then, nonstoichiometry has little effect on the creep behavior of polycrystalline CoO.

FeO

Wüstite ($Fe_{1-x}O$) also has the NaCl structure and it has been reported[36] that slip occurs on the $\{110\}\langle1\bar{1}0\rangle$ slip systems. This oxide exists only as a hyperstoichiometric (oxygen excess) phase at atmospheric pressure. The value of x may be varied in the range 0.05 < x < 0.15 by equilibration with H_2/H_2O or CO/CO_2 gas mixtures. As with most of the oxides discussed here, some controversy surrounds identification of the defect structure for wüstite. The various possibilities were summarized by Hed and Kofstad[37] who concluded that the best structural model involves formation of complex defects consisting of two iron vacancies and an iron interstitial atom either neutral as $(V_{Fe}Fe_iV_{Fe})$ or singly charged as $(V_{Fe}Fe_iV_{Fe})'$.

The compression creep of wüstite has been investigated by Ilschner, Reppich, and Riecke[14] and by Reppich.[15] Ilschner et al[14] used specimens having a coarse, columnar grained structure (0.1 - 0.3 mm av. grain dia, no length given) whose density was nearly theoretical. These were prepared by casting of molten FeO. Specimens were equilibrated and tested in CO/CO_2 gas mixtures to maintain a given composition during creep. In the temperature range 1000 - 1300°C, for values of x between 0.054 and 0.11, and at a stress of 355 psi the activation energy for steady-state creep was found to be 78 $\pm$ 5 kcal/mol, independent of x, as shown in Fig. 4. Over the same composition range at 1000°C under stresses of 310 psi to 3000 psi the creep rates fit a power law dependence where n = 4.2 $\pm$ 0.1. The creep rates were found to vary linearly with x over the entire range of temperature, stress, and x mentioned above.

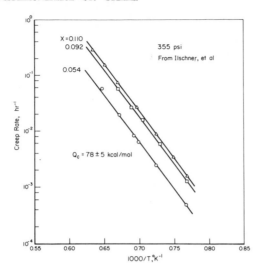

Fig. 4.   Temperature dependence of the creep rate in polycrystalline
          $Fe_{1-x}O$ for different compositions, x.[14]

Reppich[15] investigated the compression stress-strain behavior
and creep deformation of monocrystalline and polycrystalline FeO.
Work was performed in CO/CO_2 gas mixtures using polycrystals similar
to those employed in Ref. 14, electron-beam zone grown polycrystals
having coarse elongated grains with the long axis parallel to <100>
and parallel to the compression direction, and single crystals
grown by electron-beam zone refining or by the Czochralski technique.
These crystals were oriented with the <100> parallel to the compression axis.

The single crystals were found to be more plastic than the
coarse grained polycrystals, with creep rates a factor of 10 greater
than values for polycrystals tested under the same conditions. For
the single crystal tests, two equally favored conjugate {110}<1$\bar{1}$0>
slip systems were activated in every test, in agreement with results
on CoO[12] crystals tested parallel to the <100> direction. The steady-
state creep rate of the single crystals showed the same temperature
and stress dependence as was found by Ilschner et al for polycrystals.
The dependence of creep rate on x was not determined for single
crystals, but the authors assumed there would be no difference from
the polycrystalline results.

Reppich also observed sigmoidal creep curves in both single and
polycrystalline specimens. The sigmoidal shape (or the yield drop
in the compression tests) appeared only when x > 0.08, and was most
pronounced after annealing at 400°C. Evidently Ilschner, et al did
not observe sigmoidal creep curves because they did not preanneal
at 400°C before the creep test. Reppich did not determine Q_c for

the transition creep rate, $\dot{\varepsilon}_2$, but reported that compression and creep experiments gave similar results and that the strain rate dependence of the lower yield stress, $\sigma \propto \dot{\varepsilon}^{1/n}$, was n = 4, the same as the stress dependence for the creep rate.

Reppich performed detailed experiments to characterize the yield point (sigmoidal creep) phenomena. On the basis of these he concluded that the effect is caused by dislocation loops left behind at the sites of Fe_3O_4 precipitates after they had redissolved upon heating to the creep temperature. The loops initially harden the crystal, since the initial glide dislocations must cut through them. However, repeated cutting forms numerous jogs in the loops and point defects in the immediate vicinity and these cause the loops to annihilate rapidly by climb once deformation begins. This leads to a softening of the crystals and an accompanying decrease in flow stress or increase in creep rate. However, the evidence does not rule out the possibility that dislocation pinning by a complex defect such as one involving Fe^{3+}-ions might be responsible. The effect is strongest after annealing below the stability region for FeO (<570°C) and such a defect-dislocation interaction may be sufficiently strong that the pinning is retained at higher temperatures.

In the case where steady state creep was found, Reppich observed no evidence of grain boundary sliding. However, even if there had been, the large grain size prohibited grain boundary sliding from contributing significantly to the creep rate. These observations and the similarity of the creep behavior of the single and poly-crystalline materials (except for their relative strength) indicate that the same creep mechanism is operative in both. The stress dependence of the creep rate and the observation of a pronounced polygonized substructure in the polycrystalline material suggest that a recovery mechanism such as dislocation climb controls the creep rate. No substructure studies were made on the single crystals.

Examination of Table 1 shows that the activation energy for diffusion of iron in FeO is 29.7 kcal/mol, considerably below the value of 78 kcal/mol found for creep. The diffusion of oxygen in wüstite has not been studied, but it is expected to be slower than iron diffusion by analogy with CoO and NiO. The activation energies for oxygen diffusion in these oxides is in the range 60-100 kcal/mol, and it is expected that the creep activation energy for wüstite would be closer to that for oxygen diffusion than for iron diffusion. This raises the question of the meaning of the composition dependence. As for CoO, the creep rate of FeO was found to increase with increasing x, thus eliminating diffusion of oxygen via vacancies in the oxygen sublattice as the rate controlling process, and leading to the suggestion that oxygen diffusion occurs via an interstitial mechanism or through a defect complex such as $(V_{Fe}V_OV_{Fe})$.

TiO₂

Rutile crystallizes in the cassiterite (SnO₂) structure[38] with the titanium ions in a body centered tetragonal arrangement and pairs of oxygen atoms separating titanium neighbors along the alternate diagonals of (001) planes. The oxide exists as a hypo-stoichiometric phase down to $TiO_{1.98}$ [39]. There is much conflicting data concerning the defect structure of this compound. Kofstad[40] has proposed that the defect structure involves divalent oxygen vacancies predominating at low temperatures and high oxygen pressures and tri- and tetravalent interstitial titanium ions predominating at low pressures and high temperatures. At least two glide systems are operative in rutile, $\{101\}<\bar{1}01>$ and $\{\bar{1}10\}[001]$.[16,41,42]

The high-temperature steady state creep characteristics of rutile (TiO_{2-x}) have been studied by Hirthe and Brittain[16] and by Farb, Johnson, and Gibbs.[17] Both investigations were on rutile single crystals obtained from the same source. The former investi-gators studied creep in compression on specimens oriented for slip on the $\{\bar{1}10\}[001]$ system while the latter workers deformed their specimens in a four point bending jig with the "c" axis parallel to the length of the beam and the "a" axis perpendicular to the faces of the specimens. In this case glide occurred on the $\{101\}<\bar{1}01>$ system. The bending experiments were conducted in N₂ or O₂ atmospheres, while the compression study used air, air-helium mixtures, and vacuum.

The temperature dependences for creep of TiO₂ determined in the two studies are compared in Fig. 5. The results of Hirthe and Brittain have been interpolated to a stress of 6400 psi. These investigators found the activation energy for creep to vary with composition, decreasing with increasing x from Q_c = 67 kcal/mol for near stoichiometric rutile (equilibrated in air) to 40 kcal/mol for creep in a vacuum of 10^5 torr over the temperature range of 827 to 1027°C. Q_c was independent of applied stress in the range of 2030 to 12,000 psi. The stress dependence varied from n = 1.5 in air to n = 1.9 in vacuum. Farb, *et al*, obtained much different results at slightly higher temperatures, 1100 to 1230°C, but in the same stress range, 5,550 to 8,820 psi. A relative strength difference between specimens taken from different boules could be attributed to a lower cation impurity content in the softer crystals. Only the results from the softer, purer boules will be discussed.

As seen in Table 2, the values of Q_c are in the range of 115 to 162 kcal/mol for creep in an O₂ atmosphere and 135 to 179 kcal/mol in a N₂ atmosphere, and the values of Q_c decrease with decreasing stress. Farb, *et al*, fitted their data to an exponential stress dependence of the creep rate, $\dot{\varepsilon} \propto \exp(B\sigma)$, and the values of B are given in Table 2. For comparison, their data for 1 atm oxygen

at 1103°, replotted, gives n = 8, which is considerably higher than
the stress exponents found by Hirthe and Brittain. The one similarity
in the two investigations was that the creep strength decreased with
the degree of reduction, i.e., increasing x, as shown in Fig. 5.

Hirthe and Brittain found that extensive polygonization occurred
during creep, resulting in long, well defined, parallel subgrain
boundaries, and a similar tendency was noted by Farb, *et al*. The
spacing, ℓ, between the boundaries varied with stress as $\ell \propto 1/\sigma^{\alpha}$
where $\alpha = 2/3$ in air and 4/5 in vacuum. The presence of the well
defined boundaries suggests that dislocation climb occurred extensively
during creep, and thus creep may be climb controlled. Table 1 shows
that the diffusion rates of O and Ti are not very different in this
temperature range and both have activation energies similar to that
observed by Hirthe and Brittain for creep in air. Therefore, diffusion
of either, or more likely, both species could control the creep rate.
Haul and Dumbgen[8] found that the diffusion of oxygen was independent
of p_{O_2}. Thus the effect of atmosphere on Q_c observed by Hirthe
and Brittain could be caused by the Ti diffusion contribution to the
creep rate. The Q_c values reported by Farb, *et al*, are so different
from the diffusion results that no rationale can be made. It is
possible that this is caused by the different slip systems operative in
the two studies. Nadeau[43] has pointed out the {110}[001] slip system
appears to be harder than the (101)<$\overline{1}$01>. However, according to Fig. 5
above about 1100°C the reverse could be true.

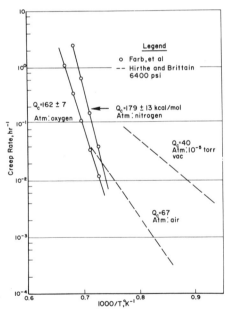

Fig. 5.   Temperature dependence of the creep rate in TiO_{2-x} single
crystals in compression[16] and bending.[17]

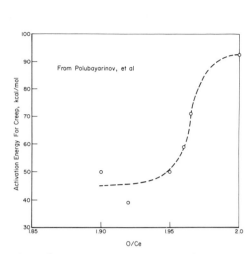

Fig. 6. Composition dependence of the activation energy in CeO$_{2-x}$.[11]

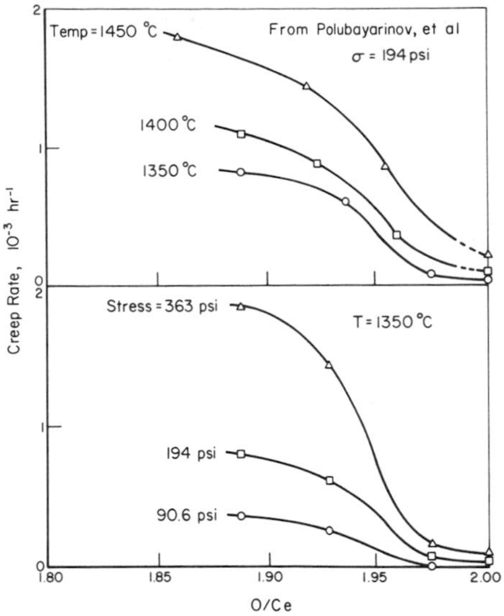

Fig. 7. Composition dependence of the creep rate of CeO$_{2-x}$ at various temperatures and stresses.[11]

CeO$_2$

Cerium dioxide has the cubic fluorite crystal structure in which the cerium atoms are arranged on a fcc-lattice and the oxygen atoms are located at distances of 1/4 and 3/4 along the unit cell diagonals. When nonstoichiometric, it is oxygen-deficient. Bevan and Kordis[44] suggested that above about 1000°C, the single phase region extends from CeO$_{2.00}$ to CeO$_{1.72}$. The nature of the defect structure for CeO$_{2-x}$ is a matter of controversy. Kevane[45] suggested that the primary point defects are oxygen vacancies. In a more recent paper, Kofstad and Hed[46] proposed that doubly charged oxygen vacancies predominate under high oxygen pressures, close to the stoichiometric composition, while single and doubly ionized cerium interstitials, Ce$_i$ and Ce$_i''$ are the dominant defects for values of x > 0.005.

The creep behavior of pressed and sintered polycrystalline CeO$_{2-x}$ has been investigated by Polubayarinov and co-workers.[11] The specimens were 99.7% pure, with 4% total porosity, no open porosity, and an average grain size of 15 µm. Four point bend tests were performed in air and in He in the temperature range 1350 to 1450°C and stress range 91 to 363 psi. The results should be viewed with some caution, as it appears that the specimens were equilibrated to the desired composition before creep, then tested in an air or helium atmosphere under conditions which might have caused the

composition to vary during creep, changing it from the reported
composition.[47] As shown in Fig. 6, the activation energy for creep
of CeO_{2-x} at a stress of 194 psi varied from 92.5 to 39 kcal/mol
as the O/Ce ratio was decreased from 2.00 to 1.90. The creep rates
increased with increasing x in the region of $0 \gtrless x \gtrless 0.14$ as shown
in Fig. 7. At 1350°C the stress dependence of the creep rate
varied from $n = 1.2$ to 2.2 as the O/Ce ratio increased from 1.90
to 2.00.

The stress dependence of the creep rate and the fine grained
microstructure suggests that the primary creep mechanism is either
Nabarro-Herring creep or a grain boundary creep mechanism. No
metallographic studies were made to investigate the presence of
grain boundary sliding. As seen in Table 1, cerium diffusivities
have not been measured in CeO_2, and oxygen diffusion rates have
been studied by a questionable technique. Thus, comparisons between
creep activation energies and diffusional activation energies cannot
be made at this time.

UO_2

Uranium dioxide crystallizes in the cubic fluorite structure
and has a homogeneity range extending over tens of atomic percent
above 1000°C.[48] Hypostoichiometric compositions are not readily
obtained below 1700°C.[49] Therefore, studies of physical properties,
including creep and diffusion, have been performed on specimens
containing an oxygen excess. The predominant point defects are
believed to consist of oxygen interstitials and vacancies, perhaps
in complexes of two interstitials and an oxygen vacancy.[50] The defect
structure on the uranium sublattice has not been clearly established,
but it is likely that uranium vacancies occur as a minority defect
in hyperstoichiometric compositions.

The influence of nonstoichiometry on creep in compression of
UO_2 single crystals has been studied by the authors.[18] It was found
that both the temperature and the stress dependence varied with O/U
ratio. At a stress of 10,000 psi, creep activation energies
decreased from 135 to 57 kcal/mol as the O/U ratio increased from
 2.0001 to 2.062. The activation energies are plotted as a function
of $log_{10}x$ in Fig. 8. (Results of other creep studies are tabulated
there in [59-61,63,64,71,72]). These values are in good agreement with
activation energies determined in other studies where the composition
has been controlled. In most cases, however, the O/U ratio has not
been treated as a variable and activation energies obtained for
so-called "stoichiometric" UO_2 have a range of values from 71 to
130 kcal/mol, all of which are listed in Fig. 8. For stoichiometric
UO_2 Armstrong et al[51] obtained an average activation energy for
creep in bending of 118 ± 9 kcal/mol in the stress range 3600 to
1800 psi and the temperature range of 1350 to 1685°C. This is close to
the value found by Seltzer, et al[18] for "stoichiometric" UO_2.

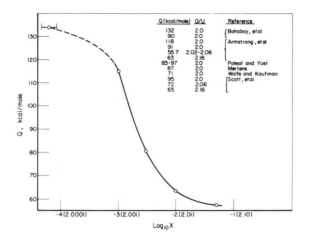

Fig. 8.   Composition dependence of the creep activation energy in UO_{2+x} single crystals.[18]

The creep activation energies for "stoichiometric" UO_2 which have values of about 130-135 kcal/mol are some 20 Kcal/mol higher than the highest reported uranium self-duffusion activation energies, and are much higher than those for oxygen diffusion (see Tables 1 and 2, and Ref. 52).   A correction in Q_c for the temperature dependence of the elastic modulus[54] amounts to less than 2 kcal/mol. The values of 55-70 Kcal/mol for creep of highly hyperstoichiometric UO_2 are considerably lower than the uranium self-diffusion activation energies reported in this O/U region by Hawkins and Alcock.[9]   More recently, however, Lay[55] has reported an activation energy for uranium diffusion of 55 kcal/mol determined from sintering experiments for material with an O/U ratio of 2.08.

A stress exponent of n = 3.3 was obtained from studies on bending creep of stoichiometric single crystals having various orientations.[51]   For the compression experiments at 1100°C, n varied from 7 for stoichiometric UO_2 to about 17 for O/U > 2.001[18] (Fig. 9). At higher temperatures, 1300 and 1400°C, n varied from 7 for stoichiometric UO_2 to 10 for higher O/U ratios.[18]   At 1100°C, the reproducibility of the data was checked by changing the O/U ratio from ~2.001 to 2.01 for a specimen used to determine the stress dependence, and then continuing the stress change tests at O/U = 2.01.   The results are given by the filled triangles at O/U = 2.01 in Fig. 9.   The agreement with the data obtained without changing the O/U is very good.   Seltzer, et al,[18] have also shown that $\dot{\varepsilon} \propto x^2$ in the region of constant stress exponent, for O/U > 2.001, where x is given by UO_{2+x}.   This x^2 dependence has been observed in several diffusion and sintering studies[55-57] where uranium diffusion rates were found to increase with increasing oxygen excess.   This dependence was predicted by Lidiard[58] on the basis that uranium

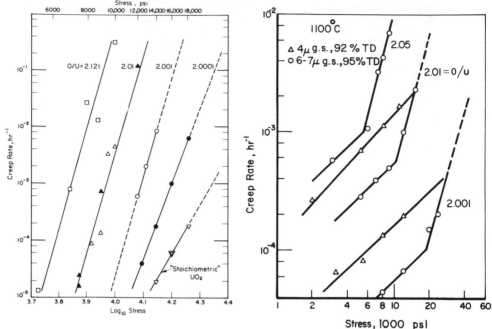

Fig. 9. Stress dependence of the
creep rate for different
O/U ratios in UO_{2+x} single
crystals.[18]

Fig. 10. Stress dependence of the
creep rate in poly-
crystalline UO_{2+x} for two
grain sizes at different
O/U ratios.[18]

diffuses via vacancies in the U sublattice as a U^{4+} species. Thus,
it appears that creep in UO_2 single crystals is governed by a
uranium diffusion controlled mechanism. The decrease of Q_c with
increasing O/U is possibly due to the diminishing importance of
thermally created uranium defects.

The influence of nonstoichiometry on creep in polycrystalline
UO_2 has been studied by Seltzer, Clauer and Wilcox[18] and Armstrong
and Irvine.[19] Seltzer, *et al*, investigated two grain sizes, 4 and
6-7 μm in compression over the range 2000 to 24,000 psi at 1100°C
on material having 95 and 92% of theoretical density. The dependence
of the creep rate on stress and O/U is shown in Fig. 10. At low
stresses the creep rate increases linearly with stress, but at high
stresses there is a power law dependence, with n = 4. This behavior
has often been observed in stoichiometric UO_2.[59-61] In the power
law region, the creep rate varies with oxygen excess, x, as
$\dot{\varepsilon} \propto x^{1.75}$. This is close to the $\dot{\varepsilon} \propto x^2$ dependence observed for single
crystals, and suggests the the creep mechanisms are the same. In

the linear, low stress region, $\dot{\varepsilon} \propto x$. This dependence of creep rate
on x in the low stress region was first observed by Armstrong and
Irvine[19] for 6 μm grain size material at 1300°C.

The transition from linear stress-to power law-dependence in
Fig. 10 occurs at widely different stresses. This is because of the
different dependence of the creep rates on x in the two regions.
The transition stress also depends on temperature when Q_c is different
in the two regions. In early studies of stoichiometric UO_2,[59,60]
Q_c was the same in both regions and the transition stress depended
only on grain size.[59,62] In the power law region the creep rate is
essentially independent of grain size, but in the low stress reg on
the creep rate depends on grain size, d, as $\dot{\varepsilon} \propto 1/d^2$. This increase
in creep rate with decreasing grain size can be seen in Fig. 10.

Microstructural changes resulting from creep of UO_2 have been
observed in compression[60] and bending[19,59-64] experiments. In the
low stress region Poteat and Yust[60] noted grain boundary sliding and
intergranular voids created from the formation and growth of chains
of pores on the grain boundaries. Armstrong, et al[63] also observed
grain boundary sliding and separation in their bending creep experiments. In addition, in the high stress region the ductility decreased
and the structure contained many intergranular cracks.

Armstrong and Irvine[19] found that stoichiometric UO_2 exhibited
intercrystalline cracking much more readily than $UO_{2.015}$. Grain
boundary migration occurred during creep of both stoichiometric and
nonstoichiometric specimens. Grain boundary separation was observed
in $UO_{2.012}$ specimens, but the fissures so formed had not linked
together to form fully developed intergranular cracks. From these
observations, it is clear that grain boundary sliding occurs during
creep of polycrystalline UO_2, and that failure due to intercrystalline
cracking is a distinct possibility for stoichiometric UO_2 subjected
to high stresses.

In the high stress region, the stress and composition dependence
of the creep rate are similar for single crystals and polycrystalline
material. The creep rate is controlled by uranium diffusion. There
are not sufficient data on the stress dependence in the power law
region for high O/U ratios to determine whether or not the stress
dependence in polycrystalline UO_2 will increase as O/U increases.
Also, the composition dependence of Q_c for polycrystalline UO_2 is
not known.

At low stresses and fine grain sizes, grain boundary creep
mechanisms predominate. The low activation energies obtained by
Armstrong and Irvine[19] suggest that grain boundary diffusion is
important. Thus, creep could occur by either grain boundary diffusion
creep,[34] or by grain boundary sliding governed by grain boundary
diffusion.[31,35]

GENERAL DISCUSSION AND SUMMARY

It is clear that a complete interpretation of the observed effects of stoichiometric variations on creep of oxides cannot be made until more complete data are available on the influence of stoichiometry and foreign atom concentrations on self-diffusion of the oxide constituents. This is especially true for the slower diffusing species which is usually expected to control diffusion creep mechanisms. In addition, if the identity of the minority defect is unknown, the dependence of creep rate on p_{O_2} or concentration of majority defect is usually of little help in determining whether creep is diffusion controlled. For example, in FeO_{1-x} and $Co_{1-x}O$ the Q_C values for creep in the power-law stress region are much higher than the cation diffusion activation energy and are in the range of that for oxygen. However, the dependence of oxygen diffusion on x or p_{O_2} is unknown and the p_{O_2} dependence of the creep rate is opposite to that expected for a mechanism involving simple oxygen vacnacies. Since the identity of probable oxygen defects is not known, it cannot be concluded with certainty that creep is diffusion controlled. Furthermore, it is not possible to separate effects on the stoichiometry dependence of the creep rate contributed by diffusion from those contributed by changes in the lattice resistance to dislocation glide, such as ordering into superlattices, clustering, and the formation of complex defects.

Therefore, any discussion of diffusion controlled creep mechanisms must usually be very tentative. A possible exception may be creep of UO_2 in the power-law stress region. Here the dependence of activation energy and creep rate on O/U ratio are both consistent with other data available for uranium diffusion. However, the diffusion data of Hawkins and Alcock[9] predict the opposite dependence of Q_C on O/U ratio to that actually observed.

A general discussion follows on the various aspects of the influence of stoichiometry on creep of the oxides reviewed. It is, of necessity, quite general since most creep mechanisms, especially those associated with the power-law region of creep, are difficult to differentiate on the basis of Q_C and stress dependence of the creep rate alone. These must be accompanied by investigation of the microstructural changes during creep, including the dislocation substructure. In the low stress region in polycrystalline materials, the Nabarro-Herring model predicts a definite creep rate if the self-diffusion coefficients of both species in the oxide are known. The other creep models contain disposable parameters and are again difficult to differentiate without detailed microstructural studies.

High Stresses

In FeO and UO_2 the activation energy for creep and the dependence of Q_C and the creep rate on composition are the same for both the

single and polycrystals in the power-law stress region. For FeO the
stress dependence of $\dot{\varepsilon}$ is also the same. Therefore, the creep
mechanisms are probably the same in the single and polycrystalline
FeO in this stress region. However, in UO_{2-x}, n = 4 to 5 independent
of x for polycrystalline material, but in single crystals n varies
from 7 for stoichiometric UO_2 up to 17 for large x at 1100°C. Thus
the stress dependences are similar for stoichiometric UO_2 (the n for
polycrystalline UO_2 could be as high as 6 to 7) but they diverge as
x increases. Therefore, although they share the same temperature
and composition dependence of the thermally activated process, e.g.,
uranium diffusion, the creep mechanisms are possibly quite different.
In the polycrystalline UO_2 the creep behavior is similar to that
observed in the high stress region in many polycrystalline materials,
but in UO_2 single crystals the high stress dependence indicates that
a different creep mechanism or combination of mechanisms become
operative. For example, the increase in the uranium defect concentration with increasing x increases the uranium diffusion rate, and
therefore the creep rate, orders of magnitude above what it is in
stoichiometric UO_2. This allows factors such as those affecting
dislocation glide velocity, or previously unimportant barriers to
dislocation motion, to assume importance in controlling the creep
rate. What these factors may be is not yet known. Another factor
which could affect the creep behavior is the influence of stoichiometry on the operative slip system. Yust and McHargue[65] found that
UO_2 single crystals oriented for single slip on $\{100\}\langle110\rangle$ showed
slip on this system only in stoichiometric crystals. In hyperstoichiometric crystals slip occurred on a less faborably oriented
slip plane.

In polycrystalline $Co_{1-x}O$, Q_c is nearer that for oxygen diffusion
than for cobalt diffusion, but is still below the oxygen diffusion
activation energy. The creep rates increase with increasing p_{O_2} in
both CoO and FeO but appear to vary linearly with x, or as
$\dot{\varepsilon} \propto p_{O_2}{}^{0.2}$ in FeO, while in CoO the x and p_{O_2} dependence are greater
than this at 1 atm O_2 but decrease with decreasing p_{O_2}. The pre-
dominant cation defects are different for the two systems, being
neutral or singly charged complex defects in FeO, and singly and
doubly charged cobalt vacancies in CoO. Therefore, the minority
oxygen defect may be different in each case. The creep behavior
appears to be diffusion controlled in the high stress region and it
is probable that this corresponds to a dislocation climb controlled
process.

Low Stresses

The near linear stress-creep rate dependence region in single
crystals of TiO_2 [16] is not usually observed in single crystals.
Here, grain boundary creep mechanisms invoked for polycrystalline
materials can be excluded. The subgrain Nabarro-Herring creep model
presented by Weertman[66] is linear in stress only if the subgrain

size is independent of stress. The viscous glide models for
creep[26-28] and Nabarro's[67] dislocation climb model also predict
a linear rate only if the dislocation density is independent of
stress. Since creep at reasonable rates in single crystals must
occur by dislocation motion, and the dislocations are both annihi-
lating each other and leaving the crystal during creep, the dislocation multiplication rate must equal the annihilation rate over
a range of stress. Furthermore, if the applied stress is not the
same as the effective stress, or if there is a critical stress for
glide or climb, then the slope, n, of a log ε vs. log σ plot, where
σ is the applied stress, would be greater than unity even though
the climb or glide velocity varied linearly with the effective stress.

In TiO_2, n was slightly greater than one, and the substructure
size, i.e., the polygonized wall spacing, was stress dependent.[16]
However, if one inserts the measured stress dependence of the wall
spacing into Weertman's model,[66] the theoretical stress dependence
of the creep rate is larger than that observed. Thus the specific
model relating the observed stress dependence of the creep rate to
that observed for the substructure is not clear. A possible reason
why the temperature and stress dependence of creep were so different
in the two TiO_2 creep studies is that different types of slip
systems were operative. The long parallel polygonized boundaries
observed by Hirthe and Brittain were numerous and well defined showing
that most slip occurred on one system. Farb, et al,[17] report only
a trend towards polygonization. Hence, the dislocation substructures
developed during creep are different and this could certainly influence the relative creep behavior. Although Hirthe and Brittain's[16]
results are in accord with a diffusion activation energy of Ti, O,
or both, the high Q_c values measured by Farb, et al, cannot be
explained by diffusion.

In the fine grained polycrystalline materials, e.g., UO_2, CeO_2,
and CoO, the stress dependence of the creep rate decreases at low
stresses, whereas in coarse grained FeO this was not observed. This
is the expected behavior if grain boundary or Nabarro-Herring creep
mechanisms govern creep at low stresses. The grain boundary creep
mechanisms may also exhibit a decrease in Q_c relative to that
observed in the power-law stress region, if they are grain boundary
diffusion controlled. This was observed in CoO. However, a grain
size dependence of the creep rate must also be determined before
grain boundary creep mechanisms are invoked, since TiO_2 single
crystals have also shown a low stress dependence of the creep rate.
More needs to be done on polycrystalline UO_2 to compare Q_c in the
low stress region with that observed at high stresses.

Sigmoidal Creep

Sigmoidal creep curves have been reported for CoO,[12] FeO,[15] TiO$_2$,[17] and UO$_2$.[18,51] All of these observations were for single crystals except for coarse grained FeO. In UO$_{2-x}$ and Fe$_{1-x}$O the creep curves at small deviations from stoichiometry have the usual initial transient followed by steady state creep stages. However, at higher values of x the creep curves become sigmoidal. Sigmoidal creep curves are associated with effects which (1) cause dislocations to be strongly pinned before loading, or (2) make dislocation glide initially difficult. The first case has been treated by various dislocation dynamics models.[21-24] The suggestion of Reppich[15] concerning sigmoidal creep in FeO corresponds to the second case. A similar argument could be applied to CoO having Co$_3$O$_4$ precipitates at low temperatures that dissolve upon heating to creep temperatures ≳900°C. However, it is difficult to imagine that the debris of dislocation loops left behind by the dissolving precipitates would not disappear by climb during an anneal at high temperatures, such as is often observed for radiation damage.[69] Reppich found that a long, high temperature anneal on undeformed crystals (4 days at 1250°C) had no effect on the yield point. Sigmoidal creep in compression of UO$_2$ was observed for O/U > 2.01.[18] In the experiment performed at 1100°C where the O/U ratio was changed during the creep test, a sigmoidal creep curve was not observed at UO$_{2.0001}$, but did appear after the O/U was changed to 2.01. Here the incubation period for creep must be associated with the change in O/U ratio and pinning effects which occur at 1100°C. Possible causes for this are pinning by atmospheres of charged defects or by the formation of nonglissile stacking faults.[70]

In stoichiometric UO$_2$ and in TiO$_2$, sigmoidal creep curves were observed in bending experiments,[17,51] but not in compression.[16,18] Thus, these sigmoidal creep curves are probably due to the method of testing and do not necessarily reflect a strong dislocation pinning effect, as do the compression tests.

In summary, the study of the effect of nonstoichiometry on creep of oxides is only in its very early stages. Much needs to be done in the areas of self-diffusion and defect structure studies of the slower moving defect to complement the creep studies. Also the creep investigations should be systematically extended to study the influence of ordering on the sublattices, antiphase boundaries, and complex defect structures on creep behavior. In metallic materials these structural features can have a significant influence on high temperature creep. Further research on creep of oxides should also include, where possible, detailed studies of the dislocation structure evolved during creep. Although it is possible in some cases to identify the defect whose diffusion controls creep, it is still not possible to identify specific dislocation creep processes. Microstructural studies will assist in resolving this problem.

ACKNOWLEDGMENT

The authors wish to acknowledge the support of the U.S. Army Research Office-Durham, under Grant DA-ARO-D-31-124-G1002.

REFERENCES

1. F. A. Kroger, The Chemistry of Imperfect Crystals. North Holland Publishing Company, Amsterdam. 1964.
2. S. Anderson, B. Collén, U. Kuylensierna, and M. Magnéli, Acta Chem. Scand., 11 1641 (1957).
3. I. V. Vinokurov and V. A. Ioffe, Soc. Phys. Solid State, 11 207 (1969).
4. R. E. Carter and F. D. Richardson, J. Metals, 6 1244 (1954).
5. W. K. Chen and R. A. Jackson, J. Phys. Chem. Solids, 30 1309 (1969).
6. L. Himmel, N. F. Mehl and C. E. Birchenall, J. Metals, 5 827 (1953).
7. D. A. Venkatu and L. E. Poteat, Mater. Sci. Eng., 5 258 (1969/1970).
8. R. Haul and G. Dumbgen, J. Phys. Chem. Solids, 26 1 (1965).
9. R. J. Hawkins and C. B. Alcock, J. Nucl. Mat., 26 112 (1968).
10. J. F. Marin and P. Contamin, J. Nucl. Mat., 30 16 (1969).
11. D. N. Polubayarinov, E. Ya. Shapiro, V. S. Bakunov, and F. A. Akopov, Inorganic Materials, 2 336 (1966).
12. A. H. Clauer, M. S. Seltzer, and B. A. Wilcox, submitted to J. Mat. Sci.
13. A. H. Clauer, M. S. Seltzer, and B. A. Wilcox, to be published.
14. B. Ilschner, B. Reppich, and E. Riecke, Faraday Society Disc., 38 243 (1964).
15. B. Reppich, Phys. Stat. Sol., 20 69 (1967).
16. W. M. Hirthe and J. O. Brittain, J. Amer. Ceram. Soc., 46 411 (1963).
17. N. E. Farb, O. W. Johnson, and P. Gibbs, J. Appl. Phys., 36 1746 (1965).
18. M. S. Seltzer, A. H. Clauer and B. A. Wilcox, Battelle Report No. BMI-1886, Battelle Mem. Inst. Columbus, Ohio, 1970.
19. W. M. Armstrong and W. R. Irvine, J. Nucl. Mat., 9 121 (1963).
20. B. Fisher and D. S. Tannhauser, J. Electrochem. Soc., 111 1194 (1964).
21. P. Haasen; p. 701 in Dislocation Dynamics. Ed. by A. R. Rosenfield, G. T. Hahn, A. L. Bement, and R. I. Jaffee, McGraw-Hill Book Company, New York, 1968.
22. E. Peissker, P. Haasen and H. Alexander, Phil. Mag., 7 1279 (1961).
23. G. A. Webster, Phil. Mag., 14 775 (1966).
24. J. Gilman, J. Appl. Phys., 36 2772 (1965).
25. B. Reppich, P. Haasen and B. Ilschner, Acta Met., 12 1283 (1964).
26. C. R. Barrett and W. D. Nix, Acta Met., 13 1247 (1965).
27. N. F. Mott, Creep and Fracture of Metals at Elevated Temperatures, HMSO, London (1956) p. 21.
28. J. Weertman, Trans. AIME 218 207 (1960).

29. J. D. Eshelby, C. W. A. Newey, P. L. Pratt, and A. B. Lidiard, Phil. Mag., 3 75 (1958).
30. K. N. Strafford and H. Gartside, J. Mat. Sci., 4 760 (1969).
31. R. C. Gifkins and K. U. Snowden, Nature, 212 916 (1966); and J. Amer. Ceram. Soc., 51 69 (1968).
32. F. R. N. Nabarro, p. 75 in Report on Conference on Strength of Solids, Phys. Soc. of London, 1948.
33. C. Herring, J. Appl. Phys., 21 437 (1950).
34. R. L. Coble, J. Appl. Phys., 34 1679 (1963).
35. R. Raj and M. F. Ashby, Met. Trans. 2 1113 (1970, and M. F. Ashby, R. Raj and R. C. Gifkins, Scripta Met., 4 737 (1970).
36. G. Vagnard and J. Manenc, Compt. Rend., 255 104 (1962).
37. P. Kofstad and A. Z. Hed, J. Electrochem. Soc., 115 102 (1968).
38. R. W. G. Wyckoff, Crystal Structures, John Wiley and Sons, New York, 1963.
39. M. E. Straumanis, T. Ejima and W. J. James, Acta Cryst., 14 493 (1961).
40. P. Kofstad, J. Less Common Metals, 13 635 (1967).
41. J. B. Wachtman, Jr., and L. H. Maxwell, J. Amer. Ceram. Soc., 40 377 (1957).
42. K. H. G. Ashbee and R. E. Smallman, J. Amer. Ceram. Soc., 46 211 (1963).
43. J. S. Nadeau, Report No. 67-C-243, General Electric Research and Development Center, Schenectady, New York, June, 1967.
44. D. J. M. Bevan and J. Kordis, J. Inorg. Nucl. Chem., 26 1509 (1964).
45. C. J. Kevane, Phys. Rev., 133 1431 (1964).
46. P. K. Kofstad and A. Z. Hed, J. Amer. Ceram. Soc., 50 681 (1967).
47. F. A. Akopov and D. N. Polubayarinov, Ogerepory [4] 37 (1965).
48. L. Lynds, W. A. Yound, J. S. Mohl, and G. G. Libowitz; p. 58 in Advances in Chem. Series No. 39, Washington, D. C. (1963).
49 E. A. Aitken, H. C. Brassfield, and R. E. Fryxell, p. 435 in Thermodynamics, IAEA, Vol. 2, Vienna (1966).
50. B. T. M. Willis, Proc. British Ceram. Soc., 1 9 (1964).
51. W. M. Armstrong, A. R. Causey and W. R. Sturrock, J. Nucl. Mat., 19 42 (1966).
52. J. Belle, J. Nucl. Mat., 30 3 (1969).
53. R. J. Hawkins and C. B. Alcock, J. Nucl. Mat., 26 112 (1968).
54. C. R. Barrett, A. J. Ardell, and O. D. Sherby, Trans. AIME, 230 200 (1964).
55. K. W. Lay, General Electric Research and Development Center, Report No. 70-C-064, February, 1970.
56. J. M. Marin, H. Michaud, and P. Contamin, Compt. Rend., 264 1633 (1967).
57. Hj. Matzke, J. Nucl. Mat., 30 26 (1969).
58. A. B. Lidiard, J. Nucl. Mat., 19 106 (1966).
59. R. A. Wolfe and S. F. Kaufman, Report No. WAPD-TM-587, October 1967.
60. L. E. Poteat and C. S. Yust; p. 646 in Ceramic Microstructures. Ed. by R. M. Fulrath and J. A. Pask, John Wiley and Sons, New York, 1968.

61. P. E. Bohaboy, R. R. Asamoto, and A. E. Conti, Report No. GEAP-10054, May, 1969.
62. M. S. Seltzer, A. H. Clauer and B. A. Wilcox, J. Nucl. Mat., 34 351 (1970).
63. W. M. Armstrong, W. R. Irvine and R. H. Martinson, J. Nucl. Mat., 7, 133 (1962).
64. W. M. Armstrong and W. R. Irvine, J. Nucl. Mat., 12 261 (1964).
65. C. S. Yust and C. J. McHargue, Bull. Amer. Ceram. Soc., 49 851 (1970).
66. J. Weertman, Trans. ASM, 61 681 (1968).
67. F. R. N. Nabarro, Phil. Mag., 16 231 (1967).
68. A. H. Cottrell, Dislocation and Plastic Flow in Crystals, Oxford Univ. Press, Oxford, 1953.
69. A. D.Whapham and B. E. Sheldon, Phil. Mag., 12 1179 (1965).
70. C. Ronchi and H. Blank, Nuclear Metallurgy, 17 175 (1970).
71. R. Scott, A. R. Hall and J. J. Williams, J. Nucl. Mat., 1 39 (1959).
72. P. R. Mettens, "The Creep Strength of Polycrystalline Uranium Dioxide", Thesis, Rensselear Polytechnic Institute, Troy, New York, 1964.

DISCUSSION

C. B. Alcock (University of Toronto):   Those who work in the thermo-dynamic properties of nonstoichiometric oxides now believe that microdomains of different composition from the matrix exist in sub-stances with gross departures from stoichiometry. Is it known how the presence of microdomains should effect the creep properties of nonstoichiometric oxides?
Authors: It would be expected that, if anything, they would increase the creep strength. Their presence would be similar to that of com-pletely coherent precipitates or clusters and they would therefore interact with moving dislocations. If their size and distribution were appropriate, this interaction would tend to pin the dislocations and thus require higher stresses to move them.

J. A. Pask (University of California at Berkeley):   The data for UO_2 indicating a grain size dependence at low stresses where the stress exponent is 1, and no grain size dependence at higher stresses where the stress exponent is >1, is a beautiful example of a match with the expectations of the theoretically developed relationships. Diffusion-controlled mechanisms are expected to have a grain size dependence and dislocation motion-controlled mechanisms are not.

PANEL ON COMMERCIAL REFRACTORIES IN SEVERE ENVIRONMENTS:

INTRODUCTION

G. O. Harrell, Moderator

North Carolina State University

Raleigh, North Carolina

Commercial refractories are represented broadly by a wide variety of materials, ranging from relatively simple compositions to highly complex multicomponent systems. Their behavior in severe environments is governed by the same basic chemical and physical principles as other materials. But the interaction of the many components with environments that are generally multifaceted presents a much more complex situation than for many other materials. The improvements that have been made in these materials involved not only materials research but iterative analyses involving raw material characterizations, processing effects, property tests, design and construction techniques, and careful interpretation of service failures. The dramatic progress in commercial refractory developments in the past 10-20 years has been made possible by this approach, and especially by the use of modern analylical tools and techniques and by close cooperation between users and suppliers.

The difficulty of summarizing the behavior of commercial refrac-tories within these few papers can be readily realized when one considers that the industry is comprised of more than 150 companies offering more than 4500 different brands. Many are similar, but their small differences may be important and required for specific applications. While the refractory trends in the United States have been largely influenced by the corrosive metals industry – over 72% of all production is used in the production of ferrous and non-efrrous metals – correspondingly severe environments are encountered in many other industrial applications; resulting in a large total spectrum of conditions. Frequently each situation almost becomes a special case.

A review of all environments is impossible for our purpose but a discussion of a few might illustrate the scope. A wide variety of gases and vapors are encountered, e.g., water, chlorine, sulfur, fluorine, alkalies, etc. Frequently, the confined process may be highly reducing for part of the cycle and oxidizing thereafter; several species of gases may be coexisting. Molten environments are particularly severe for chemical attack and sometimes the charge changes from acidic to basic during a process. Coupled with the above are the ever-present physical problems of abrasion; thermal, mechanical, and structural spalling; varying stress-strain conditions; expansion and contraction; and temperature gradients. Further complexity comes from restraints imposed by construction techniques and economics.

Understanding the nature of refractories helps clarify their service behavior in severe environments. They are truly composites, with most of them containing a wide variety of impurities which interact with and/or modify their several major components; phase boundary effects become predominant in service. Most refractories contain large amounts of porosity and a wide distribution of grain sizes. Interactions with many environments frequently cause compositions, space concentrations, and properties to change constantly in service. Corresponding changes in structural and chemical behavior can therefore be expected. It would be safe to say that under most operating conditions refractories are thermodynamically unstable. Following kinetic laws, small increases in operating temperatures can drastically increase reactions. Compounding the refractory to provide kinetic barriers is a primary approach used to increase service life. New design concepts and manipulation of operating conditions are often helpful.

Research to improve refractories can not ignore design and construction effects. Numerous refractory suppliers and users point out that it is not sufficient to improve the brick intrinsically if the changes made impair the value when assembled in a furnace lining. The refractory must be in harmony with the environment, both initially and as cyclic and aging effects occur. It is needful for the researcher to be familiar with the details of operating service conditions. It also often means a re-examination of design installation techniques when drastic changes in refractory properties are made. Most operators strive for a "balanced lining life" in which all parts wear at about the same rate for maximum life. It is then not surprising to find a furnace lined with several different kinds of refractories, and to learn that their interactions with each other must also be considered.

A final word about available refractories for severe environments involves economics. The science and technology is available in many cases to improve existing refractories, such as up-grading the

chemical purity. But it cannot be practiced unless the furnace
operator is repaid through longer service life, increased output, and
substantial savings in costs for furnace rebuilds. Fortunately,
American industry often has the enlightened approach of "it isn't
what it costs, it's what it's worth", e.g., increases in the availability of high purity magnesia resulted in a 2-3 fold increase in
refractory cost, but correspondingly decreased refractory cost per
ton of steel produced and lessened furnace down-time. Furnace life
has been dramatically increased in a short 10 year span from around
200 heats to over 1,000 heats. Better refractories for severe
environments must ultimately meet this "test of the market place".

CHEMICAL PROPERTIES OF REFRACTORIES

S. C. Carniglia

Kaiser Aluminum & Chemical Corporation

Pleasanton, California

ABSTRACT

*The immense variety of chemical aspects of refractories appli-
cations is sampled by descriptions of a few MgO and Al$_2$O$_3$ types
and of their interactions with severe use environments found in the
steel, glass, and other process industries. Some chemical origins
of mechanical behavior and corrosion lifetime are illustrated.*

INTRODUCTION

Commercial refractories are employed primarily in the management
of high temperature heat and in the containment of hot and corrosive
chemical agents; hence their chemical stability in the service
environment is a first requisite. Whether "primary" or not, many
industrial applications of refractories also entail a considerable
load – bearing function in addition to that of a barrier. One immediately becomes interested in the appropriate mechanical properties,
and usually those at high temperatures. These properties too have
their origins in chemistry.

Refractories, in facing their chemical adversaries, undergo
compositional and microstructural alterations which may deteriorate
their mechanical and other performance characteristics long before
they may become perforated by corrosion. These materials are generally
not thermodynamically stable in their service environments, and
their development has been in large part a matter of erecting kinetic
barriers to deteriorative changes that are, ultimately, inevitable.

Commercial refractories are assuredly all microcomposites, if
not macrocomposites in addition. Nearly all of them contain some
quantity of siliceous material, distributed in boundaries and/or
interstices among the crystals of the major phase(s) if not a substantial component of those crystals themselves. Nearly all commercial refractories are porous, from a small to a very high degree.
If we arbitrarily exclude those materials that are substantially
glassy as well as those made by fusion casting, then more than a
very small percentage of porosity means also the occurrence of
connections among pores, hence permeability. These are common
elements of character with which we have to deal in assessing the
properties of refractories and their interactions with the
environments of the high temperature process industries.

The major phases of the most widely used oxidic refractories
include "simple" oxides such as ZrO_2, MgO, CaO, SiO_2, Al_2O_3, Cr_2O_3 and
a few more; complex metal oxides such as spinels; "simple" SiO_2
compounds such as forsterite, zircon, and mullite; then a broad
array of more complex silicate compounds. The interstitial phases
possible, containing the above elements plus a host of metal oxide
impurities, not to mention phosphates, sulfates, fluorides, etc.,
number in the thousands. This treatment will necessarily select
only a few cases, illustrative of the breadth of problems encountered.
The discussion will emphasize some relationships between chemistry
and performance; other panelists (Ruh, Smothers) are expected to
complement this point of view through other emphases.

CHEMISTRY AND MECHANICAL BEHAVIOR

Construction of refractory linings out of individual bricks
accomplishes a large number of useful purposes; but it also produces
extreme local loads due to imperfections of fit. In approaching
the problem of mechanical performance, Davies[1] reports use of
photoelastic stress measurement methods and the utilization of low
temperature strength and high temperature creep data in the analysis
of stress redistribution.

Setting a limit of 0.3% cumulative compressive creep in 50
cycles of a blast furnace stove, Davies presents curves of allowable
loading vs. temperature for three types of Al_2O_3+SiO_2-containing
brick, as in Fig. 1. These data imply differences in high-temperature
plasticity that have little to do with the properties of the most
refractory component, mullite. These bricks contain roughly 55, 40,
and 25 w/o, respectively, of other silicate species less refractory
than mullite. The temperatures of equal creep stress fall almost
linearly at the rate of $100°C$ per 10% change in Al_2O_3 content.
This is a much greater rate of fall than is seen in the melting
point of the Al_2O_3-SiO_2 system; the Al_2O_3-SiO_2-Fe_2O_3 system appears
to agree much more closely. It is, in fact, this complex and volumi-
nous impurity system that is responsible for creep.

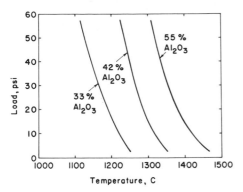

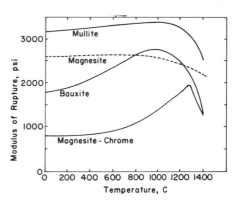

Fig. 1. Allowable load vs temperature for three types of
Al_2O_3–SiO_2 bricks.[1]

Fig. 2. Strength vs temperature
for several mineral–based
refractories[2].

Jackson and Laming[2] give modulus of rupture vs temperature curves
(Fig. 2) for several mineral-based refractories. The two alumina-
type materials, bauxite and mullite, peak at almost exactly the
same temperature, 1000°C, in spite of the fact that the major phases
melt at least 250°C apart. The break in the curve for a natural
magnesite occurs at 1000°C as well. Micrographs show a much smaller
quantity of interstitial material in the magnesite than in these
aluminous refractories; but the similarity in softening behavior
suggests remarkable similarities in impurity minimum melting points
in spite of the widely differing major phases.

The addition of chrome ore to magnesite does not decrease the
siliceous impurity content; but Cr_2O_3 in the MgO–CaO–SiO_2–Fe_2O_3–Al_2O_3
system raises the minimum melting point by about 200°C and increases
the surface tension of the liquid, hence decreasing the penetration
of crystallite boundaries by the liquid phase. Accordingly, the
strength of the magnesite+chrome refractory (Fig. 2) peaks 300°C
higher than does that of magnesite.

The increase in strength with increasing temperature below 1300°C
for MgO+Cr_2O_3 and below 1000°C for Al_2O_3 (bauxite) is striking.
Matters of boundary structure, coherence, thermal expansion, elasticity,
and anisotropy as well as plasticity need to be investigated in
order to understand crack propagation and interruption as a function
of temperature. Neely, Boyer, and Martinek[3] point out some differences
in behavior of MgO+Cr_2O_3 refractories that relate to the stage of
manufacture at which the components are reacted, i.e., to different
nonequilibrium states of the material when overall chemical content
may be very similar.

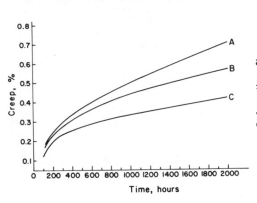

Fig. 3. Compressive creep of (A)
94.0%, (B) 99.5%, and (6)
99.9% pure Al$_2$O$_3$ refractories at 1250°C, 100 psi.[4]

Fig. 4. Compressive creep of a
99.5% Al$_2$O$_3$ refractory
at 1400°C.[5]

Wosinski, Shay, and Edminster[4] studied compressive creep of
alumina refractories of 94.0, 99.5, and 99.9% purity, at 1250°C.
Their data at 100 psi load are smoothed in the curves in Fig. 3.
Analytical data for the three materials are given in Table 1.

Sorting out the effects of porosity and grain size from those
of impurities in comparing materials "B" and "C" is impossible from
these data alone. On the other hand, materials "A" and "B" can be
compared and the high creep rate of the former attributed unequivocally
to its high silica, sodium, and iron content. Phase diagrams indicate
that particulate TiO$_2$·Al$_2$O$_3$ and possibly mullite should be in the
interstitial regions, accompanied by an Na$_2$O-SiO$_2$-Fe$_2$O$_3$-Al$_2$O$_3$ system
whose incipient melting could be at very low temperatures. An

Table 1. Chemistry of Alumina Refractories of Figure 3.[4]

	A	B	C
Al$_2$O$_3$, w/o	94.0	99.5	99.9
SiO$_2$, w/o	3.8	0.07	0.05
Fe$_2$O$_3$, w/o	0.2	--	0.01
TiO$_2$, w/o	1.1	--	--
Na$_2$O, w/o	0.3	0.25	0.04
Σ Others, w/o	0.1	0.08	--
Porosity, %	10.	19.	4.
Grain Size, μm	2-150	1-150	2-6

intricate knowledge of the solution content of the impurity elements in alumina would be necessary before the actual composition of the interstitial silicate phase and its melting point could be estimated.

In dealing with small quantities of impurities it is clearly impractical to work always from first principles: chemical analyses are imprecise, the distributions of elements among phases are not accurately predictable nor accurately measurable, and the amounts of minor phases are quite difficult to estimate petrographically. Hence working from composition to properties is both costly and precarious. It is often more directly revealing to use sensitive measures of temperature-and stress-dependent mechanical behavior as the primary tools, backing these with characterizational investigations to whatever degree of rigor in chemical correlation may seem required. It is evident that a high degree of rigor will not produce great practical reward in a 40% Al_2O_3 brick; but at the $\gtrsim$98% Al_2O_3 level or the $\gtrsim$98% MgO level a fairly fundamental understanding of the role of impurity chemistry may well pay off.

Such is the nature of a study by Trostel[5] of the compressive creep of a 99.5%-Al_2O_3 refractory, characterized by 23% porosity and a wide range of "grain" sizes from < 1 to 1500 μm. Trostel notes that the coarsest "grains" were not single crystals; but they included very large crystallites. The impurities included 0.07% SiO_2 and 0.25% Na_2O.

Trostel's strain measuring sensitivity was about 1/10 that of Wosinski, *et al*[4], so that secondary creep at 1200°C could not be distinguished even up to 400 psi loading. Figure 4 is his log-log plot of data at 1400°C. In Fig. 5 he analyzes temperature – dependent data from 1300° to 1450°C at 50 psi for the activation energy. A combined expression showing all variables (applicable over only limited ranges) is

$$\dot{\varepsilon} = k\sigma^{1.0} t^{1/2} \exp(-184,000/RT). \tag{1}$$

This value of Q compares variously with those of Warshaw and Norton,[6] Passmore and Vasilos,[7] and Fryer and Roberts[8] for "pure" alumina ceramics given in Table 2. Trostel's activation energy of secondary creep, which is here constant over the full temperature range studied, seems to reflect the vacancy diffusion kinetics of aluminum oxide.

By constrast, Kreglo and Smothers[9] report data on 58, 72 and 78% Al_2O_3 refractories in the temperature range 1320-1430°C and in the range of 20-80 psi compressive load. Their secondary creep – rate equation has the form,

$$\dot{\varepsilon} = k\sigma^{1.0} t^{1.0} \exp(-Q/RT). \tag{2}$$

Table 2. Activation Energy for Creep of Al_2O_3 Ceramics

Ref.	Alumina	Temperature,°C	Q,kcal/mole
6	3-13μ, dense	1600-1800	130.
6	50-100 dense	1600-1800	185.
7	fine, dense	1350-1500	142.
8	fine, 8% porosity	1385-1675	120.
8	fine, 35% porosity	1385-1675	185.

Figure 6 gives the analysis of activation energies. The 58% Al_2O_3
material shows a low value of 78 kcal/mole, constant over the entire
temperature range studied. The other two materials show two
characteristic activation energies.

These bricks were all in nonequilibrium states before testing.
Cristobalite and α-Al_2O_3, both present initially in all three microstructures, reacted rapidly at 1480°C to the exhaustion of the
former, while α-Al_2O_3 remained present in all compositions along
with mullite. Another crystalline species was recognized in all
three bricks, presumably derived from a siliceous complex containing
Al, Fe, Ca, Mg, and alkali metal oxides. Since this phase in the
58% Al_2O_3 brick was shown to disappear almost instantly above 1400°C
and to diminish rapidly above 1320°C, presence of a liquid phase
at all test temperatures appear certain; but liquefaction *per se*
cannot be determining the creep rate or its activation energy,
because the latter would appear to be very high and parting failure
would be expected at very low strains. Processes occurring in the
network of solid phases must be invoked to explain the stability of
secondary creep and its activation energy.

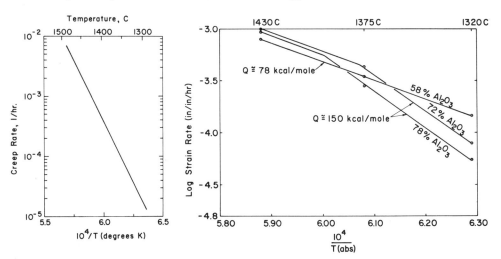

Fig. 5. Compressive creep   Fig. 6. Compressive creep of three
         of 99.5%-Al_2O_3 mullite refractories vs
         refractory vs temper-   temperature, 25 psi.[9]
         ature, 50 psi.[5]

Let us consider a deformation model which includes two processes
in series: (a) sliding along crystallite boundaries, and (b) deformation of crystallites, i.e., boundary migration. In *stable* secondary
creep these two processes are sufficient: no void formation accompanies
the deformation. The two processes are also necessary, and series-
connected, in that "a" cannot occur appreciably without either "b"
occurring or else voids forming.

Let us further assume, as Carniglia[10] has done previously, that
shear deformation along a crystallite boundary at low temperatures
is an easier, more rapid process and one of lower activation energy
than is the deformation of a crystal. Then at low temperatures,
process "b" which requires lattice diffusion will be rate-controlling.
In some higher temperature range, process "b" will be quite able to
keep up with the rate of grain boundary sliding, and process "a"
will be rate-controlling. The model is thus consistant with two -
branched creep curves such as those of Fig. 6.

Let us detail the above model somewhat further as a postulate for
refractories. As a simple example, let us compose a refractory of
three "phases", as represented in Figure 7b (contrasted with a "pure"
oxide ceramic depicted in Fig. 7a): (1) a major, highly refractory
crystalline phase; (2) intercrystalline boundaries that are "direct
bonded", yet of different chemical composition from that of the
major phase interior; and (3) a low melting interstitial phase. A
final case will have to be recognized, in which phase "1" is discontinuous, phase "2" is hence absent, and phase "3" is a continuous
network surrounding all refractory grains, as in Fig. 7c.

In a system such as in Fig. 7b, many grain interstices that
would otherwise have to deform by process "b" will be in part more

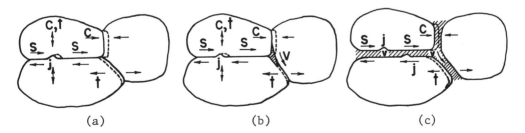

         (a)   (b)   (c)

Fig. 7. Schematic of "phases" and deformation processes in refrac-
        tories (a) "pure" ceramic compact, (b) "direct bonded" re-
        fractory, and (c) "silicate bonded" refractory. Grain boundary
        shear [s] produces compressive [c] and tensile [t] stresses at
        jogs [j] and at boundaries oblique to the applied load; these
        stresses can be relieved (a) only by grain deformation; (b)
        in part by plastic or viscous deformation [v] of the low melting
        interstitial phase; (c) entirely by plastic or viscous flow [v]
        of the low melting phase which bathes all refractory grains.

easily stress relieved by viscous flow in phase "3"; while other
areas as well as jogs in "clean" grain boundaries will still require
process "b" for stable deformation. The overall rate of grain
accommodation by process "b" will thus be greater at any given temperature in the presence of phase "3" than in its absence; but at
low temperatures the activation energy of diffusion in the granular
phase will still apply. Because the strain rate due to process "b"
is higher when a low melting interstitial phase is present, the
process "b" kinetic curve will be displaced to lower temperatures.
The composition of phase "2" is affected by the presence of low
melting impurities, hence process "a" will be more rapid and of
lower activation energy in an impure refractory than in a "pure"
ceramic compact (compare both branches of Fig. 8b with Fig. 8a).

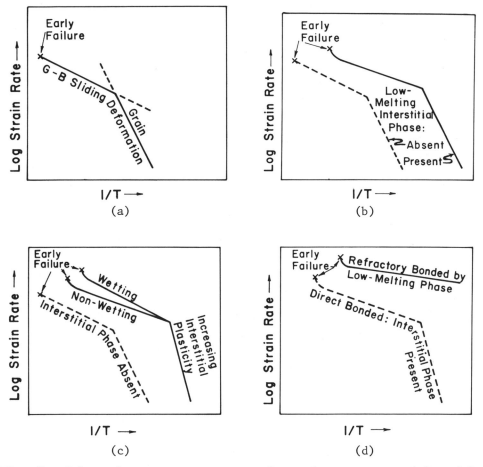

Fig. 8.   Schematic creep rate curves for refractory material models
          (a) "pure" ceramic compact; (b) "direct bonded" refractory;
          (c) progressive processes in "direct bonded" refractory;
          (d) "silicate bonded" refractory.

To the extent that melting or softening phenomena in phase "3" take place over a range of temperatures instead of sharply, the *apparent* activation energy of process "b" may be increased over that in the pure major phase (Fig. 8c). Also, to the extent that the area of intercrystalline "direct bonding" is progressively diminished with increasing temperature by wetting and intrusion of liquid phase "3", the *apparent* activation energy of process "a" may be increased over that for sliding of boundaries of constant area (Fig. 8c).

If phase "1" is discontinuous, then neither process "a" nor "b" will occur. Both will be replaced by the viscous deformation of the low melting, continuous intercrystalline phase "3" (Fig. 8d). The single resulting activation energy is presumed to be lower than that which would apply if phase "2" were present.

With these patterns in hand, let us re-examine Kreglo's curves (Fig. 6).[9] In the 58% Al_2O_3 brick there is known to be progressive melting over the entire temperature range studied; therefore the linear curve should represent the upper branch of Fig. 8b or the "wetting" branch of Fig. 8c. The melting point (below 1320°C) signals the transition from process "b" to "a".

Alternatively, there may be inappreciable direct bonding, with the solid curve of Fig. 8d applying. The two-branched curve for the 72% brick appears to present both branches of Fig. 8b or 8c, the interstitial phase beginning to melt at about 1375°C. The highest purity brick presents the same two branches, the transition (viz., minimum m.p. of the interstitial phase) being about 1390°C. Since the major or most refractory phases are principally mullite and secondarily α-Al_2O_3, it is probable that the activation energy of the lower temperature branches, ~150 kcal/mole, applies to diffusion processes in mullite. Bakunov *et al*[11] obtained the value Q = 123 kcal/mole for bending creep of a 99.5$^+$% pure, 95$^+$% dense mullite ceramic above 1500°C.

Figure 9 is a collection of tensile creep data obtained for magnesia refractories in the range 1300-1450°C by Busby and Carter.[12] One material giving an exceptionally high apparent activation energy (labelled "2-2-2") also gave early parting failure at 1350°C; both characteristics can be attributed to a liquid phase of large quantity. The chemistry, including a high content of silica (1.75%), a low Ca:Si mole ratio (1.2), 1.9% Fe_2O_3, and 0.16% B_2O_3, is in accord (see Table 3). The curve labelled "1-1-1" represents similar impurity proportions but in lesser quantity (1.0% SiO_2), and the material was more densely compacted; early parting failure was staved off to 1450°C.

Among the remaining periclases, the absolute value of creep rate decreases in the temperature range studied with increasing Ca:Si

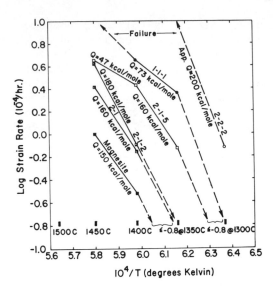

Fig. 9.   Tensile creep rates of six MgO refractories vs temperature, 7 psi load.[12]

ratio; but the activation energies are another matter.   Another matter is also in evidence when one searches for an intrinsic property of high purity MgO as a basis of comparison.

Vasilos, Mitchell, and Spriggs[13] published data on the bending creep of 1-3 μm, 99.8% pure, 99.8% dense polycrystalline magnesia, measured between 1180 and 1260°C.   Their calculated value of the

Table 3.   Chemistry of Magnesia Refractories of Figure 9.[12]

Wt.	Periclases					Magnesite
	2-2-2	1-1-1	2-1-5	2-1-2	2-1	
MgO	93.2	95.8	91.4	94.4	97.2	94.3
CaO	2.0	1.2	1.9	2.1	1.7	3.3
SiO_2	1.75	1.0	0.9	0.9	0.5	1.8
Fe_2O_3	1.9	1.3	4.9	1.4	0.1	0.5
Al_2O_3	1.1	0.5	0.7	0.9	0.2	0.1
Cr_2O_3	0.02	0.006	0.02	0.15	0.3	0.01
B_2O_3	0.16	0.13	0.15	0.10	0.05	0.03
$\Sigma SiO_2 + R_2O_3$	4.93	2.94	6.67	3.45	1.15	2.44
Ca:Si (moles)	1.2	1.3	2.3	2.5	3.6	2.0

activation energy was 74 kcal/mole. When their published raw data
are replotted in a repeat of their calculation, a linear array is
obtained giving Q = 104 kcal/mole. Coble[14] published data which
show considerably more scatter but which give an estimated value
Q = 123 kcal/mole. Vasilos et al[13] referenced previous measurements
of diffusion - controlled phenomena of several kinds in MgO, whose
activation energies ranged from 60 to 112 kcal/mole with no modal
value in view.

Interpretation of Busby and Carter's data[12] thus has to stand
without an unequivocal reference to the intrinsic behavior of MgO.
There is a tendency visible in the low temperature dashed extrapolations in Fig. 9 toward values of Q in the vicinity of 180
kcal/mole; and relatively high values are also seen in the upper
temperature ranges of all materials of high Ca:Si ratio. Considering
the temperatures, these are probably completely solid systems; and
considering the extent of MgO-MgO bonding in the high Ca:Si
materials,[12] it would seem that 150 < Q < 180 kcal/mole is in fact
describing the deformation of magnesia crystals in these solids.
Thus the steepest curves of Fig. 9 appear to conform to the steep
branch of Fig. 8b or 8c.

The refractory "2-1-5", whose curve turns over above 1400°C,
is high in Fe_2O_3 (4.9%); and so it is probable that the decreased
slope represents progressive melting above 1400°C. This branch
thus corresponds to the upper branch of 8b or 8c, and the low activation energy of 47 kcal/mole corresponds to boundary sliding in
this MgO. There is no clear basis for selection between "wetting"
and "nonwetting" (Fig. 8c) from the data, but progressive boundary
penetration by the liquid is likely.

In the "1-1-1" material between 1350 and 1400°C (where
Q = 73 kcal/mole), it seems that the "wetting" curve of Fig. 8c is
represented because in the next 50°C interval parting failure
occurs. Considering the low Ca:Si ratio and the amount of B_2O_3
present (0.13%), the arguments of Taylor, Ford, and White[15] favor
this: material "1-1-1" most likely shows initial melting around 1350°C
and penetration of grain boundaries above this temperature.

Finally, Kreglo and Smothers[16] studied a single periclase
refractory in compressive creep over the temperature range
1200-1430°C. Their Arrhenius plot is given here as Fig. 10. The
data are quite linear, and calculation of the activation energy
gives a value Q $\cong$ 33 kcal/mole which is exceptionally low. The
model of Fig. 8d is suggested: a discontinuous refractory phase
separated by a wetting glass whose softening point is at or below
1200°C. The published chemistry of this brick[16] does not appear in
accord (CaO 0.9%, SiO_2 0.5%, Al_2O_3 0.1%, Fe_2O_3 0.2%, Ca:Si = 1.8)
but Kreglo in private communication confirmed the additional presence
of 0.3% B_2O_3 and confirmed this microstructure in every detail.

Thus, it appears that the high temperature mechanical properties of refractories can be rationalized with their chemistry. Let us now examine the chemistry of a few corrosion phenomena.

CHEMISTRY AND DURABILITY

Osborn[17] has paid tribute to the contributions to the efforts in the determination of silicate phase diagrams. White[18] has also published an account of both the importance and the results of such efforts. White put the objectives of refractory material design for chemical durability into these direct terms:

"Controlling the composition to avoid the formation of low melting eutectics and to minimize the amount of liquid phase formed at high temperatures in service;

"Controlling the geometrical distribution of the melts formed, in such a way that their effect is minimized."

Considerations of these objectives in Basic Oxygen Steelmaking refractories were discussed in a companion paper (this volume, p. 57).

Basic brick containing both MgO and Cr_2O_3, so extensively used in open hearth steelmaking since the 1930's, is another example. Of White's collection of diagrams, two are presented here as Figs. 11 and 12. Figure 11 is a small pseudo-ternary portion of the MgO–CaO–SiO_2–Fe_2O_3–Cr_2O_3 liquidus, illustrating the rise in melting point occasioned by substituting chrome oxide for iron oxide in the system. Figure 12 illustrates the degree to which Cr_2O_3 reduces the wetting tendency of iron - containing silicate melts on MgO surfaces at 1550°C, a property directly related to the prevalence of direct crystal bonding and hence to penetrability by iron-rich slags and softening of the refractory.

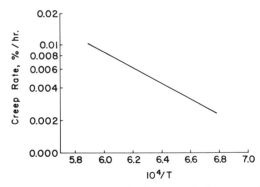

Fig. 10.   Compressive creep rate of a periclase refractory vs temperature, 25 psi load.[16]

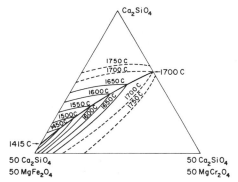

Fig. 11. Liquidus in a pseudo-ternary part of the MgO-CaO-SiO$_2$-Fe$_2$O$_3$-Cr$_2$O$_3$ phase system.[18]

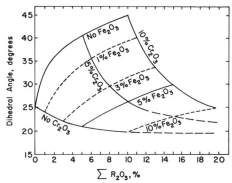

Fig. 12. Effect of Cr$_2$O$_3$ and Fe$_2$O$_3$ on dihedral angle between MgO grains in periclase-monticellite refractories, 1550°C.[18]

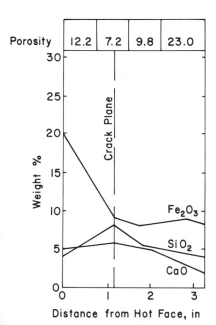

Fig. 13. Distributions of Fe$_2$O$_3$, CaO, and SiO$_2$ in used open-hearth roof brick.[19]

Fig. 14. Periclase (89% MgO) bricks after 2 years' service in glass checker port: top three courses.[22]

Figure 13 illustrates the distributions of Fe_2O_3, CaO, and SiO_2 in an MgO+Cr_2O_3- based roof brick after service.[19] Iron oxides travel predominantly by solid state diffusion, as solution components of magnesia, chrome, and spinel phases. Lime and silica move from the hot face, on the other hand, as components of a liquid phase occupying voids, boundaries, and crystal interstices. The long range liquid transport is primarily through the connected porosity, wherein the silicate moves down the temperature gradient until it freezes.

The greatest concentration of silicates lies just above the freezing plane: here, in the presence of a silica - rich liquid medium, densification proceeds as can be seen from the post - service porosity levels indicated in the several zones in Fig. 13. Broad cracks often appear, parallel to the hot face and in the liquid - rich zone or in the freezing plane just behind it; and often the 0.5-2 in. thick slab of brick isolated by such a crack will break off, leaving a new cycle of penetration, mass transport, and "slabbing" or "peeling" to follow. Much progress has been made against this cycle by adjustments of the chrome, silica, and iron oxide contents of brick and by the adoption of very high firing temperatures ($\geq 1700°C$). These in combination have encouraged formation of the resistant "direct" crystal bond.[20,21]

In several other major uses, magnesia or magnesia+chrome brick seemingly perform only the passive function of conducting hot gases or as a heat exchanger, having to cope just with entrained particulates. Some examples are the roofs of reverberatory furnaces, blast furnace stoves, and glass tank checkers. However, none of these turns out to be passive. The particulates (and splashes) may be molten minerals or glass, the hot condensible vapors are SiO_2, Na_2O or K_2O, SO_2, SO_3, etc. The exposure is inevitably to liquid-forming chemicals. Take the glass furnace checker for example. The specimens described by Van Dreser and Cook,[22] after two years' service (Fig. 14), were substantially altered. In another material under test at the same time, the first five courses of brick were not present to be photographed. The corrosive agents in this case were SiO_2, CaO, Na_2O, SO_3, CO, CO_2, H_2O and O_2, plus Al_2O_3, MgO, and As_2O_5; at a top temperature of 1400°C.

The severity of brick deterioration appeared to be proportional to the amount of recrystallization of the primary phase occurring in the hot brick.[22] This, in turn, varied more with the major phase composition: MgO, Cr_2O_3, Mg_2SiO_4, etc., than with the bonding or interstitial medium. In contrast to the open hearth steelmaking environment where MgO+Cr_2O_3 showed up so well, here the simple magnesia bricks excelled. Magnesia refractories of low Ca:Si ratio performed better in glass checkers than did those of high Ca:Si ratio which are preferred in basic steelmaking.   Spinel-bonded high purity periclase performed exceptionally well.

"Alumina" refractories encompass a spectrum of products from natural fireclays of as little as 30% Al_2O_3 content, to mullite ($3Al_2O_3 \cdot 2SiO_2$) and mullite - clay materials, bauxite (natural Al_2O_3) and bauxite - enriched compositions, and finally to refractories based on beneficiated or refined bauxite or even synthetic Al_2O_3. Economic pressures favor the use of mineral and beneficiated mineral products over synthetic aluminas where the requirements will permit. But the former products are all relatively high in silicate content, and there are several industrial processing situations wherein appreciable silicates cannot be tolerated.   For these the processed alumina refractories are a necessity.

Molten Al and its alloys, especially those containing Mg, attack silicates with great rapidity by reduction.   This not only deteriorates vessel linings but also contaminates the metal with Si.   The aluminum industry effected major savings by advancing from the superduty fireclays to refractories made from refined bauxite, running about 90% Al_2O_3.   The melting of glass is, on the other hand, considerably more demanding.   One even finds, for some specialty glasses, the use of platinum and iridium lined vessels.   Glasses produced on a massive scale must turn to cheaper vessels than these; but the cost of their installation, the cost of down time for repair, and the problem of contamination of the product demand that refractory linings of glass - melting tanks be of "technical ceramic" quality.   Here it is common for synthetic alumina refractories to run 95% Al_2O_3 or higher.

Glass corrosion mechanisms in alumina refractories are essentially grain boundary liquefaction mechanisms.   For a change of viewpoint, let us examine a mechanical aspect of glass tank corrosion which was first identified in 1924 by Scholes[23] and has been discussed recently by Begley and Herndon.[24]   This is termed, "upward drilling".

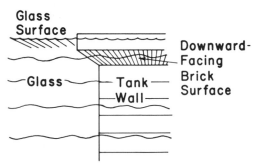

Fig. 15.   Schematic of downward-facing glass tank brick subject to "upward drilling".[24]

Fig. 16.   High-alumina glass tank brick subjected to "upward
           drilling" by soda-lime glass (Ref. 24, by permission).[24]

     "Upward drilling" is an accelerated corrosion, by molten soda -
lime glass, of refractories that expose a downward - facing surface
to the liquid. A typical geometrical configuration is presented
in Figure 15. This occurs in the construction of channels, for
example. But a downward - facing brick is not necessary: a pore,
crack, or joint which is closed above and open below, and immersed
in the glass, will suffice. The phenomenon requires further,
(a) movement of the glass, and (b) gas bubbles in the glass.

     The phenomenon is associated with a gas bubble which becomes
trapped against the downward - facing refractory surface, and
around which the glass flows. The bubble creates an interference
with laminar flow such that turbulence brings fresh alkali constantly to it. It is thought that oscillatory movement of the
bubble stirs the alkali into the film between it and the brick,
where reaction proceeds. Thus the trapped bubble functions as a
turbulence - generator, or "pump". Experiments have confirmed some
of the expected characteristics, for example:[24] (1) small (1/8-1/4 in.)
bubbles drill more rapidly than large (> 1 in.) bubbles; (2) drilling
rates in cracks or crevices increase with increasing crevice width;
and (3) "upward drilling" rates in various refractories correlate
with rates of "metal line" (i.e., meniscus) corrosion. The recession
of downward - facing surfaces may be several times faster than that
of others. A photograph of a test specimen illustrating the results
is presented in Fig. 16, supplied by Begley and Herndon.[24]

   Efforts are under way to characterize different refractories
as to their resistance to the "upward - drilling" effect. One
would anticipate a potential here for identifying those factors
which separately contribute to reaction kinetics, and for developing
a predictive capability from first principles. Since the phenomenon
so aptly points up the role of kinetic barriers in the protection
of refractories, wherein thermodynamic stability does not exist,
its description is appropriate to close this review: precisely on
the theme on which it began.

REFERENCES

1. W. Davies, Refractories J. 46 (6) 8 (1970).
2. B. Jackson and J. Laming, Refractories J. 45 (11) 328 (1969).
3. J. E. Neely, W. H. Boyer and C. A. Martinek, Jr., Am. Ceram.
      Soc. Bull. 49 (8) 710 (1970).
4. J. F. Wosinski, G. C. Shay and W. H. Edminster, Am. Ceram. Soc.
      Bull. 48 (5) 540 (1969).
5. L. J. Trostel, Jr., Am. Ceram. Soc. Bull. 48 (6) 601 (1969).
6. S. I. Warshaw and F. H. Norton, J. Am. Ceram. Soc. 45 479 (1962).
7. E. M. Passmore and T. Vasilos, J. Am. Ceram. Soc. 49 166 (1966).
8. G. M. Fryer and J. P. Roberts, Proc. Brit. Ceram. Soc. 6 225 (1966).
9. J. R. Kreglo and W. J. Smothers, J. Metals 19 (7) 20 (1967).
10. S. C. Carniglia; pp. 425-71 in Materials Science Research, Vol. 3
      Ed. by W. W. Kriegel and Hayne Palmour III, Plenum Press,
      New York, 1966.
11. V. S. Bakunov, E. S. Lukin, and D. N. Poluboyarinov, Tr. Mosk.
      Khim.-Tekhnol. Inst. 50 216 (1966).
12. T. S. Busby and M. Carter, Trans. Brit. Ceram. Soc. 68 (5) 205
      (1969).
13. T. Vasilos, J. B. Mitchell and R. M. Spriggs, J. Am. Ceram. Soc.
      47 203 (1964).
14. R. L. Coble; p. 706 in High - Strength Materials. Ed. by V. F.
      Zackay, John Wiley and Sons, Inc., New York, 1965.
15. M. I. Taylor, W. F. Ford, and J. White, Trans. Brit. Ceram. Soc.
      68 (4) 173 (1969).
16. J. R. Kreglo and W. J. Smothers, J. Am. Ceram. Soc. 50 457 (1967).
17. E. F. Osborn, Edward Orton Jr. memorial lecture, 72nd Annual
      Meeting, American Ceramic Society, Philadelphia, May 4, 1970.
18. J. White; p. 77 in Refractory Materials. Ed. by A. M. Alper,
      Academic Press, New York, 1970.
19. All - Basic Furnace Sub-Committee Report, J. Iron Steel Inst.
      185 304 (1957).
20. M. L. Van Dreser and W. H. Boyer, J. Am. Ceram. Soc. 46 257 (1963).
21. M. L. Van Dreser, Blast Furnace and Steel Plant, April, 1964.
22. M. L. Van Dreser and R. H. Cook, Am. Ceram. Soc. Bull. 40 (2)
      68 (1961).
23. S. R. Scholes, Glass Ind. 5 (9) 161 (1924).
24. E. R. Begley and P. O. Herndon, Am. Ceram. Soc. Bull. 49 (7) 633
      (1970).

PHYSICAL PROPERTIES OF REFRACTORIES

Edwin Ruh

Harbison-Walker Refractories Company

Pittsburgh, Pennsylvania

ABSTRACT

High temperature physical corrosion and resistant properties of refractories and their measurements are briefly discussed. Laboratory simulations of service conditions provide information about specific properties which must be advantageously chosen in combination in developing and/or selecting refractories capable of meeting actual service applications.

INTRODUCTION

The subject of this conference "Ceramic in Severe Environments" typifies the refractories industry today. Virtually every furnace lined with refractories in this country is being pushed at higher temperatures or greater production rates than ever before, creating more severe service conditions. Refractories are unique materials. They are unlike most other ceramics in that they contain both coarse and fine grained material which may be totally dissimilar in composition and character, but yet bonded together to form the refractory which performs a service for industry. To illustrate, three severe environments where refractories are used have been selected.

OPEN HEARTH ROOF REFRACTORIES

All of the present-day steelmaking processes present extreme environmental conditions. To single out one of these processes, Fig. 1 shows two bricks after a period of service in an open hearth furnace roof. The magnesia-chrome brick at the left represents a comparatively new development in basic refractories, that of direct bonding.[1] When

(a)   (b)

Fig. 1.   Basic refractories after service in an open hearth roof,
          (a) direct bonded magnesia-chrome refractory (b) burned
          magnesia-chrome refractory not exhibiting direct bonding.

refractories are exposed to the atmosphere and slag conditions in
an open hearth roof, they are subjected to concentrations of iron
oxide, silica and lime. A chemical analysis of the zones from the
hot face to the cold face of the brick reveals a concentration of
CaO and SiO_2 ~2-3 in. from the hot face.   Figure 2 shows a
typical plot of these oxide concentrations versus depth in inches.
It has been found the direct bonded brick have superior physical
properties and perform better in the severe wear areas of the open
hearth roof.   The photomicrograph (Fig. 3) indicates the direct
contact between the dissimilar large chrome ore grains and the

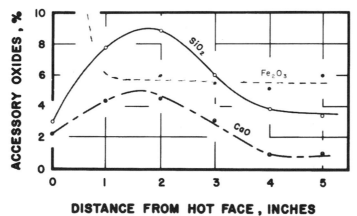

Fig. 2.   Distribution of accessory oxides in a magnesia-chrome
          refractory after service in an open hearth roof.

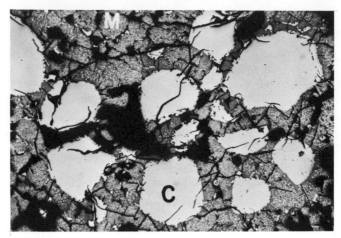

Fig. 3.   Photomicrograph of direct bonded magnesia-chrome refractory.
Note the direct contact of the chrome ore "C" and the
magnesia "M".

magnesia. This structure is commonly referred to as direct bonding.
Refractories with this type of bonding exhibit very superior
performance in long time load tests as shown in Fig.4.

GLASS TANK REGENERATOR REFRACTORIES

The second service which might be considered a severe environment is that of refractories in the checker chamber of glass tank
regenerators.   Since these regenerators are usually operated
on a 20 min. cycle, it is obvious that the brick, particularly
in the top, are subjected to extreme thermal shock conditions. Also,
the checkers are subjected to batch carry-over in the gases, thus

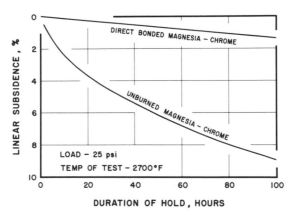

Fig. 4.   Long time load test of direct bonded magnesia-chrome and
unburned magnesia-chrome refractory.

Fig. 5. Two refractory brick after removal from service in a
         glass tank regenerator.

providing chemical contamination as well as thermal shock.[2] In
fact, conditions from top to bottom in the checker chamber vary
so greatly, that it is advantageous to divide the checker setting
into three zones, each sone being characterized by its own set of
conditions. Characterization includes evaluation of refractory

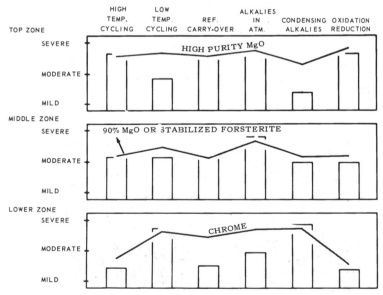

Fig. 6. Profiles of refractory brick properties for glass tank
         regenerator checker service.

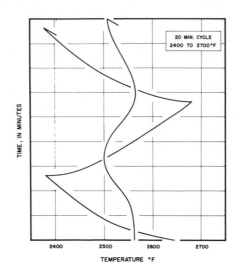

Fig. 7.   Temperature variation in a magnesia refractory during
          cycling in the simulated regenerator furnace.

performance in terms of (1) high temperature cycling, (2) low
temperature cycling, (3) batch carry-over, (4) alkalies in the
atmosphere, (5) condensation of alkalies, and (6) oxidation and
reduction. This list is by no means exhaustive.

     Figure 5 shows two refractories after service in a glass tank
regenerator. Figure 6 illustrates the brick property profiles that
would be desirable for these three checker zones. To simulate this
checker service, a laboratory furnace has been constructed which
will automatically cycle sample regenerator settings between any
two temperatures that may be desired, thus representing an attempt
to simulate actual service.   Figure 7 illustrates the

Fig. 8.   Two chrome-magnesia refractory brick after removal from
          service in a copper converter end wall.

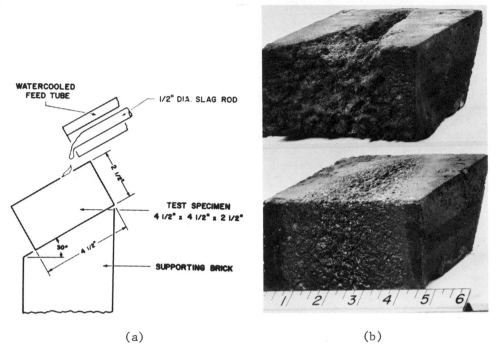

(a)   (b)

Fig. 9.   Drip slag corrosion test for copper converter service.
          (a) schematic of furnace, (b) refractories after test
          (upper) conventional brick, (lower) brick especially
          developed for this service.

kind of cycle which is achieved in this experimental furnace. Work
in this area has lead to recommendations which have materially
increased the life of glass tank checkers by providing the right
refractory for each of the three zones.

COPPER CONVERTER REFRACTORIES

     The third example of refractory materials under severe environ-
ments may be found in the copper converter. Figure 8 shows two
magnesia-chrome refractories which were removed from the end wall
of a copper converter. In this application, slag attack may be
considered one of the prime modes of refractory wear. To simulate
the slag corrosion of refractories, a drip slag test has been
devised. In this test, $\frac{1}{2}$ in. dia slag rods are fed through a
water cooled jacket so that as the slag enters the furnace, it
melts and drips on the refractory test specimen. Figure 9a illustrates a schematic of this test and Fig. 9b shows two brick of
different compositions after being subjected to the same slag.
Through the use of this test, special chrome-magnesia refractory
compositions have been developed which have extended the life of
refractories in the copper converter.

SUMMARY

       As stated at the outset, the foregoing examples were selected
to highlight the complexities associated with the production,
development and service of refractories. Developmental studies
cannot be profitably conducted in production furnaces. Thus, there
are many tests in the development laboratory which may be utilized
to simulate refractory conditions in service. Frequently, it is
necessary to use a series of tests, each one of which evaluates
one factor in the service life of refractories. Through the use
of these tests, it is possible to make recommendations for actual
field service. Ultimately, the proof lies in the actual application.

REFERENCES

1. Ben Davies and Frank H. Walther, Amer. Ceram. Soc. 47 (3) (1964).
2. Frank H. Walther and John Kivala, Ceramic Industry, November 1963.

ADDITIONAL READING

C. Burton Clark and J. Spotts McDowell, J. Metals 11 (2) 119-124
       (1959).
O. M. Wicken and H. A. Freeman, "Comparisons of Refractory Exposures
       In Ferrous and Non-Ferrous Operations", Presented at the
       Symposium-Pyrometallurgical Processes in Non-Ferrous Metallurgy,
       The First Operating Metallurgy Conference, November 29, 1965,
       Pittsburgh, Pennsylvania.
C. Burton Clark and J. Spotts McDowell. Amer. Ceram. Soc. Bull.
       42 (7) 404-408 (1963).

RELATION OF PROPERTIES OF REFRACTORIES TO BOF USE

       W. J. Smothers

       Bethlehem Steel Corporation

       Bethlehem, Pennsylvania

                              ABSTRACT

       Use of a single brick type for the initial complete lining
permits determination of the wear pattern of a BOF furnace. Im-

*provements in refractories must be made to obtain balanced wear
by increasing lining thickness in the critical areas, or when
maximum permissible thickness from a design standpoint is used,
by upgrading brick quality. The essential criteria for evaluation
of brick quality have been established, and are now being used to
produce superior brick. For almost two years, 200-ton BOF units
thus equipped have averaged over 1000 heats per lining.*

      Our goal is to obtain the lowest possible refractories cost
per ton of steel produced. Balanced lining wear is therefore a
major objective. Basic-oxygen-furnace (BOF) refractories fall
into three main process categories: (1) pitch-bonded, (2) tempered
and (3) burned impregnated. Our procedure in startup of a BOF
shop is to line the vessels with essentially one class of brick that
has the best proven performance, namely burned, pitch-impregnated
brick with high magnesia content. A wear pattern similar to that
shown in Fig. 1 is often obtained. Location of the high wear areas
indicates where improvements in refractories must be made to obtain
balanced wear. On a short range approach, the more expensive
burned brick in low wear areas could be replaced by lower cost
tempered or pitch-bonded brick. Also, some improvement in balancing
lining wear may be made by increasing lining thickness in the critical
areas. Thickness can be increased, however, only as long as steelmaking volume requirements and metal bath depths are not altered
significantly. On a longer range approach, through laboratory and
plant tests, criteria may be established to select the best quality
of refractories within a given class. For example, brick used in
critical areas of the furnace, such as the charge pad, are of the
burned-impregnated type and should have the following properties:
(1) high hot strength, (2) high resistance to thermal shock and
spalling, and (3) good resistance to slag attack.

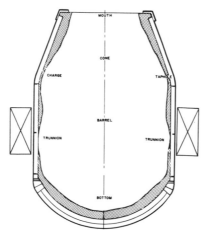

Fig. 1.   Typical wear pattern – Lackawanna BOF

Historically, the hot strength of burned impregnated brick with high magnesia content has increased through the efforts of the refractory manufacturers. By means of our hot compression strength test at 1540°C (2800°F), we found that in the early days of our operations some brick had a strength of only 300 psi. With this brick, we obtained only 400 heats in a BOF before a hole developed in the charge pad (Table 1). As hot strength increased to around 2400 psi, we obtained 815 heats before the first hole appeared. We find that the best indication of brick quality in a furnace is the number of heats on the furnace before the first hole appears in the working lining. Hot strength, then, is one of the important properties of these BOF brick. It should be emphasized, however, that there is an upper level of strength, above which other properties of the brick suffer; e.g., resistance to thermal shock may decrease.

We have developed a laboratory slag test which gives us results that can be correlated with those obtained in the BOFs. Corrosion of brick can be determined quantitatively and the depth of slag penetration measured in the laboratory test. Observations made on used brick from a charge pad showed that carbon content was not entirely uniform through the remaining thickness of brick. Through microscopic observation of the first 1/4 in. thickness of several brick at the hot face, we found that slag penetration was a function of carbon content of the brick (Fig. 2). Slag penetration decreased to a minimum as a carbon level of 3% was approached. If the burned impregnated brick retains more than 3% carbon, the slag shows minimum penetration and therefore the slag cannot wet and attack the MgO grains of the high magnesia brick.

Table 1. Correlation of hot strength with charge pad life.

Brick Brand	Compressive Strength (psi)	Heats to First Hole in Charge Pad
A	300	400
B	800	440
C	2400	815

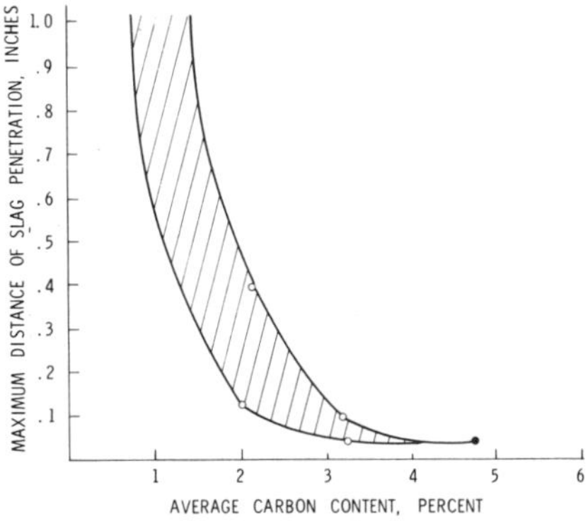

Fig. 2. Effect of carbon content on slag penetration.

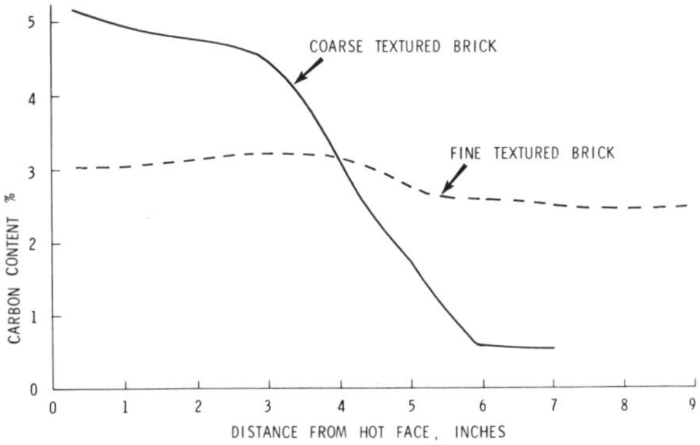

Fig. 3. Effect of brick structure on retained carbon.

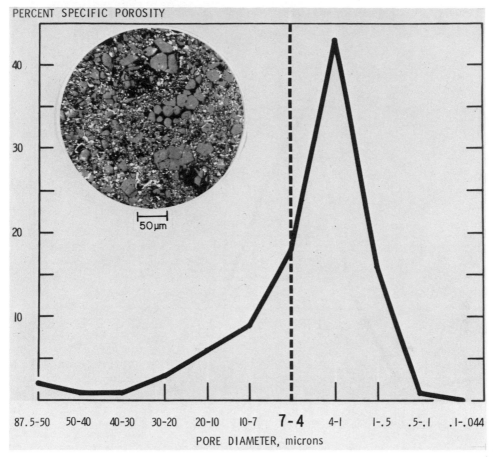

Fig. 4.   Ignited pore-size distribution and coked microstructure of
          burned pitch-impregnated periclase brick characterized
          by predominantly fine pores filled with fine, well-dispersed
          carbon crystallites. Carbon (white), Pores (black), MgO (gray).

     One important characteristic which controls the retention of
carbon is the brick texture or microstructure. A coarse textured
burned brick removed from a BOF (Fig. 3) showed a high carbon content
at the hot face, but only a small amount of carbon retained at the
cold face so that when 4 in. of this brick would be worn away, the
carbon content would quickly drop below 3% to permit slag penetration.
A fine textured burned brick, however, retained approximately 3%
carbon throughout its length. Use of the fine textured brick, therefore, will assist in giving longer life in the BOF.

     In additional work, we have found that pitch-impregnated
refractories with a coarse texture and large pores have poorly distributed coarse-grained carbon crystallites. As pore size decreases,
carbon crystallite size also decreases and the carbon becomes more

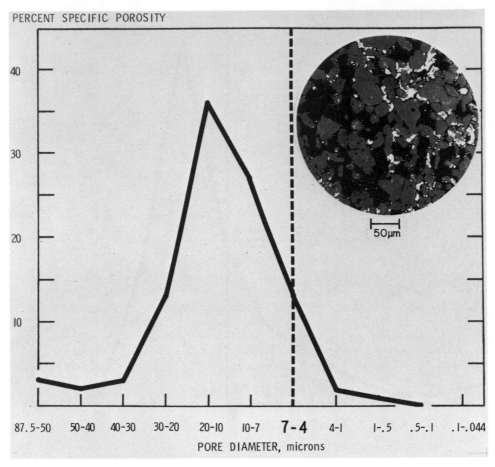

Fig. 5. Ignited pore-size distribution and coked microstructure
         of the brick characterized by predominantly coarse pores
         partially filled with coarse, poorly-dispersed carbon
         crystallites.

evenly distributed throughout the matrix structure. We find that
when the carbon is fine grained and evenly distributed, not as much
carbon in the brick is required to minimize slag penetration.

     Quantitative measurements of pore size distribution show that
when the majority of pores lie in the 1-4 μm range with a unimodal
distribution, the microstructure shows the desirable fine matrix
pores filled with fine-textured, well-dispersed carbon crystallites
(Fig. 4). A burned impregnated brick with this microstructure gives
good service life in the BOF. At the other extreme, a unimodal
distribution with a maximum of 10-20 μm shows large matrix pores
partially filled with coarse-textured, poorly-dispersed carbon
crystallites (Fig. 5); this brick would show poorer service life.

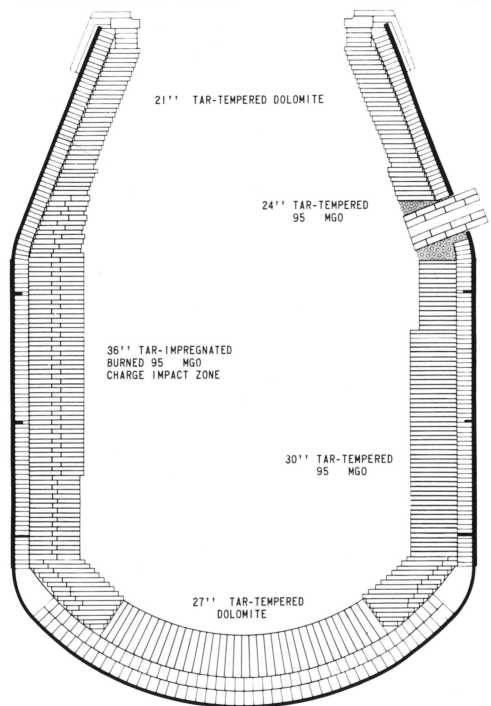

Fig. 6. Sparrows point lining configuration.

Through proper characterization of BOF brick, we can then specify what we would call a balanced brick to have the desired combination of strength, carbon content and microstructure.

A typical BOF lining configuration of our vessels at our Sparrows Point plant is shown in Fig. 6 to illustrate the principle of balancing lining wear by either increasing lining thickness or changing brick type in selected areas.

In summary, as an example of our selection of the best brick and refractory practices, our Sparrows Point BOF shop has for now almost two years averaged over 1000 heats per lining on 200-ton furnaces.

DISCUSSION

C. B. Alcock (University of Toronto): It has been suggested that the cost per ton of finished steel has such a small contribution from refractories cost that improvements on the best materials presently available would have negligible effect.

Does the panel agree with this view, and if so, how can the basic scientist help the refractories industries? Would improvements in processing and analysis of materials be most valuable?

W. J. Smothers: Sufficient improvements have been and continue to be made on refractories to reduce their costs in operation. Enough money can be saved that steel companies are interested in further improvements in refractories and in their use. Not only is it important to have purer materials, but their assemblage into suitable shapes having the necessary physical properties is important.

S. C. Carniglia: The further reduction in refractories cost per ton of steel produced is not so much the motivation as are matters of capital investment, down-time, and lining replacement and maintenance costs. The challenge of increasing lining lifetime remains, and while efforts to meet this include considerations of overall purity, the detailed chemistry of brick as affected by specific impurities, additives, and their relative amounts deserves at least equal attention. Each industry, and each process within each industry, presents unique factors in regard to refractory cost and durability that are being approached in the same rational way.

C. B. Alcock: I am delighted to hear that the panel stresses the importance of the potential contribution of the chemist over that of the physicist!

COMPATIBILITY OF BIOCERAMICS WITH THE PHYSIOLOGICAL ENVIRONMENT

Samuel F. Hulbert, Jerome J. Klawitter and Ralph B. Leonard

Clemson University

Clemson, South Carolina

ABSTRACT

When prosthetic materials are placed in the body, consideration must be given to (1) the effect of the physiological environment upon the prosthetic metal, medical polymer, or bioceramic, and (2) the effect of the prosthetic material and its corrosion or degradation products upon the fluids and tissues of the surrounding environment. The interaction which occurs between biomedical materials and the physiological environment is discussed with emphasis on the mechanism of biocorrosion, biodegradation, and wear as well as on the toxicology of implant materials. Since bone and soft tissue invade certain physiologically acceptable porous ceramic materials, there is reason for optimism that ceramics will become very useful materials of construction for orthopedic and oral appliances.

INTRODUCTION

Ceramics have produced for the chemical, steel and glass industries a variety of products to be used in severe environments such as high temperature reducing atmospheres or corrosive liquids. The advent of the nuclear reactor presented the ceramist with even more extreme conditions. The materials had to survive high radiation fluxes, extreme temperature gradients and corrosive liquids yet remain dimensionally and mechanically stable and develop no cracks or pinholes and, in addition, had to conform to these requirements for ten to fifteen years.

The space age presented the ceramist with the exotic problems of heat shields for space crafts. A new aspect was included in this problem, namely, human life itself depended upon the ceramic performing successfully.

Now a completely new challenge has arisen for the ceramist, the field of bioceramics. Briefly, this entails the use of ceramics as surgical implants within a living body. This environment is obviously different from a nuclear reactor or outer space, but is equally or possibly more hostile. Even more important this application directly involves the welfare of human beings and therefore the ceramic implant must not fail for the duration of the life of the patient.

The use of various materials as surgical implants is not new. One can find early attempts at bodily repair by implantation as far back as the pre-Christian era. It was not until the late nineteenth century that medical knowledge had progressed to the point that serious attempts at bodily repair by use of materials could be considered. The field of materials science, however, had not yet developed and implants by surgeons included such materials as brass and copper. As might be expected, such implants were doomed to failure by extensive corrosion.

Modern medicine has found it imperative to use a wide range of materials to repair defects in the body. Advancements in polymer chemistry have supplied the surgeon with a wide selection of materials such as knit dacron arterial grafts and plastic spheres or discs for artificial heart valves. Modern metallurgy has developed new alloys from which a wide variety of orthopedic devices are fabricated ranging from plates and screws for fractured bones to artificial hip joints. Although metals have performed admirably well as orthopedic appliances, they suffer from corrosion in the exceedingly hostile environment of the body.

There are many instances where an implant must be left within the body for the duration of the life of the patient. Such cases include artificial hip joints, replacement of missing segments of structural bone or the replacement of facial bones due to disease or trauma. The search for a suitable material for such applications has come to the field of ceramics.

No material placed within a living body can be considered to be completely inert. The only material which is completely compatible with the physiological environment is autogenous tissue. However, by their very nature, ceramics do not suffer from corrosion as do metals. Extensive progress in ceranic science has led to a variety of materials with chemical, physical, and mechanical properties which make them candidates for long-term implants within a living body.

DETERMINING COMPATIBILITY OF CERAMICS

The Hostile Biological Environments

Materials engineers and polymer chemists have produced a variety of materials to be used in corrosive environments. Usually the chemistry of a given environment is specifically known and the material can be designed accordingly. The environment of a living body, however, is quite different. This environment is an aqueous solution of approximately 1 M NaCl containing organic acids, proteins, enzymes, biological macromolecules, electrolytes and dissolved oxygen, nitrogen and carbon dioxide. The pH of the fluid is about 7.35, but drops to 5.3 to 5.6 upon injury to the tissue and returns to normal in about ten days.[1] In addition to its complex chemical make-up the activities of the various dissolved species are constantly changing. It is not surprising that long term metallic implants are apt to fail due to corrosion.

Deterioration of man-made materials in biological environments is most certainly due to a great many factors. The body seems to "digest" many polymeric materials and it is still not known what role enzymes and other body constituents play in the degradation of materials. Polymers are particularly susceptible to changes which lead to failure. Such changes can be rearrangement of the polymer itself such as breaking of cross-links or outright chemical degradation. Oppenheimer[2] detected carbon-14 in the urine of rats after intramuscular implantation of polystyrene, polyethylene, and polymethylmethacrylate. The carbon-14 had to have been freed from the main chain of the polystyrene and polyethylene and from the ester side chain of the methacrylate, emphatically demonstrating the actual metabolization of polymers by the body. Chemical analyses of various polymer implants which have shown deterioration indicate cleavage of the polymer chain. Both oxidation degeneration and hydrolysis have been described as probable mechanisms of polymer degradation in implants. The reader is reminded that all these rather drastic changes occur at body temperature.

One would expect metals and alloys to experience problems of corrosion. It is noteworthy to observe that alloys specifically developed for other highly corrosive and reactive environments often are badly attacked within a living body.

Ferguson, Laing and Hodge[3] implanted a wide variety of metals and alloys in rabbit muscle and then examined the surrounding muscle tissue by spectrochemical analysis; their results showed that metallic elements appear in the tissues around metal implants in concentrations which are significantly greater than normal, no matter how noncorrosive the implants are in other environments.

The stainless steels depend upon a closely adherent oxide surface layer for their resistance to corrosion. This layer in turn depends upon the continuous presence of oxygen in the environment. In the body tissues it is quite common to have varying oxygenation, and thus stainless steel implants are quite susceptible to corrosion. Colangelo and Greene[4] examined 53 **orthopedic** implants of Type 316 stainless after they had been removed from humans; 24 were corroded. If multicomponent devices are used, an electrolytic cell is obtained and corrosion is almost certain; in the previous study it was found that 21 out of 23 were corroded.

The body removes the constituent metal ions from such alloys as Hastelloy, Inconel, and Elgiloy. The situation is made worse by the fact that implants are usually subjected to repeated stress, particularly if the implant is connected to the skeletal system, and hence stress corrosion and eventual failure is apt to occur.

The Ceramic - Tissue Interface

It should be emphasized again that no foreign material placed within a living body is ever completely compatible. The only substances which enjoy complete compatibility are those manufactured by the body itself (autogenous); any other substance is recognized as foreign, initiating some type of biological reaction. Therefore, the goal in the selection of materials for implant devices is to choose those which will perform the desired function with a minimum of adverse biological reaction.

One must consider two aspects of an implanted material: (1) the effect of the biological environment on the material and, (2) the effect of the material on the biological environment. Even if the former effect is minimal and the material retains its physical and mechanical integrity, the latter aspect is of utmost importance. If the material causes a change in the surrounding tissue or physiological fluid, the result may be pain, disfunction or loss of an organ or limb or even eventual death of the patient. It is obviously of extreme importance, therefore, to take great care in the evaluation of the living tissues surrounding an implanted material.

Connective tissue, especially bone, directly contacts present metallic bone fixation plates, screws, rods and artificial hip joints.[5] The most important research effort in bioceramics is to develop a ceramic to replace missing or damaged segments of bones. Whereas the present metallic implants are generally considered temporary, the ceramic bone implants would be expected to last for the duration of the life of the patient. Hence the interest in the bone-ceramic interface.

Bone is a remarkable tissue. It is constantly rebuilding itself with new, stronger bone and even builds in new layers in
response to new stresses placed upon the skeletal system.[6] If
nothing interferes with this growth potential of bone, it has the
capacity to grow into an implant material of the proper design and
to effect a strong mechanical bond.

Muscular tissue may come into contact with an implanted
material in several instances. If facial bones are replaced with
ceramic, for example, one side of the implanted piece may be against
the remaining facial bones, while muscular tissue is closed over the
opposite side. Muscular tissue does not regenerate like bone, but
rather repairs itself by producing a fibrous tissue. This fibrous
tissue, however, can also be effective in attaching existing muscles
to implanted materials of the proper design.

It is often necessary to have a permanent opening through the
skin (epithelial tissue) and outer muscle layer of the body to the
internal organs. The most important need for such a device at
present is to attach a patient to an artificial kidney. Presently
no suitable device has been developed, but it will be in direct
contact with both muscular and epithelial tissue.

Presently, amputees must attach their artificial limb to a
remaining stump by a series of belts, straps and buckles, a method
which has not changed in concept for many centuries. Research is
presently in progress to develop a permanently attached artificial
limb. A device would be implanted directly into the remaining bone
stump and exit through the muscle and skin to the exterior where
the artificial limb would be attached. The materials used in such
a device would be in direct contact with connective, muscular and
epithelial tissue.

These few examples show that an implanted material may come
into direct contact with a wide variety of tissues. However,
human values insist that the material must not cause any adverse
biological response.

Types of Tissue Response

The various responses of living tissue to an implanted material
range from small temporary changes in adjacent cells to macroscopic
reconstruction or degradation of tissue.[7] The consequences of the
various responses can range from the insignificant to those that
compromise the well-being or even the life of the patient. A full
discussion of tissue response would require hundreds of pages, but
a brief description of the most important types will help to introduce the reader to the phenomena that must be considered in the
evaluation of an implanted material.

Phagocytosis. The area adjacent to an implant may contain debris of two types: (1) inorganic - small pieces may break or scale off from the implant itself or (2) biological - the implant may damage adjacent tissue leading to a wide variety of cellular debris. Bacteria and organic and inorganic foreign matter may be removed by a process called phagocytosis, which is simply the ingestion of the material by a cell. If the cell cannot digest the material, it often simply holds it and thus isolates the particle. When the cell dies, another cell picks up the indigestible material.

The primary ingestors of bacteria are the polymorphonuclear cells, especially the neutrophilic leukocytes. A great number of these cells will be present around an implant if infection is present. The monoculear cells are proficient at picking up cellular debris and nonbiological foreign matter. The monocyte performs this function in the bloodstream while the macrophage resides in body tissues. In an implant site where there is trauma and necrosis one would expect to find a large number of macrophages or monocytes engulfing damaged tissue. They would also be present around an implant that is undergoing continuous degradation, hydrolysis or corrosion. If an object is too large to be engulfed by a single cell, the mononuclear cells will coalesce to form foreign body giant cells.

One final topic should be mentioned, and that is the phenomena of chemotaxis. If the previously mentioned cells move toward an area, it is described as positive chemotaxis and if they move away from an area, it is negative chemotaxis. The mechanism of chemotaxis is still being debated, but suffice it to say that some ceramic implants cause a positive reaction and some a negative reaction. The type and degree of chemotaxis produced by a ceramic implant may be important to its success or failure. If no phagocytic cells are present, the implant may fail because of infection; if too many are present, normal healing of the tissues may be impaired.

By observing the types of cells around a material after it has been implanted for a given length of time, one is able to gain some idea as to its "compatibility" with the surrounding tissue.

Immune Responses. The immune response is a complicated reaction through which a body is able to recognize unfamiliar substances such as proteins as foreign and to render them harmless. The unfamiliar substances are termed antigens and the body proceeds to produce antibodies to destroy them.

The immune response is responsible for the rejection of organs transplanted from one person to another. This rejection occurs because the body is able to recognize the protein of the transplanted organ as "foreign". All of the specific cells which are responsible

for this immune response and their roles are not completely understood, but the most important cells in the process are lymphocytes,
mast cells and eosinophilic leukocytes. Increased populations of
the aforementioned is an indication that an immune response is
occurring.

An immune response does not directly occur with a ceramic
implant because the ceramic is not made of protein. A ceramic implant can cause an immune response in an indirect manner by
releasing ions or radicals into the surrounding tissue, which, in
turn, react with various proteins and denature them. The changes
wrought in the proteins may be very slight, but the immune response
is so specific that the body produces antibodies to destroy these
proteins. The results can obviously be catastrophic.

Abscesses. An abscess is a localized collection of pus and is
composed of dead cells which have been phagocitizing bacteria and
necrotic tissue. This condition may be due to infection or it may
be due to chemical or mechanical trauma caused by the implant to
the surrounding tissue. If the abscess is due to the latter case,
the normal healing process is greatly hindered, if not completely
stopped. An abscess around an implant is a severe form of rejection.

Neoplasms. A neoplasm is a growth comprised of an abnormal
collection of cells whose growth is uncoordinated with that of normal
tissue. It may or may not be cancerous. A large number of studies
have been done with a wide variety of materials. The exact mechanism
of the formation of a neoplasm is not known and many conflicting
results have been published. Needless to say, the formation of a
neoplasm at an implant site is cause for serious concern.

Poisons. Any substance can be poisonous to a living body if
present in large enough concentrations. Substances such as lead,
mercury, arsenic and beryllium are known poisons and implants containing them are not considered. There is a vast list of metals such
as iron, aluminum, chromium, vanadium, cobalt, etc. about which
little is known concerning tissue response. If a ceramic released
its metallic ions into the surrounding tissue, they may do several
things. They may denature proteins and trigger the immune responses
as mentioned earlier or they may collect in an organ like the liver
or kidneys and cause damage. An increased concentration of a
metallic ion may change the metabolism of the body and cause anoxia,
cardiac damage, renal failure, etc. Such types of tissue response
are exceedingly difficult to detect and may go completely undetected
by a physician.

Inflammation. In normal wound healing there is an initial,
acute inflammation which subsides shortly. It is characterized by
a dilation of blood vessels, edema, and an increased concentration

of white blood cells. A chronic inflammation, however, is a much
more serious reaction and is of prolonged duration. Very high
concentrations of leukocytes appears and an abnormal granulation
tissue forms. Repeated episodes of necrosis occur and scar tissue
may form. This results in continuous pain and loss of function with
little or no healing.

Evaluation of Tissue Response

The question of the compatibility of a material is so complex
that it can only be tested by actual implantation in a laboratory
animal. Small pellets of a material to be tested are implanted in
the bone, muscle or skin and the animals maintained for various
lengths of time. When the animal is sacrificed, the implant is removed intact with the surrounding tissue.

The tissue and the implant contained within it go through a
standard histological sequence of fixation in formalin and dehydration
in ethyl alcohol. Ordinary soft tissue sections can be embedded in
paraffin and thin sections cut with a microtome for microscopic
examination. However, when a ceramic is embedded in either bone or
soft tissue, this method cannot be used because the knife cannot
cut the hard ceramic. A method has been devised where the tissue
is infiltrated with unpolymerized methylmethacrylate (MMC) and then
placed in partially polymerized MMC where it is permitted to poly-
merize to a solid block. Sections ~75 μm can then be cut
on a precision diamond saw.[8]

Staining of the thin sections is required in order to make the
cellular structure visible. The staining of tissue is a common
histological technique and many different stains exist. Paragon, a
commercial polychromatic stain, was found to give good results for
the sections embedded in plastic. This resulted in the nuclei
staining blue-gray with the cytoplasm a paler blue-gray; erythrocytes,
the elastica of blood vessels and collagen fibers stained a bright
red; cartilage was blue; and osteoid tissue pink. Following the
staining, the sections are mounted on glass slides and examined
under the microscope. This allows one to determine the response of
the tissue to the implant, as outlined in the previous section.

One of the primary interests in bioceramics is their use as a
replacement for missing bone segments. If the ceramic contains
~150μm dia pores, one observes bone tissue growing into the pores
thus resulting in a strong mechanical bond. The initial unmineralized
bone tissue which is formed is called osteoid; it later calcifies into hard osseous tissue. One cannot easily distinguish the difference
between mineralized and unmineralized tissue under the microscope.

In order to determine the degree of mineralization of the tissue one uses the techniques of microradiography and the electron beam microprobe.

Microradiography[9] is fundamentally the same as clinical radiography, i.e., x-rays are transmitted through a specimen and the differences in absorption are recorded on an x-ray sensitive film. The term microradiography refers to those techniques which utilize a high resolution film and allow for microscopic examination of the resulting radiograph.

The thin sections containing the implant are held tightly against the film in a special holder and exposed to x-rays. The resulting radiograph is examined under a microscope where the degree of mineralization of the osteoid tissue within the pores can be ascertained by the radiopacity.

Another method of determining the degree of mineralization is with the electron beam microprobe. Thin sections to be examined are mounted on flat glass discs and a film of carbon deposited for electrical conduction. Each section of the bone implant contains a portion of the adjacent cortical bone as well as the implant. Analyses for calcium and phosphorous are first performed on the adjacent natural bone in order to establish the calcium and phosphorous intensities indicative of fully mineralized bone. This technique provides a standard for mineralized bone which has been subjected to the same previous experimental procedures as the areas of interest within the pores of the ceramic. Calcium and phosphorous analyses are then made of the tissue within the pores of the ceramic.

POROUS CERAMICS TO ATTACH PROSTHESES TO THE MUSCULO-SKELETAL SYSTEM

The most promising use of a porous ceramic material is the possibility of bone (or soft tissue) ingrowth and direct attachment of the prosthesis to the musculo-skeletal system. If bone growth into a porous structure is to be obtained, the pores must be large enough to host the development of the organic and inorganic constituents of bone as well as the bone cells. Furthermore, if bone is to form and exist inward past a depth of 100µm, which is the extent of canalicular effectiveness, blood vessels must also penetrate into the ceramic.[6]

To provide sufficient room for the cellular and extracellular constituents of bone, as well as a small blood vessel, pores of at least 50µm in dia appear to be necessary. The pores of the ceramic must interconnect to allow the ingrown blood vessels to anastomose freely with one another. A good blood supply is extremely important

for it supplies the calcium and phosphorous necessary for mineralization of the organic matrix as well as nourishment for the ingrown
tissue.

The use of an open pored ceramic represents a material which
closely resembles the structure of cancellous bone. Once ingrown
with a vascular connective tissue, the similarities between the
porous ceramic and cancellous bone become more evident. In the
natural case, cancellous bone is converted to compact bone by the
successive deposition of lamellae of new bone resulting in a series
of primary osteons. In like manner, a porous ceramic ingrown with
a vascular connective tissue provides a latticework and a cellular
environment conductive to appositional bone formation.

The analogy between the implanted ceramic and cancellous bone
not only provides a biologically comprehensible basis for anticipat-
ing bone formation but also provides further design criteria for the
ceramic. Once ingrown with a vascular tissue, the appositional
deposition of bone should initiate a series of primary osteons. To
allow for full development of the osteons, pore sizes of approximately 200µm must be provided.

The overall pore structure must be taken into consideration as
well as the individual pore size. In the most general sense, an
open pored ceramic material can be considered as a series of
individual pores connected to one another by openings which are
smaller than the pores themselves. The interconnection pore size
will control the ingrowth of bone. For example, a ceramic with pore
sizes of about 100µm would be of little value for bone ingrowth if
the interconnection pore size was only a few microns. Since the
development of a porous ceramic which totally eliminates restrictive
pore interconnections is highly unlikely, careful attention must be
given the interconnection pore size.

As the size of the pores is increased, the overall possibility
of bone growth should also increase. Larger pores provide more room
and a less tortuous path for the ingrowth of a vascular tissue and
subsequent mineralization. However, a second factor also enters
into the picture. The size of the pores can not be increased
indefinitely without decreasing the strength of the material to a
point where it would be useless from a practical point of view.

EXPERIMENTAL

Fabrication of Porous Ceramics

Porosity was introduced into a calcium aluminate material by
firing fine grained Al_2O_3 with granular $CaCO_3$ at 1450°C (2650°F) for

24 hr.[8] The desired pore size distributions were achieved by controlling the grain size of the $CaCO_3$. During the reaction the $CaCO_3$ decomposed into CaO and CO_2, and subsequent reaction of the CaO and Al_2O_3 resulted in an open pored calcium aluminate material, with a pore size distribution approximately the same as the carbonate grain size distribution.

The calcium aluminate pellets were produced by reacting 41 w/o $CaCO_3$ with 59 w/o Al_2O_3. Decomposition of the carbonate during firing resulted in a mixture of 27 w/o CaO and 73 w/o Al_2O_3. The $CaO-Al_2O_3$ phase diagram predicts that if chemical equilibrium were obtained the pellets would be composed of two calcium aluminate phases, $CaO \cdot Al_2O_3$ and $3CaO \cdot 5Al_2O_3$. The x-ray diffraction analysis of the pellets identified the presence of the two calcium aluminate equilibrium phases and gave no evidence of unreacted Al_2O_3 or CaO. The x-ray diffraction data were identical for type I through type V pellets.

Table 1. Ceramic Pore Structure Designations

Calcium Carbonate Grain Size, μm	Pore Structure Type	Percent Pore Interconnections >100μm
<45	I	---
45 - 75	II	---
75 - 100	III	20
100 - 150	IV	37
150 - 200	V	63

Identification of only the equilibrium phases indicated that the reaction had approached chemical equilibrium and the percentages of the $CaO \cdot Al_2O_3$ and $3CaO \cdot 5Al_2O_3$ phases were approximated by the equilibrium values of 20% $CaO \cdot Al_2O_3$ and 80% $3CaO \cdot 5Al_2O_3$, as obtained from the phase diagram. Similarly, porous $CaTiO_3$ and $CaZrO_3$ can be produced by reaction of granular calcium carbonate with TiO_2 or ZrO_2.

Another method of producing porous ceramics that also appears promising is by a foaming technique. This method merely entails mixing a very fine ceramic powder ($\sim<1\mu$ m) with an organic binder, adding a commercial foaming agent and mechanically whipping the mixture.[10] The resulting foam chemically hardens and is subsequently fired. The organic constituents burn out leaving behind a ceramic with interconnected pores. Various ceramics are being processed by this method and the results for foamed alumina are very promising. Pellets of foam-alumina implanted into dog femora showed complete mineralized bone throughout the 1/4" pellet.

Implantation of Ceramic Specimens

Pellets of various pore structures of calcium aluminate were
implanted into the midshaft region of the femurs of adult dogs. The
pellet specimens were implanted for periods of 4, 11, and 22 weeks.
This time progression allowed for the evaluation of the initial
rapid ingrowth of tissue, the slower long term changes in tissue ingrowth and provided sufficient time to examine for a chronic
inflammatory response. Implantation in a long bone gave a site
exhibiting the availability of the normal potential bone regenerative
areas--specifically the periosteum, the endosteum and the osteogenic
capabilities of the mesenchymal tissues of the blood vessel walls and
primitive reticular cells of the bone marrow. This implant site also
placed the implant specimens within a stressed portion of the skeletal system.

In addition to the calcium aluminate bone implants, porous and
nonporous ceramics have been implanted in soft tissue. Discs of
three ceramics, calcium aluminate ($CaO \cdot Al_2O_3$), calcium titanate
($CaO \cdot TiO_2$) and calcium zirconate ($CaO \cdot ZrO_2$) were implanted into each
of the following sites in rabbits: (1) the masseter muscle of the
jaw, (2) the muscles of the thigh, (3) the paravertebral muscles in
the back, and (4) the connective tissue superficial to and adjacent
to the external oblique muscles of the abdomen.[11]

After the prescribed implant periods, the pellet specimens
were removed and the type of tissue ingrowth analyzed using
histological tissue sections, microradiography and electron microprobe analyses. The tissue ingrowth was related to the pore structure of the various implants.

RESULTS OF CERAMIC IMPLANTS

Ceramic Bone Implants[8]

Microscopic analysis of the implanted pellets in bone revealed
the ingrowth of soft tissues and osseous tissues. Evaluation of the
histological sections identified the majority of soft tissue ingrowth
to be a vascular fibrous tissue, although limited areas of myeloid
tissue infiltration were observed. The osseous tissue ingrowth consisted of both mineralized bone and unmineralized bone (osteoid).
Tissue growth into the implant pellets was found to be dependent on
pore structure. Table 2 presents the type of tissues observed within the pores of the Type I through Type V implants during the 22
week duration of the experiment.

Analysis of the Type I implant showed no evidence of living
tissue ingrowth. The Type II implants were found to be completely
infiltrated with a vascular fibrous tissue but provided no evidence

Table 2.   Tissue Growth Into the Type I Through Type V Implants*

	Implant Type				
Tissue Ingrowth	I	II	III	IV	V
Fibrous	0	+	+	+	+
Osteoid	0	0	+	+	+
Mineralized Bone	0	0	0	+	+

*0= ingrowth of tissue not observed; += ingrowth of tissue observed.

of osseous tissue ingrowth. Analysis of the Type II implants revealed a limited penetration of mineralized bone, ∿50µm deep, into
the open surface pores of the pellets. This limited degree of
mineralized bone penetration was not considered indicative of
mineralized bone ingrowth. Fibrous tissue ingrowth and the initiation of osteoid development was found throughout the inner pore
volume of the Type III pellets. Analysis of the Type IV implants
revealed mineralized bone ingrowth to a maximum depth of ∿500µm
after 22 weeks. Osteoid and fibrous tissue were found throughout
the inner pore volumes of the Type IV pellets. The osteoid tissue
within Type IV implants was observed to have developed a lamellar
structure similar to the Haversian systems of cortical bone.
Analysis of the Type V implants showed the deepest penetrations of
mineralized bone, reaching inward to a depth of ∿1500µm after 22
weeks. Fibrous tissue and highly organized osteoid tissue were also
found within the Type V implants. In all the implant specimens an
osteoid seam ∿50µm thick was found separating the calcium aluminate
ceramic from the mineralized bone both adjacent to the pellets and
in the areas of mineralized bone ingrowth. The depth of mineralized
bone ingrowth is presented as a function of implant type and implant time in Table 3.

Table 3.   Depth of Mineralized Bone Ingrowth

Implant	4-Weeks	11-Weeks	22-Weeks
Type I	no ingrowth	------	------
Type II	no ingrowth	no ingrowth	no ingrowth
Type III	50µm	50µm	50µm
Type IV	--------	200µm	500µm
Type V	--------	600µm	1500µm

The minimum interconnection pore size ranges found necessary for mineralized bone, osteoid and fibrous tissue ingrowth are summarized in Table 4.

Table 4.   Minimum Interconnection Pore Size Range Necessary For
           Mineralized Bone, Osteoid and Fibrous Tissue Ingrowth

Tissue	Interconnection Pore Size Range
Mineralized Bone	100µm
Osteoid	40 - 100µm
Fibrous	5 - 15 µm

Microscopic examination of the tissues within and surrounding the calcium aluminate implant specimens showed no obvious signs of an inflammatory response. During the analysis particular attention was paid to areas of tissue ingrowth containing blood vessels, where one would expect to find evidence of a cellular exudate if an inflammatory reaction had taken place. Careful examination revealed only a few giant cells within each tissue section. The areas containing the giant cells did not show a local accumulation of inflammatory cells. The limited number of giant cells observed and the lack of local inflammation were felt to be indicative that an adverse foreign-body reaction had not taken place.

Even though the calcium aluminate implants gave no evidence of inducing an inflammatory or foreign-body reaction, the overall compatibility of the implants with the musculo-skeletal system must consider the osteoid tissue found within and surrounding the implants. Large areas of osteoid, as found within and surrounding the implants, are unmineralized and indicative that some factor or factors have interfered with the mineralization process.

A plausible theory suggested by the experimental evidence and the results of previous investigations is that hydration of the calcium aluminate resulting in an increase in pH and a retardation in mineralization due to aberrations in collagen development.

The presence of the calcium aluminate material did not evoke an inflammatory response despite the fact that it was known not to be totally inert and was found to have degraded during the implant periods. This observation helps substantiate the concept that tissue tolerable materials can be produced from ceramic materials if careful attention is paid to the toxicity of their constituent ions. The incompatibility of the calcium aluminate implants was evidenced by the retardation in mineralization which, in light of the experimental evidence, appeared to be the result of an

interaction of the calcium aluminate material with the mineralization
process. It therefore appears that the combination of CaO and
Al_2O_3 which provided the tissue tolerability was also responsible
for the retarded mineralization.

In the final analysis it is difficult to arbitrarily classify
any implant material as compatible or noncompatible in the strictest
sense of the words. The calcium aluminate material used for the
purpose of this research was considered, in the most general sense,
as being tissue tolerable but not totally compatible with the
musculoskeletal system.

Ceramic Implants in Soft Tissue[11]

In this study, coin-shaped discs of about 15 mm dia. were
implanted into the masseter muscle, the muscles of the thigh, and
connective tissue at various sites in rabbits. A series of non-
porous discs were made of TiO_2 , Al_2O_3, $CaO \cdot Al_2O_3$, $CaO \cdot TiO_2$, and
$CaO \cdot ZrO_2$. After six months all materials were tolerated well,
except for a very few atypical implants. The acute, post surgical
inflammatory responses of the experimental animals were compared by
observing the implant site for redness and swelling. The duration
of this response was longer in some animals than in others; however,
no definite pattern could be shown, except that the acute inflammatory period in all animals receiving implants was longer than that
of the controls, who merely received an incision. Histological
sections showed that each implant impervious disc had been sealed
off by a layer of fibrous connective tissue. This layer was
generally of variable thickness but showed no signs of infection,
chronic inflammation, or toxic effects. The encapsulation was
attributed to the movement of the rigid disc in soft, living tissue
which caused an observed irritation, followed by proliferation of
granulation tissue. The granulation tissue was later remodeled
into the fibrous encapsulation.

A second set of porous discs were implanted using $CaAl_2O_4$,
$CaTiO_3$, and $CaZrO_3$. These ceramic discs all contained a system of
random, interconnecting pores with diameters on the order of
100-150μm. All of the porous ceramics showed a shorter period of
post-surgical inflammation, although still longer than the controls.
Histological examinations showed that these porous discs had much
thinner fibrous encapsulations, and about half of the implants
appeared to have no surrounding fibrous layer at all. The pores
throughout the implants were filled with connective tissue and
blood vessels.

There were a few pores which contained fragments of blood
clots at three months, and these were greatly diminished but still

present at nine months. There was no evidence of any toxic reaction, or chronic irritation of any kind, except in a very few samples. Immediately after incision, blood clots and fibrin networks permeate the porous implant. This network is continuous with the walls of the incision, and as inflammation progresses, the clot is remodeled into fibrous connective tissue. The implant is held firmly to the tissue rather than sliding and scraping as the muscles contract, and the mechanical effect of abrasion is greatly reduced.

The porous implants healed more quickly and demonstrated a thinner fibrous encapsulation than the impervious implants due to the early and continued adherence of surrounding tissue. Hydration was not evident in $CaAl_2O_4$ in the nine month rabbit muscle implants and may be a problem only in bone where it appears to interfere with calcification.

The ceramic implants of $CaAl_2O_4$, $CaTiO_3$, and $CaZrO_3$ evinced no signs of a toxic, immune or carcinogenic response after nine months in muscle and connective tissue of rabbits, and were well tolerated after an initial, acute inflammation, as demonstrated by the absence of inflammatory cells and the normal morphology of the cells present.

This investigation shows that ceramics should find practical applications in prosthetic devices employed in soft tissue.

FUTURE USES OF BIOCERAMICS

The most outstanding advantage that bioceramics have over other materials is their great chemical stability in the extreme environment of the living body. It must be made clear that bioceramics will not totally displace other biomaterials. Some applications require the flexibility of polymers while others may require the high strength and toughness of metals.

For long term implantation, however, ceramics seem to hold the greatest promise. Furthermore, the porous ceramics have the ability to allow bone or soft tissue ingrowth thus producing a strong mechanical bond. However, if an implant is to be only a temporary device, one would not desire to have tissue ingrowth and a metal implant may be more desirable.

Let us look briefly at some possible applications of bioceramics. The most obvious application is the replacement of missing bone segments.   For example, facial bones may be lost due to cancer or trauma and presently nothing exists to rebuild the structure of the face. Although initial applications of bioceramics to this problem will be the replacement of rather small missing pieces of bone, eventually it is hoped to be able to replace an entire jawbone.

It is well known to dentists that the stability of lower dentures depends upon an adequate alveolar ridge. The alveolar ridge is the bony area on the upper surface of the jawbone and serves to support the teeth. In many patients this bony ridge is resorbed by the body after the teeth are extracted. This leaves only soft tissues to support the dentures with the result that the lateral instability makes the dentures useless. Various materials have been used in attempts to rebuild the alveolar ridge but none have been as successful as the ceramic implants which are being investigated. An oral surgeon surgically exposes the alveolar ridge, inserts a piece of porous ceramic and closes the soft tissue over it. The bone grows into the ceramic and thus anchors it in place. This artificial alveolar ridge then gives a firm base on which the dentures sit thus allowing the patient to wear normal dentures.

Behind the ear is an area of spongy bone called the mastoid. If this becomes infected, a surgeon performs a mastoidectomy and removes the spongy bone. This results in a cavity which cannot be covered with the existing soft tissue. This cavity becomes continuously infected throughout the life of the patient and is a rather serious medical problem. Surgeons have tried many methods in attempts to obliterate this cavity with only limited success. Muscle pedicle flaps tend to atrophy while fat, cartilage or bone chip implants tend to resorb. This is an ideal problem for the use of ceramics. No mechanical stress is involved in this area; all the ceramic must do is to fill the cavity and support the soft tissue closed over it. Presently several forms of ceramics are being tested for use in mastoid cavity obliteration. The most useful form of ceramic seems to be chips or spheres which are packed into the cavity. The soft tissues which are closed over the area will grow into the cavity and anchor the ceramic pieces. Experiments are presently being conducted in this area.

Lastly, ceramics may be used in conjunction with metals. Many patients have hips rendered unusable by various diseases. The present procedure is to replace the head of the femur with a metal ball with a stem that is driven down into the shaft of the femur. This is essentially a force-fit and often works loose after a few years. This is a major operation and it would be of tremendous benefit to the patient if a hip joint could be inserted that would last him the duration of his life. Presently, work is in progress to put a porous ceramic coating on the metal stem of the hip prothesis so that bone ingrowth will securely anchor it.

These are just a few brief examples of how bioceramics can help mankind in a new and unique fashion, by withstanding the most severe environment of all, the living body.

ACKNOWLEDGMENT

This investigation was supported by grant AM #13272 from the National Institute of Arthritis and Metabolic Diseases.

REFERENCES

1. A. White, P. Handler, and E. Smith, Principles of Biochemistry, 3rd Edt., McGraw Hill Book Co , 1964.
2. B. S. Oppenheimer, E. T. Oppenheimer, A. P. Stout, and I. Danishefsky, Cancer Research 15 333 (1955).
3. P. G. Laing, A. B. Ferguson, and E. S. Hodge, J. Biomed. Mater. Res., 1 135 (1967).
4. V. J. Colangelo, and N. D. Greene, J. Biomed. Mater. Res. 3 247 (1969).
5. C. O. Bechtol, A. B. Ferguson, and P. G. Laing, editors, Metals and Engineering in Bone and Joint Surgery, The Williams and Wilkins Co., Baltimore, 1959.
6. A. W. Ham, Histology, 6th Edt., J. B. Lippincott Co., Philadelphia, 1969.
7. S. L. Robbins, Pathology, 3rd Edt., W. B. Saunders Co., Philadelphia, 1967.
8. J. J. Klawitter, "A Basic Investigation of Bone Growth Into a Porous Ceramic Material"; Ph.D. Thesis, Clemson University, Clemson, S.C. (1970).
9. J. Jowsey, et al., J. of Bone and Joint Surg., 47A (4) (1965).
10. The Mearl Corp., Roselle Park, N.J.
11. S. J. Morrison, "A Basic Investigation Into the Compatibility of Ceramics with Soft Tissue"; M.S. Thesis, Clemson University, Clemson, S.C. (1970).

DISCUSSION

D. J. Godfrey (Admiralty Materials Laboratories): I don't suppose silicon nitride has been evaluated for this application, but it would appear to have several advantages which would merit examination of its bio-compatibility. It can be a very strong material, in some forms it is readily formed to precise and complex shapes, and it is very inert to nearly all aqueous corrodents. Our work is showing excellent performance as a sea water environment bearing material, and it can be made in forms which have appreciable permeable porosity, which have been reinforced with strong fibres to yield materials tougher than bone.

Authors: We are not aware of the utilization of silicon nitride as a material of construction for any present day artificial organs. Also, we are unaware of any testing of its compatibility with the physiological environment. We concur that it's compatibility with the physiological environment ought to be investigated because of its chemical inertness.

STATIC FATIGUE IN GLASSES AND ALUMINA

J. E. Burke, R. H. Doremus, W. B. Hillig and A. M. Turkalo

General Electric Company

Schenectady, New York

ABSTRACT

Experimental results on delayed failure of FN borosilicate glass fused silica, and polycrystalline alumina are presented and compared with previous results on soda-lime silicate glasses. For higher stresses the results for the glasses fit on a universal curve; at lower stresses the reduced time for failure for FN and fused silica becomes longer than for soda-lime glass. The alumina data show a different slope than those for the glasses, implying a different corrosion reaction.

Static fatigue data define conditions for proof tests for ceramic parts subject to load. A plot of log stress against log failure time is particularly valuable for this definition, and allows one to calculate the time and stress for a convenient proof test.

INTRODUCTION

A glass or brittle ceramic fails after being stressed for some time at a lower level than the short-time failure stress. This reduction of strength, known as delayed failure or static fatigue, results from reaction of the glass or ceramic with atmospheric water, because no such reduction occurs in a dry atmosphere. Static fatigue in a soda-lime-silicate glass was studied by Mould and Southwick,[1] who found that all their results fit on a single curve on a log $t/t_{1/2}$ vs. S/S_N plot, where t is the time to failure at applied stress S, S_N is the breaking stress at liquid nitrogen temperature, and $t_{1/2}$ is the time to failure at $S_N/2$. At the temperature of liquid

435

nitrogen the reaction of glass and water is negligible, so that S_N represents the strength of unreacted glass. Wiederhorn[2] has studied the velocity of crack growth in glass under a stress; his results are closely related to static fatigue.

In this paper we report the results of delayed fracture experiments on three different materials; FN (7052) borosilicate glass, fused silica, and 96% polycrystalline alumina. The data for the glasses fit on the same curve as those of the soda-lime glass on a log $t/t_{1/2}$ vs. S/S_N plot at higher stresses, while the alumina data fit on another line. Explanations for these results are considered.

The static fatigue data can be used to define certain minimum design limits for delayed failure of brittle materials. Proof tests at certain stresses and times are useful in defining these limits. The amount of data needed to define the limits of delayed failure is discussed.

EXPERIMENTAL METHODS

The strength at 78°K (S_N) was measured on rods 3 to 4 mm dia and 5 to 7 cm long in a four-point bend test with an appropriate testing jig in an Instron tensile machine. The delayed failure measurements were made in a multiple station static load apparatus. A load designed to give a stress in the range 0.4 to 0.8 of S_N in a four-point bend test was applied to rods 2.5 to 4 mm in diameter and 7 to 8 cm long. The actual stress was calculated from the dimensions of the rods and the geometry of the apparatus. A beam span of 1.3 cm was used in both delayed failure and liquid nitrogen tests. The tests were carried out in an air-conditioned room with relative humidity ~50% at ~23°C.

Rods of General Electric FN (7052) glass (approximate composition SiO_2 66%, B_2O_3 24%, Al_2O_3 3%, Na_2O 4%, K_2O 3%) and of Amersil fused silica were centerless ground with a 220 grit (600 grit for the silica) Carborundum wheel to give uniform surface damage. The 96% alumina rods (Wesgo AL 500) were given three different treatments: (1) as received; (2) centerless ground, cleaned with acid, fired in air for several different temperature cycles; (3) processed as in (2) but not ground.

RESULTS

One hundred glass samples were tested at each loading level. The median log time to fracture at a certain load can be used to determine the variation of fracture time with load. However, a new method of treating the data gave considerably more information.

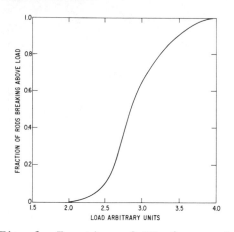

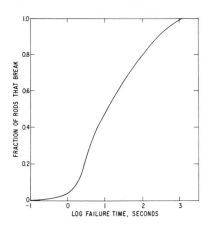

Fig. 1. Fraction of FN glass rods that break above a certain load at 76°K.

Fig. 2. Fraction of FN glass rods that break after a certain time when stressed at 50% of the median liquid nitrogen strength at room temperature.

The method depends upon the assumption that the characteristic distribution of strengths is found in the 100 samples, and that this distribution is related to the distribution in failure times at a particular stress. For example, for samples of FN glass tested at liquid nitrogen temperature with a median fracture load of 2.85, 95% fracture above a load of 2.50, 80% above 2.63, 70% above 2.70, etc. as shown in Fig. 1. Thus when the samples are loaded at room temperature to 50% of their median liquid nitrogen strength (1.42), some samples are actually loaded to more than 50% of this strength and others to less. The distribution of log failure times at a load of 50% of the median liquid nitrogen strength is shown in Fig. 2. Then the log failure time above which 90% of the samples fail (0.25) can be taken as the log time for failure at a load of 1.42 with a liquid nitrogen strength of 2.50, or a ratio of 56.8%. In this way the curve in Fig. 3 for FN glass was found from measurements at seven different stress levels.

Results on delayed fracture of fused silica rods held at five different loads and calculated as described in the last paragraph are shown in Fig. 4. At liquid nitrogen temperature the strength distribution was close to gaussian in the fracture load. The median strength at 78°K was 15,000 ± 800 psi. Results on delayed fracture of rods of polycrystalline alumina are summarized in Fig. 5. The slopes of the lines on the S/S_N against log failure time plot are about the same for samples treated in different ways,

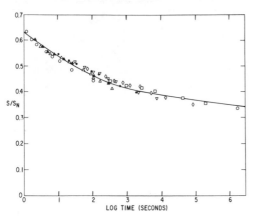

Fig. 3. Log failure time at room temperature for FN glass rods held at different fractions S/S_N of the median liquid nitrogen stength; S/S_N, 0, 0.55; ▲, 0.525; △ 0.50; ●, 0.475; ▽, 0.45; ◇, 0.425; □, 0.40.

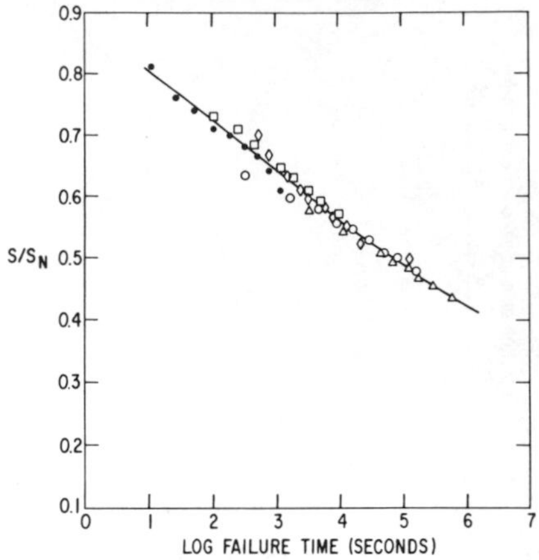

Fig. 4. Log failure time at room temperature for fused silica rods held at different fractions S/S_N of the liquid nitrogen strength: ●, 0.7; □, 0.65; ◇, 0.6; 0, 0.55; △, 0.5.

although the times to failure at $S/S_N = \frac{1}{2}$ were different. The data on the ground and unground processed samples indicate that the specimens with the highest value of S_N failed in shorter times than specimens of lower S_N, when tested at the same fraction of their liquid nitrogen temperature strength. This behavior is in qualitative agreement with that found by Mould and Southwick[1] for soda lime glass.

Mould and Southwick found that all their data for soda lime glass with different surface treatments fitted on the plot of reduced stress, S/S_N, against log reduced failure time log $t/t_{1/2}$, where $t_{1/2}$ is the time to failure at $S_N/2$. Their data are compared with ours[2] on such a plot in Fig. 6. The results for FN glass and fused silica fall on the same line as for the soda lime glass at $S/S_N > .5$, but the FN data deviate below this stress. Since the slope of the linear portion is related to the stress dependence of the corrosion reaction,[3] this correlation suggests that the rate limiting reaction in the delayed fracture of the different glasses is the same, probably being the hydrolysis of silicon-oxygen bounds. The alumina data show a lower slope, indicating a different rate limiting reaction.

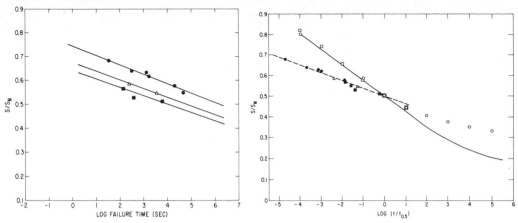

Fig. 5. Log failure time at room temperature for polycrystalline alumina rods held at different fractions S/S_N of the liquid nitrogen strength. ●, as received; Δ, ground and processed; and ■, processed only.

Fig. 6. Log reduced failure time $t/t_{1/2}$, where $t_{1/2}$ is the time to failure at $S_N/2$, for different materials held at different fractions S/S_N of the liquid nitrogen strength. Continuous line, soda-lime glass[1]; □, fused silica; 0, FN borosilicate glass; ●, Δ, ■, alumina as-received, ground and processed, and processed only.

ENGINEERING DESIGN

While the conventional plot of S/S_N vs. log t or log $t/t_{1/2}$ is satisfactory for discussions of mechanism of static fatigue, a relationship between S and t is needed for engineering purposes. The most straightforward approach is to measure S_N and $t_{0.5}$ for the structural element in question, and then construct a graph of S vs. t from the universal curve, or a similar curve obtained from specimens of simple geometry. Unfortunately such measurements are inconvenient, and in many cases substantially impossible. Simpler approaches are needed.

In principle, $t_{1/2}$ should be related to S_N, and from Mould and Southwick's data one can deduce empirically that $t_{1/2} = bS_N^{-m}$ where m = 6.5. However, their data are limited to values of S_N ranging from 10,000 to 20,000 psi, and the exponent is surprisingly high. More experimental data from specimens with a wider range of S_N values are needed before such a relationship is used for design purposes.

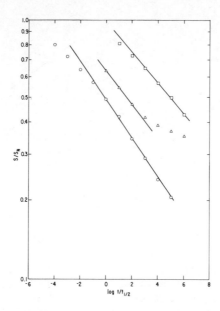

Fig. 7. Log failure time as a function of log stress, where stress
is plotted as the fraction of liquid nitrogen strength:
□, fused silica; Δ, FN borosilicate glass. Log $t/t_{1/2}$ for
soda-lime glass; O, universal curve of Mould and Southwick.

The universal curve can be approximated up to intermediate
times by a straight line:

$$\log t = -p^{S/S_N} + \log t_{1/2} + K' = -p^{S/S_N} + K \qquad (1)$$

where p = 13. If S_N is determined for the structural element, the
constant K can be evaluated for the maximum time and S/S_N that the
piece will be required to support. From the K a minimum value of
S/S_N can be determined for any convenient short time, and a proof
test can be performed to eliminate pieces that will not support
that stress. In this approach a difficult measurement of S_N must
be made, since from Eq. 1 a graph of log t vs. S will have a slope
which is dependent upon S_N.

Charles[4] and Mould and Southwick[1] have observed that many static
fatigue data yield a straight line on a log-log plot at values of
S/S_N less than about 0.6. In Fig. 7 the fit is good for Mould and
Southwick's data on soda-lime glass and our data for fused silica.
Neither set of results shows an indication of a static fatigue
limit when plotted in this way. The data for FN glass do show some
indication of a static fatigue limit. The straight line portion
of this curve follows a power law of the type

$$\log t = -n \log S + \log t_{1/2} + n \log S_N + c = -n \log S + C \qquad (2)$$

where n is about 13. This relationship cannot hold at short times since the maximum failure stress is S_N, but represents the data well at the longer times where prediction is desired.

This log-log plot is more convenient than that of Eq. 1 for design and prediction purposes, since all curves of log t vs. log S have the same slope, and are not influenced by different values of S_N. If n = 13, one can compute, for example, that if a glass piece is to support a stress of 3000 psi for 10^6 seconds (about 4 months), C is at least 51.2. With this value of C the piece should support a proof test of at least 7500 psi for 10 sec. All weaker specimens can be eliminated by such a test. If the value of n is uncertain, it can be found by measuring the time of fracture for specimens of simple geometry at two or three different stresses.

In the above treatment it is assumed that Eq. 2 fits the static fatigue behavior of the specimens at all stresses. If the failure time at low stresses becomes longer than predicted by Eq. 2, as it does for the FN glass in Fig. 7, then there is a factor of safety over the safe time predicted by the proof test. At short times the proof stress derived from Eq. 2 may be above the expected breaking stress, as shown in Fig. 7. Again the proof test is conservative.

In the above method the stresses in the samples to be tested must be known. If a stress analysis is difficult, specimens can be held at two or three different loads until failure, defining the static fatigue curve of Eq. 1 in terms of load instead of actual stress. Then a proof test can be defined as before, using this curve as a guide.

CONCLUSIONS

On the universal curve in which S/S_N is plotted against log $t/t_{0.5}$ all glasses fit on the same straight line above $S/S_N = 0.5$. At lower stresses data for a borosilicate glass deviate to give longer failure times at a high stress than the stress at which deviations for a soda-lime glass appear. Polycrystalline alumina shows a different slope on the universal plot, probably because the reaction of water with the alumina is different than with silicate glasses. Static fatigue data can define conditions for proof tests for glass parts subject to load; a plot of log stress against log failure time is particularly convenient for this definition.

ACKNOWLEDGMENTS

Part of this work was supported by ONR Contract N 00014-68-C-0126. V. J. DeCarlo assisted with the experimental work.

REFERENCES

1. R. E. Mould and R. D. Southwick, J. Am. Ceram. Soc. 42 582 (1959).
2. S. M. Wiederhorn, J. Am. Ceram. Soc. 50 407 (1967); 54 (1970).
3. R. J. Charles and W. B. Hillig; p. 511 in Symposium on Mechanical Strength in Glass. Union Scientifique Continentale de Verre, Charbroi, Belgium, 1962.
4. R. J. Charles, J. Appl. Phys. 29 1549 (1958).

DISCUSSION

Peter R. Kosting (U. S. Army Material Command, Retired): In view of the interest in predicting conditions for proof testing, for engineering purposes it would be beneficial to indicate also the values for 99% probability of failure and for no failure.

Authors: If one has available the spread of fracture stresses for the component for which prediction is being made, it would indeed be desirable to indicate the 99% confidence limit or some similar value. The data presented in this paper were for abraded or damaged laboratory specimens, and the spread in fracture strength can refer only to that particular set of specimens, and cannot be taken to be characteristic of a glass of that composition. The point of the strength extrapolation technique proposed was to permit a proof test of a specific specimen to be made at short times to predict whether it itself would support lower stresses for a longer time.

To provide a confidence limit for a production piece, a statistically significant measure of the distribution of strengths to be expected from the manufacturing process would have to be obtained. We know of no way to get this from laboratory specimens - because what one is really observing is the spread due to variability in surface perfection of pieces actually made in the process of concern. Given this distribution, which could be obtained either at liquid nitrogen temperature or, by a more extensive series of tests at room temperature one could construct a series of stress vs time to fracture curves. At a given stress, these curves would display increasingly shorter fracture times as the required confidence level was increased.

The point of our suggested proof testing procedure, involving the assumption that $\log t = -n \log S + C$, was to permit a proof test on our individual specimen without the arduous work of constructing curves from real specimens to take into account the distribution of surface flaws and strength.

S. M. Wiederhorn (National Bureau of Standards): If one considers data on abraded glasses or crack velocity studies there appears to be no static fatigue limit for silica glass. Tests on fibers of silica by Proctor and Mallinder, however, do show a fatigue limit which suggests that fatigue limit studies might better be studied on unabraded materials such as fibers or chemically polished glass. *Authors:* We agree most heartily. Most of the data upon which the existance of a static fatigue limit is postulated comes from specimens having liquid nitrogen temperature strengths ranging from 10,000 to 20,000 psi. Taking 0.15 S_N as a reasonable static fatigue limit, this results predict long term glass strength of 1500 to 3000 psi, and these and exposure times of months under stress may be needed to observe fracture at these stresses.

From the data of Mould and Southwick one can deduce (and you have) that $t_{0.5}$ is inversely proportional to the 6.5 power of S_N. Assuming that $t_{0.15}$ (the presumed static fatigue limit) is similarly proportional to S_N, one can further deduce that specimens stronger than about 100,000 psi should fail in times of a few seconds at loads of 15% S_N if no static fatigue limit exists. Hence, the use of strong or even pristine specimens should permit the determination of whether the concept of a static fatigue limit is correct in much shorter testing times. One might expect some experimental difficulty because the scatter in strength in strong specimens will be greater than in predamaged ones. Perhaps preliminary proof testing would permit the weakest specimens to be rejected before the static fatigue tests are conducted.

N. J. Kreidl (University of Missouri): Is not log-log plot leading to a dangerous extrapolation of "no limiting strength", e.g., for silica?
Authors: None of the existing data on abraded specimens lead to the conclusion that there is a static fatigue limit for fused silica. To the best of our knowledge a static fatigue limit for fused silica has not been established. In any case, the "dangerous" prediction would be to count on one if it does not exist, since one might then falsely predict a nonexistant load bearing capability.

H. Palmour III (N. C. State University): To what degree would cyclic variations either in stress or in corrosion agent (water) in the real world of engineering applications influence the interpretations and extrapolations of these steady state data?
Authors: Delayed fracture by static fatigue invokes only the concept that the rate of corrosion is accelerated by the application of a stress, and does not consider whether this stress is cyclic or not. Since it is predicted that the rate increases exponentially with stress, one would conclude that only the highest cyclic stresses would contribute to failure. A cyclic stress with a given maximum

value would be less effective in causing failure than a steady state
stress of the same maximum value. It is of course conceivable that
a cyclic stress could produce other effects not contemplated by
the standard theory, but we know of no experimental evidence that
cyclic stressing (of glass) is as deleterious as constant stressing.

John B. Wachtman, Jr. (National Bureau of Standards): For the purpose
of designing a proof test to insure a desired survival time at a
given stress what is the advantage of the log-log plot as opposed
to the semilog plot? The resulting proof test stress should be
the same for any method of plotting provided the curve does pass
through the data.
Authors: It is a question of experimental feasibility. Assuming
that the correct relationship between stress and failure time is
that presented by Eq. 1, then the slope of a curve in which S is
plotted against log t is $-p/S_N$. Knowing a desired value of S and t
for design purposes, one can predict proof testing conditions at
short times only if S_N is known, because the slope of a log t vs S
curve is dependent upon S_N. But S_N is difficult to determine.

If the relationship expressed by Eq. 2 is equally good (and it
appears to be so from data available), all plots of log S vs log t
have the same slope - the effect of both S_N and $t_{1/2}$ on the time
to fracture appear only in the intercept constant.[2] Hence, a line
of slope n = -13 drawn through the values of S and t desired will
predict the value of S which must be sustained at the proof testing
time. No knowledge of S_N is needed. The difference is also a
fundamental one. The expressions of Eq. 1 and Eq. 2 are not
equivalent - they are however so similar that one cannot decide
which (if either) is correct from currently available data.

THE BEHAVIOR OF REACTIVE CERAMICS IN ATMOSPHERIC ENVIRONMENTS

Andre Accary

Centre D'Etudes Nucleaires de Saclay

France

ABSTRACT

Using UC as an example, the reaction of highly reactive carbides was studied in different environments containing water. The aim was to elucidate the atmosphere corrosion mechanism which was found to be that of stress corrosion. Previous data were reinterpreted according to this mechanism.

THE CORROSION OF CERAMICS

When a ceramic is exposed to a corrosive atmosphere one of several phenomena can take place. The reaction products can adhere to the surface of the ceramic material and form a protective layer through which the reactive species must migrate for the corrosion reaction to proceed. Usually, this migration is slow and the corrosion rate decreases and becomes eventually negligible as the protective layer grows thicker. Due to the volume difference between the corrosion products and the reacting ceramic, stresses are generated across the interface between the corrosion layer and the ceramic. Because of the lack of placticity of that layer when the stress level is high enough, i.e., after the corrosion layer has reached a sufficient thickness, the layer fractures and usually peels off as a powder. The powder consists primarily of the reaction products, e.g., the hydration of calcia produces $Ca(OH)_2$ as a fine powder[1,2] and the calcia fractures as such when undergoing reaction. The fracturing has been ascribed to a "wedging" effect of the $Ca(OH)_2$ formed, but could possibly be associated with the perturbation of the stress pattern in the calcia when successive layers are reacted.

445

The calcia is never stress free because of the formation conditions (sintering of CaO powder or solidification of liquid CaO).

There are other examples where the amount of corrosion product is very small and does not result in the formation of a corrosion product layer but where the initial ceramic cracks or even powders. A well known example of such a phenomenon is the so called molybdenum disilicide "pest",[3,4] in which the disilicide exhibits[5,6] a very high oxidation resistance at temperatures in excess of 600°C but is completely transformed into a powder within 50 to 100 hr between 400°C and 600°C without more than 0.6 w/o of the initial silicide being oxidized. A similar behavior is observed for the actinide carbides when exposed to the atmosphere at room temperature for several days. The corrosion of the uranium carbides is very important in the handling and storage of nuclear fuels and has been studied in great detail at the Commissariat à l'Energie Atomique.[5,6] This paper summarizes these studies.

MATERIALS STUDIED

The materials studied were "industrial monocarbides" containing two phases, a nearly stoechiometric UC as the major phase with uranium metal in the samples containing <4.8 w/o C. These materials are usually called hypostoechiometric carbides. The uranium metal is located at the grain boundaries (Fig. 1). The materials with >4.8 w/o C, usually called hyperstoechiometric carbides, contained UC_2 which was found at the grain boundaries as well as in the grains where it forms a Wiedmanstätten structure (Fig. 2).

Samples were prepared by two methods, electron beam melting[7] and arc melting.[8] The characteristics of the two products are summarized in Table 1 which shows that the electron beam melted carbides are lower in impurities and have a larger grain size.

SAMPLE FABRICATION

The samples used were small cylinders 13mm in dia with a height of 40 mm (samples A) or of 10 mm (samples B). The two height-to-diameter ratios were chosen in order to provide two different surface to volume ratios.

The samples were machined from cast slugs. In order to remove as completely as possible the surface imperfections they were surface finished by careful grinding using a resin bonded Carborundum wheel. One of the flat faces of the B samples received a metallographic polish to allow microscopic study of the sample surface as the corrosion proceeded.

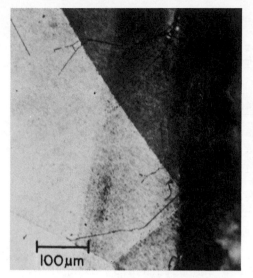

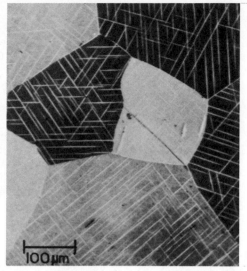

Fig. 1. Hypostoechiometric carbide. Uranium is visible
in the grain boundaries.
Micrograph plane is per-
pendicular to the surface;
a few microcracks run from
the surface.

Fig. 2. Hyperstoechiometric
carbide. The white
lines are Wiedmanstätten
platlets of UC_2 in UC.

Table 1. Carbide Characteristics

Melting Method	Carbon Content* (w/o)	Grain Size (µm)	Impurities Content (p.p.m.)		
			N_2	H_2	O_2
Electron beam	4.5 to 5.5	200-1000	50-250	1-30	100-500
Arc	4.7 to 5.9	100-400	450-1000	1-10	500-1000

*The composition of a given sample is homogeneous and known to an
accuracy of ±0.02%.

ATMOSPHERES USED

From prior experience it was known that water, either as
liquid[9] or as atmospheric moisture, was one factor in the atmospheric
corrosion of the uranium carbides. Other possible agents were oxygen
or nitrogen. In order to explore a range of atmospheres, six
environments were selected. Those with a water content lower than
that of the laboratory were controlled by means of dessicants. The

Table 2. Water Vapor Pressures above the Drying Agents Used.

Drying Agent	Water Content (mg/1)	Reference
P_2O_5	2.5×10^5	Morley*
$Mg(Cl_4)_2 \cdot 2H_2O$ mixed with $Mg(ClO_4)_2 \cdot 4H_2O$	2×10^3	Pascal†
Ca SO_4	5×10^3	Bower†

*Handbook of Chemistry and Physics 43rd ed. (1961).
†Pascal-Nouveau Traite de Chimie Minerale Tome IV (ed. Masson).

two drying agents used and the corresponding water vapor pressures are shown in Table 2. A very dry atmosphere was obtained by using P_2O_5 (water vapor content $\sim 10^2$ v.p.m.) and Mg $(ClO_4)_2 \cdot 2H_2O$ to obtain an atmosphere with ~ 1 v.p.m. of water. The six atmospheres selected were:

Atmosphere I - air saturated with water at 25°C
Atmosphere II - laboratory atmosphere at 25°C
Atmosphere III - laboratory air confined in the presence of
        $Mg(ClO_4)_2 \cdot 2H_2O$
Atmosphere IV - argon in the presence of $Mg(ClO_4)_2 \cdot 2H_2O$
Atmosphere V - vacuum (10^2 torr) in the presence of
        $Mg(ClO_4)_2 \cdot 2H_2O$
Atmosphere VI - vacuum (10^2 torr) in the presence of P_2O_5

Atmospheres III-VI were "dry"; III contained oxygen and nitrogen while IV-VI did not. Atmospheres V and VI differed from III in that the oxygen and nitrogen pressures were very low.

EXPERIMENTAL PROCEDURE

The samples, except those for atmosphere II, were kept in a closed vessel either with some liquid water for atmosphere I or the dessicant for atmospheres III to VI. They were periodically removed from the vessel for examination which consisted of (1) visual examination and weight change measurements for the A samples, and (2) visual and microscopic examination and weight change for the B samples.

The samples exposed to atmosphere I were of the type A and were examined by microfractography after having been fractured near their middle in a small press using a three point bend technique. A single examination for each specimen was possible.[6] Samples exposed to

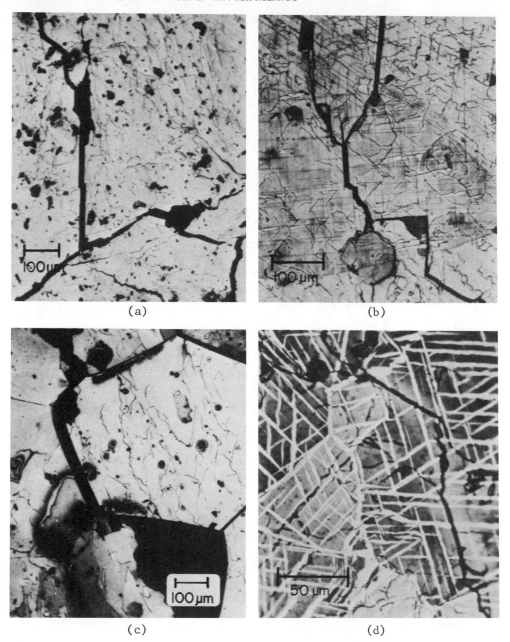

(a)

(b)

(c)

(d)

Fig. 3. Uranium carbide corroded by laboratory air (a) hyperstoechiometric (4.92 w/o C, E.B. melted) 80 days exposure,
unetched, point reactions shows as dark dots; (b) hyperstoechiometric (5.0 w/o C, E.B. melted) 80 days exposure,
etched; (c) hypostoechiometric (4.7 w/o C, E.B. melted)
60 days exposure, etched; (d) hyperstoechiometric (5.7 w/o C,
arc melted) 120 days exposure, etched.

atmospheres III-VI were, in fact, corrosion tested in a cyclic manner
consisting of long exposures to the controlled atmosphere and very
short exposures to the laboratory atmosphere during the time necessary
for examination. The total duration of the study of samples exposed
to atmospheres II-VI was 8-10 months. The samples subjected to
microfractography (atmosphere I) were ruptured after 1-3 days.

EXPERIMENTAL RESULTS

Atmosphere II: Laboratory Air at 25°C

On the electron beam (E.B.) melted samples, corrosion is visible
on the polished surface of the B samples when they have been exposed
from eight days to a month. After a long period, two systems of
cracks appear on the surface of the samples as exemplified in Fig. 3a
which shows both macro and microcracks with some point reactions.
Figure 3b (a hyperstoechiometric sample) shows intragranular cracking.
The macrocracks run parallel to the UC_2 platlet and the microcracks
seem to be independent of the platlets. Figure 3c (a hypostoechiometric sample) shows intergranular macrocracks with the microcracks
running through the grains.

For the A samples macrocracks are visible after 2-3 months of
corrosion. The samples exhibit a slight swelling (1% linear after
8 mos). No significant quantity of powder was produced and a weight
grain of $15-20 \times 10^4$ g/g was observed after 8-10 mos. The weight changes
with time are presented in Fig. 4. There is no significant difference
between the samples A and B.

On the arc melted samples the first traces of corrosion appeared
after 1/2 - 2 mos. In the case of hyperstoechiometric samples
(Fig. 3d) the macrocracks are not as obviously parallel to the UC_2
platlets as for the E.B. melted material. The swelling is slightly
lower than for E.B. melted material and the weight gains are also
lower in most cases (Fig. 4).

Atmosphere III: Laboratory Air in Presence of $Mg(ClO_4)_2 \cdot 2H_2O$

For the E.B. melted material, no evidence of corrosion was
visible under the microscope before 1/2 - 1 mo. of exposure. After
8 mos. the polished surface of a sample is exemplified by Fig. 5a
for an hyperstoechiometric material, the corrosion is localized to
the UC_2 platlets and their vicinity with little point corrosion.

The surface of an hypostoechiometric sample after 3 mos. is
depicted by Fig. 5b; the corrosion is restricted to the grain boundary
U metal and certain grains. No cracking is visible. The weight
decreases as shown in Fig. 4. They were lower than 10×10^5 g/g after
10 mos. and very scattered; no relationship between the corrosion rate
and the shape of the samples can be established (samples A or B).

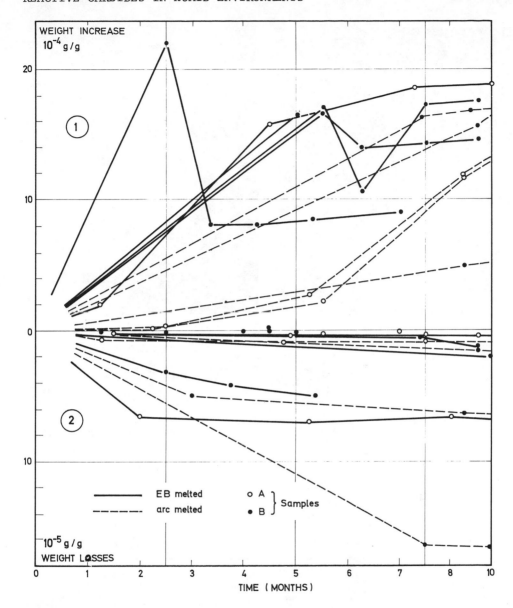

Fig. 4. Weight changes of samples exposed to curves (1) laboratory
         air and curves (2) air + $Mg(ClO_4)_2 \cdot 2H_2O$.

     For the arc melted samples the appearance after 6 mos is the one
shown on Fig. 5c where point corrosion is visible as well as a little
microcracking (barely visible). No swelling is observed and the
weight loss is of the same order of magnitude as for the E.B. melted
material.

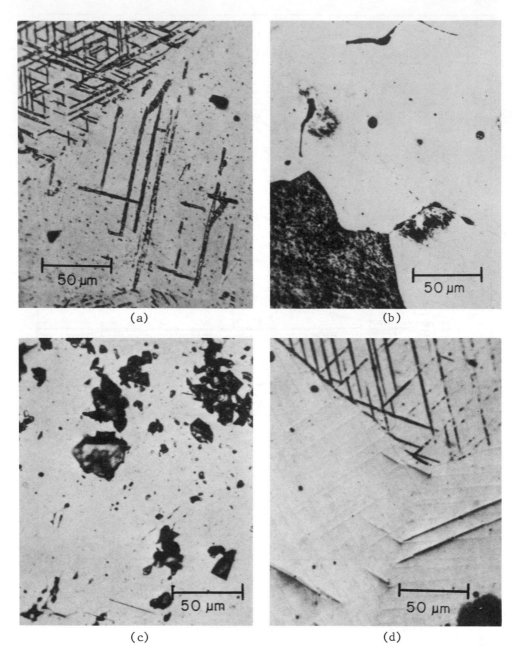

Fig. 5. Uranium carbide corroded in air and argon equilibrated with
Mg(ClO$_4$)$_2 \cdot$2H$_2$0, unetched (a) hyperstoechiometric (E.B. melted)
exposed 6 mos., air; (b) hypostoechiometric (E.B. melted)
exposed 3 mos., air; (c) hyperstoechiometric (arc melted)
exposed 6 mos., air, (d) hyperstoechiometric (E.B. melted)
exposed 7 mos., argon.

Atmosphere IV: Argon with $Mg(ClO_4)_2 \cdot 2H_2O$

The corrosion after 7 mos. for the hyperstoechiometric material is represented by Fig. 5d: no cracking is visible and the attack is restricted to the UC_2 platlets and to a few points. The weight losses are $<5 \times 10^5$ g/g in 8 mos. In the case of the arc melted material, the corrosion is limited to a very slight point corrosion. No clear differences can be made between the E.B. melted and the arc melted samples.

Atmosphere V: Vacuum with $Mg(ClO_4)_2 \cdot 2H_2O$

The corrosion was not significantly different for the E.B. melted and the arc melted material, neither was there a descernible difference between the hyperstoechiometric and the hypostoechiometric samples. The appearance of a polished surface after two months of exposure is typified by Fig. 6a which shows surface corrosion but no macrocracking; however, there is some very slight microcracking. After six months the aspect was as shown in Fig. 6b where the surface corrosion has involved large areas. The x-ray examination of the "corrosion products" showed no crystalline species other than UC. All lost weight which was as great as 40×10^4 g/g after 2 mos. (Fig. 7). The weight loss versus time curves fell into two families without any known reason.

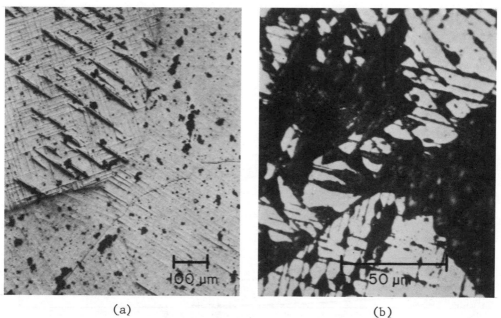

(a)   (b)

Fig. 6. Hyperstoechiometric uranium carbide specimens corroded in vacuum (10^{-2}mm Hg) equilibrated with $Mg(ClO_4)_2 \cdot 2H_2O$ (a) E.B. melted, exposed 2 mos.; (b) arc melted, exposed 4 mos.

Atmosphere VI: Vacuum and P_2O_5

No surface reaction or cracking was observed for either the E.B. or for the arc melted samples whether they were hyperstoechiometric or hypostoechiometric.   The only change observed was a very slight weight loss, $<6.10 \times 10^5$ g/g (Fig. 8).

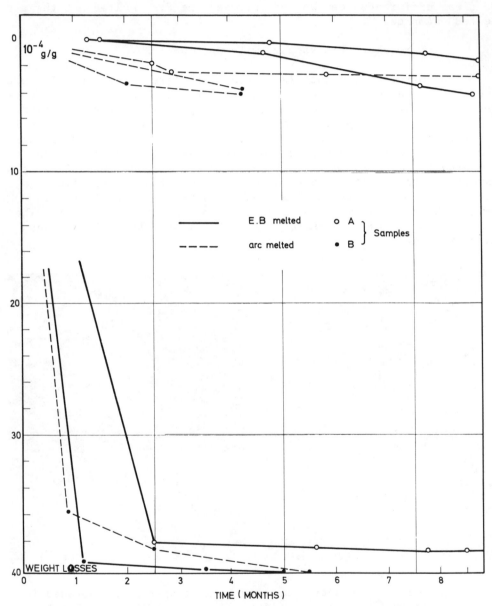

Fig. 7. Weight losses of samples exposed to vacuum (10^{-2}mm Hg) equilibrated $Mg(ClO_4)_2 \cdot 2H_2O$.

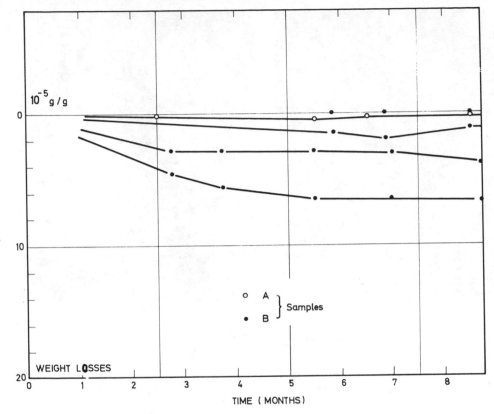

Fig. 8.   Weight losses of E.B. melted samples exposed to vacuum
          $(10^{-2}$ mm Hg) equilibrated with P_2O_5.

Atmosphere I: Air Saturated with Water at 25°C

  Examples of fracture surface of samples "ruptured" after 1-3
days of exposure are represented by Fig. 9.   A typical cleavage
fracture with "river patterns" running along the direction of propagation of the crack is shown in Fig. 9a; perpendicular to that
direction, interference rings show that some corrosion product
has accumulated in different thickness along the direction of propagation of the crack.   Figure 9b shows an interference ring system
with uniform colors in steps.   Each step was characteristic of a
given uniform corrosion product thickness.   Figure 9c shows another
example of this "step thickness variation" of the corrosion products
on the crack surface.   Similar fracture patterns have been observed
in the hydration of calcia single crystals[2] and ascribed to a wedging
effect.

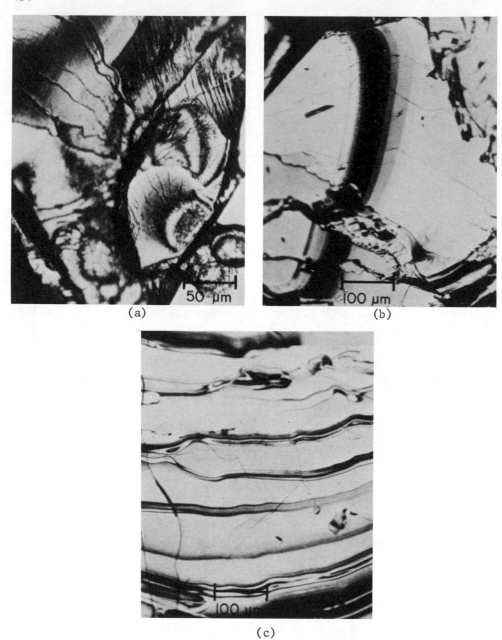

Fig. 9. Uranium carbide specimens (E.B. melted) corroded in H_2O-
         saturated air, 25°C, (a) hyperstoechiometric, exposed then
         fractured; (b) stoechiometric, exposed then fractured;
         (c) same material as in (b), interference fringes show
         step motion of crack.

INTERPRETATION OF THE EXPERIMENTAL RESULTS

The major experimental facts can be summarized as follows:

1. There is a significant crack formation only for the atmospheres with a high water vapor content. The crack pattern consists of a microcrack and a macrocrack network. When macrocracking is observed weight gain and swelling are also present. If no macrocracking can be observed the samples undergo a very slight weight loss.

2. For the moderately dry atmospheres, the absence of a gas other than the water vapor enhances the corrosion.

3. If air is replaced by argon in a moderately dry atmosphere the corrosion is decreased only slightly.

4. A "very dry" vacuum or a moderately dry gaseous atmosphere lead to about the same very low corrosion rate.

5. When the material cracks during corrosion, the crack surface exhibits interference colors which show that the crack propagated in successive steps.

The above facts are consistent with stress corrosion cracking produced by water vapor. The experimental facts as stated above (Nos. 1 to 5) will now be discussed in terms of this mechanism.

Since no cracking was observed without a sufficient water content in the corroding media, it is clear that the major corroding agent is water. Oxygen can possibly lead to some corrosion but as the comparison of the results reported for dry air and dry argon show that the corrosion does not lead to cracking.

The uniform thickness of a step shows that the exposure time to corrosion is uniform for any single step; the formation of a step is therefore practically instantaneous, i.e., the crack propagates in "jerks". This type of crack propagation suggests that its mechanism is that of stress corrosion as has been suggested for calcia.[2] The mechanism of that process has been analyzed previously by Hillig and Charles:[10] when a material under corrosion is stressed, the crack tips act as stress concentrators and thus, because of the higher energy level of the material under high stress the activation energy for the corrosion reaction is decreased and its thermodynamical driving force increased. Both phenomena enhance the corrosion rate. This tends to reduce the radius of curvature of the interface between the material and the corrosion products at the tip of the crack. On the other hand, if the radius of curvature decreases, the activity of the surface material decreases and thus tends to decrease the corrosion rate. The radius of curvature of a crack tip is the result of the balance between these antagonistic trends and is therefore influenced by the stress level since the later influences two of the three phenomena mentioned above. This radius of curvature in turn determines if the crack will or will not propagate under the existing stress, the condition being that the Griffith criterion be satisfied.

If the stress level is high the corrosion rate will be high at the tip of the crack which will progressively sharpen until the fracture condition is attained. The crack then runs in the un-corroded material with a very sharp head and extends suddenly until the stresses are relaxed to a level too low for further propagation: a step is thus formed. The corrosion resumes under a low stress field, the tip of the crack interface will then tend to be blunted unless the stress field is increased sufficiently to restore a high stress level in the vicinity of the crack tip, this may be due to at least two mechanisms acting separately or simultaneously: (1) the accumulation at the crack tip of corrosion products with a volume greater than the one of their parent material, and (2) the variation of the stress pattern in the material, this variation being due to the propagation of other cracks.

Mechanism 1 would suggest that the width of the steps should be uniform as mechanism 2 does not. Examination of Fig. 9 does not show such uniformity and thus indicates that the mechanism 2 is almost certainly operative.

The location of corrosion at the tip of the crack is confirmed by the fact that a low pressure atmosphere enhances the corrosion rate since the transport velocity of water by diffusion through the gaseous phase in the crack increases when the pressure of the gas decreases.

REFERENCES

1. R. Collongues, Private communication.
2. R. W. Rice, J. Amer. Ceram. Soc. 52 (8) 428-436 (1969).
3. J. Berkowitz-Mattuck, Bul. Soc. Française de Céramique 77 91-98 (1967).
4. J. Berkowitz-Mattuck, TAIME 233 1093-1099 (1965).
5. P. Rousset and A. Accary, Atmosphere corrosion of Uranium Carbon Alloys, Symposium on Carbides Harwell 1963.
6. P. Rousset and A. Accary, J.M.N. 17 (2) 149-152 (1965).
7. J. Trouve, Doctoral Thesis, Orsay, 1968.
8. A. Portneuf and R. Hauser, Nuclear Meeting Rome 1963.
9. B. Mansard, Private communication.
10. W. B. Hillig and R. J. Charles; pp. 683-84 in High Strength Materials. Ed. by V. F. Zackay, John Wiley and Sons, New York, 1965.

DISCUSSION

R. F. Stoops (North Carolina State University): Is the surface cracking illustrated in your figures typical of that caused by machining or were your specimens machined under condition to accentuate surface cracking?

Author: The surface cracks illustrated are typical of those encountered in machined surfaces of uranium monocarbide.

R. W. Rice (U. S. Naval Research Laboratory): Some of your results are similar to those I have reported for CaO where corrosion cracking appeared to be due entirely to the wedging action of the corrosion product resulting from its greater volume. A distinct ring pattern is obtained since the crack runs when the wedging stress reaches the critical level. By its advance, the crack reduces the stress, and comes to rest after a short distance. It will not run again until corrosion again builds up sufficiently to repeat the process. Your situation does appear to be more complicated, but such corrosion wedging action may be quite important in tests where weight is gained.

THE PROCESS OF HYDROSTATIC COMPACTION OF B_2O_3 GLASS

N. Mizouchi[*] and A. R. Cooper

Case Western Reserve University

Cleveland, Ohio

ABSTRACT

Upon application of high pressure, glass exhibits both elastic and anelastic behavior even at room temperatures. The consequence of anelastic behavior has been recognized as an "irreversible" densification. The technique developed permitted measurement of the instantaneous volume of B_2O_3 glass under hydrostatic pressure up to 22 kbar. It was observed that (1) anelastic volume strain is highly nonlinear in pressure, (2) apparent bulk viscosity may be lower than 10^{15} poise at onset of pressure application and tends to approach infinitely large value as time advances, and (3) more than one ordering parameter is necessary to be consistent with observed compaction under pressure.

INTRODUCTION

Bridgman's early work[1-3] on the effect of high pressures on simple glasses revealed two interesting results: (1) simple glasses tend to have compressibilities that increase with pressure in contrast with most crystalline materials, and (2) there are "irreversible" increases in density that can be produced by high pressure.

Many investigators[4-9] since that time have looked at these phenomena and proposed various explanations to account for the somewhat unusual behavior. In almost all cases the studies have been on the aftereffect of high pressure on the density or index of

---

* Now at Technical Center, Owens-Illinois, Inc., Toledo, Ohio.

refraction of glass. Particular concern has been directed at the
influence of shear in affecting irreversible compaction.[5-7] While
simple oxide glasses show marked irreversible compaction on application of high pressure, the effect is in no sense confined to
these glasses and more complicated inorganic glasses[4,5] and polymeric
glasses[9] show this behavior although often to a much lesser degree.

In an effort to gain a better understanding of the effect of
high pressure on glass we have studied the dependence of the volume
of B_2O_3 glass on pressure and time, thus allowing an examination
of the compaction process itself, as well as the irreversible effects
of compaction. Our experiments were confined to hydrostatic
pressure because of its greater simplicity.

Configurational changes responsible for the irreversible density
increase are of particular interest, but the following sections
also describe the compaction process, summarizing some of the interesting results of the experiments and comparing them with
various models.

DESCRIPTION OF EXPERIMENTS

The material used for all experiments (initially anhydrous
B_2O_3) was melted at Corning Glass Works at a temperature of 1200°C
in vacuo (5mm Hg) for 12 hr. The major impurity was silica
(0.01-0.1 w/o) according to spectrographic analysis of the raw
materials; infrared analysis of H_2O content in a melted glass showed
approximately 0.05 w/o. Rod-shaped specimens of ~0.5 cm dia were
drilled from the B_2O_3 glass using kerosine as a lubricant to avoid
exposure of the glass to moisture. The specimens were cut to a
length of ~1.13 cm, polished smooth on both ends with fine silicon
carbide and diamond abrasives, annealed overnight at 270 ± 1°C, and
then cooled at a rate of 0.5°C/min and stored in a vacuum desiccator.

Compaction experiments were conducted in a modified Bridgman
apparatus manufactured by the Harwood Engineering Company. The
details are described elsewhere.[10] The pressure transmitting medium
was a 1:1 mixture of pentane and isopentane. A dilatometer was
constructed using a linear variable differential transformer (LVDT)
which monitored the length of the sample at pressure. It was
calibrated using dummy specimens (KCL crystals, commercially pure
Fe). Signals from both the LVDT and the pressure measuring manganin
coil were properly amplified and displayed on a dual pen recorder,
as indicated in the schematic diagram (Fig. 1). The biggest uncertainty in the read-out of sample length came from the variations
in room temperature which effected the amplifer output. The amplifier was maintained as nearly as possible at constant temperature;
the data were corrected for small variations in amplifier temperature. Most experiments conducted were of relatively simple design,

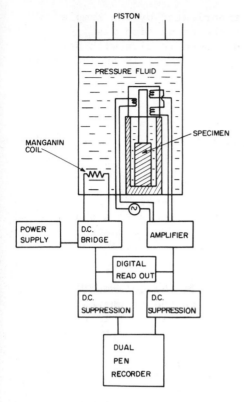

Fig. 1. Schematic diagram of experiment.

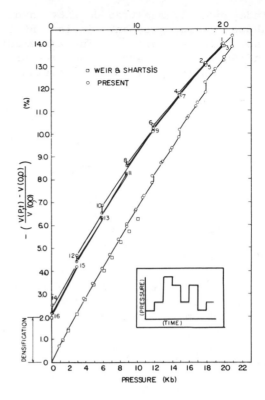

Fig. 2. Fractional volume change (%) vs applied pressure up to 21 kbar. In a multistep experiment. The numerals give the order of the pressure change after 21 kbar was attained. The inset gives schematic description of experiment.

i.e., one step[*] changes in pressure. However, on a few occasions multiple step experiments, such as shown in Fig. 2, were carried out. Figure 2 reveals that for a maximum pressure of 21 kbar, the total fractional volume change was 14%, while the irreversible changes in volume (aftereffects) were 2%. Comparison of the irreversible volume change based on the LVDT measurements with the

---

[*]Although the experiments are classified as step experiments, in reality it always required a finite period of time to change from one pressure to another. Typically, a change of one kbar could be achieved in about one minute.

independently measured density change experienced by the sample
showed close agreement, indicating that the compaction was
isotropic and uniform. The agreement of these results with Weir
and Shartsis'[11] data at low pressure (up to 5 kbar) is an indication
of the validity of our calibrations.

RESULTS AND DISCUSSION

Elastic Behavior

Changes in volume of glass occur from two separable causes:
the instantaneous changes in bond length in response to a change
in external conditions and the slower response due to changes in
the configuration or structure. We term these elastic and anelastic
respectively.

At low pressures the anelastic changes in the volume of glass
occur so slowly that for normal pressure rates the volume change
observed is practically all due to elastic behavior. This was
seen in Fig. 2 where the volume vs pressure data at short times
corresponded very closely with that predicted from Weir and
Shartsis'[11] measurements of compressibility of B_2O_3. At higher
pressures, however, a time-dependent change in volume occurs, and
it is difficult to separate this from the elastic effect. However,
it is possible to wait until the time-dependent change of volume
at pressure is very small and then quickly (~10 sec) reduce the
pressure, ΔP, by ~0.2 kbar. The dual pen recorder displays this
pressure change and the associated length change which allows the
compressibility, defined as $K = -1/V \, dV/dP$, to be calculated.
Uncertainties of the pressure change, ΔP, of ~1 bar and of the fractional length change of 6×10^6 can result in an uncertainty in
the compressibility,* $K' \simeq -1/V_o \, \Delta V/\Delta P$, of $\pm 1\%$. Results obtained
in this manner are shown in Fig. 3. Experimental precision is seen
to be ~ $\pm 2\%$. The two sets of results agree with each other and
with Weir and Shartsis' data at low pressures. This method provides
a means of extending the compressibility data to higher pressures
under the assumption that compressibility does not depend markedly
on the change in configurational volume.**

---

* For convenience, and for direct comparison with previous workers,[11]
we will often use $K' \equiv V/V_o \, K$ in place of K. V_o is a volume of
the uncompacted glass.
** This assumption is analogous to the often accepted assumption
that thermal expansion does not depend on fictive temperature.

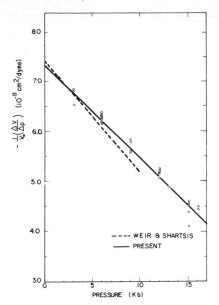

Fig. 3. Compressibility (k') vs applied pressure up to 16 kbar.

Simple Viscous Compaction

For a viscous system which is deforming slowly under the influence of a constant hydrostatic pressure, P, the Naviar-Stokes equations can be simplified by the elimination of all inertial terms and shear stresses to yield[12]

$$- \frac{(P-\hat{P})}{\eta} = \frac{1}{V} \left(\frac{\partial V}{\partial t} \right)_P \tag{1}$$

where η is the volume viscosity* and $\hat{P}$ is the internal or fictive pressure defined as the pressure with which the density of the material would be in equilibrium.

The significance of Eq. 1 is readily understood. In contradiction to usual viscous flow under shear stress where deformation proceeds until the stress is removed, under a hydrostatic stress the compaction continues only until the external pressure equals the internal (fictive) pressure. It is clear that the volume viscosity can be calculated if $\hat{P}$ and the fractional rate of change of volume, $1/V \, (\partial V/\partial t)_P$, are known.

---

* $[\lambda+2/3\mu]$ in the notation of Lamb[12] is often considered to be zero for gases and simple fluids. However, in network liquids the volume viscosity is typically of the same order as the shear viscosity, μ.

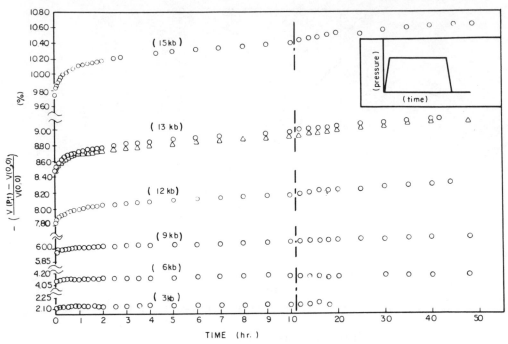

Fig. 4.   Anelastic volume change in one-step experiments at various
          pressures up to 15 kbar.

To determine the latter, one-step experiments (e.g., Fig. 4)
were carried out where the pressure was increased as rapidly as
possible (~1 kbar/min, adiabatic temp. change <3°C) to a specified
pressure and held for a period, usually ~50 hr.

The volume change expected from purely elastic deformation
could be calculated from the compressibility data in Fig. 3 which
allows a good estimate to be made of the appropriate initial volume
at pressure, $V(P,0)$.

Experiments such as those shown in Fig. 5 where volume expansion and volume compaction are carried out at the same pressure
allow an estimate of the volume "at equilibrium" with a given
pressure from the apparent limiting volumes of these curves.   From
this data a plot of $\hat{P}$ vs V_{equil} is constructed as shown schematically
in Fig. 6.   Note that at volume $V(0,0)$ equal to that of the uncompacted glass, there is a large uncertainty in the value of $\hat{P}$.   It
ranges from one atmosphere to significant negative pressures.
Davies and Jones[13] have also postulated a negative fictive pressure
for glass at room temperature and pressure.   There is also a
marked uncertainty at the maximum pressure, 15 kbar, as here we
have no way to estimate the lower bound to V_{equil}.

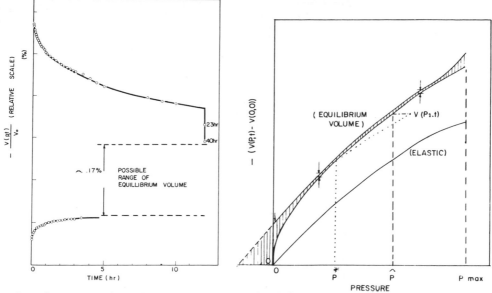

Fig. 5. Determination of "equilibrium volume" at 6 kbar. The ordinate has an arbitrary scale.

Fig. 6. Schematic diagram for determination of fictive pressure, $\overset{*}{P}$ and P. Shaded region represents uncertainty.

Two ways exist for determining the upper bound of fictive pressure. Consider a specimen with volume $V(P_1,t)$ at external pressure, P_1. By our definition, following that originally applied to the Navier-Stokes equations, the fictive pressure, P, is obtained by moving horizontally along the dashed line until the $V(\hat{P})$ line is intersected. If instead, one were to consider fictive pressure, analogous to fictive temperature, to relate only to configurational aspects of the structure, then the elastic deformation should be eliminated by proceeding along the curve (dotted) parallel to the elastic deformation. This results in a lower value of internal pressure which we term $\overset{*}{P}_o$. It is clear that the value of viscosity from Eq. 1 depends on which value of fictive pressure, $\overset{*}{P}$ or $\hat{P}$, is chosen. It is for this reason that we emphasize the difference and reiterate that we have followed the original definition here using $\hat{P}(V)$, primarily because it is more convenient at this stage.

The results of volume viscosity vs external pressure and time, t, i.e., $\eta(P,t)$, are shown in Fig. 7. Several points are worthy of note: (a) the viscosity is a sharply decreasing function of pressure at all times; (b) the viscosity is an increasing function of time at all pressures; and (c) the magnitude of viscosity values encountered range upward from those customary in the glass transition range, $10^{13} \sim 10^{14}$ poise.

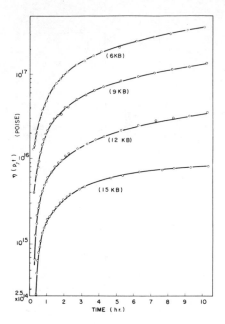

Fig. 7.   Bulk viscosity vs time at various pressures.

Kovacs Model

For the one-step compaction experiments, viscosity increases
with time are equivalent to viscosity increasing as volume decreases.
An explanation for such an occurrence in volume relaxation was
suggested by Kovacs[14] based on the recognition that free volume
decreases with compaction. A clear test for this premise exists
in the expansion experiments shown in Fig. 5. Calculations of
viscosity for these experiments invariably show viscosity increasing
with time, i.e., with increasing volume in direct contradiction
to the free volume ideas put forth by Kovacs.

Single Ordering Parameters

Equation 1 provides a single ordering parameter, the fictive
pressure, $\hat{P}$.   Implicit to this treatment is the idea that for the
description of the macroscopic volume deformation of glass all the
necessary information regarding the structure can be distilled to
a single parameter, $\hat{P}$. This is equivalent to the notion that glass
configuration can be described by a single fictive temperature.
The observation that the isothermal viscosity, η, is a function of
P and t or equivalently of P and $\hat{P}$ does not violate Eq. 1.

However, a contradiction of Eq. 1 and the idea behind it occurs
if two different viscosities, η_1 and η_2, can be found at the same

value of P and V, or P and $\hat{P}$. To test this premise we could give two glasses, equivalent initially, drastically different pressure histories and then study their volume change. A crossover point[15] where both glasses have the same volume and external conditions but have different rates of volume change would invalidate Eq. 1 because it would require viscosity to be a function of another internal parameter, i.e., more than a single ordering parameter would be required. Such crossover points have often been found in volume relaxation studies after abrupt temperature changes in glass transition regions.[16-18]

While we do not have any crossover points, we show in Fig. 8 a plot of the deformation behavior of two glasses (I and II), each having a different pressure history, but otherwise equivalent. The results are likewise inconsistent with Eq. 1 when it is recognized that irreversible thermodynamics requires that viscosity (being a conductivity for momentum) must always be greater than zero. Since systems must tend toward equilibrium at long times, $V(P_1, \infty)_I = V(P_1, \infty)_{II}$. This can occur only if 1/V dV/dt changes sign for one of the glasses, which is forbidden because P-$\hat{P}$ cannot change sign (P=$\hat{P}$ implies equilibrium) and η is always positive.

We conclude that description of the iosthermal compaction process of B_2O_3 glass requires more than one ordering parameter. This is consistent with the results and conclusions of a number of workers[15,17,19] who have studied glass transition behavior of various inorganic glasses.

Multiple Ordering Parameters

Much of the freedom possessed by generalized multiordering parameter models is present in a two-parameter model as was shown by Macedo *et al.*[15] in discussing the results of volume relaxation at glass transition. Shown in Fig. 9 are the simplest schematic (spring and dashpot) representations of one- and two-parameter systems. In the one-parameter case, Fig. 9a, the spring with compressibility K_0 gives the elastic behavior. The spring and dashpot pair in parallel give the anelastic behavior, the dashpot with viscosity η_1 provides damping and the stress on the spring with compressibility K_1 defines the ordering parameter $\overset{*}{P}$.

Thus, the single ordering parameter model in Fig. 9a possesses all the qualities of simple viscous compaction at constant pressure. The stress acting on the dashpot is the difference between the external and internal stress. Hence the rate of change of volume is given by

$$- \frac{1}{V} \left(\frac{\partial V}{\partial t} \right)_P = \frac{1}{\eta_1} (P - \overset{*}{P}) \qquad (2)$$

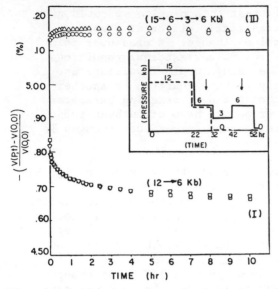

Fig. 8. Volume changes of glasses I   Fig. 9. Spring and dashpot
         and II at 6 kbar.   representation of
                                                  ordering parameters.

which is identical to Eq 1. The relation between $\overset{*}{P}$ and V is obtained
from definition of the spring compressibility as follows

$$\ln \frac{V(P)}{V(0)} = -K_1\overset{*}{P} \tag{3}$$

where $V(0)$ is the equilibrium volume at $\overset{*}{P} = 0$. Substituting for $\overset{*}{P}$:

$$-\frac{d\ln}{dt}\left(\frac{V(\overset{*}{P})}{V(0)}\right) = \frac{P}{\eta_1} + \frac{1}{K_1\eta_1}\ln[V(\overset{*}{P})/V(0)] \tag{4}$$

The solution of this equation when $\overset{*}{P}_o$ is the internal pressure at
time zero is:

$$\ln \frac{V(\overset{*}{P})}{V(0)} = -K_1P + K_1(P-\overset{*}{P}_o)e^{-t/\tau_1} \tag{5}$$

where $\tau_1 = K_1\eta_1$. Adding $K_1\overset{*}{P}_o$ to both sides to get the volume
strain at pressure gives

$$\ell n \ \frac{V(\overset{*}{P})}{V(\overset{*}{P}_o)} = -K(P-\overset{*}{P}_o)(1-e^{-t/\tau}{}_1) = -K_1 (\overset{*}{P}-\overset{*}{P}_o) \tag{6}$$

The equality on the right hand side is simply a consequence of the elastic behavior of the spring.

Using only the linear term in the series expansion for $\ell n \ (V(P)/V(P_o))$ gives the volume change as a function of time

$$V(\overset{*}{P}_t)-V(\overset{*}{P}_o) \equiv V(t)-V(t=o) \simeq -V(\overset{*}{P}_o) \ K_1(P-\overset{*}{P}_o) \ (1-e^{-t/\tau_1}) \tag{7}$$

Notice for a single parameter model the volume is a monotonic function of time.

If there are more than one pair of elements, as in Fig. 9b, then the total volume change as a function of time at constant pressure is the sum of the changes of all the elements, say n elements.

$$V(t)-V(t=o) = \sum_{i=1}^{n} [V(\overset{*}{P}_t)-V(\overset{*}{P}_o)] \simeq \sum_{i=1}^{n} - V_i \ (\overset{*}{P}_o)K_i (P-\overset{*}{P}_io)(1-e^{-t/\tau}{}_i)$$

$$\simeq \sum_{i-1}^{n} - a_i (1-e^{-t/\tau}{}_i) \tag{8}$$

The behavior does not depend explicitly on the constants $\overset{*}{P}_io$, K_i and $V_i(\overset{*}{P}_o)$ for each pair, but only on the products: a weighting factor, $a_i = V_i(P_o)K_i(P-\overset{*}{P}_io)$ and a time constant $\tau_i = K_i\eta_i$.

The number of extremum points (maxima or minima) permissible according to Eq. 8 is (n-1). Thus the fact that there has never been more than one extremum point observed during volume relaxation studies near the glass transition is consistent with the previous successes of the two-ordering parameter model.[15]

Another feature of the two-parameter model is that it naturally possesses the already noticed characteristic of apparent viscosity increasing with time. This occurs because a low viscosity, η_i, results in a correspondingly small time constant, τ_1. The small time constant processes occur earliest and hence low viscosity processes dominate. These qualitative factors suggest the importance of a quantitative test of the two-parameter model. Fig. 10 shows

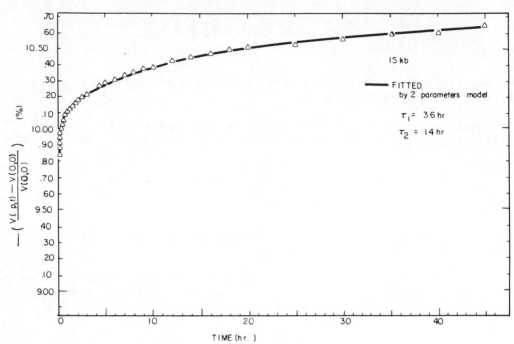

Fig. 10.   Anelastic volume change at 15 kbar. Fitted with
            2-parameter model.

the 15 kbar data from Fig. 4 compared with the prediction of Eq. 8
for the case where a_1 = 3.56 x 10^{-3}, a_2 = 5.69 x 10^{3}, τ_1 = 0.36 hr,
τ_2 = 14 hr.

While the predicted curve does not represent the best fit to
the data, they are the result of a fitting procedure whereby the
deviations in results were separately minimized for the ratios $\frac{a_1}{a_2}$, $\frac{\tau_1}{\tau_2}$.
The reasonable fit of actual results to the predictions is
obvious. Notable, however, is a deviation where the actual volume
change at short times is greater than predicted. The deviation in
fractional volume change is fairly large (~10%).

A more severe test of the two-parameter model is provided by
volume expansion experiments at room temperature of a sample previously at 18 kbar. Data ranges from 180 to 1.65 x 10^{5} sec, with an
additional point at 4.8 x 10^{7} sec (1.5 year). Shown in Fig. 11
are the results along with a two-ordering parameter fit to the data,
with a_1 = 5.45 x 10^{-4}, a_2 = 2.73 x 10^{3}, τ_1 = 4.0 hr., τ_2 = 150 hr.

The fit is reasonable but again the expansion is faster than
predicted at short times. The residual volume change predicted
for t = 4.8 x 10^{7} sec is ~1.17%, while the measured change is ~1.05%.

Thus the two-parameter model with constant properties appears to
give a reasonable description at short times, it does not accurately
predict the process. We conclude that it is premature to ascribe
fundamental significance to a two-ordering parameter model, and
it is inadvisable to extrapolate data using this premise.

It is interesting to note that $(\tau_2/\tau_1)_{15\ kbar} = 39$ and
$(\tau_2/\tau_1)_{1\ bar} = 38$. While this correspondence is perhaps coincidental,
it suggests that the pressure dependence is the same for both time
constants. There is no doubt that a multiordering parameter
(n $\geq$ 3) will supply sufficient adjustable parameters to allow fitting
all of the data (Figs. 4, 5, 10 and 11) within the experimental
uncertainty. We did not pursue this approach because of its tedium,
because it becomes more and more difficult to ascribe physical
significance to the increasing number of parameters, and most
importantly, based on experience with the room temperature recovery
data, we have no confidence that fitting such results over time
interval (t $\leq$ t < t_1) will allow prediction to times t $\gg$ t_1.

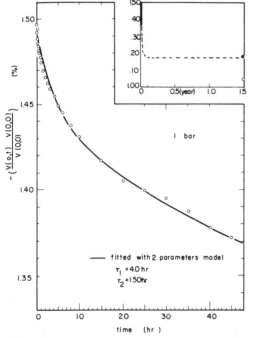

Fig. 11. Volume recovery at 1 bar.

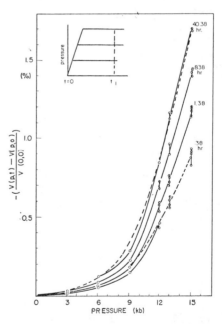

Fig. 12. Nonlinearity of anelastic
volume change in one-
step experiments at
various pressures with
time, t_1, as a parameter.

On the other hand the multiordering parameter model is attractive
conceptually for glass compaction because it suggests that a variety
of microscopic processes (each having its own time constant) are
involved in structural rearrangements of simple glasses.

Memory

A single ordering parameter only gives a measure of the
deviation of a system from equilibrium, i.e., $(P-\overset{*}{P})$. Thus, a two-
ordering parameter model is the simplest model to permit a "memory"
effect. The "memory" is contained in the ratio of the spring tension
of springs 1 and 2 (Fig. 9b). Clearly it is possible for the model
to have the same length with various ratios of the springs. Only
one bit of memory is permitted, however, as all the information
about history has to be distilled into the two tensions, $\overset{*}{P}_1$ and $\overset{*}{P}_2$.

Ordering Vector

Because of the necessity to invoke at least two ordering parameters to describe glass compaction, it is suggested that instead
of *ordering parameter*, the term *ordering vector* be used. The vector
is conceived to have an many elements as necessary to adequately
describe the properties and behavior. Each of the properties is
described by a dot-product of this vector with an appropriate
characteristic vector for that property. Notice that in Eq. 8 the
volume at time t can be thought of as the dot-product of a vector,
$\{V(P_o)K(1-e^{-t/\tau})\}$, with the vector, $(P-\overset{*}{P})$, describing the deviation
from equilibrium.

Nonlinearity in Pressure

For the anelastic deformation to be linear in *pressure*, it is
required that $a[V(P,t)-V(P,o)] = V(aP,t) - V(aP,o)$ for all time,
t, and all choices of nonnegative real number, a. That this is
not the case is seen in Fig. 12. Rather it appears that -
over the time and pressure regions we have examined - the anelastic
volume change is approximately a cubic function, i.e.,

$$a^3[V(P,t) - V(P,o)] = [V(aP,t) - V(aP,o)] \qquad (9)$$

This nonlinearity is consistent with the observations in Figs. 7,
10 and 11 that viscosities and time constants are steep functions
of pressure. If it proves that time constants all have some
pressure dependence, as indicated by Figs. 10 and 11, then it may
be possible to "linearize" the behavior in a way equivalent to that
proposed by Narayanaswamy.[19]

Multistep Experiments

Rather than to attempt a quantitative treatment of multistep experiments at this time, we prefer to attempt a reconciliation of some of the important phenomenological observations of our experiments on a qualitative basis using a two-parameter model with viscosities as steep functions of pressure.

At low pressures both η_I and η_{II} are very large but $\eta_I << \eta_{II}$ ($\simeq\infty$). Slight increases in pressure give imperceptible changes in configurational volume after short times because of small driving force and high viscosities. This gives the appearance of a threshold pressure for anelastic compaction. As pressure increases, viscosities decrease and configurational volume changes occur during the experimental times at pressure. Because of the lower viscosities, it is possible to quench only some of the configurational volume change by quickly reducing the pressure. At room pressure, expansion occurs primarily from process I with viscosity η_I and a residual compaction exists even after long times because $\eta_{II} \to \infty$.

Firmer support for this notion will exist if quantitative verification is obtained. Until then, it is perhaps better to use this speculation as a guide from which a more complete explanation can be built.

CONCLUSIONS

It has been shown that a one-parameter (single fictive pressure) model is unsatisfactory to describe compaction behavior of B_2O_3 glass under hydrostatic pressure at room temperature. A two-parameter model which is easily conceived schematically is in reasonable quantitative agreement with simple one-step experiments. Time constants are shown to be steep decreasing functions of pressure which is consistent with the nonlinear dependence of configurational volume on pressure and may offer a qualitative explanation for the observed phenomena in volume compaction of glass by hydrostatic pressure. The concept of a system with multiple ordering parameters is readily visualizable from spring and dashpot pairs in series since the extension of every spring can define a separate ordering parameter.

ACKNOWLEDGMENTS

We sincerely thank the Owens-Illinois Corporation for the financial support they gave to this work and for the encouragement and interest of various persons in their Technical Center during the course of this work.

REFERENCES

1. P. W. Bridgman, Am. J. Sci., 7 81 (1924).

2. P. W. Bridgman, Am. J. Sci., 10 359 (1925).

3. P. W. Bridgman and I. Simon, J. Appl. Phys., 24 [4] 405 (1953).

4. O. L. Anderson, J. Appl. Phys., 27 [8] 943 (1956).

5. H. M. Cohen and R. Roy, Phys. Chem. Glasses, 6 [5] 149 (1965).

6. J. D. Mackenzie, J. Am. Ceram. Soc., 46 [10] 461 (1963).

7. W. Poch, Phys. Chem. Glasses, 8 [4] 129 (1967).

8. J. Arndt and D. Stöffler, Phys. Chem. Glasses, 10 [3] 117 (1969).

9. R. M. Kimmel, Ph.D. Thesis, M. I. T. (1968).

10. F. Birch, E. C. Robertson and S. P. Clark, Jr., Ind. Eng. Chem. 49 [12] 1965 (1957).

11. C. E. Weir and L. Shartsis, J. Am. Ceram. Soc., 39 [9] 299 (1955).

12. H. Lamb, Hydrodynamics, Cambridge University Press, 1932.

13. R. O. Davies and G. O. Jones, Advances in Phys. 2, 370 (1953).

14. A. J. Kovacs, Trans. Soc. Rheology, 5, 285 (1961).

15. P. B. Macedo and A. Napolitano, J. Res. NBS 71A [3] 231 (1967).

16. R. W. Douglas and G. A. Jones, J. Soc. Glass Tech. 32, 309 (1948).

17. H. N. Ritland, J. Am. Ceram. Soc., 39 [12] 403 (1956).

18. S. Spinner and A. Napolitano, J. Res. NBS 70A [2] 147 (1966).

19. O. S. Narayanaswamy, private communication.

DISCUSSION

N. J. Kreidl (University of Missouri at Rolla): Have your interesting B_2O_3 glasses, with their remarkable temperature-volume anomalies yielding varying densities (structures), been tested for their glass transitions?
Authors: No.

R. H. Doremus (General Electric Company): The authors have shown that a model based on a single relaxation time is inadequate to describe their data. Two relaxation times are better, but still not completely satisfactory. In fact a whole series of "relaxation times" are probably operating. Would it not be more realistic to return to the actual molecular process and try to develop a better two-parameter model? As the compaction proceeds, molecular groups become increasingly entangled, so that further compaction is increasingly difficult. Perhaps some mathematical function could describe this process.
Authors: We agree with your comments. As is mentioned in the text, we feel that it is a tedious exercise to get best fit from a distribution of relaxation times. It is not justified because of the doubtful utility for extrapolation. Other mathematical functions seem to be better, e.g.,

$$\frac{V(p,t)-V(P,0)}{V(P,0)} = a\,[1-\exp{(-t/\tau)}]^{1/n}, \; n > 1$$

but at present no model justifies this. With regard to molecular description we have purposely avoided their discussion in this paper as our thoughts are incomplete.

FACTORS INFLUENCING THE USE OF CERAMIC MATERIALS IN DEEP-SUBMERGENCE APPLICATIONS

J. C. Conway

The Pennsylvania State University

State College, Pennsylvania

ABSTRACT

Analytical and experimental research conducted on ceramic materials in order to evaluate pertinent material properties in the deep-ocean environment is presented. Areas investigated include evaluation of brittle material failure criteria, determination of meaningful material properties, fortification of ceramics by rubber, ductile or induced compressive layers, modification of graphites by high temperature-high pressure infiltration, corrosion resistance of ceramic materials, and vehicle fabrication feasibility. Results of the various investigations are compiled and evaluated with regard to deep-ocean requirements and recommendations are offered for future research.

INTRODUCTION

Hulls of deep-diving vehicles, sea-bottom structures, and instrument housings to be used at great ocean depths are preferably made buoyant, so that they are independent of suspension from the surface or of positive applications of power to bring them to the surface after a submergence. External buoyant components can be added; however, they reduce serviceability. A search is on, then, for materials that are light enough and yet have the required physical properties, ease of fabrication, and acceptable fabrication cost to provide the complex pressure-resistant hulls, structures and housings of the future.

Glasses, ceramics, graphites as well as other brittle materials possess all of the required physical and mechanical properties to

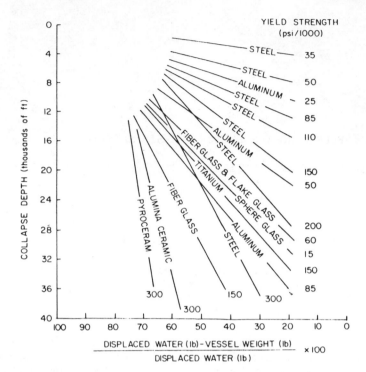

Fig. 1.   Relative strength of rib-stiffened cylinders.

make then desirable as structural materials for deep-submergence shells (Fig. 1). Factors precluding their use have included inadequate knowledge of the failure process, a lack of adequate physical and mechanical properties, sensitivity to point, impact and cyclic loading and severe fabrication difficulties.

The Ordnance Research Laboratory (ORL) of The Pennsylvania State University has conducted research concerning the use of ceramic materials in deep-submergence applications for the past 10 years.[2-5] Initial research concerned itself with basic brittle materials behavior leading to the design and fabrication of a cylindrical underwater ceramic vehicle, the Benthos vehicle. Testing of this vehicle revealed areas in which additional knowledge was needed. Further investigations conducted in the areas of design of a ceramic deep-submergence vehicle, failure criteria and properties evaluation, fortification of ceramics by protective coatings, modification of base materials and use of ceramics for corrosion protection constitute a major portion of this paper.

DESIGN OF A CERAMIC DEEP-SUMBERGENCE VEHICLE

Following preliminary tests conducted on simple monocoque cylinders and spheres and ring-stiffened cylinders fabricated from

Fig. 2.   Benthos Vehicle.

both Pyroceram 9606 and alumina ceramic, the multicomponent Benthos
vehicle, was fabricated and tested (Fig. 2).   The vehicle was composed of a titanium hemispherical nose, three ring-stiffened cylinders
and a ring-stiffened afterbody of Pyroceram 9606.   Components were
connected by joint rings designed to minimize localized bending
stresses and point loadings.

Test results indicated that although the elastic stress and
stability behavior of the vehicle was adequately predicted with
existing theory, more information was needed on material properties
and failure criteria before reliable design would be possible with
the ceramic materials.   Failure of the components occurred in two
ways.   The first failure mode, occurring at relatively low external
hydrostatic pressure, consisted of concentric cracking of the
cylinder due to point loading of the sealing surface (Fig. 3).   Subsequent measurements indicated that radial runout of 0.001 in. at
the bearing surfaces induced point loading.   A realistic understanding
of the failure mechanism was unavailable.   The second mode, occurring
at high pressure, was found in areas of high bending stress gradient
and resulted in cracking and spalling in the tensile zones and
crushing in the compressive zones (Fig. 4).   This mode of failure
could be adequately predicted by the maximum stress theory.   The
components were inherently vulnerable to impact loading, and cyclic
testing yielded conflicting results.   Corrosion resistance of the
material was not investigated.

Fig. 3. Section of cylinder Fig. 4. Failure mode in ring-stiffened
       showing concentric   Benthos cylinder at end ring.
       cracking.

FAILURE CRITERIA AND PROPERTIES EVALUATION

Following testing of the Benthos vehicle, a long-term program
was initiated to investigate the failure mechanism in the brittle
material and to evaluate available mechanical and physical properties
with regard to the failure mechanism. As expressed earlier, the
failure mode occurring in areas of high bending stress gradient
was adequately predicted by the maximum stress theory, utilizing
the measured flexural strength of the material as the design parameter. The second failure mode, that of point loading, was studied
further.

To simulate point loading in the brittle shell, plates of
various depth-to-width ratios were line-loaded (Fig. 5).[6] Longitudinal and transverse stresses were predicted by a mathematical
solution (Fig. 6) and experiments were conducted on plastic plates
at loading rates designed to force the plastic into quasi-brittle
behavior. Mathematical results indicated a biaxial stress field
near the line load consisting of high tensile transverse stresses
and high compressive longitudinal stresses for depth-to-width ratios
greater than unity. Application of the maximum principal stress
theory of failure to the critical biaxial stress zone beneath the
line load yielded fracture loads which were verified within 10%
by experimental data for each depth-to-width ratio. Crack propagation
followed fracture initiation and was photographed through a circular
polariscope with a spark camera. Isochromatic fringes, photo-
elastically recorded, revealed an extremely localized plastic zone
at the crack tip at initiation and during propagation.

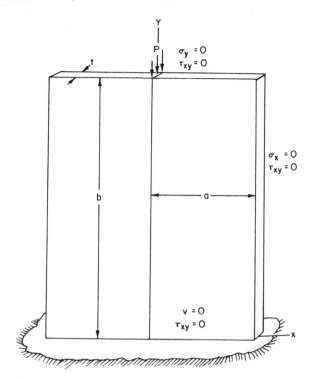

Fig. 5. Line-loaded rectangular plate.

Application of the point load data to the Benthos cylinders utilizing the maximum principal stress theory and measured material tensile strength revealed extremely low external hydrostatic failure pressures coinciding with those obtained in the Benthos cylinders. These results indicate the necessity of absolute parallelism of bearing surfaces or complete avoidance of brittle material contact in cases where two brittle components are joined.

In view of tests conducted on the Benthos vehicle and subsequent point load experiments, it seemed that the tensile or flexural strengths were the most significant measured material properties. Methods of obtaining these values are well known and will not be reiterated here. The elastic modulus and Poisson's ratio, obtained in static testing of cylinders or dynamically on beams are also useful parameters. The material compressive strength, a parameter frequently used to judge the relative merits of various brittle materials was found adequate for qualitative comparisons but of little use for design purposes.

Analytical and photoelastic investigations conducted on solid cylinders subjected to axial compressive loads showed that the stress distribution within the cylinder was extremely sensitive to the shear stress distribution and intensity on the loaded surfaces of the

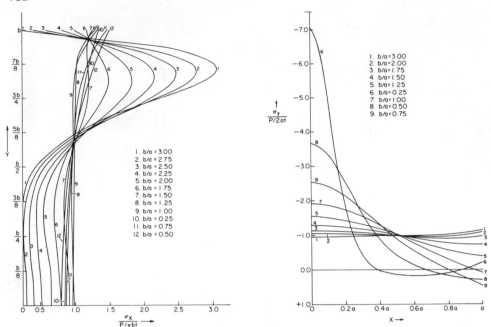

Fig. 6.   Stress distributions in terms of mean sectional stress for
various (b/a) ratios (left) horizontal stress along the
vertical axis of symmetry and (right) vertical stress
along the bottom edge of the plate.

cylinder.[1] For zero friction, a uniform axial compressive stress
and tangential and radial tensile stresses are found throughout
the cylinder. This stress state is illustrated in Fig. 7(a) where
a uniform isochromatic fringe order is found throughout the central
slice of a frozen stress cylinder with reduced end-plane frictions.
For this condition, the tensile strength of the material is the
critical material parameter and the cylinder exhibits the classical
tensile fracture. As friction on the end plane is increased shear
stresses become prominent near the edges of the loading surface as
can be seen in the corresponding frozen stress photograph of Fig. 7(b),
and the cone fracture is obtained. The measured compressive strength
for the specimen shown in Fig. 7(b) was more than double that for
the specimen shown in Fig. 7(a). In any case, the measured compressive strength is simply an indirect index, for a given boundary
condidtion, of the tensile or shear strength of the material and
cannot be used directly as a design parameter.

FORTIFICATION OF CERAMICS BY PROTECTIVE COATINGS

        Analytical and experimental studies conducted on and in con-
junction with the Benthos vehicle indicated that failure of the

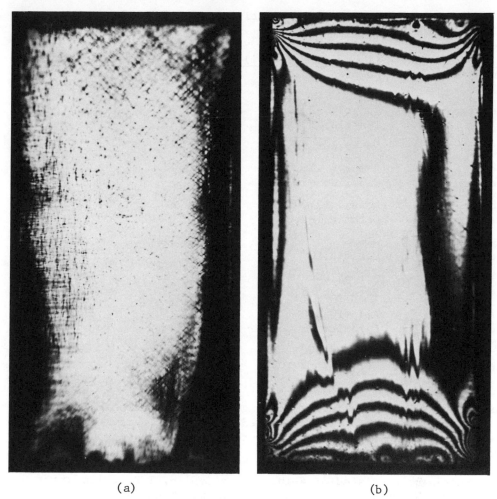

(a)   (b)

Fig. 7.   Axially loaded circular cylinders (a) negligible end
          friction with uniform stress field and (b) end friction
          with nonuniform stress field.

vehicle could be predicted in zones of high bending stress gradient
and that point loading of any portion of the structure would lead
to premature failure. Further, the occurrence of both modes of
failure could be minimized by careful design and fabrication of the
component. In order to introduce added design confidence, particularly with regard to point and impact loads, the possibility of
adding protective coatings to the base material was investigated.

     The coatings investigated were of three types; ion-exchange
induced compressive layers, compressive ceramic glazes and rubbery
or ductile coatings. Both the ion-exchange and compressive ceramic
glaze techniques resulted in external compressive layers which, it

Table 1. Evaluation of Protective Coatings in Impact.

Specimen	Surface (layer stress, psi)	Coating (in.)	Impact Energy (in. lbs)
Annealed glass	No preparation	none	155
Chemcor glass	Compr.: −30,000 Tensile: +30,000	0.005	310
Base ceramic	No preparation	none	116
Glazed ceramic	Compr.: −10,000 Tensile: +10,000	0.030	186
Rubber-coated ceramic	Lexan: +3,000 Ceramic: −3,000	0.250	310
Al-coated ceramic	Al: 0 Ceramic: 0	0.250	275
Al-coated ceramic	Al: +5,000 Ceramic: −5,000	0.250	275
Al-coated ceramic	Al: +10,000 Ceramic: −10,000	0.250	275
Al-coated ceramic	Al: +20,000 Ceramic: −20,000	0.250	275

was felt, would resist impact loading. The application of rubbery or ductile layers effectively covered the brittle ceramic with an energy absorbing layer which would prevent the peak impact stresses from reaching the specimen.

Specimens consisted of 4-in. O.D. x 3.5 in. I.D. cylinders which were impacted by a ballistic pendulum. Test results are summarized in Table 1. Increases in impact resistance exceeding 100% were found for both the Chemcor reinforced cylinders and those protected by either a rubbery or ductile layer. Compressive ceramic glazes resulted in a 60% increase. From a strength-to-weight basis, the Chemcor reinforced specimens offered maximum impact protection.

MODIFICATION OF BASE MATERIALS

Extensive investigations conducted to find an easily formable material exhibiting the inherently high strength-to-weight ratio of brittle materials led to a program to assess the influence of high-strength, high-modulus infiltrations on porous graphitic materials. Cylindrical specimens of two grades of porous graphite were infiltrated with a bronze and an aluminum alloy, the physical

Table 2. Mechanical Properties of Infiltrated Graphite.

Filter	Joint	Compr. Strength Avg. (psi)	Increase (X)	Flexural Strength Avg. (psi)	Increase (X)	Elastic Modulus Avg. (psi x 10^6)	Increase (X)	Poisson's Ratio	Density (lb/in^3)	Comp. Strength / Density (in. x 10^4)
						Graphite Grade 2020				
NONE	GREASE	10,000	--	4,750	--	1.27	--	0.23	0.0639	15.7
NONE	EPOXY	10,000	--	4,750	--	1.27	--	0.23	0.0639	15.7
AMP45	EPOXY	21,000	2.10	10,500	2.20	2.40	1.88	0.23	0.0906	26.5
AMP45	INTEGRAL	21,000	2.10	10,500	2.20	2.40	1.88	0.23	-.0906	26.5
2024AL	EPOXY	31,000	3.10	13,700	2.90	3.40	2.68	0.23	0.0794	39.0
						Graphite Grade HS82				
NONE	EPOXY	23,000	--	10,400	--	1.72	--	0.23	0.0597	38.6
AMP45	EPOXY	43,000	1.87	16,300	1.57	2.61	1.52	0.23	0.1050	43.0
2024AL	NON	47,000	2.04	19,500	1.88	4.10	2.38	0.23	0.0794	59.3

and mechanical properties of the resultant materials were determined, predictions of collapse pressure were made and the specimens were instrumented and tested to destruction. Principal strains were monitored during testing and the resultant stresses were used to verify the assumed infinite, thin-walled cylindrical solution and to determine the static elastic modulus.

Test results can be seen in Table 2. Results indicate that significant increases in both strength and elastic modulus are possible with minimal increase in density, resulting in high strength-to-weight ratios. Those ratios listed for the infiltrated graphites can be compared to corresponding values of 35.0×10^4 for 6061-T6 aluminum, 33.0×10^4 for HY 100 steel and 40×10^4 for titanium. The measured static modulus agrees well with that dynamically obtained,

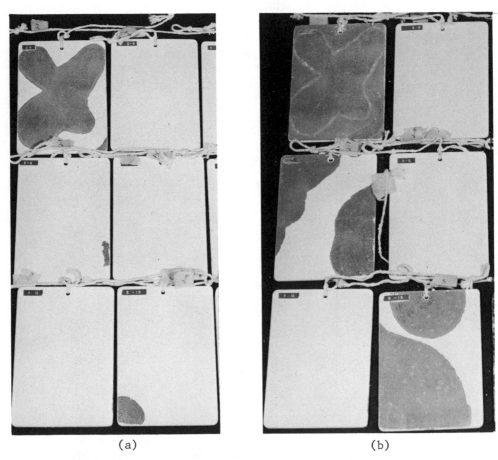

                    (a)   (b)

Fig. 8.   Corrosion of front of group 2 panels after immersion
          (a) 20 days, (b) 40 days.

and it was possible to predict collapse pressure within a reasonable percent of error by using the distortion energy theory. The ability to predict failure with a conventional failure theory is explained by a reduced brittle behavior as the ductile impregnation is added.

USE OF CERAMICS FOR CORROSION PROTECTION

Ceramics have been considered as a means of protecting currently used metallics from the corrosive deep-sea environment. Specimens consisting of 4 x 6 x 1/8 in. rectangular, 6061-T6 aluminum panels were coated on both sides with a porcelain enamel glaze. The plate edges were rounded to approximately a 1/16 in. radius to prevent chipping during the enameling cycle. After enameling, the plates were restored to a T-6 temper. Two types of enamel were tested in two thicknesses for a total of four evaluation groups. Each group consisted of 15 panels.

The panels were immersed in sea water and inspected after 20, 40, 60 and 80 days immersion. A typical progression of failure can be seen in Figs. 8(a) and 8(b). Figure 9 shows failure vs days immersion, a failure being considered as any spall area exceeding 1/8 in. in its maximum dimension. Results indicated that the technique did not offer sufficient protection in the sea water environment. Failure appeared to be due to a reduced coating

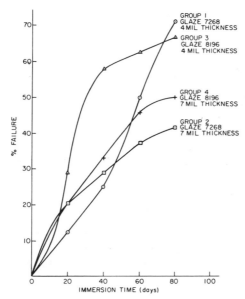

Fig. 9. Percent failure vs days immersion for test glazes of two thicknesses.

thickness at or near the 1/16 in. radius at the edge of the panel, and radii of this magnitude would certainly be encountered in under- water vehicles near joint rings, hand holes, etc. It is felt, how- ever, that prior conditioning of the metallic surface will increase the reliability of the technique, and additional research is expected in this area.

CONCLUSIONS

Ceramic and related brittle materials will find increased use in deep-submergence applications due to their high strength-to-weight ratio. Before their use becomes widespread, however, means will have to be found to design reliably with these materials. Research at ORL, both basic and applied, has been directed toward achieving this reliability in design. Some progress has been made, but much additional research must be conducted, especially in the areas of brittle material failure criteria, materials property evaluation and fortification of existing ceramics.

ACKNOWLEDGMENT

This work was supported by the Naval Ordnance Systems Command, U. S. Department of the Navy.

REFERENCES

1. J. C. Conway, "An Investigation of the Stress Distribution in a Circular Cylinder Under Static Compressive Load for Varying Boundary Conditions", M.S. Thesis, The Pennsylvania State University, 201 pp. (1963).
2. J. D. Stachiw, "Solid Glass and Ceramic External-Pressure Vessels", Tech. Rept. No. ORL-NOw 63-0209-C-2, January 1964.
3. J. D. Stachiw, "Glass and Ceramics for Underwater Structures", Ceram. Age, 80 [7] 20-23 (1964).
4. J. C. Conway. "Bethos-Deep Submergence Pyroceram Test Vehicle", ASME Publication 67-DE-5, May 1967.
5. J. C. Conway, "Structural Evaluation of a Ceramic Underwater Vehicle", Am. Ceram. Soc. Bull., 47 [11] (1968).
6. J. C. Conway, "A Study of the Factors Influencing Stress Distri- bution, Fracture Initiation and Fracture Propagation in Line- and Wedge Loaded Rectangular Plates", Ph.D. Thesis, The Pennsylvania State University, 174 pp (1968).

EFFECT OF TEMPERATURE ON ELECTRICAL CONDUCTIVITY AND

TRANSPORT MECHANISMS IN SAPPHIRE

W. J. Lackey*

North Carolina State University

Raleigh, North Carolina

ABSTRACT

The mechanisms of electrical conduction in single crystal alumina were investigated. An oxygen concentration cell was used to separate the direct current electrical conductivity into its ionic and electronic components over the temperature range 1000-1500°C. Transport number and electrical conductivity measurements were made in an oxidizing atmosphere while guarding against both surface and gas phase conduction.

The fraction of the electrical conductivity which was contributed by transport of ions decreased from approximately 0.6 at 1000°C to 0.01 to 0.02 at 1500°C. Comparison of the ionic component of conductivity with values calculated from aluminum and oxygen self-diffusion data led to the conclusion that both ion species contribute to conduction with the aluminum contribution probably being the larger of the two. At the higher temperatures conduction is predominantly electronic and probably p-type.

INTRODUCTION

Even though alumina is commonly used both electrically and mechanically, little is known concerning its electronic or diffusional material transport. As an example, it is not known with any reasonable degree of reliability whether the electrical conductivity of alumina

---

* Presently with Oak Ridge National Laboratory (Oak Ridge, Tennessee), operated by the Union Carbide Corporation for the U. S. Atomic Energy Commission.

is a result of electronic (electrons or electron holes) or ionic
transport. With this most fundamental aspect of the electrical
conductivity of alumina still unknown, it is obvious that the
mechanism of conduction is almost totally unknown. This lack of
basic information concerning transport phenomena results in an
inability to alter predictably and control properties such as
electrical resistivity and diffusional creep which are of extreme
importance in many demanding electrical and mechanical applications.

Not many years ago it was thought that charge transport in
nearly all ceramics resulted from the flow of electrons or electron
holes. However, during the past 10 years electrical conduction in
the numerous oxides listed in Table 1 has been shown to be either
completely or partially ionic in nature. Consequently, a careful
electrical conductivity study can, under favorable circumstances,
produce information on both electronic and ionic transport. The
approach used in the present investigation was to measure the conductivity of single crystal alumina and to determine experimentally,
by use of the oxygen concentration cell technique, the fraction
of the current that was carried by ions.

The general usefulness of electrochemical cells was first
realized by Kiukkola and Wagner[1] and the theoretical basis is well
established.[1-8] Briefly, the ionic transport number, t_i, is defined
by Eq. 1,

$$t_i \equiv \frac{\sigma_i}{\sigma} = \frac{\sigma_i}{\sigma_i + \sigma_e} ,$$
(1)

where σ_i = ionic component of conductivity, σ_e = electronic component
of conductivity, and σ = total conductivity. For an oxygen concentration cell the ionic transport number is given by the ratio of
the observed potential across the cell, V, to the Nernst potential,
E, which would be developed if the specimen were a completely ionic
conductor. That is,

$$t_i = V/E$$
(2)

The Nernst potential is obtained from Eq. 3.

$$E = \frac{RT}{nF} \ln(P_1/P_2),$$
(3)

Table 1. Crystalline Oxides Exhibiting Complete
or Partial Ionic Electrical Conductivity

Material[*]	Technique[†]	Material[*]	Technique[†]
Al_2O_3	A[2,9-12]	ThO_2	A[23,33]
Al_2O_3	B[13]	ThO_2	G[33]
$3Al_2O_3 \cdot 2SiO_2$	A[9]	$Th_{0.85}Ca_{0.15}O_{1.85}$	A,B,G[1]
$BaTiO_3$	A[14]	$Th_{0.85}Ca_{0.15}O_{1.85}$	A[34]
BeO	A[15,16]	ThO_2-CaO	A[21]
BeO	C[17,18]	$Th_{0.9}La_{0.1}O_{1.95}$	A[34]
CaO	A[15]	$ThO_2-La_2O_3$	A,B,G[1]
CeO_2	D[19]	$ThO_2-La_2O_3$	A[23,35]
$CeO_2+<1$ a/o CaO	E[20]	$Th_{0.87}Y_{0.13}O_{1.935}$	A,B,G[36]
$CeO_2-La_2O_3$	D[19]	$ThO_2-Y_2O_3$	A,G[23,33]
CeO_2-MgO	A[21]	$ThO_2-Y_2O_3$	A[34,37]
$CeO_2-Nd_2O_3$	D[19]	$ThO_2-Y_2O_3$	G[38]
$CeO_2-Y_2O_3$	D[19]	Y_2O_3	A[22]
Dy_2O_3	A[22]	Yb_2O_3	A[9]
Gd_2O_3	A[22]	ZrO_2	B,G[39]
HfO_2	A[23]	$Zr_{0.85}Ca_{0.15}O_{1.85}$	A,B,G[1]
HfO_2+4w/o CaO	F[24,25]	$Zr_{0.85}Ca_{0.15}O_{1.85}$	B,G[40]
MgO	A[3,9,10,15,26-28]	$Zr_{0.85}Ca_{0.15}O_{1.85}$	C[41]
$Na_2O \cdot 11Al_2O_3$	A[29-31]	$Zr_{0.86}Ca_{0.14}O_{1.86}$	C[42]
Nd_2O_3	A[9]	$Zr_{0.85}Ca_{0.15}O_{1.85}$	A[34]
$PbZrO_3-PbTiO_3$ $+1$w/o Nb_2O_3	A[32]	$ZrO_2+(CaO)$	A[9,43]
Sc_2O_3	A[9]	ZrO_2-CaO	A[21,23]
Sm_2O_3	A[9,22]	$ZrO_2-Y_2O_3$	A[35]
SrO	A[15]	ZrO_2-CeO_2-CaO	A[21]

*
 - Dashes indicate several compositions in the system were investi-
   gated.
†A - Electrochemical cell
 B - Polarization
 C - Nernst-Einstein relation
 D - Volumetric analysis of electrolytic oxygen
 E - Mass spectrometric analysis of electrolytic oxygen
   - Comparison of conductivity and permeation data
 G - Conductivity as a function of oxygen pressure

where n = number of moles of unit electrical charge transported per
mole of oxygen transported or consumed, = 4 for normally charged ions
and vacancies; P_1, P_2 = oxygen partial pressures.

    An assumption embodied in any derivation of Eq. 2 is that the
rate-controlling process of the cell reaction must be transport

through the electrolyte specimens and not exchange of oxygen between
the atmosphere and the specimen surface.[44] Kröger[2] has pointed out
that one also assumes the ions and vacancies migrating in the
electrolyte have the normal charge.

The relative merits of experimental techniques for measuring
electrical conductivity[45-47] and for determining the mechanisms of
conduction[40,48] have previously been reviewed. Also, several reviews
of electrical transport in alumina are available,[48-50] but nearly
all of the previous work was done without guarding against gas phase
conduction and in some instances without guarding against surface
conduction. It has been shown by Loup and Anthony[51,52] and others[53,54]
for conductivity studies and by Mitoff[28] for electrochemical determinations of transport numbers that unguarded measurements are of little
value when good insulators like alumina are being investigated.
Consequently, in this investigation, careful attention was given to
guarding against both gas phase and surface conduction.

EXPERIMENTAL PROCEDURE

Materials

The two sapphire disks used for the conductivity and concentration cell experiments and a specimen used for chemical analysis were
cut consecutively from a Verneuil boule. Specimen purity and
crystallographic orientation have been reported elsewhere.[48] Briefly,
the total impurity content was about 200 ppm and the longitudinal
axis of the specimens made an angle of $63 \pm 2°$ with the c axis.
Calcia-stabilized zirconia, $Zr_{0.85}Ca_{0.15}O_{1.85}$, used to calibrate the
experimental apparatus was purchased from the Zirconium Corporation
of America, Solon, Ohio.

Specimen Preparation

Specimens were surface ground flat and parallel using a 100-grit,
water-cooled diamond wheel. After thorough cleaning,[48] electrodes
were formed on the disks by air curing at 920°C type 6082 Pt paste
purchased from Hanovia Liquid Gold Division of Engelhard Industries,
Inc., East Newark, New Jersey.

Transport Number Measurements

In this investigation it was possible to guard against surface
and gas phase conduction when making the transport number and conductivity measurements. Diagrams of the apparatus are shown in
Figs. 1 and 2. The work of Mackenzie[55] was of considerable assistance
in designing the mechanical part of the cell and the electrical
circuit was patterned after the work of Mitoff.[28] A Pt wire spot

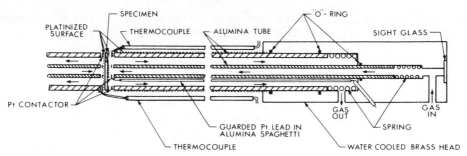

Fig. 1.   Diagram of oxygen concentration cell.

welded to the contactor of the guarded electrode served as the lead
wire which had high resistance to ground. This lead was shielded
over its entire length. As shown in Fig. 1, the outer surface of
one of the alumina tubes was platinized and was an integral part of
the guard circuit.

Oxygen and an argon-oxygen mixture were used to establish the
oxygen potential on either side of the specimen. The purity of
the argon was 99.995% and that of the oxygen was 99.998% if inert gases
and nitrogen are excluded   The total hydrocarbon content of the
oxygen was 11 molar ppm. Calibrated flowmeters were used to deliver
100% O_2 to one side of the cell and Ar-21 v/o O_2 to the other side
of the cell.

Referring to Fig. 2, a measurement was made by first having
the switch in position 1 with the electrometer functioning as a null

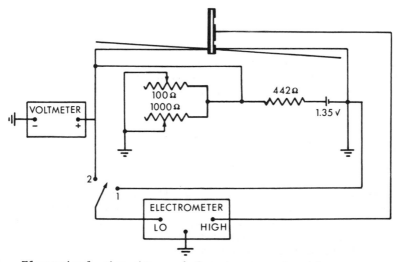

Fig. 2.   Electrical circuit used for transport number measurements.

detector. The variable resistors were adjusted until the electrometer showed the guard ring and the guarded electrode to be at the same potential. The voltage developed by the concentration cell was then measured with the voltmeter and after changing the switch to position 2 the cell voltage was measured with a calibrated electrometer which had an input impedance in excess of 10^{14} ohms. The values observed for the cell voltage with the two meters were always within 2% of each other except at voltages below about 10 mv, where the voltmeter becomes inaccurate. In all cases, the value reported was that read using the electrometer. This procedure was repeated until the cell voltage no longer varied with time.

To minimize thermoelectric potentials induced by a temperature gradient the specimen was located at the thermal center of the furnace. The temperatures read from the two thermocouples were usually in agreement to within 0.2°C which was about the sensitivity of the equipment. In no case did the thermocouples disagree by more than 1°C.

As many investigators have reported previously, the observed cell voltage increased with flow rate for low flow rates. By trial and error it was found that the cell voltage was not so influenced when the rate on each side of the cell was as large as 150 cm^3/min. This flow rate was used for all measurements.

Conductivity Measurements

After completing the transport number measurements for a given specimen, the direct-current conductivity was determined prior to removing the specimen from the oxygen concentration cell. Each of the alumina tubes shown in Fig. 1 were left in place so that the conductivity measurements were guarded against gas as well as surface conduction. The measurements were made in flowing oxygen at atmospheric pressure using a type 1644A megohm bridge manufactured by General Radio Company, West Concord, Massachusetts. A test voltage of 10 v was supplied by the bridge.

RESULTS AND DISCUSSION

Transport Numbers

The oxygen concentration cell was found to perform properly by comparison of the observed cell voltage with that expected for the completely ionic conductor, $Zr_{0.85}Ca_{0.15}O_{1.85}$. Observed voltages were within 1% of values calculated from Eq. 3. Discrepancies of this magnitude could have been caused by the 1% uncertainty in the calibration of the flowmeters.

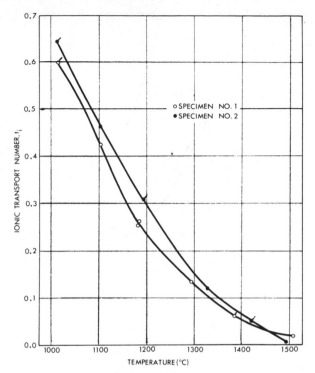

Fig. 3. Ionic transport number (t_i) of sapphire vs temperature. Cell: Pt, $O_2(0.21$ atm$)|Al_2O_3.|$Pt, $O_2(1.00$ atm$)$. The data points with the slanting dash represent experiments where the higher oxygen pressure was on the guarded side of the specimen.

The experimentally determined ionic transport numbers for the two sapphire specimens are plotted as a function of temperature in Fig. 3. The ionic transport number decreased with increasing temperature from values of about 0.6 at 1000°C to 0.01-0.02 at 1500°C.

The results of this investigation are compared with those of previous investigators of single crystal alumina in Fig. 4. Results for polycrystalline alumina[9,10,12] are not shown. With the exception of Mills,[13] who used the polarization technique, all of the data were obtained using an oxygen concentration cell. Further, in each of the concentration cell studies the atmosphere on either side of the cell was oxidizing. Although the data of Mitoff[11] were obtained with a guarded cell, these results appear questionable since they alone show an increase in ionic transport number with increasing temperature. It is speculated that at least some of the discrepancy between the results of the present investigation and those of Matsumura[28] is the result of his not using a guard ring.[56] Based on

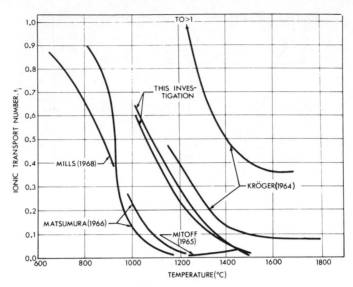

Fig. 4. Ionic transport numbers (t_i) by various investigators.

the work of Mitoff[28] with MgO, one would expect the true value of
the ionic transport number to be larger than that obtained by an
unguarded measurement. It remains, however, that although Kröger[2]
did not use a guard ring,[57] he observed larger values than in the
present investigation. Consequently, some other factor may be
equally as important in influencing the value observed for the transport number.

Since the flux-grown crystals of Kröger and the Czochralski-grown
crystal of Mills probably had the lowest density of dislocations,
it does not appear that the curves in Fig. 4 are in the order of
dislocation density. Neither Kröger, Mitoff, nor Mills
reported the impurity contents of their crystals, making an analysis
on this basis impossible.

Electrical Conductivity

The measured conductivity for the two sapphire specimens is
plotted against the reciprocal of temperature in Fig. 5. These data
are thought to represent the combined ionic and electronic conductivity.
The lines in Fig. 5 were obtained by a least-squares statistical
analysis. For a given specimen, the scatter was larger than expected.
However, the agreement from one specimen to the other is considered
good when compared to that usually observed for duplicate specimens.

The results of Moulson and Popper[54] are included in Fig. 5.
There are no other data which have been collected using the fully
guarded technique for the temperature range of interest. Unfortunately,

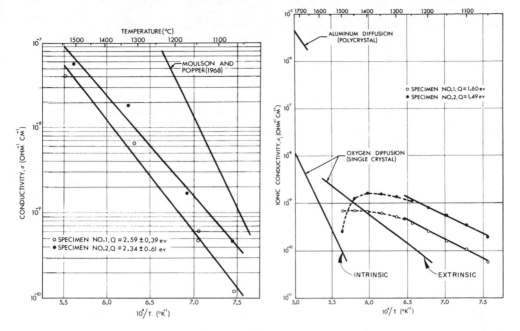

Fig. 5. Electrical conductivity (σ)
        of sapphire vs reciprocal
        temperature.

Fig. 6. Ionic conductivity (σ_i)
        of sapphire vs recipro-
        cal temperature.

Moulson and Popper did not report the purity, orientation, nor growth
technique for the crystal they investigated. They observed greater
conductivity and activation energy than in the present investigation.
Any attempt to explain the discrepancy would be highly speculative.

Ionic Conductivity

The transport number data obtained in this investigation were
used to separate the observed electrical conductivity into its
ionic and electronic components. The calculation was made at 50°C
intervals over the range 1050 to 1500°C by simply multiplying the
conductivity predicted by the least-squares equation times the transport number determined from Fig. 3.

The ionic component of the conductivity is plotted logarithmically
against reciprocal temperature in Fig. 6. At the lower temperatures,
the expected linear relationship was observed. However, at approximately 1250°C the observed ionic conductivity did not increase as
rapidly with increasing temperature as expected and actually decreased
with increasing temperature between 1400 and 1500°C. Matsumura[12]
observed the same effect. No reliable explanation of this departure
from linearity is apparent. The possibility exists that polarization

occurred at the higher temperatures, resulting in a reduction of
the observed ionic conductivity. This, however, is purely speculative.

For comparison with the observed ionic conductivity, values
calculated from the diffusion data of Oishi and Kingery[58] and
Paladino and Kingery[59] are included in Fig. 6. Since aluminum self-
diffusion data for sapphire are not available, aluminum diffusion
in polycrystalline alumina was included. One would expect aluminum
diffusion in sapphire to be equal to or less than that for polycrystalline alumina.

The activation energy observed for ionic conduction differs
significantly from that of either intrinsic aluminum or intrinsic
oxygen diffusion, as can be seen in Fig. 6. Thus, it is suggested
that the ionic conduction is the result of extrinsic aluminum or
oxygen transport. Since aluminum diffusion data are not available
in the temperature range of interest, it is not possible to establish
with certainty whether aluminum or oxygen transport is responsible
for the ionic conductivity. From Fig. 6 it appears that oxygen
transport contributes an appreciable fraction to the ionic conductivity. Assuming that aluminum diffusion is greater than that of
oxygen leads to the conclusion that both types of ions contribute
appreciably to conduction. The basis for assuming that aluminum
diffusion exceeds that of oxygen in single crystal alumina is the
fact that most thermoelectric power experiments[12,50,60] indicate that
the sign of the current carrier is negative for the low temperature
region. Since the cell measurements show the conductivity to be
predominantly ionic at low temperatures, it is concluded that the
negative sign indicates either aluminum vacancy or oxygen interstitial
diffusion. Diffusion of aluminum vacancies is the logical choice.
In addition, cation diffusion has been found to exceed oxygen
diffusion[61] in both Cr_2O_3 and Fe_2O_3, which have the same crystal
structure as alumina.

Electronic Conductivity

The electronic component of the conductivity is plotted as a
function of reciprocal temperature in Fig. 7. In the temperature
range of interest, no reliable data are available for comparison.
For experiments conducted below 1000°C, Mills[13] observed activation
energies in the range of 1.5 to 3.5 ev for specimens previously
annealed in oxygen.

The activation energies observed for the electronic conduction
appear slightly too small to correspond to half of the width of the
forbidden gap.[62] It is thus reasonable to conclude that the conduction
is extrinsic. Since the oxygen concentration cell measurements
reported here and by other investigators show the conductivity to be
predominantly electronic in nature at the higher temperatures and

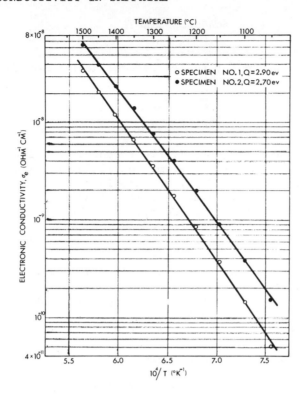

Fig. 7.   Electronic conductivity (σ_e) of sapphire vs reciprocal
          temperature.

since thermoelectric power experiments[1,2,50,60] indicate that the
current carrier is positive at the higher temperatures, it is
logical to conclude that the electronic conductivity is p-type.
Matsumura[12] reached the same conclusion. Based on a simple band
model the average activation energy observed (2.80 ev) indicates
that the acceptor level is 5.6 ev above the valence band.

CONCLUSIONS

Carefully guarded oxygen concentration cell measurements conducted in oxidizing atmospheres showed that electrical conduction in
single crystal alumina is the result of both ionic and electronic
transport. The ionic transport number decreases with increasing
temperature from approximately 0.6 at 1000°C to 0.01-0.02 at 1500°C.
It appears that both aluminum and oxygen ion-transport contribute
to the ionic component. The electronic component is probably p-type.

ACKNOWLEDGMENTS

The author wishes to thank G. S. Sheffield and Drs. R. F. Stoops and J. V. Hamme for their frequent assistance during this investigation.

REFERENCES

1. K. Kiukkola and C. Wagner, J. Electrochem. Soc. 104, 379–387 (1957).
2. F. A. Kröger, Proc. Brit. Ceram. Soc. 1 167–171 (1964).
3. S. P. Mitoff, J. Chem. Phys. 36 1383–1389 (1962).
4. C. Wagner, Z. Physik. Chem. 21 25–41 (1933).
5. C. Wagner; in International Committee of Electrochemical Thermodynamics and Kinetics, Proceedings of the Seventh Meeting.
   Butterworths Scientific Publications, London (1955).
6. C. Wagner, Z. Elektrochem. 60 4–7 (1956).
7. S. V. Karpachev and S. F. Pal'guev, Trans. Inst. Electrochem.
   1 63–72 (1961).
8. F. A. Kröger, The Chemistry of Imperfect Crystals, John F. Wiley
   and Sons, New York, 1964.
9. H. Schmalzried, Z. Physik. Chem. 38 87–102 (1963).
10. V. P. Luzgin *et al*.,Ogneupory 30, 42–44 (1965). (English transl.
    NASA TTF-10,093) National Aeronautics and Space Adminsitration,
    Washington, D. C.
11. S. P. Mitoff, Ionic Conductivity of Single Crsytal Al_2O_3,
    General Electric Research Laboratory Report 65-RL-(3964M),
    Schenectady, N. Y. (1965).
12. T. Matsumura, Can. J. Phys. 44 1685–1698 (1968).
13. J. J. Mills, Research on Electrical Conductivity and Conduction
    Mechanisms in Alumina, Technical Documentary Report ARL 68-0154,
    Wright-Patterson Air Force Base, Ohio (1968).
14. D. C. Glower and R. C. Heckman, J. Chem. Phys. 41 877–879 (1964).
15. S. F. Pal'guev and A. D. Neuimin, Soviet Phys.-Solid State 4
    629–632 (1962).
16. C. F. Cline, J. Carlberg, and H. W. Newkirk, J. Amer. Ceram.
    Soc. 50 55–56 (1967).
17. H. J. aeBruin, G. M. Watson, and C. M. Blood, J. Appl. Phys. 37
    4543–4549 (1966).
18. C. F. Cline *et al*., in Mass Transport in Oxides. National Bureau
    of Standards Special Publication 296, Superintendent of
    Documents, U. S. Government Printing Office, Washington, D. C.,
    1968.
19. A. D. Neuimin and S. F. Pal'guev, Dokl. Akad. Nauk. Tadzh.
    SSR 143 1388–1391 (1962). [Proc. Acad. Sci. Chem. Sect.,
    English transl. 143 315–318 (1962).]
20. E. L. Holverson and C. J. Kevane, J. Chem. Phys. 44 3692–3696
    (1966).

21. S. F. Pal'guev and A. D. Neuimin, Trans. Inst. Electrochem. 1
    90-96 (1961).
22. V. B. Tare and H. Schmalzried, Z. Physik. Chem. 43 30-32 (1964).
23. C. H. Steele and C. B. Alcock, Trans. Met. Soc. AIME 233 1359-1367
    (1965).
24. H. A. Johansen and J. G. Cleary, J. Electrochem. Soc. 111
    100-103 (1964).
25. A. W. Smith, F. W. Meszaros, and C. D. Amata, J. Am. Ceram. Soc.
    49 240-244 (1966).
26. H. Schmalzried, J. Chem. Phys. 33 940 (1960).
27. S. P. Mitoff, J. Chem. Phys. 33 941 (1960).
28. S. P. Mitoff, J. Chem. Phys. 41 2561-2562 (1964).
29. Ford Motor Company, Science 154 (3751) 828 (1966).
30. Ford Motor Company, Mater. Des. Eng. 64 (5) 19-20 (1966).
31. Ford Motor Company, Mater. Des. Eng. 65 (3) 14-16 (1967).
32. A. Ezis, J. G. Burt, and R. A. Krakowski, J. Am. Ceram. Soc.
    53 521-524 (1970).
33. M. F. Lasker and R. A. Rapp, (in English), Z. Physik. Chem. 49
    198-221 (1966).
34. C. B. Alcock and B.C.H. Steele; in Science of Ceramics, Vol. 2.
    Ed. by G. H. Steward. Academic Press, New York, 1965.
35. H. Peters and H. H. Möbius, Z. Physik. Chem. 209 298-309 (1958).
36. J. M. Wimmer, L. R. Bidwell, and N. M. Tallan, J. Am. Ceram.
    Soc. 50 198-201 (1967).
37. E. C. Subbarao, P. H. Sutter, and J. Hrizo, J. Am. Ceram. Soc.
    48 443-446 (1965).
38. J. E. Bauerle, J. Chem. Phys. 45 4162-4166 (1966).
39. R. W. Vest and N. M. Tallan, J. Am. Ceram. Soc. 48 472-475 (1965).
40. N. M. Tallan, R. W. Vest, and H. C. Graham; in Materials Science
    Research, Vol. 2. Ed. by H. M. Otte and S. R. Locke. Plenum
    Press, New York, 1965.
41. W. D. Kingery, J. Pappis, M. E. Doty, and D. C. Hill, J. Am.
    Ceram. Soc. 42 393-398 (1959).
42. L. A. Simpson and R. E. Carter, J. Am. Ceram. Soc. 49 139-144 (1966).
43. H. Schmalzried, Z. Elektrochem. 66 572-576 (1962).
44. L. Heyne; in Mass Transport in Oxides. Ed. by J. B. Wachtman and
    A. D. Franklin, National Bureau of Standards Special Publication
    296, Superintendent of Documents, U. S. Government Printing
    Office, Washington, D. C., 1968.
45. A. N. Gerritsen; in Handbuch der Physik., Vol. 19. Ed. by S.
    Flügge, Springer-Verlag, Berlin, 1956.
46. W. C. Dunlap, Jr.; in Methods of Experimental Physics, Vol. 6.
    Ed. by K. Lark-Horovitz and V. A. Johnson. Academic Press,
    New York, 1959.
47. J. L. Pentecost; in High Temperature Technology, Proceedings of
    an International Symposium on High Temperature Technology,
    Stanford Research Institute, Butterworth, Inc., Washington, D.C.
    1964.

48. W. J. Lackey, Electronic and Ionic Conductivity in Alumina,
        Ph.D. Thesis, North Carolina State University, Raleigh, N.C.
        University Microfilms, Ann Arbor, Michigan, 1970.
49. J. Cohen, Am. Ceram. Soc. Bull. 38 441-446 (1959).
50. J. A. Champion, Proc. Brit. Ceram. Soc. 10 51-62 (1968).
51. J. P. Loup and A. D. Anthony, Rev. Hautes Temp. Refractaires 1
        15-20 (1964).
52. J. P. Loup and A. D. Anthony, Ibid. 193-199 (1964).
53. D. W. Peters, L. Feinstein, and C. Peltzer, J. Chem. Phys. 42
        2345-2346 (1965).
54. A. J. Moulson and P. Popper, Proc. Brit. Ceram. Soc. 10 41-50
        (1968).
55. J. D. Mackenzie, J. Am. Ceram. Soc. 47 211-214 (1964).
56. T. Matsumura, personal communication, 1967.
57. F. A. Kröger, personal communication, 1969.
58. Y. Oishi and W. D. Kingery, J. Chem. Phys. 33 480-486 (1960).
59. A. E. Paladina and W. D. Kingery, J. Chem. Phys. 37 957-962 (1962).
60. S. Dasgupta and J. Hart, J. Appl. Phys. 16 725-726 (1965).
61. C. E. Birchenall, in Mass Transport in Oxides. Ed. by J. B. Wacht-
        man and A. D. Franklin, National Bureau of Standards Special
        Publication 296, Superintendent of Documents, U. S. Government
        Printing Office, Washington, D. C. 1968.
62. P. J. Harrop, J. Appl. Phys. 16, 729-730 (1965).

DISCUSSION

R. H. Doremus (General Electric Co.): The Nernst relation as
written (Eq. 3) assumes very special electrode reactions. Different
electrode reactions would give various potentials, perhaps resulting
from impurities such as water in the gas or ions in the alumina.
This is the most probable reason for different results by different
investigators. Also the ionic conductivity could be influenced by
impurities, such as aklalis, in the alumina.
Author: Even though most water was removed from the gases by passing
them through P_2O_5, one cannot definitely exclude the possibility that
side reactions might occur as suggested earlier by Davies [M. O. Davies,
J. Chem. Phys. 38 2047-55 (1963)]. However, since Schmalzried's
(ref. 9) results were independent of gas type for a given oxygen
potential, the best evidence is that such side reactions do not occur.
In the current investigation there was no evidence for the transport
of impurity ions or abnormally charged aluminum or oxygen ions
although one cannot exclude this possibility.

H. Palmour III (N. C. State University): Could the gas content
[e.g., H_2O] of these flame grown crystals be causing activated
complex(es) which thermally decompose, thus altering the ionic component of transport at the higher temperature?
Author: I do not believe that the thermal behavior of impurity-defect
complexes would lead to a reduction of ionic transport with increasing
temperature. It is possible, however, that impurity ions present
could be sufficiently mobile to contribute to ionic transport.

ELECTRODE AND INSULATION MATERIALS IN MAGNETOHYDRODYNAMIC GENERATORS

Larry L. Fehrenbacher and Norman M. Tallan

Aerospace Research Laboratories

Wright-Patterson Air Force Base, Ohio

ABSTRACT

The operation and service environment of open and closed cycle MHD generators is described. The performance characteristics of candidate electrode and insulation materials and their peculiar limitations with respect to conductivity, mechanical and electro-chemical erosion, thermal shock resistance, maximum allowable surface temperature, compositional stability (polarization and contamination), thermionic emission, and overall service life are discussed. Special emphasis is given to problems associated with the performance of various ceramic insulators and electrodes used in open cycle generators. Recommendations for research studies on specific ceramic compositions and systems showing potential as MHD generator channel materials are offered.

INTRODUCTION

The dramatic increase in requirements for electrical power in the United States and other countries has stimulated considerable research in the development of new systems to satisfy these demands. With predictions indicating that annual peak loads will double every 10 years in the U.S.,[1] many new technological approaches and concepts for necessary improvement in the generation, storage and transmission of electrical power are under consideration. Of prime interest is the increase in the efficiency of converting the primary energy source (usually heat) to electricity. Since conventional steam turbo-machinery using combustion of fossil fuels have reached an upper limit of ~40% efficiency, other energy conversion schemes have become the subject of many and varied research efforts.[2]

One such approach, magnetohydrodynamics (MHD), appears very attractive since it is capable of extracting electricity directly from the higher temperature region (1600°C - 2500°C) of the fossil fuel combustion. The MHD process is specially suited for increasing the operating temperature range of thermal power plants and offers the potential for substantial improvement in efficiency to a 55 - 60% level.[3] The efficiency of MHD energy conversion is strongly dependent on temperature and its thermodynamic potential is only fully realized at maximum temperatures. Since heat source temperatures may be as high as 3500°C, the materials used, in a MHD generator become one of the difficult and demanding engineering requirements. Before presenting a review of materials used a brief summary of MHD principles, generator developments, and current status of MHD for practical power generation will be given to serve as background information for the materials engineer.

Status of MHD Power Generation

The principle of MHD generation is the same as that for a conventional rotating generator; that is, an electrical voltage is produced when a moving conductor intersects a magnetic force field (Faraday effect). In contrast to the solid conductor generator, MHD uses a fluid (normally a gas) conductor that passes through a magnetic field to create electricity. The nature of the moving conductor is the most basic difference between the two types of generators. Even with the seeding of the fluid with a salt of a low ionization potential such as K_2SO_4 or K_2CO_3, a typical high temperature gas contains only 10^{12}-10^{13} free electrons per cm^3 in contrast to 10^{22}-10^{23} for a metallic conductor.

In contrast to common electrical generators, the combination of low electron density, high mobility of charged particles and high magnetic field give rise to the Hall effect. This phenomenon and its role in determining generator operating conditions can be best explained by the following set of simple Ohm's law equations. Assuming uniform flow in a two dimensional channel and ignoring ion-slip and pressure-diffusion effects,[4] the component current flows can be represented by

$$j_y = \frac{\sigma}{1 + (\omega\tau)^2} (E_y - UB + \omega\tau E_x) \tag{1}$$

$$j_x = \frac{\sigma}{1 + (\omega\tau)^2} [E_x - \omega\tau(E_y - UB)] \tag{2}$$

where j_y = Faraday current density (normal to gas flow direction),
σ = gas conductivity, j_x = axial current density (Hall current),
E_y = Faraday voltage, E_x = axial voltage (Hall voltage), U = gas
velocity, ω = cyclotron frequency = eB/m, B = magnetic field,
$\tau = 1/\omega_c$ (ω_c = collision frequency), and m = mass of electron.

    The axial current j_x shown in Eq. 2 is a consequence of the
Hall effect. Since the concentration of electrons is low, the
frequency of collision of electrons with the neutral and positive
ions may be reduced to the extent (large $\omega\tau$) that the net current
vector is deflected away from the normal (y) Faraday direction. In
other words, there is an electric field and current (Hall component)
in the direction of the gas flow. The angular distance between the
electric field vector E and current density vector j is known as
the Hall angle θ. The ω (cyclotron frequency) of an electron in
the magnetic field is influenced directly by the field strength and
the collision frequency by the number of gaseous particles (gas
pressure). Thus, the Hall angle and associated voltages and currents
can of course by varied by adjusting gas conditions and magnetic
fields. This versatility has spawned the development of various
MHD channel configurations to optimize performance at different Hall
angle conditions. The generator chamber has the requirements of
containing the moving fluid and transferring the voltages and
currents generated in the fluid to the load resistances. Current
carrying electrodes are usually rectangular or cylindrically shaped
with the walls consisting of alternating layers of electrodes and
insulators.

    The two most basic types of configurations are the rectangular
shaped Faraday and Hall generators which are illustrated in Figs. 1
and 2.[5] (Circular chambers are most suitable for supersonic flows).
In the Faraday generator the Hall angle should be minimized while the
Hall device naturally favors large Hall angles. Other basic designs

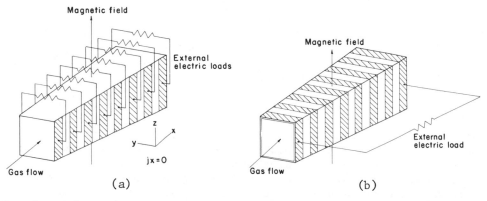

(a)   (b)

Fig. 1.   Schematics of generator channels (a) segmented Faraday and
          (b) rectangular Hall.

which are under study and appear promising are the diagonal conducting
wall (DCW) for fossil fuel energy conversion[6] and the disc generators
for large Hall angle plasma conditions.[7] As is readily evident
from Figs. 1 and 2, a MHD generator consists of simple, static
construction and is free of highly stressed, moving parts. Other
inherent advantages of MHD are: (1) the conversion process depends
on a volume force of magnetic origin as opposed to the surface
effect control of turbo-machinery advancing its potential for large,
single units capable of high power levels; (2) higher conversion
efficiency reduces the thermal and air pollution for the same power
outputs and (3) the capability of direct current power at any
voltage and current plus a rapid stop-start operation.[5] Although
several laboratory scale MHD devices and schemes have been investigated, MHD power generation systems can be divided into two primary
categories, open-cycle and closed-cycle.

Open-cycle systems involve alkali salt seeded combustion gases
of coal, oil, or natural gas which act as the working fluid in the
generator with the spent gases being exhausted to the atmosphere.[8]
The closed cycle systems incorporate a heat exchanger driven by a
combustion process which, in turn, heats a working fluid before it
passes through the generator.[5] The conducting fluid may either be
an inert gas like Ar or He seeded with Cs (plasma closed-cycle MHD)
or metal in a liquid or vapor form (liquid metal MHD) which is
returned to the heat exchanger by a compressor or pump. Closed
cycle research has been oriented towards coupling a MHD generator
with high temperature nuclear reactors. Upper temperature capability
of present reactors (1500°K gas cooled reactor) is unfortunately,
too low to provide 2000-2200°K temperatures desired for the generators.[9]
The actual exploitation of closed cycle reactor-generator systems
will most likely be realized in short term, high megawatt power
applications for military or space use.[10] Such systems are not
beyond the feasibility stage and will require extensive R and D.

The technical feasibility of open-cycle MHD has been readily
demonstrated and the primary emphasis today is on MHD-topped steam
plants. This development is highlighted by a 75 Mw steam topping
plant presently under   construction in Moscow.[11] A simplified
block diagram of a MHD-topped steam power plant is shown in Fig. 2.
Combustion products of coal and preheated air doped with 1%K-salt
are passed to the generator channel with an inlet temperature of
2400°C where the electrical energy is extracted. This d.c. power
can be converted to a.c. by solid state inverters for local power
distribution. The conversion process is completely terminated at
a low temperature limit of 1600°C and the hot gases are slowed down
by a diffuser and a portion of the remaining heat is then extracted
by the regenerative air heater for feedback to the coal combustor.
The remaining heat is used for the normal steam turbine cycle.
Recovery of K salt, (sulfur) ash, and nitrogen as nitric acid is

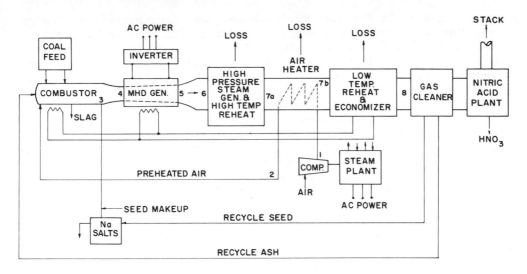

Fig. 2. Schematic of MHD topped-steam power plant.

accomplished before the final gases are exhausted out the stack.
These final recovery steps improve the economy of the process and
markedly reduce air pollution.

Since the electrode and insulator materials of the MHD channel
are exposed to high temperature, high velocity erosive gases of
the conducting plasma, these materials have been the limiting
factors in the achievement of long time reliable MHD power generation.
This paper will not treat the details of complex inter-relationship
between plasma properties and electrodes but will discuss the general
performance requirements of channel electrode and insulation
materials, and a few promising areas for materials research with
special emphasis on the open-cycle environment.

MHD CHANNEL MATERIALS

Electrodes

An electrode for long term open-cycle power generation should
be capable of efficiently transferring current densities of
1-5 amps/cm^2 for a minimum lifetime of 10,000 hrs.[12] The technical
performance and property criteria required of electrode materials
in order to satisfy the objectives of MHD power generation are:
(1) electrical conductivities greater than 1 mho/cm at the
electrode-lead contact; (2) resistance to mechanical and electrochemical erosion by hot contaminated gases and electrical discharge
arcs; (3) good thermionic (electron) emission; (4) thermal shock
resistance and (5) chemical phase stability (low vaporization rate,
chemical reaction and oxidation resistance). No electrode is

presently available for high performance, year-long operation. Design
innovations and special operating procedures will have to be implemented to realize this goal at least in the near future. However,
knowledge of the limitation of present electrode material candidates
helps to identify possible compositional and/or design improvements.

The study and evaluation of electrodes can be simplified by
treating separately the three major classes of materials; metals,
intermetallics and ceramics. In general, the cold surfaces of
water cooled metal electrodes promote high thermal and electrical
losses as well as alkali seed condensation which causes erratic
and degenerative electrical performance.[13] The hot face temperatures
of Cu electrodes must not exceed 500°C. This limitation represents
a loss of 1½% thermal efficiency in comparison to 1800°K electrode
face temperatures. At these low temperatures, the electron emission
of the electrodes is low and non-uniform, producing high voltage
drops which, in turn, creates severe discharge and arcing problems.[14]
This arcing phenomenon coupled with seed condensation (occurring at
1000°C and below) leads to severe pitting and electrochemical
erosion of the Cu and subsequent power losses. Refractory metals,
e.g., W, capable of operating at 1500°K temperatures must be protected from oxidation by noble metal (e.g., Pt, Pt-Rh and Ir)
coatings which are not economical for large scale generators.
Refractory metals (Hf, W, Nb, Mo) and their alloys even experienced
erosion and oxidation in Ar closed cycle plasmas between 1000°C
and 2000°C. Mo based alloys with additions of Hf, Zr, Ti, and Nb
performed well as a result of the formation of stable complex oxides
on the exposed surfaces.[16]

Some intermetallic compounds possess excellent high temperature
properties but in general are not satisfactory for the oxidizing
environment of open-cycle systems. However, ZrC and Nb C (m.p.
~3500°C) appear to be excellent candidates for the seeded Ar plasma
closed-cycle atmosphere.[15] They exhibited very low recession and
vaporization rates, outstanding thermal shock resistance, suitable
electrical properties and good resistance to K-attack. SiC has also
shown excellent oxidation, erosion and thermal shock resistance up
to 1900°K (SiO_2 coating limit). Its poor thermionic emission was
improved by doping with Mo, Ti, and Cr.[15] Oxidation of boride
materials have also been studied in combustion gas flows. Only a
mixture of ZrB_2 and yttrium boride gave interesting results due to
the formation of a stable cubic ZrO_2 coating.[17]

Since electrode and generator efficiency is substantially
improved the higher the temperature, the search for electronically
conducting ceramic electrodes has been extensive. Although many
ceramics inherently offer oxidation and chemical erosion resistance,
they are quite sensitive to thermal shock fracture and electrolysis
of material due to ionic conduction at high temperatures and O_2

pressures. Special design arrangements and engineering innovations
must be implemented to overcome these limitations for 2000°C open-
cycle temperatures. An excellent discussion of various design
approaches for utilizing ceramic electrodes and insulators in MHD
channels is given by Yerouchalmi.[18] The results of channel flow
tests and experimental studies on some of the more promising ceramic
materials and their specific drawbacks will be discussed.

Ideally, ceramic electrodes should be total electron conductors
to prevent the degenerative effects of polarization during current
flow. Most of the good electron conducting ceramic materials such
as Sr doped $LaCoO_3$, $PrCoO_3$, In doped SnO_2, and Li doped NiO do not
possess high enough melting points.[19]

Lanthanum chromite exhibits high electronic conductivity even
at room temperature (1 mho/m) but dissociates rapidly at temperatures
of 1300-1500°C with resultant vaporization of chromic oxide according
to

$$2LaCrO_3 \rightarrow La_2CrO_3 + CrO_3 \text{ (gas)} \tag{3}$$

Additions of SrO to give $La_{1-x} Sr_x CrO_3$ solid solutions greatly
improve hardness and strength. Electrodes made with 16 m/o SrO
dopant showed good K-seed resistance but the volatility of chromia
still limited its maximum use temperature to 1800°K.[12] The electrical
properties of $LaCrO_3$ and $LaZrO_3$ combinations have been measured.[18]
Calcia additions to $LaCrO_3$ have also shown good conductivities.[20]
A 50-50 w/o of $LaCrO_3$ and ($.25$ $Gd_2O_3-.75$ ZrO_2) has been successfully
tested for 30 hr in a 8 Mw rig.[21]

Of the refractory oxide compounds with melting points above
2000°C, the cubic fluorite oxides such as stabilized zirconia, ceria,
and hafnia have received the most attention. Several different
cubic stabilized zirconia (calcia and yttria doped) compositions
(CSZ) have been tested for electrode performance in MHD channel
environments with varying success. The CSZ materials show outstanding
mechanical and chemical erosion resistance (inert to K-seed) but
polarization (ionic conductivity effect) and thermal shock (low
thermal conductivity) are inherent problems that require special
design considerations.[12] Twelve mole percent Y_2O_3 stabilized ZrO_2
has been used at current densities of 2 amps/cm^2 with almost no
power loss due to polarization as long as sufficient oxygen was
present at the metal lead-zirconia cathode interface.[22] The rate
and degree of electrolysis increased with current density and
decreasing partial pressure of oxygen producing eventual grain
boundary precipitation of metal atoms and intergranular crack
formation.

The properties of several other rare earth oxide-zirconia systems have been studied in an effort to improve electronic conductivity while maintaining chemical and mechanical stability. The heavy rare earth type "C" (body centered cubic-thallium oxide structure) sesquioxides all form cubic fluorite (anion-vacant) solid solutions ($M_xZr_{1-x}O_{2-x/2}$) with 6-20 m/o additions. The lighter (Types A and B) rare earth oxides tend to form ordered "pyrochlore" ($M_2Zr_2O_7$) crystal structures.[23]

CSZ all show high ($t_i = 0.99$) ionic conductivities as a result of enhanced oxygen vacancy concentrations. In contrast, the ceria-zirconia systems exhibit high electronic conductivity and consequently have been studied as MHD electrode candidates in some detail.[24,25] One ZrO_2-CeO_2 combination yielded steady current-voltage behavior for over 30 hrs in an 8 Mw test fixture.[21] The $Ce_2Zr_2O_7$ pyrochlore compound and CeO_2-ZrO_2 rich solid solutions are reported to be unstable in changing O_2 atmospheres.[26] A detailed d.c. study by Casselton of the 20 m/o CeO_2-ZrO_2 composition revealed that the material deviated from ohmic behavior even at low current densities showing a sharp rise in current with slight increases in voltage.[22]

This phenomenon was attributed to current channeling and could not be prevented at current densities greater than 0.1 amps/cm^2 regardless of the oxygen partial pressure or temperature. Although the current channeling was not structurally damaging to the CeO_2-ZrO_2 body, it may induce temperature gradients severe enough to melt the anode electrode at MHD generator current densities. Praseodymium zirconate has shown much better stability than cerium zirconate in reducing and oxidizing atmospheres. Indium additions to $Pr_2Zr_2O_7$ have improved the electrical conductivity and these compositions are under further investigation.[12]

Insulators

The development of reliable insulation ceramics may be the greatest challenge. They, too, must exhibit excellent mechanical, chemical and thermal stability while maintaining high electrical resistivities (10^2 > gas phase) and dielectric strengths. The penetration of and the reaction with the alkali seeding compound is the single most detrimental factor, causing electrical leakage, arcing and material disintegration, thus gravely affecting generator performance. Insulator materials can be divided into two categories: (1) refractory oxides (Al_2O_3, MgO, BeO) and mixed oxide compounds (zirconates); (2) refractory ceramic concretes.

An erosion study of MgO insulation brick indicated that the K-salts chemically attacked and reacted at grain boundaries, promoting grain growth and structural damage of the ceramic.[27] Fused

MgO and Al_2O_3 insulators also show reactions with seeded combustion gases. Strontium zirconate, $SrHfO_3$, and $CaHfO_3$ possessed good insulating properties and seeding agent resistance.[20,28] Despite their high melting points (2700–2800°C for $SrHfO_3$ and $SrZrO_3$ and 2400°C for $CaZrO_3$),[29] wall temperatures above 1900°K are reported detrimental to material stability.[28]

Refractory ceramic concretes have been based on Al_2O_3 and MgO with the host grains bonded by refractory cements such as mono-calcium aluminate, aluminophosphate, magnesium chloride, magnesium hydroxide, etc. Upper use temperatures are 1600–1650°C and 1800–1900°C for alumina and magnesia concretes, respectively. Advantages of ceramic concretes are ease of fabrication of relatively large sizes into complex shapes and the ability to fill spaces in mosaic or modular constructions. However, the porous nature of the concretes increases their susceptibility to alkali impurity contami-nation and electrochemical erosion. A very comprehensive Russian paper discussed (1) the requirements of oxide concrete fillers on the basis of the purity of the starting material and granular com-position and (2) the kinetics and mechanisms of hardening reactions of various mixtures.[30] The electrical properties (volume resistivity, breakdown voltage, and surface flashover voltage) as a function of temperature and gaseous media are given.

Thermal insulation linings or coatings required for MHD generator parts (connective shaft of steam generator, hot blowout air duct, and exit cone of combustion chamber) have also suffered from potash condensation when the thermal operation (cool down) is disturbed.

ELECTRICAL PROPERTY STUDIES

Commensurate with the development of better ceramic oxide electrodes is the accurate measurement of their bulk electrical properties at high temperatures. The objective of a complete characterization program should be the determination of total con-ductivity, transference (ionic and electronic) number, concentration and mobility of charge carriers, and the defects responsible for the charge carriers as a function of temperature and oxygen partial pressure over the service environmental range of pressures and temperatures. Our initial research was concentrated on the CeO_2–ZrO_2 binary system (most widely studied electronically con-ducting MHD oxide electrode material). An electrical conductivity-thermogravimetric-crystal structure study has recently been completed by the authors on 22 m/o CeO_2–ZrO_2 sintered samples to 1400°C as a function of P_{O_2}.[31]

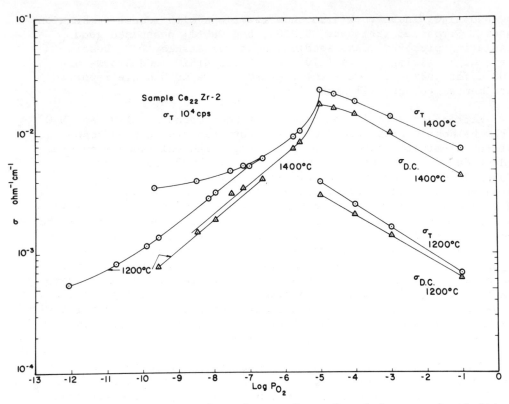

Fig. 3.   Oxygen pressure dependence of total and d.c. conductivities
for 22 m/o CeO$_2$–ZrO$_2$.

The oxygen pressure dependence of total (a.c.) and electronic
(d.c. polarization technique)[31] electrical conductivities at 1200°C
and 1400°C isotherms is shown in Fig. 3. Of note is the n-type
conduction region at higher oxygen pressures and a transition to
p-type regime at lower pressures. The conductivity maximum shifted
to higher oxygen pressures with temperature, being located at
$10^{6.6}$ atm at 1200°C and 10^5 atm at 1400°C. Assuming that the
platinum paste electrodes were blocking to ion flow, the proportion
of current (transference number) carried by ions was calculated
from the equation

$$t_i = \frac{\sigma_{a.c.} - \sigma_{d.c.}}{\sigma_{a.c.}} \qquad (4)$$

where $\sigma_{d.c.} = \sigma_{electronic}$. At 1200°C and 1400°C these calculations
gave t_i values of (0.23 for $P_{O_2}=10^1$ atm, 0.12 for 10^4 atm) and
(0.32 for $P_{O_2}=10^1$ atm, 0.23 for 10^4 atm), respectively.

Since complete blocking of the ionic component at the Pt electrodes is seldom realized experimentally at high oxygen partial pressures due to gas phase conduction through the electrode, a gaseous electrochemical cell of the type (Pt) $P_1O_2/CeO_2-ZrO_2/P_2O_2$ (Pt) was used to measure transference numbers at 1200°C in the high pressure region. The ionic transference number is given by

$$t_i = E_m/E_i \tag{5}$$

where E_m is the measured open circuit voltage (high resistance electrometer) and E_i is the theoretical voltage expected of a total oxygen ion conductor

$$E_i = \frac{2.3RT}{4F} \ \log P_1(O_2)_1/P_2(O_2)_2 \tag{6}$$
$$(F = Faraday).$$

In contrast to the polarization results, the cell measurements indicated that the conductivity was almost 100% electronic with t_i values of 0.011, 0.018, and 0.025 for P_{O_2} couples of $10^0/10^{1.99}$, $10^{1.99}/10^{2.85}$ and $10^{2.85}/10^{4.05}$ atms, respectively. The almost total absence of ionic contribution to the conductivity agreed very well with Palquev and VolchenKovas' results[32] ($t_i < 1\%$ in air) but conflicted with Millet and Guillous'[33] findings ($t_i = .53$ for $P_{O_2} = 10^{0.5}$ atm) on almost the same composition (25 m/o CeO_2-ZrO_2) at approximately 1000°C.

If parasitic neutral diffusion of O_2 (through pore structure, microcracks and/or around seals) is large enough that the required oxygen pressures are not maintained at the two sample electrode surfaces, voltage readings will be lower than the true values. Thus, experimental errors inherent in both techniques yield higher apparent t_e (lower t_i) numbers. Repetitive measurements on dense CeO_2-ZrO_2 samples in the gaseous oxygen partial pressure cell would be necessary to determine the real transference number.

The presence of a conductivity maximum and transition to apparently p-type conduction with decreasing oxygen partial pressures coupled with very long equilibration times in the conductivity transition region suggested the possibility of major atomic rearrangement and diffusion controlled crystal structure changes. Hence, thermogravimetric and x-ray diffraction analyses were conducted in various O_2 atmospheres to assess the effect of structural changes. The weight losses as a function of O_2 partial pressure at 1200°C (Fig. 4) were very slight in the high P_{O_2} region (10^1 to 10^5 atm) in comparison to the dramatic weight loss that occurred in changing from $10^{-4.84}$ P_{O_2} atm

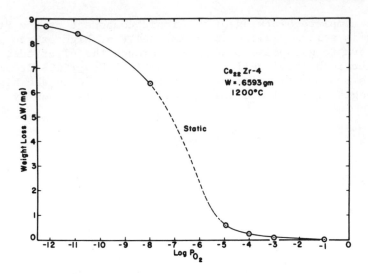

Fig. 4.   Weight change as a function of pressure at 1200°C for
          22 m/o CeO$_2$-ZrO$_2$.

mixture to $10^{-7.96}$ atm P$_{O_2}$. Assuming that all the cerium ions were
in Ce^{+4} state at the highest oxygen partial pressure (10^1) and all
the electrons released by the loss of oxygen from the lattice reduced
the Ce^{+4} ions to Ce^{+3} ions, the concentration of Ce^{+3} cations with
respect to total cerium content was 7% at P$_{O_2}$ = $10^{-4.84}$ atm, and
70.5% at $10^{-7.96}$ atm P$_{O_2}$. These large weight losses were consistent
with the weight gains of the 2000°C He sintered CeO$_2$-ZrO$_2$ samples
that were oxidized at 800°C prior to making measurements. The
black as-sintered specimens gave cubic fluorite ZrO$_2$ x-ray diffraction
patterns at room temperature (indicative of a large percentage of
Ce^{+3} ions) converting to the tetragonal zirconia crystal structure
on oxidation.

     Room temperature and hot stage x-ray diffraction analyses of
the conductivity and thermobalance samples after testing revealed
the following structural changes. In the high pressure region, the
tetragonal zirconia solid solution appears to be the stable phase
in the 1000-1400°C temperature range as determined by x-ray
diffractometer scans of the weight change sample on heating in air
to 1520°C. The low pressure p-type region (70 to 100% Ce^{+3}) does
not crystallize in the cubic fluorite ZrO$_2$ structure as might be
expected. The x-ray diffractometer traces indicated that the
Ce$_2$Zr$_2$O$_7$ pyrochlore compound and free ZrO$_2$ phases (tetragonal) are
in equilibrium with each other at these test temperatures. Approximately 30% of free ZrO$_2$ was left since the pyrochlore compound
requires two moles of zirconia for every mole of ceria.

The slow equilibration times associated with the large weight loss and conductivity transition is understandable since the formation of the pyrochlore phase requires diffusion of the cations to achieve the long-range ordering of the Ce^{+3} and Zr^{+4} cations throughout the oxygen polyhedra. Naturally, the interpretation of the conductivity behavior in the low pressure (p-type) region is complicated by the presence of the two phases. The loss of strength and structural degradation of CeO_2-ZrO_2 bodies that occur in cyclic O_2 atmospheres and temperatures is easily related to these reversible phase reactions.

The rapid equilibration and electron conduction of the high pressure tetragonal (CeO_2-ZrO_2) solid solution can be readily explained on the basis of the following reaction:

$$2\ Ce^{+4} + O_1^x \rightleftharpoons \tfrac{1}{2}O_2\ (g) + V_O^{\bullet\bullet} + 2Ce^{+3} \tag{7}$$

where $V_O^{\bullet\bullet}$ is a fully ionized oxygen vacancy with a net positive charge of two. The two electrons given up by the oxygen vacancy donor ($V_O^{\bullet\bullet}$) are trapped at Ce^{+4} acceptor sites effectively changing the cerium valence to Ce^{+3}. Electron conduction then proceeds by activated hopping of the electrons from Ce^{+3} sites to Ce^{+4} ions. Since oxygen diffusion is very rapid in anion vacant ZrO_2 solid solutions and no cation diffusion is required with this model, equilibration rates on changing O_2 pressures are expectedly high.

Although the CeO_2-ZrO_2 binary is essentially an electronic conductor over its entire composition range, the conductivity is marginal for rear metal lead contact-electrode temperatures. Also the increasing tendency for reduction of Ce^{+4} with temperature may induce destructive phase changes at hot face electrode temperatures of 1800-2000°C even at the high oxygen pressures of open-cycle operation.

Improvements in the magnitude of the conductivity and crystal structure stability by small additions of Y_2O_3 and Sc_2O_3 to a ZrO_2-CeO_2 rich base composition are currently being investigated. Hopefully, the lower valent Y^{+3} and Sc^{+3} cations will enhance the overall conductivity slightly without impairing the electronic transference number to any degree and increase the stability of the Ce^{+4} ion in the lattice (better decomposition resistance). An extensive electrolyte study by Steele et al.[34] indicated that mixed ceria-zirconia-scandia compositions gave high conductivities with small ionic contributions at 1000°C covering a range of oxygen pressures from air down to 10^{-15} atmospheres.

The yttria and scandia dopants may be better than indium oxide additions to the praseodymia-zirconia composition. Other oxide systems that appear to offer promise for 2000°K MHD electrodes are ThO_2 and Y_2O_3 hosts doped with small concentrations of donor materials. Oxides of limited substitutional solubility possessing stable higher valence states and donor levels close to the conduction band of the host oxides are the most favorable candidates. Additions of V_2O_3, Cr_2O_3, Nb_2O_3, and Ta_2O_3 additions might satisfy these criteria for thoria at high temperatures. Zirconia and thoria doped Y_2O_3 also offers interesting possibilities.

SUMMARY

Although the ultimate in MHD electrodes has not been realized yet, some of the inherent limitations of ceramics such as thermal shock, vaporization, mechanical erosion and polarization will have to be minimized by engineering design in order to increase their life expectancy. Poor thermal stress resistance has prompted the development of small segmented ceramic electrodes (peg wall or mosaic construction) and/or the use of thin layers or plasma sprayed coatings supported by water cooled metal fixtures. The erosion and vaporization problems, for example, have inspired the concept of *in situ* replenishment by seeding the plasma with small amounts of the electrode material in powder form.[35] Polarization of ionic conducting stabilized zirconia materials has been inhibited by maintaining a continuous oxygen supply to the cathodes through elaborate design arrangements.

Ceramic electrode materials with high electronic conductivity capable of stable operation at 2000°C or better will probably be fulfilled by (1) zirconia based compositions heavily doped with variable valence rare earth oxides such as ceria and praseodymia and (2) thorium and yttrium oxides lightly doped with oxides possessing stable, higher altervalent cations. The concentrations of these various oxide additions should minimize the degradation of bulk electrical properties due to impurity contamination. Insulation problems may be reduced by application of thin, dense coatings over porous insulation of the same material. Another method that offers promise for MHD topping plant generators is the continual maintenance of ceramic insulation and electrodes at high temperatures by bleeding air at low flows from the air heater when the generator is not operating.

The most likely near term solution for long life electrodes may result from the development of arc electrodes.[36] This concept involves recessing the hot surface of the electrode (refractory metal) from the main channel wall and protecting it with an inert gas such as argon. With proper duct design, an arc column can be created to transfer current from the plasma to the electrodes.

ACKNOWLEDGMENTS

*A special thanks is rendered to Kenneth Cramer for his stimu-
lating discussions, critique, and encouragement in preparing the
manuscript. The authors are also grateful to Gene Charles for
performing the hot stage x-ray diffraction analyses.*

REFERENCES

1. A. Kusko, IEEE Spectrum 5 (4) 75–80 (1968).
2. M. Petrick, IEEE Spectrum 2 (3) 137–151 (1965).
3. R. Rosa and F. Hals, Industrial Research, 69–72, June (1968).
4. G. Sutton and A. Sherman; pp. 340–88 in Engineering Magnetohydrodynamics. McGraw-Hill, Inc., New York, 1965.
5. Ibid, pp. 471–528.
6. J. Dicks *et al.* pp. 16–28, Preprint Volume, Eleventh Symposium on
   Engineering Aspects of Magnetohydrohynamic. Calif. Inst.
   Technology, March 24–26, 1970.
7. J. Klepeis and J. Louis; pp. 62–64, Ibid.
8. R. Rosa *et al.*, MHD Power Generation Status and Prospects for
   Open-Cycle Systems, "Avco Everett AMP 295, November 1969.
9. M. Petrick, IEEE Spectrum 2 3 137–151 (1965).
10. R. Rosa and J. Louis, "Position Paper on Closed Cycle MHD Power
    Generation," Avco Everett Research Lab.
11. Joint ENEA/IAWA International Liaison Group on MHD Electrical
    Power Generation," MHD Electrical Power Generation," The
    1969 Status Report.
12. P. Curtis *et al.* "Electrodes and Insulators for Open-Cycle MHD
    Generators," Preprint, Symposium on Magnetohydrohynamic
    Electrical Power Generator, Warsaw, Poland, 24–30 July (1968).
13. E. Robinson, J. Ballad, and J. Yerrell "The Performance of Metal
    Electrodes Under Various Open Cycle MHD Conditions," Ibid.
14. J. Koester *et al.*; pp. 54–60 in Preprint Volume, Eleventh MHD
    Symposium. California Institute of Technology, Pasadena,
    Calif., 1970.
15. V. Gordon *et al.*, "High Melting Oxide and Carbide Materials for
    MHD Generator Electrode Walls, Preprint, Warsaw Symposium (1968).
16. A. Borison, "The Behavior of High Melting Metallic Materials in
    an Argon Plasma Flow," Preprint, Warsaw Symposium (1968).
17. T. Morikawa, *et al.*, "Study on Semi-Hot Wall Duct for MHD Generator",
    Preprint, Warsaw Symposium (1968).
18. D. Yerouchalmi, "Etudes de Materiaux Destines a Equiper des
    Tuyeres de Conversion Magnetohydrodynamique de L'Energie
    Thermique des Gaz de Combusion en Energie Electrique", Thesis,
    Docteur-Ingenieur, De L'Universite de Paris, France (1970).
19. C. Tedmon, H. Spacil, and S. Mitoff, J. Electrochem. Soc., 116
    (9) 1170–1175 (1969).

20. Translation of "The Institute of High Temperatures of the USSR
    Academy of Sciences," "The Most Important Results of Scientific
    Research in 1969," Nauka Press, Moscow, 1970, 62 pp.
21. D. Yerouchalmi, "Hot Electrodes for Open Cycle MHD Generators,"
    Preprint Volume, Eighth Symposium on Engineering Aspects of
    Magnetohydrodynamics, Stanford, Calif., March 28-30, 1967.
22. R. Casselton, "The Nature and Consequence of Current Blackening
    in Stabilized Zirconia," Preprint, Warsaw Symposium (1968).
23. L. Fehrenbacher, L. Jacobson, and C. Lynch, in Proceedings of
    the Fourth Rare Earth Research Conference. Gordon and Breach,
    New York. April 22-25, 1964.
24. D. Meadowcroft, "The Electrical Resistivity of Some Castable
    Zirconias," to be published in J. Mat. Science.
25. R. Chapman, D. Meadowcroft, and A. Walkden, "Properties Relevant
    to MHD Applications of Some Pyrochlore Structured Zirconates
    and Stannates," to be published.
26. A. Kuznetsov, E. Keler and Fan Fu-k'ang, Zhurnal Prikladnoi
    Khimii, 38 (2) 233-241 (1965).
27. G. Kuezinski, R. Parrot, and D. Yerouchalmi "Study of Corrosion
    in Magnesia Crystals by Atkaline Salts," Preprint, Warsaw
    Symposium (1968).
28. R. Wang and D. Yerouchalmi, Rev. Hautes Temper et Refrat. t. 3,
    205-214 (1966).
29. M. Foex, "Study of Oxides Having Extremely Good Refractory
    Properties and Capable of Being Used in MHD Electrodes,"
    Preprint, Warsaw Symposium (1968).
30. A. Romanov, "Refractory Concretes as Electrically Insulating
    Material for MHD Generator Ducts," Warsaw Symposium (1968).
31. L. Fehrenbacher, Ph.D. Thesis, "Electrical Conductivity and
    Defect Structure of Ceria- Zirconia Solid Solutions, University
    of Illinois, Urbana, Illinois, Feb. (1969).
32. S. Pal'guev and Z. Volchenkova, "Electrical Conductivity and
    Transport Numbers in the $CeO-ZrO_2$ System," Translation
    Russian Journal of Physical Chemistry, 34 (2) 211-13 (1960).
33. M. Asquiedge *et al.*, "Physics-Chemical Properties of Binary
    Solid Electrolytes," Translated from Rev. Hautes Temper
    et Refract. 6 35-44 (1969).
34. B. Steele, B. Powell, and P. Moody, Proc. Brit. Ceram. Soc.,
    10 87-102 (1968).
35. J. Teno, "Long Duration Electrodes for Open Cycle MHD Generators,"
    Eighth MHD Symposium, Stanford, (1967).
36. J. Yerrell and E. Robinson, "The Development of Arc Electrodes
    for Open-Cycle MHD Use," Warsaw Symposium (1968).

DISCUSSION

H. S. Bennett (National Bureau of Standards): Is my understanding correct that materials for MHD closed-cycle generators are at present more promising than those for MHD open-cycle generators?
Authors: Yes.
H. S. Bennett: Which of the following two plants which produce the same amount of power would be least harmful to the environment? (air pollution and thermal pollution) Plant A: conventional fossil fuel, Plant B: MHD open-cycle, using fossil fuel.
Authors: Plant B.

H. Palmour III (North Carolina State University): Will the current MHD generator designs perform well under the strongly cyclic thermal and electric loads imposed by real world power demands?
Authors: Several current channel designs and related material systems appear capable of handling cyclic, high level thermal-electrical power loadings although they have never been tested in very large (>100 MW) MHD generators. A more critical problem for topping of central base power generators is long life stability of electrodes and insulators. Thermal, mechanical and electrochemical erosion that often result from the singular or combined effects of vaporization, oxidation, arcing, and seed interaction usually lead to structural degradation of the channel materials. To minimize the probability of thermal stress fracture and spalling, however, the electrodes and insulators usually contain considerable porosity (20% or so) which increases the likelihood of seed penetration and electrochemical destruction.

C. B. Alcock (University of Toronto): Have studies been made of the effects of sulphur on the chemical stabilities of the proposed oxide electrodes and conductors for open cycle operation?
Authors: The influence of sulphur on the chemical stabilities of the oxide ternary systems proposed in this paper has not been determined. However, the electrochemical erosion of the closely related $Ce_2Zr_2O_7$ system and other rare earth zirconate and stannate compounds in molten K_2SO_4 seed was measured by the General Electric Hirst Research Centre in England. Cerium zirconate was the least eroded zirconate cathode material, with the zirconates performing better than the stannates. SnO_2 was the least affected anode with $Ce_2Zr_2O_7$ probably showing the best overall resistance to K_2SO_4 erosion. Hence, oxide additions to improve its poor oxidation resistance may prove quite beneficial.

S. G. Ampian (U. S. Bureau of Mines): Has any effort been made to either recycle the seed material and/or reduce the SO_x problem by the formation of K_2SO_4 compounds?
Authors: Atmospheric contamination and the cost of potassium seed material dictate that it must be recovered and reinjected into the MHD combustion plasma. Thus, a considerable portion of research and analysis has concentrated on seed recovery and chemistry,

especially in England. In order to design MHD seed handling systems
for seeded exhaust gases, a knowledge is required of the gas stream
chemistry, the condensation temperatures of the various seed compounds, their respective compositions and how the components and
impurities in the gas stream affect the seed condensation rates.

With sulphur containing fuels such as oil or coal, pure K_2SO_4
is used as the seeding agent and is the principal condensate even
though KOH is present in greater quantities in the gas stream. Small
amounts of impurity sulfates of Na, Ca, Mg, Ni, Fe, and Cr are also
found with K_2SO_4. Techniques for separating the ash slag of coal
from the sulfate condensate and purifying it for recycling have also
been developed. If sulphur free natural gas fuel is used for open
cycle operation, K_2SO_4 is replaced by K_2CO_3 seed and the carbonate
is the principal potassium recovery material. To insure that exhaust
contamination is within safe limits, processes such as electrostatic
precipitation and gas filtration have been tested experimentally
to remove seed particles and fly ash that are carried as smoke or
fume with electrostatic precipitation offering the most efficient,
economical removal.

INTERACTION BETWEEN HIGH ENERGY RADIATION AND GLASSES

Norbert J. Kreidl

University of Missouri

Rolla, Missouri

ABSTRACT

Effects of high energy radiation depend upon the dominance of particle radiation, primarily manifested in structural changes, or ionizing electronic changes; in practice radiations and damages are mixed. Neutron fluxes of 10^{20} nvt significantly alter crystalline and vitreous silica. High intensity γ and x irradiation induces impurity and intrinsic color and spin centers in silica and complex oxide glasses, which can be differentiated and described in considerable detail. Induced changes can be used to prevent discoloration or, conversely, to read dosage. Such changes include absorption, luminescence and thermoluminescence. Space charges can lead to fracture when surfaces are violated. The understanding and use of other changes are in their infancy. Relatively low energy (light) in high intensity can cause mechanical damage. Thresholds are crucial in modern laser technology.

INTRODUCTION

The nature of the effects of high energy radiation depends upon the relative dominance of particle radiation (structural changes), or ionizing radiation (electronic changes). In practice both radiations and effects are usually mixed. Large doses are necessary to damage quartz or pure silica glass, and multicomponent oxide glasses as well. Directional bonds with fixed length in these significantly covalent materials discourage lattice diffusion, and defects are complex, usually associated with impurities in silica and nonbridging oxygen often accompanied by alkali in multicomponent oxide glasses.[1,2]

521

At the same time these conditions allow the selection of some glasses
as particularly resistant materials in severe radiation environments,
e.g., hydrogen containing synthetic silica and cerium doped optical
glasses. Detailed reviews were published by Lell *et al.*[3] and
Bishay.[4]

PARTICLE RADIATION

The basic processes involved in damage suffered by glasses under
particle radiation are *collision, charge interaction,* and in specific
interactions, *transmutation,* including fission. The damage may
extend beyond the area of direct irradiation. In the case of
collisions one refers to these areas as spikes, which may cause dis-
order, superlattices, phase changes, thermal defects, etc.

Structural Effects in Silica

Fast neutrons (>100 ev), in particular, have been observed to
affect quartz starting at 2×10^{19} nvt (=n/cm^2 integrated flux) and
saturating at about 2×10^{20} nvt. The density decreases by about
15%. Lattice parameters, refractive index and other properties[8,9]
change correspondingly; and diffusion of diffraction lines, loss in
birefringence as well as loss in rotation power indicate conversion
to a glass-like substance.[10,12,15-17] This end product is appropriately
called metamict silica, the term "metamict" having been coined
(1914) for minerals having regionally and externally conditioned
glass-like properties. This lower density of metamict silica is,
however, still about 3% higher than that of fused silica.

Moreover, and curiously, fused silica irradiated by 2×10^{20} nvt
fast neutron flux increases ~3% in density thus saturating to the
density of irradiated quartz.[10,11] Thus quartz, other crystalline
forms of silica, and fused silica appear to convert to the same type
of densest glass, metamict silica, upon saturating neutron irradiation
(Fig. 1). Its difference from ordinary SiO_2 glass seems to be in
the distribution of bond angles (the mean shifts from 142° to 138°)
and not in coordination or basic structure,[18] which appears to be
supported by the correlation of ir and Raman band changes.[9] One can
only conclude that neutron damage might involve both high and low
density zones. It appears plausible that in the case of the originally ordered, denser quartz the prevailing trend is toward lower
density.

Below saturation (<5×10^{19} nvt), damaged quartz is inhomogeneous,
exhibiting characteristic spikes,[12] and may anneal back to the
ordered condition.[19] N-irradiated silicas show higher sensitivity
to ionizing radiation indicating the creation of precursor defects

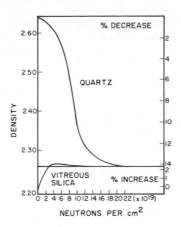

Fig. 1.   Change in density with neutron irradiation of crystalline
          quartz and fused silica (after Primak *et al.* 1955).[7]

(oxygen vacancies).[20] Interferometric studies also show induced
surface changes in n-bombarded SiO_2 which may become serious enough
to cause the surface layer to shrink and "turn up".[13,14]

Strong surface changes, such as low index layers, are the consequence of high intensity bombardment with particles (krypton,
hydrogen, deuterium, helium).[14,21-23] Heat capacity and thermal
conductivity also decrease in irradiated quartz and increase in
irradiated silica glass, converging to the same metamict silica
values.[24] Reference may be made to Lell *et al.*[3] for additional
properties and details.

Transmutation Effects

The most serious particle damage to conventional glasses is
the transmutation (capture of thermal neutrons) of boron to the
smaller Li-ion, leading to collapse.[25] Boron containing glasses,
such as the common borosilicate glasses, must not be used in a
neutron flux environment.

Glasses may, of course, become "hot" due to specific nuclear
reactions leading to radioactive daughter species. One must
consider specific glass components in neutron environments, particularly K^+, Ca^{++}, Zn^{++}, As^{3+}, Ce^{3+}, Co^{3+}.[26]

EFFECTS OF IONIZING RADIATION ON SILICA

Impurity Related Centers

Most of the effects of ionizing radiation on silica are related
to impurities; they were first discovered and, later, best understood. Originally, the impurity was often not known to be crucial,

or even present. A typical case is that of germanium. The induced electronic effects can be classified by charge as hole or electron trap centers, as evidenced by optical or magnetic spin resonance studies, and by their association with a specific impurity.

   Germanium centers. Germanium, a neighbor of silicon in the periodic system, can easily substitute and is frequently present as an impurity in SiO_2. The high electron affinity of Ge makes it active as an electron trap:

$$Ge^{4+} + e^- \rightarrow Ge^{4+,-*} \text{ or } (Ge^{4+} O_4)^0 \rightarrow (Ge^{4+,-}O_4)^-$$

If Al^{3+} is also present (a few ppm have been found in nearly all quartz with charge compensated frequently by an alkali ion, e.g. Li^+), this Al^{3+} may become a corresponding hole trap:

$$Al^{3+} + e^+ \rightarrow Al^{3+,+} \text{ or } (Al^{3+}O_4)^- \rightarrow (Al^{3+,+}O_4)^0.$$

This change in the charge of the $(AlO_4)^-$ group allows the compensating alkali (e.g., Li^+) to migrate towards capture by the new germanium center. The total reaction may be written as:

$$Li^+ (AlO_4)^- + (GeO_4)^0 + e^- + e^+ \rightarrow (Al^{e+}O_4)^0 + Li^+ (Ge^{e-}O_4)^-.$$

The visible result is discoloration. These centers, as well as the corresponding ones not involving alkali migration at low temperatures [e.g., $(Ge^{e-}O_4)^-$ without the neighboring Li^+] have all been established through elegant optical and spin resonance studies.[27-30] The description given here summarizes Mackey's model.[30] Similar interpretations apply to the case of other alkali additions.

   Aluminum centers. A basic model of radiation induced centers in Ge free SiO_2 glasses with Al impurity, is:

$$Li^+ (Al^{3+}O_4)^- \rightarrow (Al^{3+,+}O_4)^0 + Li^0.$$

The original impurity defect is usually charge-compensated by alkali, say Li^+. However, on closer study centers revealed by optical and spin resonance studies show a lack of correlation and a complexity much beyond the simple model.

---

*The designations $Ge^{4+,-}$ or $Al^{3+,+}$ etc. are sometimes used instead of the conventional Ge^{3+}, Al^{4+} to symbolize the preservation of the original surrounding (of Ge^{4+}, Al^{3+}, etc.) leading to specific electronic states.

Assuming an oscillator strength of 1, one can calculate the number of centers from the optical absorption coefficient and the half band width.[36] The bands between 1.0 and 6.0 ev thus can be associated with the analytical aluminum content. (70 ppm vs $N \gtrsim 10^{16}$ cm^{-3}) of quartz. By heating and reirradiating, it is possible to show in which cases heat sensitive precursor defects affect the induced electron or hole traps.[37] By doing this for both optical and paramagnetic spin centers, assignments can be compared and the meager correlation that is found indicates the complexity of the Al centers. At least one optical band, its relation to spin centers definitely associated with Al (and its sensitivity to heating) still are best explained by the $Al^{3+,+}$ hole center first proposed by O'Brien.[39]

Smoky quartz, a natural material, shows the bands at 4600 and 6200 Å associated with damaged silica. Amethyst most likely contains Fe in the place of Al.[28 b] If alkali is replaced by hydrogen, no coloration is observed. This seems to fit the notion that hydroxyl or hydrogen-bonded protons will not easily migrate.

Other impurities in SiO_2. The iron centers in amethyst, citrine and synthetics are more complex.[40-42] Titanium is involved in the color center of rose quartz which initially is not caused by radiation, but on irradiation, converts to a brown color center involving titanium and aluminum, alkali and/or hydrogen.[43-45] Hydrogen strongly affects defect centers, irradiation processes, and effects. It can become manifest and distinguishable as to detailed arrangements in ir spectra.[28 c,d] Reference is made also to the following section on "intrinsic" centers.

Intrinsic Defects in Silica

When about 100 times the dose causing induced impurity defects ($\sim 10^9$ r) is applied to purest silica, "intrinsic" defects are detected, some admittedly involving hydrogen also. Assignments are based primarily on electron spin resonance (ESR) interpretations.

The most important induced centers are [31,32,38] (1) E_1, corresponding with a 215 nm band, an oxygen vacancy with one (or three) trapped electron(s) (Fig. 2a) and (2) E'_2, corresponding with a 230 nm band, a silicon oxygen vacancy pair, perhaps with an associated hydrogen impurity (Fig. 2b).

Oxygen vacancies can be bridging or nonbridging, the latter particularly in more complex systems and the related P_2O_5. The defect electron states can be combined affecting neighbor orbitals,[35] using an extensive quantitative estimation of the g-tensor from ESR data. The method allows one to "see" only odd numbers (1-3) of trapped electrons (g < 2.0023, g > 2.0023).

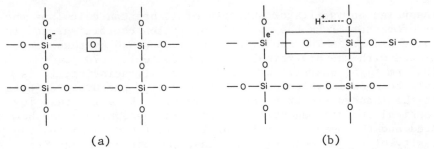

(a)   (b)

Fig. 2.   Models of induced centers (a) E'$_1$ center (b) E'$_2$ center.

Defect structures of SiO$_2$ are interesting and important parameters in effects of SiO$_2$ on the human organism, since they affect
the redoxy system of the reaction chain in "cell breathing".[46]
Intrinsic (and impurity) defects are accompanied by characteristic
luminescences which have been used as petrological tools.[47]

MULTICOMPONENT GLASSES

Effects of Ionizing Radiation on Alkali Containing Oxide Glasses

In spite of the complexity of possible effects of high intensity
radiation on multicomponent glasses it has become possible to identify,
correlate, and interpret definite induced optical absorption bands
that seem associated with alkali and its oxygen neighbor, frequently
a nonbridging oxygen (i.e., one having only one highly charged
neighbor, e.g. Si). Typically, three bands may appear, e.g., (1)
at 620 nm, the position being independent of the species of alkali,
the intensity increasing from Li to Cs, (2) at 415 - 490 nm, the
position shifting from Li to Cs, and (3) at 300 nm, independent of
alkali, the position changing with concentration.

These centers are now usually assigned to hole traps at nonbridging oxygens associated ("type 2") or not associated ("type 1")
with the modifier (alkali) cation.[48,49] The corresponding discoloration damages the performance of optical glasses. Under
reducing conditions sodium silicate glasses show greater induced
visible absorption when irradiated and observed below 300°K than
when observed above 300°K. Bishay[4] suggests convincingly that a
positive hole center as well as an electron trap center contribute
to the 2 ev absorption, the hole center being much more temperature
stable. The electron trap could be a nonbridging oxygen vacancy,
destabilized at higher temperature by Na$^+$ capturing the electron
and moving away. At lower temperatures and when Na$^+$ is replaced
by the less deformable and less mobile K$^+$, Cs$^+$, Rb$^+$ this process is
inhibited and the absorption persists at higher temperatures. The
decrease of absorption under low dosage in the high energy uv range

is associated with the behavior of nonbridging oxygen.[5] The water
content should be considered carefully, since it was found that as
little as 0.01% OH strongly enhances the induced hole centers.[82]

As aluminum approaches or exceeds the alkali in concentration,
specific centers of the type discussed for silica glass may develop.
Optical centers induced in borosilicate glasses were first described
by Levy.[52]

Two types of ESR signals are characteristic of alkali silicate
glasses:[53,61] (1) a broad, asymmetric signal g = 2.0012, width 19G,
was assigned to nonbridging oxygen and, with caution, related to the
optical bands at 620 and 400 (since it is also found in aluminosilicate glasses some nonbridging oxygen was postulated in them) and
(2) a narrow, sharp signal was assigned to a trapped electron,
related to Weeks' E centers (see Figs. 2a and b), i.e., electrons
trapped in oxygen vacancies. A third six-line signal which is found
only in aluminosilicate glasses can be removed by heat treatment,
and is considered a hole center similar to that found in SiO_2 containing $Al+R^+$ impurities; i.e., $Al(O_4)^{e+}$, although its absence in
some aluminosilicates has not been explained. Spin centers (1) and
(2) are also found in borosilicate glasses.[54,55]

The correlation of spin and optical centers is not as good as
one may wish, perhaps indicating that spin signals demand specific
centers involving hole trap, nonbridging oxygen and alkali.[61]

Although reference is made here to Tucker's[48] first suggestion
and Bishay's[49] extensive recent analysis, it is instructive to indicate briefly that the hole trap assignment was strongly supported
by the fact that the demonstrated radiation reaction of cerium
additions, ($Ce^{3+} \rightarrow Ce^{3+,+} + e^-$) which creates holes suppresses the
base blase centers and, therefore, the assigned hole trap character.
The reverse reaction suppresses electron traps. This method was
worked out by Stroud.[56]

A very important role in the analysis and assignment of centers
is the study of their annihilation("bleaching") rate by heat or
irradiation in their own energy band. This subject has been studied
extensively by Levy, including the important case of cerium.[50] [52]
Doping optical glasses with cerium in a certain oxidation-reduction
equilibrium thus prevents the discoloration due to (1) and (2) centers
in the base glass, and Ce-doped glasses are used under the name of
"protected optical glasses" for optics in "hot" areas.[61]

The principles described thus far may be illustrated by a few
examples for glasses other than silicate glasses. Induced absorption
bands in Li to Rb diborate glasses are shown in Fig. 3 according to
Bishay[4] and Yokota[57] who had still considered as electron traps what

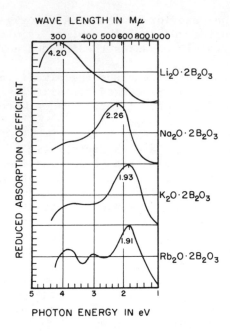

Fig. 3.   Induced absorption bands in alkali diborate glasses
          (Yokota).[57]

are now recognized as hole traps for reasons explained by Arafa
et al.[58]   Spin centers in irradiated alkali borate glasses permit
the recognition of considerable structural detail. A hole trap can
be associated with an oxygen shared between a 4 and a 3 coordinated
boron by the computer analysis of g and hyperfine tensors for
experiments with glassy and crystalline $Li_2O.4B_2O_3$ whose structure
is known.   In potassium borate glasses Arafa and Bishay[59] discern
3 hole-trap and 2 electron-trap centers and can locate a hole trap
at a diborate group.   For further details on borate glasses see
the work of Bishay.[4]

    Such detailed assignments not only help the understanding of
irradiation effects but are extremely valuable in connection with
spin resonance based structural analyses of nonirradiated borate
glasses, which are among the most successful achievements in the
investigation of noncrystalline solids, such as those by Bray's
group at Brown University.

    In alkali phosphate glasses hole trap centers related to those
in other alkali glasses are found.   Because of the transparency of
phosphate glasses in the ultraviolet, Kreidl and Hensler[60-62] used
them first as models to explain the effect of cerium and found
that the suppressed centers seemed to be hole centers, just as is
now recognized for all alkali glasses.

The detailed information from the spin and optical signals of these hole centers allowed Beekenkamp, *et al.*[64] to assign them to specific oxygen locations characterized as between two phosphorus network forming cations, between one phosphorus and a modifying low valency cation, having only one phosphorus neighbor affixed by a double bond, etc. Various investigators[33,64-66] were able to construct rather detailed bond models for irradiated atom groups in these glasses. Weeks *et al.*[31-38] for instance, were able to make an assignment as definite as a nonbridging oxygen vacancy plus trapped electron defect representable by an sp hybridization, with a variation in O-P defect angle depending upon $M_2O:P_2O_5$ ratio in various glasses, indicative of definite configurations. Among other non-silicates, calcium aluminate glasses were irradiated by Yanisheviskii *et al.*[67] Under high pressure, peaks tend to shift to shorter wave lengths.[68,69]

Although physical properties of multicomponent glasses can be manipulated by radiation, unlike plastics, few glass materials have been treated with industrial beneficiation in mind. Some studies were made at St. Gobain, mostly by LeClerc[109] and Paymal.[106] The most recent publication on radiation strengthening of conventional glasses are by Sil'vestrovich & Plisko.[70]

Transition Elements in Complex Glasses

The effect of cerium as a hole trap competing successfully with the damaging process affecting the coloration of the base glass (described in the preceding section) proved to be only one example of the behavior of many transition elements and related species with variable valency. As a matter of fact, the author and his associates studied and found these same effects in phosphate glasses during the course of developing and producing optical glasses immune against high gamma doses.[60] We found probable hole traps for additives such as Mn, Co, Cu, Fe, Ni, V, reduced titanium (Ti^{3+}) etc., each with a characteristic radiation coloration corresponding most likely to a reasonably new species since the ligand oxygen environment does not follow the new electronic state as it would in in a chemical synthesis. Details on recent work were given by Bishay.[4] In two cases, that of cobalt[62] and of bismuth [6,63] the radiation induced colors were, indeed, applied to radiation dosimetry with reasonable success. Cerium in lithium glasses performs as a scintillation counter.[109-111]

Silver Glasses. Radiophotoluminescence.

In phosphate glasses, Ag^+ may, as is now recognized, trap a hole, the resulting $Ag^{+,+}$ emitting a characteristic orange fluorescence in linear dependence on integrated gamma dose, thus serving well in a system of radiophotoluminescence dosimetry designed by Schulmann *et al.*[71] About 1945, the author provided a glass developed

to study silver solubility, composed of 50% $Al(PO_3)$, 25% $Ba(PO_3)_2$,
25% $K(PO_3)$ with an addition of 8% $Ag(PO_3)$. This glass was more
recently modified for improved energy independence by the introduction of light elements (Li, Mg) replacing Ba and K. Details on
radiophotoluminescence glass dosimetry were summarized by Kreidl
and Hensler.[61]

Effects on Electroluminescence and Mössbauer Effect

Effects of high energy radiations on other luminescence
phenomena are also being observed, for instance on Mössbauer
emmission spectra and electroluminescence of terbium doped glass
by Vbron'ko *et al.*[88]

Lead Glasses. Shielding Windows.

In addition to the hole traps characterizing oxide glasses
under high-intensity high-energy gamma irradiation, a characteristic
induced center (1.5 ev) is associated with Pb^{2+} in lead glasses.[7]
Coloration and fading have been studied by Barker *et al.*[72,73] One
observes unusually rapid bleaching of induced centers at room
temperature in ordinary light environment. This property and the
high cross section of lead for x and γ radiation are the bases of
the wide use of high lead (no alkali) glasses as radiation shielding
windows,[61] without the need to resort to cerium protection, in
moderately severe environments. The useful correlation of radiation
data with ir, x-ray and other data in the difficult evaluation of
lead glass structures has been reviewed most recently by Bishay.[4]

EFFECTS ON REFRACTIVE INDEX

From first principles an effect on refractive index must be
expected via induced absorption bands at the low energy side of
which the refraction must increase sharply, while it must decrease
sharply at the high energy side ("anomalous dispersion"). For
optical glasses, the observed effect under about 10 rad is of
the order of 2×10^4, sufficient to disturb optical performance.
Neutron irradiation will, as described earlier, affect the refraction
as the density changes. Surface indices are changed by ion
bombardment.[21,22]

DAMAGE BY ELECTRONS-SPACE CHARGES

Extremely large doses (1×10^{16} e/cm^2) of (1.2 Mev) electrons
are needed to cause the formation of color centers resulting in a
faint blue tint in the extremely pure (<0.2 ppm impurity, except
OH which may contribute to radiation resistance) synthetic silica

glass (SiO_2 from $SiCl_4$ + H_2)). The discoloration penetrates only
~0.09 cm beneath the surface corresponding to the range of 1.2 Mev
electrons.[75] Again, the more serious damage found by the same
authors in fused and crystalline quartz must be ascribed to the
type of induced impurity centers discussed in the preceding
sections.

In infrared transmitting materials used in guidance systems,
Mev electrons from the environment cause luminescence which represents damaging noise. Downing *et al.*[76] studied these effects from
an engineering viewpoint, but fundamental processes have not been
well established.

Electrons may induce long-lived (6 mo.) space charges, which
may show a complex gradient on the surface where ionized air may
provide conduction.[77]

Space charges, introduced by electrons and other radiative
environment may discharge from any disturbance, e.g., a mechanical
lesion at the surface, in the form of lightening or tree shaped
patterns.[77-81] Hardtke[80,81] refers to his beautiful collection
of discharge photographs as a "botanical collection."

In an orbital environment ("worst-case orbit" in Rogers'
terminology[83]) the charge induced in a year would be one order of
magnitude larger[25] than that observed to cause destructive discharges
from surface flaws. This suggests extreme care on surfaces exposed
to severe orbital environment. Silica windows can be used to
protect the more sensitive glasses needed in the optics of a
system.[84]

RADIOTHERMOLUMINESCENCE

The important phenomenon of thermoluminescence induced by high
energy radiation (*radiothermoluminescence*) is less well understood.
Minute impurities greatly affect the process. An electron may fall
into an empty defect level of an Al impurity in a silica glass
from which thermal energy may lift it to the conduction band from
which it may fall back to its original position.[95] Thermoluminescence
induced by radiation in quartz is now applied at Oxford's Research
Laboratory of Archaeology to dating and authenticating ceramics.[86]
Thermoluminescence radiation dosimetry has become a field of intense
investigation since about 1960. A radiothermoluminescent glass
has been proposed by Ginther,[87] however, the most active radiothermoluminescence dosimetry programs involve crystalline ceramics
such as the Mn-doped LiF first proposed by Schulmann and more
recently described by Fowler and Attix[89] and Schulmann *et al.*[71]

Effects of High Intensities of Light Laser Damage to Glass

Ultraviolet radiation has been observed to cause changes in glasses in specific cases.[90-92] But, even lower energy (visible and infrared) radiation can damage glass in high intensities. Soon after the inception of the glass laser in 1961, damage to the lasing glass as well as to accessory glass components by laser radiation was observed and, eventually, has become the main limiting factor for the development and the usefulness of high power devices. Besides numerous specialist references[97-105] to laser damage and its control, two comprehensive sources of detailed information are the special (1970) publications of ASTM[107] and NBS.[108] It should be stated that in addition to high energy radiation, high intensity radiation of light energy can cause filamentary damage to glasses by a self focusing process[93,94,96] which causes rapid increases in local temperature and plasma formation. Destruction may also be initiated at the surface[95] and at the interface of inclusions (as small as 1 μm) particularly platinum[105] from modern furnace components now used in the manufacture of pure technical glasses. The mechanism was described by Bliss.[93] Elimination of these latter sources of damage is expected (see e.g., ref. 95) to push the damage threshold from 10 beyond 100 joules/cm^2. Most of these experiences were obtained with Nd-doped silicate glasses and may have to be supplemented with new experience if and when other dopants(e.g., erbium) or other base glasses should become significant.

Laser damage can be utilized to grind, cut or drill glass.[112] Cracks (often time-delayed), covered bubbles, craters and non-crystalline deposits from vapors emerging in smoke rings accompany these processes and their control is the task of a new development in glass working.

Reference is made at this point to the following paper which will deal with laser effects on ceramics exclusively.

REFERENCES

1. D. Billington and J. J. Crawford, Radiation Damage in Solids, Princeton University Press, 1961.
2. G. J. Dienes, J. Phys. Chem. Solids, 13 272-278 (1960).
3. E. Lell, N. J. Kreidl, and J. R. Hensler, pp. 3-94 in Progress in Ceramic Science. Ed. by J. E. Burke, Pergamon, New York, 1966.
4. A. Bishay, J. Non-Cryst. Solids, 54-114 (1970).
5. A. Bishay and I. Gomoa, Phys. Chem. Glass 9 (1968).
6. A. Bishay, Phys. Chem. Glass 2 (2) 33-38 (1961).
7. A. M. Bishay and M. Maklad, Phys. Chem. Glass 7 149 (1966).
8. A. E. Clark and R. E. Strakna, Phys. Chem. Glass 3 121 (1962).

9. R. H. Stolen, J. T. Krause and C. R. Kurkjian, Faraday Conf. Vitreous State (1970). To be published 1971.

10. W. Primak, L. Fuchs and P. Day, J. Amer. Ceram. Soc. 38 (4) 135-139 (1955).

11. W. Primak, L. Fuchs and P. Day, Phys. Rev. 92 1064-1065 (1953).

12. W. Primak, Phys. Rev. 110 1240 (1958).

13. W. Primak, J. Appl. Phys. 35 1342-1347 (1964).

14. W. Primak, Oral Communication (1965).

15. M. Wittels, Phys. Rev. 89 656-657 (1953).

16. M. Wittels, Phil. Mag. 2 1445-1461 (1957).

17. M. Wittels and F. A. Sherill, Phys. Rev. 93 1117-1118 (1954).

18. I. Simon, Phys. Rev. 103 1587-1588 (1956).

19. E. V. Kolontsova and I. V. Telegina, Sov. Phys. Sol. State 7 (9) (1965).

20. C. M. Nelson and J. H. Crawford, Jr., J. Phys. Chem. Solids 13 290-305 (1960).

21. R. L. Hines, J. Appl. Phys. 28 587-591 (1957).

22. R. L. Hines and R. Arndt, Phys. Rev. 119 623-633 (1960).

23. A. Leadbetter and J. A. Morrison, Phys. Chem. Glass 4 188-192 (1963).

24. A. Cohen, J. Appl. Phys. 29 591-593 (1958).

25. W. A. Hanning "Selection of optical materials for use in radiation and orbital environment," presented at the Annual Meeting, Soc. Photo-Optical Instrumentation Eng., St. Louis, Mo. (1966).

26. M. Swerdlow and R. F. Geller, NBS 5272, May 1957.

27. A. J. Cohen and H. L. Smith, J. Chem. Phys. 28 401-405 (1958).

28. A. Kats (a) Verres et Refr. 12 191 (1958), (b) Philips Res. Rept. 11 113 (1956), (c) Ibid, 1962 17 133, 201, (d) (with Haven and Stevels) Phys. Chem. Glass 1 94 (1960), (e) Ibid, 1962 3 69, (f) (with J. Stevels) Philips Res. Rept. 11 115 (1956).

29. A. Halperin and H. E. Ralph, J. Chem. Phys. 39 63 (1963).

30. J. Mackey, J. Chem. Phys. 39 74 (1963).

31. R. A. Weeks and C. M. Nelson, J. Amer. Ceram. Soc. 43 399 (1960).

32. R. A. Weeks and C. M. Nelson, J. Appl. Phys. 31 1955 (1960).

33. R. A. Weeks and E. Lell, J. Appl. Phys. 35 1932 (1964).

34. R. A. Weeks, Phys. Rev. 130 570 (1963).

35. R. A. Weeks, "Oxygen vacancies", presented at the 2nd meeting on Point Defects in Non-Metallic Solids, British Ceramic Society, Sussex, England, Sept. 26-28, 1966.

36. R. A. Weeks and E. Sonder, P. 869 in Paramagnetic Resonance, Vol. 2 (W. Low, Ed.) Academic Press, New YOrk, 1963.

37. R. Weeks, J. Amer. Ceram. Soc. 53 (4) 176-179 (1970).

38. R. A. Weeks and P. J. Bray, J. Chem. Phys. 48 5 (1968).

39. M. C. O'Brien, Proc. Roy. Soc. A231 404-14 (1955).

40. G. Lehmann and W. Moore, J. Chem. Phys. 44 (5) 1741 (1966).

41. T. I. Barry and W. Moore, Sci. 144 289 (1964).

42. D. R. Hutton, Phys. Lett. 12 310 (1964).

43. W. R. Hechler, Titania in SiO_2, Thesis, Hamburg, 1962.

44. K. Przibram, Luminescence. Pergamon Press, 1962.
45. P. M. Wright, J. A. Weil, T. Buch, J. H. Anderson, Nature 197 246 (1963).
46. C. Boose, K. Robock and W. Klosterkoetter, Naturwissensch. 51 (16) 385 (1964).
47. J. V. Smith and R. C. Stenstrom, J. Geol. 73 (4) 627 (1965).
48. R. F. Tucker, Advances in Glass Tech. 1 103 (1962).
49. D. M. Yudin, Sov. Phys. Sol. State 7 (6) 1399 (1965).
50. P. Levy, J. Chem. Phys. 23 764-765 (1955).
51. P. Levy, Phys. Chem. Sol. 13 287-295 (1960).
52. P. Levy, J. Amer. Ceram. Soc. 43 389-395 (1960).
53. J. S. Van Wieringen and A. Kats, Philips Res. Rept. 12 432 (1957).
54. S. Lee and J. P. Bray, J. Chem. Phys. 39 2863 (1963).
55. G. O. Karapetyan, 3rd All-Union Conf. on the Glassy State [Engl. Translation, p. 319], Consultants Bureau, New York, 1960.
56. J. S. Stroud, J. Chem. Phys. 37 836 (1962).
57. R. Yokota, Phys. Rev. 95 1145 (1954).
58. S. Arafa and A. Bishay, VIII International Congress on Glass, Soc. Glass Tech., London, 1968.
59. S. Arafa and A. Bishay, J. Amer. Ceram. Soc. 53 (12) 390-396 (1970).
60. N. J. Kreidl and J. R. Hensler, J. Amer. Ceram. Soc. 38 423-432 (1955).
61. N. J. Kreidl and J. R. Hensler, P. 217-241 in Modern Materials, Vol. 1, Ed. by H. Hauser, Academic Press, 1958.
62. N. J. Kreidl and G. E. Blair, Nucleonics 14 56-60, 82-83 (1956).
63. L. D. Pye, J. R. Hensler and A. W. Snyder, Oral Communication (1964).
64. P. Beekenkamp, H. J. VanDyk and J. M. Stevels, Proc. VII Intern. Congress on Glass, Brussels, 1965, Gordon and Breach, Publishers, New York, 1966.
65. J. W. H. Schreurs and R. F. Tucker, p. 616 in Proc. Intern. Conf. Phys. Non-Cryst. Solids, 1964, North Holland, Amsterdam, 1965.
66. J. W. H. Schreurs and R. F. Tucker, J. Chem. Phys. 47 818 (1965).
67. V. M. Yanishevskii, T. A. Siporov, V. A. Tyulkin, Neorg. Mat. 516 1159-1160 (1969).
68. T. Yamamoto, S. Sakka, M. Tashiro, J. Non-Cryst. Sol. 1 (6) 441-454 (1969).
69. T. L. Purcell and R. A. Weeks, Phys. Chem. Glass 10 (5) 198-208 (1969).
70. S. I. Sil'vestrovich and T. A. Plisko, Steklo i Ker. 25 12-15 (1968)
71. J. H. Schulmann, R. D. Kirk and E. J. West, p. 3 in Luminescence Dosimetry. Ed. by F. H. Attix, AEC Symposium Series No. 8, 1967. [Available Nat. Bur. Standards].
72. R. S. Barker, D. A. Richardson, E. A. G. McKonkey and R. E. Yeadon, Nature 188 1181 (1960).
73. R. S. Barker, T. K. Karrison, D. A. Richardson, and R. Rimmer, Nature 191 374 (1961).
74. I. H. Malitson and V. J. Dodge, Annual Meeting Optical Soc. of America, 1965.

75. G. A. Haynes and W. E. Miller NASA Rept. TN-D2620. March, 1965.

76. R. G. Downing, F. T. Snively and W. K. Van Latta, AFML-TR-66-224, 1966.

77. B. Gross, Phys. Rev. 107 368 (1957).

78. F. J. Campbell, pp. 19-22 in Proc. 17th Annual Power Sources Conference (1963).

79. E. Lell and R. A. Weeks, Bull. Am. Phys. Soc. II 8 340 (1963).

80. F. C. Hardtke, Phys. Rev. Letters 9 339-361 (1962).

81. F. C. Hardtke and K. R. Ferguson, pp. 369-381 in Proc. 11th Conf. Hot Labs. and Equipment, New York (1963).

82. M. Maklad and N. J. Kreidl, paper submitted to IX International Congress on Glass, 1971.

83. S. C. Rogers, IEEE Trans. Nucl. Sci. NS 10 97 (1963).

84. J. B. Rittenhouse and J. B. Singletary, Space Mat. Handbook, supplement to 2nd ed. "Space Materials Experience" NASA SP 3025-ML-TDR64-40-SUPP1., 1966.

85. S. Cohen, Verres et Refr. 20 (5) 336-342 (1966).

86. R. B. Mazess and D. W. Zimmerman, Science 152 3720 347 (1966).

87. R. Ginther, U. S. Patent 3,396,120. 6 August, 1968.

88. Yu Vbron'ku, A. Kaminsk, V. Osiko, JETP Letters 2 (10) 294 (1965).

89. J. F. Fowler and F. H. Attix in Luminescence Dosimetry. (F. H. Attix, ed.) Academic Press, New York, 1966).

90. K. Ooka and T. Kishii, Yogyo Kyokai Shi 76 (869) 4-10 (1968).

91. K. Ooka and T. Kishii, J. Mat. Sci. 4 (12) 1039 (1969).

92. K. Ooka and T. Kishii, Yogyo Kyokai Shi 78 171-173 (1970); J. Non-cryst. Sol. 3 344 (1970).

93. E. S. Bliss, ASTM Special Tech. Publ. 469 1970 (see also articles by F. W. Quell and others in this issue).

94. C. G. Young, Proc. IEEE 57 1267 (1969).

95. J. E. Swain, P. 69 in ASTM Special Tech. Publ. 469, 1970.

96. R. R. Alfano and S. L. Shapiro, Phys. Rev. Lett. 24 592 (1970).

97. F. Davit and M. Soulie, Compt. Rend., 261 3567 (1965).

98. O. Olness (a) J. Appl. Phys. 39 (1) 6-8 (1968); (b) Appl. Phys. Lett. 8 283 (1966).

99. V. K. Vladimirov, Fiz. Tverd. Tela 9 (10) 2804 (1967).

100. C. R. Guiliano, Appl. Phys. Lett. 5 (17) 137 (1964).

101. N. M. Kroll, J. Appl. Phys. 36 34 (1965).

102. G. H. Conners and R. A. Thomson, J. Appl. Phys. 37 3434 (1966).

103. A. Wassermann, Appl. Phys. Lett. 10 132 (1967).

104. C. E. Garmire and C. H. Townes, Phys. Rev. Lett. 13 479 (1964).

105. R. F. Woodcock, G. A. Granitsas, G. Silverberg, (a) NONR rep. 4656 (00) 1 July-31 Des., 1965, ARPA 306; (b) Final Report, Aug., 1969.

106. J. Paymal (a) Verres et Refr. 15 259, 341, (1961); (b) ibid, 16 20, 100 (1962); (c) Canadian Patent 688,944; (d) J. Amer. Ceram. Soc. 43 430 (1960), 47 548 (1964).

107. Damage in laser glass. ASTM Special Rech. Rept. 469, 1970.

108. Laser Materials. NBS Technical Note 514, 1970.

109. P. LeClerc, Bull. d'Inform. Sci. et Tech. du Commissariat a l'Energie Atomique No. 98.

110. J. Paymal (a) with P. LeClerc, M. Bonnaud and S. de Bonnery, Selected topics in radiation dosimetry, Intern. Atom. Energy Comm. Vienna 1961; (b) with M. Bonnaud, Silicates Ind. (1) 1-15 (1962); (c) with P. LeClerc, J. Amer. Ceram. Soc. 47 (11) 548-554 (1964).

111. R. J. Ginther (a) Colloquium Certain Aspects Dosimetrie des Rayonnem. Agence Intern. Energie Atom., Vienna, 7-11 June 1960; (b) IRE Trans. Nucl. Sci. 5 (3) 92-95 (1958).

112. J. R. Shewell, Paper presented at Amer. Ceram. Soc. Glass Div. Fall Meeting, Bedford Springs. 1970.

113. G. Hass and W. R. Hunter, Appl. Opt. 9 2101 1970.

114. G. A. Haynes and W. E. Miller, NASA-TN-D 2620, 1965.

115. (a) A. R. Bailey and P. D. Townsend, Optics & Laser Technology, 117, August 1970.
        (b) P. D. Townsend, ibid. 65, May 1970.

APPENDIX

Interesting effects in the *uv* range have been reported very recently.[113] If vitreous silica obtained by hydrolysis of $SiCl_4$ is exposed to 1 Mev electrons ($10^{15}/cm^2$) the uv transmittance is reduced below 300 nm, and pronounced absorption at 200-220 nm is produced. Even greater losses occur at higher doses, and a loss of the order of 1% can be expected at 10^{14} e/cm^2. Silica fused from quartz, and, unlike hydrolyzed silica, free from significant amounts of (OH) groups shows much less initial absorption at near 150 nm but its impurity content is cause for greater absorption losses under 10^{15} e/cm^2 than with hydrolyzed silica. Radiation with uv from a mercury-quartz lamp can restore some (but not all) of the lost transmittance,[113,114] as does heating at ~1500°C for hours. One may explain this by the assumption that electron bombardment not only fills but also creates traps. At least in some cases the heat-bleached glass is more sensitive to additional radiation; protons (3Mev) cause similar induced centers to form.[113]

Another recent set of information refers to processing of glass surfaces by ion beams.[115]

DISCUSSION

A. R. Cooper (Case Western Reserve University): Your illustration of metamict SiO_2 (Fig. 1) seemed to suggest the neutron radiated fused SiO_2 is more "ordered" than unirradiated SiO_2. Would you comment on this apparent increase in order?

Author: I did not wish to suggest that metamict SiO_2 is more ordered, on the contrary, it may be less ordered.

DAMAGE TO CERAMICS FROM HIGH INTENSITY Q SWITCHED LASERS

Herbert S. Bennett

National Bureau of Standards

Washington, D.C.

ABSTRACT

One important factor limiting the advance of high power laser technology is the failure of laser materials due to optically induced damage. Examples of surface damage, extrinsic damage produced by inclusions, and intrinsic bulk damage (beam trapping) are presented. A model to treat metallic inclusions which absorb an appreciable amount of incident laser radiation in glass laser rods is formulated and used to estimate thermal stresses and changes in refractive indices due to the thermal stress field. The feasibility of optical techniques to detect incipient damage sites also is discussed.

INTRODUCTION

One of the important factors which is limiting the advance of high power laser technology is the failure or degradation of laser materials due to optically induced damage. In the past, the empirical approach to laser materials has been adequate. That is, the damage or degradation to laser materials has been circumvented by equipment design or by a cut-and-try improvement in the laser materials. Damage and degradation as used above are distinguished in that damage refers to a localized phenomenon which occurs during the course of a very few pulses whereas degradation implies the much more gradual and general deterioration of the laser performance over many pulses. There may of course be some overlap. Only the phenomenon of damage in laser materials will be discussed in this paper. Also, the term laser materials refers in this paper to the more common high performance laser materials such as Cr-doped Al_2O_3

(ruby), Nd-doped laser glass, and Nd-doped yttrium aluminum garnet
(YAG). This problem of optically induced damage now is an obstacle
to increasing the performance characteristics of high-power-pulsed
solid state lasers. It appears that further improvement in laser
materials will require a well integrated and coordinated effort
between the laser materials manufacturer and the user.

Before laser radiation can produce damage, sufficient energy
must be absorbed to initiate some kind of failure mechanism. The
ways in which the laser energy is absorbed to produce damage are
numerous. Some mechanisms require imperfections in the material
that depend upon properties subject to control during the manufacturing process, (e.g., impurity atoms, color centers, or inclusions). Others result from more intrinsic properties and constitute fundamental limitations on the performance of the material.

A large number of variables influence the damage process under
laser radiation. For example, the ambient temperature, the pulse
duration, the manner in which the population of the energy levels
of the lasing ion are inverted, the laser beam diameter, the laser
frequency, and the history of the laser material during manufacturer
all may be significant in particular instances of laser damage.[1]
This number of variables makes meaningful comparisons of various
experimental findings with one another very difficult. It also
complicates attempts to understand the relationships among experimental results, theories, and operational laser systems. In
addition, the probability that a given mechanism occurs depends very
much upon the point of laser operation in this space of many parameters which may vary over decades in value.

In this paper, a brief review of the major causes for damage
in laser materials is presented. In particular, examples of each
type of damage are cited. Then a model to treat absorbing inclusions in laser materials is outlined. Finally, the model is used
to estimate the thermal stresses, the changes in refractive indices,
and the lens effect, and to examine the feasibility of using optical
techniques to detect incipient damage sites before they cause
damage.

SUMMARY OF DAMAGE MECHANISMS

When an inclusion absorbs appreciable energy at the laser frequency, its temperature may rise sufficiently to produce tensile
stresses in the surrounding host material which exceed the breaking
strength of the host. This mechanism is referred to as extrinsic
bulk damage. The resulting internal fractures may either take the
form of an almost perfect disk or a star burst. The discoid fractures are usually a few optical wavelengths thick, and of the order

of millimeters in diameter. They tend to have the normal to their
plane more or less parallel to the laser beam. The star burst
fractures are usually isolated and assume the form of a chaotic-
crushed volume of several millimeters in size.

There appear to be two possible mechanisms by which surface
damage arises; plasma-induced mechanical damage and thermal erosion.
Most researchers agree that plasma formation is present in both
mechanisms, that the state of the surface is important in generating and maintaining the plasma, and that thermal heating by
absorption of the laser light at the surface controls the plasma
initiation. Lubin[2] proposes that a near-surface plasma is formed
so rapidly that local regions of the glass experience an explosive
impact by shock waves from the expanding plasma. The local region
then undergoes Hertzian compression, with the build-up of tensile
forces circumferentially about the impact area. Ring cracks may
form if the tensile stresses are large enough. When the compression
wave reflects from nearby surfaces, and return as tensile waves to
the impact area, sections of this area may be ejected. The thermal
erosion process occurs when the initiated plasma locally evaporates
and erodes the laser material by the bombardment of the surface
with thermally excited ions. It is very likely that the eroded and
evaporated material may contribute to the laser energy absorption
and hence to the growth and extinction of the local plasma. Hence
the erosion process is probably more sensitive to the chemical composition and structure of the surface than the mechanical process.
The terms pitting, crazing, and crenelation usually are used to
describe the features of surface damage. The damage areas may have
linear dimensions from less than a micron to a few millimeters.

Intrinsic bulk damage (beam trapping) in laser materials is
initiated by some mechanism which focuses the laser beam down to a
diameter of the order of microns. This self-focussing mechanism
increases the energy density until the threshold for a fundamental
damage mechanism in an initially, pure nonabsorbing material is
reached. This fundamental mechanism is probably electron avalanche
breakdown and/or electron heating by multiphoton absorption processes.[3] The dominant mechanism which initiates the self-focussing
for a particular example is very sensitive to the pulse duration.
For pulses longer than a nanosecond, the mechanism is most likely
the electrostrictive effect (stimulated near-forward Brillouin
scattering) in which acoustic waves travel transverse to the laser
beam and produce changes in the refractive index. For picosecond
laser pulses the mechanism may be the quadratic Kerr effect or the
combination of absorptive heating and the intrinsic temperature
dependence of the refractive index $(\partial n/\partial T)\rho > 0$ for constant density
ρ. Self-focussing damage tracks are described as filaments which
are continuous or composed of a trail of very small fracture star
bursts. These filamentary tracks may be several centimeters long

and 1-3μm dia. The distance between the very small fractures
(bright spots) is ∿1μm.

Detailed theories exist which treat each of the above mechanisms separately and for a particular range of the laser parameters.
The following two sections contain a theoretical study of the
extrinsic bulk damage due to inclusions.

A MODEL FOR ABSORBING CENTERS

One of the severe problems encountered in high-power-solid-
state laser systems is the thermal damage to laser rods and optical
elements. One such type of damage is thought to arise from metallic
or dielectric inclusions. Such inclusions may absorb an appreciable
amount of incident laser radiation and thereby may undergo thermal
expansion. This produces major stresses within the host material.
Estimating such thermal properties requires the consideration of
solutions to the heat diffusion equation and the thermal stress
equations with appropriate boundary conditions.

Two models to represent the behavior of absorbing centers in
laser materials have been proposed recently.[4,5] The two models
have several features in common; however, they differ slightly in
the boundary conditions at the absorbing center-host interface.
Also, the respective researchers have used their models to answer
different questions. The model of Ref. 5 is outlined below and some
of its important predictions are given in the section on Results and
Conclusions. In particular, the optical path length change for a
probing light ray passing near the absorbing center, the radial and
tangential stress components, and the changes of the refractive
index for radially polarized and tangentially polarized light due
to the thermal stress field have been computed. The dependence of
the maximum value of the tensile stress upon the size of the inclusion and upon the physical properties of the host have been examined.
The feasibility of using optical techniques to detect metallic and
dielectric inclusions in laser materials before they cause damage
also has been studied. These computations suggest that the use of
laser pulse widths of the order of microseconds or longer may be
more promising for the detection of small incipient absorbing centers than the use of nanosecond pulse widths. In addition, the
lens effect has been estimated.

The model for absorbing centers in laser materials contains
many physical assumptions which are necessary to render the problem
solvable. The major assumptions are summarized here.

a)   The absorbing inclusion is a sphere of radius r_0 and is always in good thermal contact with the host. The effects of shape and orientation to the incident radiation are neglected.

b)   The host material is isotropic, continuous and of infinite extend. It also is initially at an ambient temperature T_0 and free from all stresses and strains. Because the energy content of the incident radiation is finite, the latter statement requires the temperature to be T_0 at infinity and all stresses and strains to vanish at infinity. The distribution and nature of microcracks and optical imperfections are not treated in the model.

c)   The linear-thermal-elastic equations are assumed to give a reasonable description of the processes which ultimately may lead to catastrophic damage. They are valid only when a local temperature exists and when distances are larger than atomic dimensions ($\sim 10^8$ cm). A relaxation time t_r for the definition of a local temperature T is approximately the reciprocal of a characteristic vibration frequency of the material. These relaxation times t_r for Pt, Sb, and Al_2O_3 inclusions and for the laser glasses are about 10^{13} to 10^{12} sec. Hence, the equations are physically meaningful only when times t are much larger than 10^{12} sec.

d)   It is assumed that the radiation of heat by the center-host interface and by the heated glass close to the absorbing center may be neglected in the thermal-elastic equations. The laser beams studied in Ref. 4 and Ref. 5 contain energy fluxes at least 10^4 times greater than the energy flux produced by a black body at 600°C. The calculations of Ref. 5 also show that the energy flux due to thermal conduction greatly exceeds the energy flux due to radiation for times <1 sec. Because the temperature is close to the ambient temperature whenever the pulse width of the laser beam is <1 msec and whenever the time after cessation of the laser beam is >1 sec, the long time behavior is not in the region of practical interest for detecting damage centers before they cause damage. Hence, it is assumed that all times are <1 sec.

e)   The linear-thermal-elastic equations contain a coupling term and an inertial term. These two terms may be neglected whenever the three characteristic times which occur in the absorbing center-host system satisfy a set of inequalities. These times are the following. The pulse width τ of the incident radiation determines in part the rapidity of heat generation. The characteristic relaxation time for temperature equilibration (thermal diffusion) is $t_T \sim (r^2/a^2)$, where r is the radial distance from the center of the inclusion and a^2 is the thermal diffusivity. The characteristic mechanical time required for the production of stress waves is $t_M \sim (r/v)$, where v is the speed of propagation of elastic waves. Boley and Weiner have demonstrated that when $\tau \gg t_M$ and $t_T \gg t_M$, then the coupling and inertia terms are small compared to the other terms

in the equations.[6] The above inequalities are reduced to inequalities containing the pulse width τ and the radial distance r;
namely: $\tau \gg 2.8 \times 10^6$ (sec/cm)r and $r \gg 6.7 \times 10^7$ cm. It is assumed
that these inequalities are satisfied and thereby that the coupling
and inertial terms may be neglected.

   f) The absorbing center-host interface is assumed to be an
adhesive boundary. An adhesive boundary maintains the thermal contact between the absorbing center and the host and obtains when the
following three conditions are valid: (1) the hydrostatic pressure
in the absorbing center equals the negative of the radial stress
component in the host at $\underline{r} = r_o \hat{r}$; (2) the tangential stress components are discontinuous across the interface; and (3) the radial
displacement vector is continuous across the interface.

   In summary, the following inequalities describe the regions for
which the model is physically meaningful:

$$10^3 > \tau > 10^{-12} \text{ sec}, \quad 1 > t > 10^{-12} \text{ sec},$$

$$r > 6.7 \times 10^{-7} \text{ cm}, \quad \text{and } \tau > 2.8 \times 10^6 \text{ (sec/cm) } r$$

RESULTS AND CONCLUSIONS

   The numerical results which the model developed in Ref. 5
predicts are summarized here. Among the many input parameters, the
absorptance $A(\lambda,T)$ of the center-host interface is perhaps most
sensitive to the initial thermal contact and surface conditions of
the inclusion and host. The wavelength is λ. The numerical results
are given for the case in which $A(\lambda,T) = 1$. This presents no
additional problem because the temperature $T_h(r,t)$, the optical path
length change ΔL for a probing light ray passing near the inclusion,
the stress components σ_{rr} and $\sigma_{\theta\theta} = \sigma_{\psi\psi}$ for the spherical coordinates
r, θ, and ψ, and the changes in the refractive index due to the
thermal stress field for radially polarized light Δn_r and for tangentially polarized light Δn_θ, are all directly proportional to the
absorptance. The subscript h refers to properties of the host
material.

   The model is used to study three questions. First, how the
maximum tangential-tensile stress varies as a function of the
radius of the spherical inclusion for a fixed energy density E_L and
pulse width of the laser beam τ. Second, how the maximum of the
tangential-tensile stress varies as a function of the thermal conductivity and the thermal expansion coefficient of the host. And
third, if the maximum temperature of the inclusion is limited to a
fixed value, $T_h(r_o,\tau) = 600°C$, how the parameters of the incident
laser beam should be varied to increase the probability of detecting

by optical techniques a small incipient absorbing center before it causes damage.

The bulk thermal, elastic, and optical properties for the laser material used in these calculations are representative of two Nd-doped laser glasses manufactured domestically. The inclusions are Pt, Sb, and polycrystalline Al_2O_3 absorbing spheres.

The variation of the maximum tensile stress $\sigma_{\psi\psi}$ (max-tensile) as a function of r_o for a laser beam having an energy density $E_L = 20$ (J/cm^2) and a pulse width $\tau = 30$nsec is studied. When $10^{-4}cm > r_o > 5x10^{-5}cm$, the maximum tensile stress exceeds by as much as a factor of two the theoretical strength of the glass $(6x10^9 N/m^2)$. The calculations show that Sb and highly absorbing Al_2O_3 inclusions are more likely to cause damage than Pt. The maximum tensile stresses for Sb and Al_2O_3 inclusions are about 20% larger than those for Pt.

These results demonstrate that submicron-sized inclusions have the greatest probability to cause damage in laser glass hosts. Experimental measurements performed by industrial researchers[7] have verified tentatively this theoretical result. Very large $(r_o > 10^{-4}cm)$ and very small $(r_o < 5x10^{-5}cm)$ inclusions are not likely to produce damage.

Intuitive arguments exist to explain why the temperature $T_h(r_o,t)$ and the thermal stress have maxima at some $r_o = r(max \ T_h)$ for fixed E_L and τ and decrease for values of r_o greater than $r(max \ T_h)$ and less than $r(max \ T_h)$. As $[r_o/r(max \ T_h)]$ becomes >1, the volume increases much faster than the surface area of the sphere. The inclusion receives less energy per unit volume. Therefore, the maximum surface temperature at the end of the pulse decreases. As $[r_o/r(max \ T_h)]$ becomes <1, the equilibration time (r_o^2/a_h^2) approaches zero. Then the surface temperature at the end of the pulse cannot deviate much from the equilibrium temperature due to the extremely short equilibration time. Because the thermal stresses are functions of $T_h(r_o,\tau)$ they also exhibit a similar qualitative behavior as functions of r_o.

The maximum tensile stress as a function of thermal conductivity K_h and the thermal expansion coefficient α_h is also studied for $r_o = 5x10^5cm$, $\tau = 30$ nsec, and $E_L = 20$ (J/cm^2). It is found that increasing K_h from 0.008 to 0.04$(W/cm°C)$ with all remaining properties kept constant decreases the maximum tensile stress from about $9x10^9$ to $4x10^9$ (N/m^2). Again, all the other properties of the host are kept the same except for the thermal expansion coefficient. The maximum tensile stress is studied then as a function of α_h. The maximum tensile stresses are found to have minimum values for $10x10^6 > \alpha_h > 8x10^6$ $°C^{-1}$. Values of α_h outside this range yield larger

stresses. They also are slowly varying functions of the thermal
expansion coefficient. Increasing α_h by a factor of three produces
only a 10% variation in the maximum tensile stress. Hence, the
thermal conductivity influences greatly the behavior near the region
of maximum tensile stress and the thermal expansion coefficient
plays only a minor role in the value of the maximum tensile stress.
Among the many possible glass laser hosts, the expansion coefficients
vary by about an order of magnitude from 10^{-6} to about 10×10^{-6} $°C^{-1}$
and all the remaining bulk elastic and thermal properties vary by only
small amounts. The latter variations usually do not exceed a factor
of two. Hence, extensive research on altering substantially the composition of present laser glasses to increase the damage threshold
due to inclusions is not warranted by the predictions of this model.
It suggests that probably at the best, a factor of two increases in
damage threshold due to inclusions could be achieved. *

The model also shows that the maximum value for the tensile
stress does not depend upon the Young's modulus E in a straightforward manner as some researchers have suggested. The Lamé constants are the independent elasticity variables and changes should
be discussed in terms of them. If it were possible to alter only
the Young's modulus, then the stresses would increase in a monotonic
fashion with increasing values of E. But such changes are not
possible in practice. The model predicts that glasses with higher E
values are not necessarily more resistant to damage.

The optimum parameters of the incident laser beam to be used for
detection of incipient damage centers before they cause damage are
examined for the case in which the maximum temperature is 600°C at
the cessation of the laser pulse. The computations indicate that
pulse widths of the order of μsec or longer heat the center more
slowly and thereby produce larger optical path length changes for a
probing light ray near the inclusion. For example, a Pt sphere with
radius $r_0 = 10^{-4}$ cm produces a path length change of about 2×10^{-6} cm for
$\tau = 100 \mu sec$. Researchers[8] using microsecond pulses and holography
have detected 5×10^{-3} cm inclusions without its causing damage. It is
not known yet whether this method can be improved sufficiently to
detect 10^{-4} cm inclusions.

Finally, the lens effect due to heated regions of the host is
estimated. Whenever the refractive index increases with temperature,
$(dn_h/dT_h) > 0$, and even though the inclusion does not produce damage
at its site the heated region of the host surrounding the inclusion
might focus the same laser pulse or a succeeding laser pulse which
occurs after sufficiently short times and before the heated region
cools. The lens effect is estimated by considering the focal length
f of a spherical lens having a radius r and mean refractive index n_1.
The refractive index of the host is n_h and because $(dn_h/dT_h) > 0$;
$n_1 > n_h$. Some numerical examples for platinum inclusions are cited.

---

* This statement does not include possible chemical changes.

When $r_o = 10^4$ cm, $E_L = 20$ (J/cm^2), and $\tau = 30$ nsec, then an effective focal length due to the lens effect occurs at $\tau \sim 3$ μsec and $f \sim 14$ cm. But the tensile stress at $t = \tau = 30$ nsec exceeds the theoretical strength of the host before the lens effect becomes most important. Consider now another example for Pt in which the maximum tensile stress is less than the theoretical breaking strength of the glass host. When $r_o = 10^4$ cm, $E_L = 0.33$ (J/cm^2), $\tau = 30$ nsec, and $T_h(r_o, \tau) = 600°$C, the effective focal length attains a value of about 10^4 cm for times $t \sim 3\mu$sec.

Hence, the model predicts that the lens effect arising from heated inclusions probably does not cause damage. Those cases for which the maximum tensile stress is less than the theoretical tensile strength of the glass have minimum effective focal lengths which are much greater than any dimensions of Nd-doped glass elements used in present laser systems. In those cases for which the minimum effective focal length is comparable to the size of the host, the tensile stress exceeds the theoretical strength of the glass and probably causes damage before the lens effect becomes large enough to heat another inclusion or to initiate an intrinsic damage mechanism such as self-focussing.

In conclusion, the model enables one to answer qualitatively many questions concerning laser damage due to absorbing centers and concerning the detection of such centers before they cause damage.

REFERENCES

1. E. S. Bliss; p. 9 in Damage in Laser Glass. Ed. by A. G. Glass and A. H. Guenther. ASTM Special Technical Publication 469, Philadelphia, Pennsylvania, 1969.
2. M. Lubin, private communication.
3. Y. Raizer, Soviet Physics Uspekhi 8 650 (1969).
4. R. W. Hopper and D. R. Uhlman, J. Appl. Phys. 41 4023 (1970).
5. H. S. Bennett, J. Appl. Phys. 42 619 (1971).
6. B. Boley and J. Weiner, Theory of Thermal Stresses. John Wiley and Sons, Inc., New York, 1960.
7. R. Beck and Haynes Lee, private communication.
8. Elias Snitzer, private communication.

DISCUSSION

L. M. Gold (U. S. Army Frankford Arsenal): Please elaborate on the assumed conditions of compatability at the particle-host interface.

Author: The boundary conditions are:
  a)   The hydrostatic pressure,
       $P_c = -1/3 \, (\sigma_{rr,c} + \sigma_{\theta\theta,c} + \sigma_{\psi\psi,c}) = -\sigma_{rr,c}$, in the
       absorbing center at $\underset{\sim}{r} = r_o \, \hat{r}$; i.e., $\sigma_{rr,c}(r_o,t) = \sigma_{rr,h}(r_o,t)$.
  b)   The tangential stress components are discontinuous across·
       the interface.
  c)   The radial displacement vector $u \, \hat{r}$ is continuous across the
       interface; that is, $u_c(r_o,t) = u_h(r_o,t)$.

EFFECTS OF ULTRAVIOLET RADIATION ON LATTICE IMPERFECTIONS IN PYROLYTIC BORON NITRIDE

John D. Buckley

NASA Langley Research Center

Hampton, Virginia

and

James A. Cooley

Martin-Marietta Corporation

Denver, Colorado

ABSTRACT

Pyrolytic boron nitride was exposed to 310 equivalent sun hours of ultraviolet radiation in a space environment simulator. Lattice parameter comparisons show a definite increase in lattice imperfections in the crystal structure resulting from the ultraviolet irradiation.

INTRODUCTION

Spacecraft temperature control is accomplished by regulation of the radiative interchange of thermal energy between the vehicle and its low-density environment. In order to maintain the temperature of spacecraft subsystems within operating limits, passive thermal control coatings are frequently applied to vehicle surfaces. In cases where a net transfer of energy to the environment is necessary, white thermal control paints having low ratios of solar absorptance to infrared emittance are used.

A requirement for a thermal control coating is that its optical and radiative properties must remain stable when subjected to long periods of exposure to a space environment. Unfortunately, many white

547

thermal control coatings are so affected by exposure to solar ultraviolet radiation and/or indigenous charged particles that their
optical properties change and they no longer control temperature
within desired or even tolerable limits. The primary cause of these
optical property changes is the damage induced in the pigments by
radiation. No white thermal control coating yet formulated is really
optically stable in both electromagnetic and charged particle
space radiation environments. The objective of this investigation,
therefore, was to evaluate pyrolytic boron nitride as a pigment for
a thermal control coating and to identify radiation damage using
X-ray diffraction techniques.

SPECIMENS AND MATERIAL

Pyrolytic boron nitride (produced by Union Carbide Corp.) having
a density of 2.2 g/cm^3 and a purity greater than 99.99% was used in
this investigation.[1] Specimens approximately 2.54 x 2.54 x 0.63 cm
were cut from a bar of pyrolytic boron nitride initially 7.62 x 2.54 x
0.63 cm. [Solid boron nitride would have to be ground into a fine
powder before it could be incorporated into a carrier and used as a
coating pigment. The type of grinding would affect purity and,
depending on the type and level of the impurity, could have an
additional detrimental influence on the thermal control coating.]

EQUIPMENT AND PROCEDURE

The pyrolytic BN was tested for ultraviolet radiation degradation
in a small vacuum chamber evacuated by cryosorption and ion pumping
techniques. A single test specimen was mounted against a metal heat
sink kept at 298°K by flowing H_2O. The heat sink with specimen was
moved and positioned vertically in a silica glass tube which was
part of the vacuum system. A standard Gier-Dunkle integrating sphere
attached to a Beckman DK-2 spectrophotometer was positioned over
the tube so the specimen was centered in the sphere. In this way
spectral reflectance data were obtained directly on the front face
of the specimen *in situ.*

The data reported are corrected for reflectance differences
between specimens exposed in air and through the silica tube. The
solar simulator incorporated a 2.5 kw Xe compact-arc lamp with a
mirror and lens system to focus the radiation from the lamp on the
chamber window; it produced radiation equivalent to the intensity
of three suns.

X-ray diffraction measurements were made in a General Electric
XRD-5 diffractometer using Ni-filtered Cu K$_\alpha$ radiation. Specimens
were scanned before and after exposure to the ultraviolet radiation

to observe changes in lattice parameter of the BN resulting from
radiation damage.

Qualitative analysis was accomplished by fast scanning the
surface of a solid specimen at 2Θ = 2°/min over the range
15°$\geqslant 2\Theta \geqslant$75°; "d" spacings were compared with those suggested for BN
by the vendor[1] and the ASTM card file.[2] Lattice parameters were
determined by scanning over the major peak (002) of the BN pattern
at a speed of 0.2°/min. Angular locations of this peak before and
after exposure to radiation were determined from the bisectors of
the half-height lines.

RESULTS AND DISCUSSION

Figure 1 shows specimens as-received and exposed to 310 equivalent sun hours (ESH) in the Martin-Marietta solar simulator. The
discoloration (light brown) of the exposed surface suggests that it
is not as resistant to ultraviolet radiation as the ZnO - K_2SiO_3
pigments (e. g., Z93) now used as thermal control coatings (Fig. 2).

Comparison of fast scanned patterns of as-received pyrolytic
BN and the specimen exposed to 310 ESH of ultraviolet radiation indicated variation in "d" spacing due to radiation damage. A slow scan
of the major (002) BN peak revealed distinct differences in peak
location and geometry before and after exposure to ultraviolet
radiation (Fig. 3). The midpoint of the half-height line is
2Θ = 26.02° for the as-received material, corresponding to a "d"
spacing of 3.42 Å. This value is within the range of "d" spacings
commonly found for pyrolytic BN.[1] The comparable value for exposed
specimen is 2Θ = 26.25°, representing an apparent decrease in
crystal lattice "d" spacing from 3.42 Å to 3.39 Å. Evidently, the
BN crystal lattice contracted as a result of exposure to ultraviolet

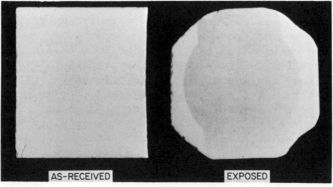

AS-RECEIVED   EXPOSED

Fig. 1. Pyrolytic BN specimens as received and after exposure to
310 ESH of uv radiation.

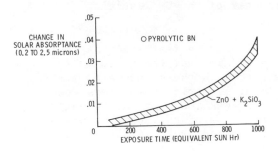

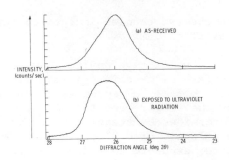

Fig. 2. Ultraviolet solar absorptance data on pyrolytic BN and ZnO-K₂SiO₃.

Fig. 3. X-ray diffraction patterns for (002) plane of pyrolytic BN in (a) the as-received condition and (b) after exposure to uv radiation for 310 ESH.

radiation. The increased area and rounded apex of the (002) X-ray peak of the exposed specimen (Fig. 3b) when compared to the sharper apex of the as-received BN pattern (Fig. 3a) may be indicative of decreasing order in the periodic arrangement of atoms. This decrease in long-range order (periodic arrangement) in irradiated BN is attributed to an increase in crystal imperfections produced by the ultraviolet radiation. The good thermal conductivity of boron nitride suggests that the damage (brown discoloration) observed on the surface of the specimen was due to radiation rather than to heating produced by the ultraviolet source.[4]

The significance of the X-ray patterns (Fig. 3) is that pyrolytic boron nitride, a high temperature refractory candidate for white thermal control coatings, has shown a measureable change in crystal structure resulting from exposure to the ultraviolet component of simulated solar radiation. This crystallographic change has not been observed in other refractory oxides making up most of the candidate white thermal control coating pigments.[5] An explanation for the absence of observable crystallographic change in refractory oxide ceramics subjected to ultraviolet radiation is attributed to the ease with which the highly electronegative oxygen in air is absorbed into irradiated oxygen-poor refractory oxide ceramics when removed from the solar simulator. Since BN does not absorb nitrogen and heat (as do the oxide ceramic materials) after exposure and removal from the solar simulator these crystallographic changes provide an alternate interpretation of damage mechanisms occurring on refractory white thermal control coatings when subjected to ultraviolet rays from the sum.

CONCLUDING REMARKS

Changes observed in the diffraction angle and geometry of the (002) X-ray peak of pyrolytic boron nitride shows that ultraviolet radiation produces imperfections in its crystal lattice. These imperfections are believed to produce a decrease in the interplanar "d" spacing resulting in lattice contraction in the surface layers of irradiated pyrolytic BN. This sensitivity to radiation damage makes pyrolytic boron nitride unsuitable as a pigment for thermal control coating.

REFERENCES

1. M. Brasche and D. Schiff, in Design Eng., 78-81, Feb. 1964.
2. American Society for Testing Materials "ASTM X-Ray Diffraction Card on Boron Nitride 9-12." (Data obtained from Hadfields Ltds., Sheffield and A.E.R.E., Harwell.)
3. G. A. Zerlaut and L. Node, "Development of Space-Stable Thermal-Control Coatings". Rept. No. IITRI U6002-63, April 1968.
4. T. A. Ingles and P. Popper pp. 144-67 in Special Ceramics. Ed. by P. Popper. Academic Press, New York, 1960.
5. W. S. Slemp,(NASA Langley Research Center), Personal communication Nov. 1970.

DISCUSSION

R. N. Katz (Army Materials and Mechanics Research Center): Which type of pyrolytic BN did you use, anisotropic or isotropic?
Authors: Anisotropic.

I. Ahmad (U. S. Army Watervliet Arsenal): Could the brown color on the uv-irradiated BN specimen be due to the formation of boron by photodecomposition of BN?
Authors: Yes.
I. Ahmad: What was the temperature of the surface during irradiation?
Authors: Approximately 25°C.

G. C. Kuczynski (University of Notre Dame): In the X-ray diagram (Fig. 3b) there is a suggestion that there are two poorly resolved peaks very close to each other. These two peaks can belong to two different phases (including ordering). In view of this diagram (b) can be interpreted as follows: the phase characterized by the higher angle peak is enhanced by uv radiation, resulting in a broad peak where two peaks are not resolved.

Authors: We are not aware of any X-ray evidence supporting the existence of nearby peaks attributable to second phases or related to ordering in BN. Thus we have interpreted the observed changes in X-ray patterns in terms of a real change in crystal structure resulting from radiation damage.

S. W. Bradstreet (Consultant - Dayton, Ohio): BN is a layered, hexagonal, highly anisotropic structure comparable in many ways to graphite. In the ideal crystal, alternate atoms around a regular hexagon are B and N, so each lies in a trigonal planar array with three neighbors of the other kind, presumably by a complete sharing in saturated covalent (and probably hydridized) Sp^2 electrons. In the ideal lattice, the unused electrons of boron atoms form saturated covalent pairing with alternate N atoms in the plane above or below in such a way that the rule of 8 is satisfied, and unlike carbon in graphite, no electrons for conduction are left over in the plane.

This ideal structure is seldom achieved. The strong B-B bonds may form interfering pairs or trigonal units for which interplanar nitrogen atoms or planar nitrogen position vacancies become stable. Thus in hot-pressed BN, pyrolytic BN, and other achievable solids, (as with polycrystalline and pyrolytic graphite) the ideal indexing of one layer with respect to the layer above it or below it is seldom achieved.

On such real structures, the added local photon energy of uv energy will be most notable on those atoms which prevent the BN from achieving the ideal (i.e., the least energetic) configuration. Thus $\bar{N}^3$ ions may be converted to N° atoms (which are very small and may be accepted at interplanar hexagon centers without strain). B-B bonds will not be broken by such moderate energies, but will now form domains in which electron densities are sufficiently high to act as optical absorbing regions, with consequent discoloration and measurable increase in surface emittance. (The ideal BN lattice is surprisingly transparent to uv, in the planar directions, and penetration effects may be several hundred $\overset{\circ}{A}$ deep).

The effect of irradiation, then, will be (at ordinary temperatures) to allow the BN planes to approach each other more closely where the (larger) N^{-X} atoms are lost or converted to more positive valancies; there will thus be a reduction in *average* interplanar spacing as those regions in which interfering nitrogen interstitials are released or made smaller. Since some interplanes are not affected, line broadening will be observed.

On remedial treatment (exposure to heat and/or air) oxidation of some B-B bonds will recreate transparency and lowered emittance, but original sharp peak d-spacing will not be recovered, since a wide variation in interplanar spacings will still exist.

In substance, the effect is similar to the thermal neutron irradiation of pyrolytic graphite above 1100°C. Changes in BN a-spacing should be followed by electron diffraction.

THE USE OF COMPUTER EXPERIMENTS TO PREDICT RADIATION

EFFECTS IN SOLID MATERIALS

J. R. Beeler, Jr.[*]

North Carolina State University

Raleigh, North Carolina

ABSTRACT

Prediction of radiation effects requires an analysis of defect annealing in addition to the analysis of the primary damage state directly produced by irradiation. The transient and steady states associated with defect production and annealing depend upon the particular types of defect configurations possible under irradiation conditions. Many of the important defect configurations are much different in character than those which can be produced thermally. Experience has shown that the analysis of radiation effects must be based on discrete particle interaction rather than continuum models, providing one of the main reasons for using the computer experiment method in radiation effects prediction. The methods used to simulate atomic collision cascades in solids and to simulate the crucial initial stages in the subsequent defect annealing process are described and illustrated with results from computer experiment simulations. Special emphasis is placed upon the influence of ceramic-like precipitates on defect production and annealing in reactor structural materials.

INTRODUCTION

This article describes computer simulation of irradiation induced crystal lattice defect production and of the annealing process for these defects. Particular emphasis is placed on simulation of collision cascades in BeO and on simulations which describe the

---

[*] Currently Battelle Visting Professor, Ohio State University.

the effects of ceramic precipitates on irradiation damage-anneal processes in metallic systems.   The discussion is limited to topics concerned with neutron irradiation.   Prominence is given to defect aggregation mechanisms and properties which bear on the nucleation, growth and stability of voids in metallic systems.

It will be helpful to begin by outlining the general nature of a computer simulation approach for the problem at hand.   To a great extent computer simulation of neutron irradiation induced defect production and annealing is an empirical method.   First of all, experimentally measured neutron cross sections are used, in a straightforward Monte Carlo simulation of fast neutron slowing down, to compute (1) the number density of high-energy atoms displaced as a direct result of neutron collisions, and (2) the energy spectrum of these high-energy atoms, which are called primary knock-on atoms (PKA).   A schematic description of PKA production by a fast neutron appears in Fig. 1.   Each PKA immediately initiates an intricately branched sequence of atomic collisions (collision cascade).   In most cases, each cascade leaves an aftermath of crystal lattice defects many of which are isolated vacancies and interstitials, but an important fraction of them are made up of (a) vacancy and (b) interstitial aggregates.   In addition, there are defect aggregates comprised of impurity atoms combined with either vacancies or interstitials.

The second step is collision cascade simulation for selected PKA energies over the entire PKA energy range and on the basis of experimental data as well. Specifically, the outcome of each collision in the cascade simulation is decided on the basis of a semiempirical atomic interaction function.   Normally, the elastic constants and phonon dispersion data are used to fix the slope and curvature of the atomic interaction energy, as a function of the separation

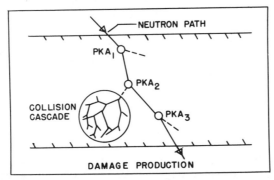

Fig. 1.   Schematic of PKA production by a neutron collision chain in a finite sample, and the initiation of atomic collision cascades.   The insert is a greatly magnified view of the collision cascade initiated by PKA_2.   The distance between PKA production points is ~1cm.   A typical cascade region is about 100Å dia.

distance between two atoms. The strength of the interaction is set
up on the basis of available data for the sublimation, stacking
fault, surface and the vacancy formation energies. Shock wave data
can be used to characterize close atom encounters, and rare gas
atom collision data are used to characterize very close atom
encounters in which the inner electron shells of the colliding atoms
overlap. For the most part, cascade simulation is performed on
the basis of classical mechanics theory for elastic collisions.
Inelastic collision energy loss corrections are introduced, however,
whenever the geometric and energetic conditions for a significant
inelastic collision interaction occur.

The third major step in an overall irradiation effects simulation
is that of defect annealing. Annealing simulation results depend
sensitively on the initial instantaneous defect distribution at
the start of the annealing process. Simulation of PKA production
by fast neutrons (already described) provides a statistical description
of the instantaneous spatial distribution of atomic collision cascade
centers produced by neutrons having a given energy distribution.
These cascade centers are the points at which the PKA are produced.
In addition, it also provides a statistical description of the energy
of the PKA which initiated a cascade at a given center. The information provided by collision cascade simulation completes the picture
of neutron irradiation induced crystal lattice defect production.
Namely, cascade simulation gives a description of the number of
defects and the distribution of defects by a cascade initiated by
a PKA with a given energy. This atomic level description of the
instantaneous defect state produced by neutron irradiation is the
input information needed for the simulation of defect annealing.
Annealing simulation describes changes in the defect distribution
as a function of time between successive random additions of new
defect production by subsequent neutron collisions in the region
concerned. The irradiation damage-anneal process can be thought
of as an indefinitely extended sequence of randomly superposed
fast-quench and ageing operations. The fast quench corresponds to
localized supersaturated defect production by a cascade during a
time interval of about 10^{-14} sec. The ageing time is the time between
successive neutron collisions in the local region concerned and is,
therefore, a random variable. In this context, the local region
concerned is $\sim$100 Å dia. Figure 2 is a schematic outline of the
process described above.

The use of these computer techniques for irradiation effects
prediction in engineering design work came about largely because
the particular effects observed are highly sensitive to the details
of the irradiation history and the microscopic state of the irradiated
material. This is a natural consequence of the nature of the neutron
irradiation process. Analysis of this process requires a description
of particular collision sequence combinations for both neutron and
atom collision chains. The latter case requires a good description of

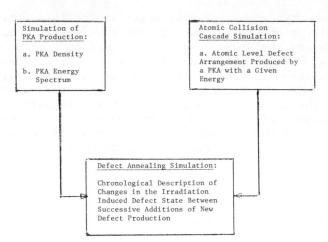

Fig. 2.   Diagram of the interrelations among the three principal
          types of computer simulations used to describe irradiation
          effects in crystals.

collision trajectories. The only way yet known for handling such
problems is computer simulation. This is true because there is
(1) no practicable mathematics for handling complex combinational
problems and (2) no practicable mathematics for handling many-body
problems dealing with arrays of discrete particles which are not
highly symmetrical and/or periodic.

     As mentioned above, collision cascade simulation is done for
both reasons. If one traces a pencil point along the various collision chain branches on a cascade trajectory map (see Fig. 6), it
is at once evident that a description of these simultaneously
evolving trajectories presents insuperable permutational and combinational problems in a conventional approach. These problems are
insuperable because there is no way to ask for and to execute
particular decision steps in a conventional calculation. But when
one can follow each atom along and "see" when its path begins to
interact with the paths of each other atom, as does the computer
program, the problem reduces to properly queueing energetic atoms
to take turns in moving ahead, on the basis of the one that has
the largest velocity being given first place in the queue. This
simple scan-and-queue operation solves the permutational and combinational problems.   At the same time, it also solves the problems
of handling the geometrically asymmetrical shape of the collision
cascade and irreversibility and mechanical locking aspects in the
energetic and defect structure senses, respectively. Finally, having
replaced analytical mathematical techniques with simple scan-and-queue
procedures in order to dispense with the combinatorial, asymmetry,
and irreversibility problems, one finds computer simulation to be
the quickest way to execute the practicable techniques.

PKA DENSITY AND ENERGY SPECTRUM CALCULATIONS

Given the type of target nucleus and neutron energy, the possible outcomes of a particular neutron-nucleus collision are described statistically by the relative magnitudes of the neutron reaction cross sections at the energy in question. The probability for some type of reaction to occur per unit path length is given by the total neutron reaction cross section. By using (1) this cross section to select distances between successive collisions with nuclei and (2) the relative magnitudes of the cross sections for the various reaction types as the basis for selection of a particular type of reaction at each collision, it is possible to describe the production and the energy spectrum of PKA by fast neutrons. Since the outcome of each event is described statistically by the experimental data at hand (the cross sections and material composition), it is natural to use a Monte Carlo (statistical sampling) technique to simulate neutron collision histories.

The analog Monte Carlo method is perhaps the most flexible and convenient method for doing a PKA energy spectrum and density calculation (abbreviated PKA spectrum calculation) for a finite sample.[3,4] Given the energy spectrum and angular distribution of neutrons incident at the sample surface, a Monte Carlo method is used to trace out neutron collision histories in the finite sample. A schematic description of the problem appears in Fig. 2. Materials containing up to 10 different chemical elements can be treated in such a calculation. Cross sections for elastic and inelastic neutron scattering and neutron absorption are used to describe the relative frequencies of the different collision modes between a neutron and the atomic nuclei in the sample. The PKA energy spectrum and density distribution are obtained in histogram form by tallying the energy transfers to struck nuclei and the spatial region in which each neutron collision occurred. Since complete, energy-dependent cross section information is needed to trace out the neutron collision histories, it is also possible to tally transmutation and other absorption events of interest in each particular case, i.e., H-production, He-production, B-fission, etc. Therefore, in a Monte Carlo PKA spectrum calculation it is possible to obtain the following information on the characteristics of the PKA produced during irradiation by neutrons from a given energy spectrum and angular distribution: (1) PKA density per unit neutron flux for each chemical element in the sample (PKA yield) as a function of position in the sample; (2) PKA energy spectrum for each chemical element in the sample as a function of position in the sample; (3) transmutation element atom density per unit neutron flux as a function of position in the sample for type of transmutation element, and (4) transmutation element atom energy spectrum for each type of transmutation element.

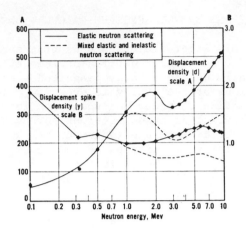

Fig. 3. Displacement density (Frenkel pair density), d, and displacement spike density (PKA density), y, per unit neutron
exposure as a function of neutron energy for a square-base
(0.28 x 6.35 cm long) specimen of α-iron.

A desirable flexibility in the use of the information obtained
from a PKA spectrum calculation is obtained if neutron histories
are traced down in energy to the point where no energy transfer to
an atom can be realized which exceeds the sublimation energy for
that atom. This procedure will give an energy distribution for
struck atoms which includes many atoms with energies so low that
they would not actually be displaced from normal sites in a perfect
crystal. Such a distribution can be renormalized easily, however,
on the basis of any appropriate "displacement energy" (or energy
distribution) model or for a particular case at hand. These models
are pertinent when one wants to account either for directional effects
or for the presence of previously produced damage.

Figure 3 illustrates the behavior of the PKA density as a
function of neutron energy in α-iron.[5] In the language of the present
article, the words "Displacement Spike density (y)" in Fig. 3 should
be read as being "PKA density (y)". The effect of inelastic neutron
collisions is also illustrated. Taking inelastic neutron scattering
into account, the general trend is for local PKA production by fast
neutrons to decrease with neutron energy in a *finite* sample of
material.

COLLISION CASCADE SIMULATION

The ideas and techniques described in the classic paper of
Gibson *et al.*[6] (called the dynamical method) form the basis for
nearly all of the current computer experiment simulation work on
crystal lattice defects. These investigators treated the production

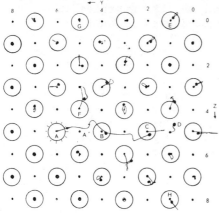

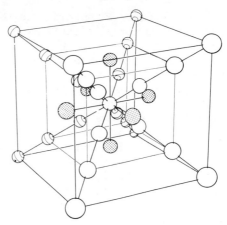

Fig. 4. Atomic orbits in Cu pro-   Fig. 5. A recombination region
        duced by the movement of a   for a vacancy and an inter-
        40 ev Cu knock-on atom   stitial in α-iron at 0°K.
        started at position A   A split interstitial
        and directed 15° above   centered at any of the non-
        the -y axis. Large circles   central positions will
        and smaller dots give   recombine with a vacancy at
        initial positions of atoms   the centered position.
        in the (100) plane and in
        the plane below, respec-
        tively. Vacancy is created
        at A; a split interstitial
        at D.

of a stable Frenkel pair[*] in copper which they simulated by ejecting
an atom from a normal lattice site with several energies ranging up
to 100 ev in each of several different directions. They thereby
determined *minimum* energy for the pair-production as a function
of the direction in which an atom is ejected from a normal lattice
site. These minimum energies are called the directional displacement
threshold energies for Frenkel pair production. The displacement
threshold energy measured experimentally is usually the average of
the directional displacement threshold energies. This classical,
many-body calculation in which the 3N Newtonian equations of motion
for N atoms in a crystallite are solved simultaneously using the
central difference approximation numerical technique for the solution
of differential equations. The forces among the atoms were derived
from a semiempirical potential energy interaction function. An
example of a displacement simulation for a 40 ev copper atom is
given in Fig. 4.

---

[*] A vacancy and an interstitial atom so positioned that the atom
    can not spontaneously recombine with the vacancy.

In reactor engineering applications, the PKA energies of greatest importance with respect to displaced atom production usually range up to 500 kev., i.e., those in this energy range produce damage which makes the dominant contribution in limiting the service life of a structural reactor component. Unfortunately, it is not practicable to simulate collision cascades initiated by PKA with energies above 0.2 kev using the dynamical method because of the computer time costs. Consequently, most cascade simulation work, being in the kev range, is based on the "branching sequence of binary collisions approximation", or briefly the "binary collision approximation".[7] The validity of this approximation is based on the fact that atomic collisions at energies above about 15 ev take place so rapidly that only the direct participants make any sensible local movement during the collision process. This is nicely illustrated in Fig. 4 for quite low energies, in fact. Except for a few special cases connected with focused collision chains, the direct participants consist of a pair of colliding atoms. In the focused collision special cases, it turns out that the collision events can be accurately described as a superposition of simultaneous binary collisions.

The validity of the binary collision approximation, however, is not alone sufficient to allow one to simulate damage production by high energy collision cascades. Also needed is a workable criterion for distinguishing between unstable and stable Frenkel pairs. Prior to the dynamical calculations of Gibson *et al.*, no such criterion existed. The dynamical method, however, allows one to introduce a well defined recombination region V_R within which a vacancy and an interstitial center site can not coexist. The recombination region is defined in terms of a collection of separation distance vectors between the centers of the two elemental defects, rather than upon a scalar radial separation distance. Any Frenkel pair characterized by a separation distance in this collection is unstable; all others are stable.

By basing Frenkel pair stability decisions on this criterion, the collision trajectory of each energetic atom can be traced in a rigid lattice structure down to an energy at which actual tracing of the trajectory shows it can not penetrate the current cage of nearest atoms surrounding it, and at which time it then becomes an interstitial atom candidate. Whether or not this interstitial atom candidate is in fact stable, with respect to the current local distribution of vacancies, is determined using the recombination region criterion. In this way, a cascade can be simulated without having to invoke a displacement threshold energy. Each energetic atom is allowed to collide about as long as it can make headway. When it becomes collisionally hemmed in as an interstitial atom, a test is made to see if it should be recombined with a local vacancy. In effect, this procedure grafts the many-body dynamical

calculation results for Frenkel pair stability onto the terminal
part of each binary collision approximation cascade branch.   Figure 5
is a perspective drawing of the three-dimensional recombination
region in α-iron computed by Erginsoy *et al.*[8] using the dynamical
method.   The separation vectors which define this region are <111>,
<200>, <222> and <333>.   Counting the central site, 31 atom sites
are involved.

Fig. 6 shows an atom trajectory map for an atomic collision
cascade in BeO initiated by a 10 kev iodine ion.[9]   Comparisons with
similar plots at 5 and 50 kev (not illustrated) show that as the
energy of the initiating particle increases, the shape of the
cascade becomes progressively more elongated.   This is typical of
collision cascades in metals as well as in ceramics.   These cascades
in BeO were among the first complete cascade simulations performed
using the binary collision approximation.   The methods developed
for cascade studies in BeO were later refined and used to make
extensive calculations for high-energy cascades in α-iron, tungsten
and copper.

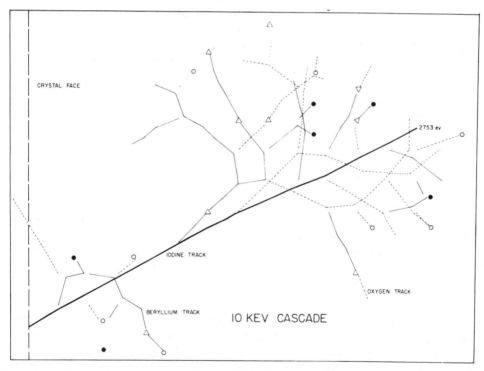

Fig. 6.   The initial part of a collision cascade initiated by a
10 kev iodine ion in BeO.   Note that the trajectory map
has been drawn for the part of the cascade during which
the iodine ion slows down from 10 kev to 2.75 kev.

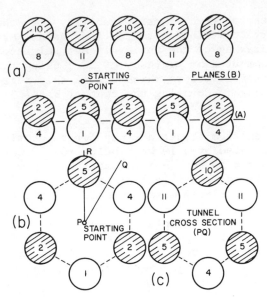

Fig. 7.   Wurtzite structure of BeO, (a) basal plane A,(b) top view
          of plane A (also the cross section of the tunnel-like
          confinement geometry for pure channeling along the
          c-axis), (c) cross section of (b) for pure channeling
          along the PQ direction between basal planes. Atoms
          ejected (started) along directions in plane A are deflected
          into a trajectory confined between basal planes. Atoms
          ejected (started) along directions in plane B remain in
          plane B.

     A description of the arrangement of atom types in the BeO
wurtzite structure is given in Fig. 7. Two main channels exist in
the wurtzite structure along which energetic atoms can penetrate
long distances. One is a hexagonal ring channel along the c-axis
and the other is an asymmetrical cross section channel parallel to
the basal planes. Confinement of energetic atom trajectories between
basal planes is prounced in BeO and has an important effect upon
the distribution of damage left by cascades in BeO. A similar
confinement of energetic trajectories between (110) planes occurs
in bcc metals.

     A schematic of the cascade structure in BeO for a 9 kev Be PKA
appears in Fig. 8. It represents the ejection of a 9 kev beryllium
atom from a normal lattice site in a randomly chosen direction.
Most of the atom displacements occurred in a planar region, parallel
to the basal plane, 4 Å thick and ~30 Å dia. The longitudinal cross
section of this region is indicated by the rectangular outline in
the figure. The arrangement of vacancies in the vacancy-rich planar
damage region is shown in Fig. 9. Figure 10 shows the small amount

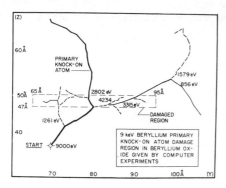

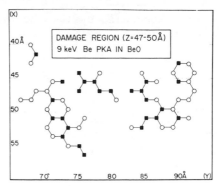

Fig. 8. Trajectories of a 9 kev beryllium PKA in BeO and the principal secondary knock-on atoms it produced.

Fig. 9. Basal plane projection of vacancy strings as outlined in Fig. 8. Filled squares are vacancies, open circles are sites from which atoms were temporarily ejected but were not sufficiently displaced to prevent relaxation onto a normal lattice site.

of damage produced in the basal plane above the damaged region. This paucity of damage was also observed in the basal plane below the central damaged basal plane. The geometry of the damaged region produced by oxygen PKA was also typically planar. Figure 11 is a map of the planar damage region produced by a 16 kev oxygen PKA in BeO.

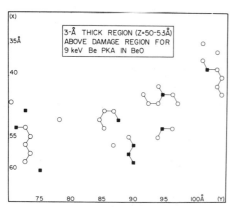

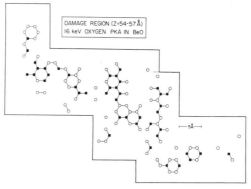

Fig. 10. Vacancy production in the 3 A-thick layer above the central damage region.

Fig. 11. Planar damage region produced by oxygen PKA in BeO.

In a monatomic system the displacement efficiency K(E) of a PKA is defined by the relation

$$g(E) = K(E)E \tag{1}$$

where g(E) is the average number of displaced atoms (stable Frenkel pairs) produced by a cascade initiated by a PKA with energy E. The displacement efficiency for a binary system is a four-element matrix defined by the relations

$$g_{ij}(E_i) = K_{ij}(E_i)E_i \tag{2}$$

where $g_{ij}(E_i)$ is the number of type-j atoms displaced in a cascade initiated by a type-i PKA with energy E_i. The K_{ij} for BeO expressed as displacement per kev are $K_{11} = 3.85$, $K_{12} = 2.02$, $K_{21} = 3.92$ and $K_{22} = 2.02$ where the index 1 signifies a beryllium atom and the index 2 signifies an oxygen atom. E_i is expressed in kev.[10] These K_{ij} were obtained by averaging the number of displacements produced by 10 cascades initiated by Be- and by O- PKA at each of several PKA energies over the range 1 - 16 kev.

The displacement efficiencies of Eq. (2) were used in conjunction with the results of a Monte Carlo PKA spectrum calculation for BeO to compute the number of Be and O atoms displaced per neutron as a function of neutron energy.[11] The results of this calculation are summarized in Fig. 12.

Figures 13 and 14 describe channeling trajectories in BeO.[1] These figures pertain to channeling parallel to the basal plane, i.e., along the direction PQ in Fig. 9. The different semi-helical

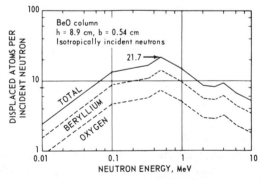

Fig. 12.   Number of displacements in a 0.54 cm BeO column as a function of neutron energy.   Isotropic incidence.

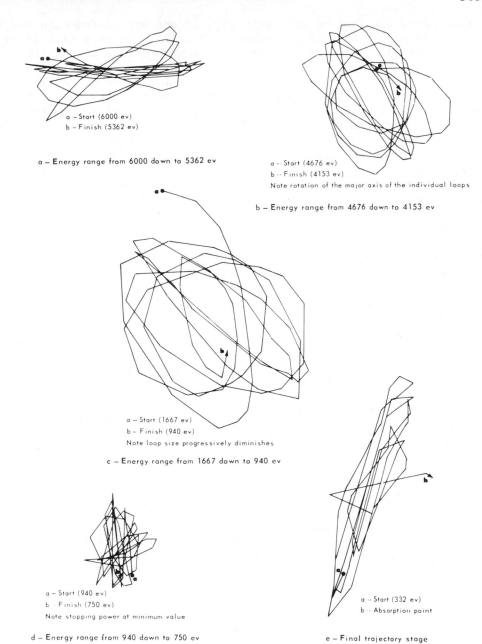

a – Start (6000 ev)
b – Finish (5362 ev)

a – Energy range from 6000 down to 5362 ev

a – Start (4676 ev)
b – Finish (4153 ev)
Note rotation of the major axis of the individual loops

b – Energy range from 4676 down to 4153 ev

a – Start (1667 ev)
b – Finish (940 ev)
Note loop size progressively diminishes

c – Energy range from 1667 down to 940 ev

a – Start (940 ev)
b – Finish (750 ev)
Note stopping power at minimum value

d – Energy range from 940 down to 750 ev

a – Start (332 ev)
b – Absorption point

e – Final trajectory stage

Fig. 13. Projected image of a channeled 6 kev oxygen atom onto a plane normal to the channel axis (see Fig. 7b). Projections (a), (b), and (d) portray pure channeling with no damage production. Projections (c) and (e) are for trajectories associated with occasional Frenkel pair production. The length of the entire channeled trajectory was 2000Å. (See Fig. 4).

trajectory cross section shapes in Fig. 13 are associated with varying degrees of energy loss as described by Fig. 14. In this example of pure channeling, the atom trajectory is confined to a tunnel-like region in the BeO crystal, i.e., it is a nearly one-dimensional trajectory. This channeling influence, the influence of regular atomic arrays to confine energetic atom trajectories to a strictly defined form, is the basis for the planar damage regions observed in the cascade simulations for both Be- and O-PKA in BeO. Energetic atom trajectories also tend to be confined between two neighboring basal planes and exhibit a nearly two-dimensional character. This confinement is called quasi-channeling[7,12] since it effectively removes only one degree of freedom whereas pure channeling effectively removes two degrees of freedom from the dimensionality of the atom trajectory. Pure channeling is a relatively rare event but quasi-channeling occurs in some degree for almost every trajectory in a cascade.[12] Each time a quasi-channeled atom is deflected back into the interplane confinement region, small groups of vacancies are produced as shown in Fig. 9 and Fig. 11. PKA trajectories which channeled along the c-axis were observed to adopt pure channeling trajectories, as a rule, and they typically produced very little damage. This important crystal structure effect (quasi-channeling) upon the distribution of cascade produced defects was first predicted by computer experiments. It has since been confirmed by field ion microscopy in W and α-Fe.[13] The pure channeling process was confirmed experimentally in an extensive series of ion bombardment experiments.[14]

Work currently in progress indicates that the channeling trajectory directions found for high energy particles in crystals are also the essential trajectories for diffusing atoms at thermal energies.[15] Although one usually thinks of a diffusing atom as moving along the close-packed line in diffusion by vacancy mechanism, closer examination shows that this is not the fundamental guiding principle; rather, it seems to move along the easiest channeling directions. Hence in α-Fe, the diffusion path is a collection of [001] and [110] segments rather than a single [111] segment. In fcc systems it is a coincidence that the close-packed direction is also the easiest channeling direction.

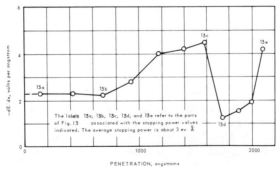

Fig. 14.   Stopping power $-dE/dx$ for a 6 keV oxygen channeling trajectory as a function of penetration distance in BeO.

SIMULATION OF DEFECT ANNEALING

Only a relatively few annealing simulation studies have been made. Aside from primitive work for BeO, none have been performed for a ceramic system. Simulation studies for metallic systems are limited to α-Fe.[5,7,16] We will briefly discuss the Monte Carlo method[2] and describe results pertaining to the influence of ceramic precipitates and pre-precipitates on defect annealing in metallic systems.

The annealing simulations for α-Fe have divided themselves into two classes, short-term and long-term annealing. The absolute time scale associated with short-term annealing depends, of course, upon the temperature. During short-term annealing, the movement of each mobile defect is generally influenced by the relative locations and orientations of all other defects in its vicinity. This circumstance is the basis for the influence which the initial deployment of defects in the primary damage state has upon the character of the annealed defect state. It is not correct to think of each mobile defect as moving about in a random walk type of excursion during the annealing process. Rather, the "first" defect to move usually has an immense effect upon the nature of the defect distribution after short-term annealing. A description of strongly interrelated defect motions can be obtained on a statistical basis if one knows the activation energy for the movement of a given defect in a particular direction as a function of how other defects are deployed in its immediate vicinity. It turns out that it is possible to obtain this type of information using defect interaction computer experiments. We will assume this information is known and continue with a description of the essential constitution of an annealing simulation calculation.

The key idea is to list each vacancy and interstitial as if it were a separate, independent defect element in an elemental defect table, even though it may be a member of a defect cluster.[2] Each elemental defect in this table is assigned a permanent iden-tification number, a type number (vacancy or interstitial), and a class index which is equal to the size of defect cluster to which it belongs. In addition, each defect carries an adjacency number, which is the identification number of one other defect adjacent to it in any cluster to which it may belong. This simple scheme is flexible enough to describe all contingencies in a vacancy-interstitial annealing process. The basic block diagram for the annealing program is given in Fig. 15. The basic time interval for the calculation is commonly assigned a value such that the jump probability for the fastest moving defect during this interval is 0.1. This assignment appears to provide a useful compromise between computational efficiency requirements and the necessity for accounting for possible multiple jumps (fluctuations) during the "average" time between jumps. During each successive time interval, each extant elemental defect

in the defect table is selected at random without replacement and
its opportunities for movement are examined. If it is physically
possible for this defect to move, a possible crystallographic jump
direction is selected at random and the migration energy for this
direction is constructed from precomputed tables of activation
energies for correlated defect movement. A random number is selected
and compared with the Boltzmann factor defined by the migration energy.
If this random number exceeds the Boltzman factor, the defect is
not moved and another defect is selected from the defect table for
examination.

     Annealing simulations for cascade produced defect states show
a short-range annealing regime of about a microsecond in which
strongly interacting defects regroup themselves, with about 90% of
them being annihilated. The survival of defects appears to correlate
with the presence of large vacancy clusters in the primary damage
state produced by the cascade. During short-term annealing the
interstitials tend to cluster and, as it appears from dynamical
simulations, transform from cluster to loop configurations at about
16 interstitials.[17] Variational calculations indicate that vacancy
loops do not start to form until about 200 vacancies are assembled
into a planar array on a close-packed plane.[18,19] This planar aspect
of the prerequisite cluster shape for subsequent loop formation
together with the tendency for cascades to leave planar damage
regions enhances vacancy loop formation relative to that which would
follow from an isotropic damage region.

     Metalloid impurity atoms importantly influence the crystal
lattice defect annealing process by trapping individual defects and
by altering the shape of the Frenkel pair recombination region.
Trapping of either a vacancy or an interstitial by a single metalloid
atom impedes the defect migration activity. Trapping by two metalloid

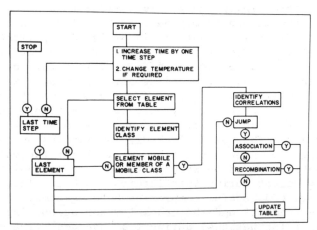

Fig. 15. Block diagram for a generalized defect annealing program.

atoms effectively stops defect migration. The size of the recombination region around a trapped vacancy is about two-thirds of the
size of the recombination region about an untrapped vacancy. This
serves to impede interstitial-vacancy recombination.

During long-term annealing most defect configurations are well-
separated from other defects during most of their migration history.
This being the case, their movement can be simulated on the basis of
the binomial distribution for jump length components along each
Cartesian axis. Only when the defect comes within correlation
range of other defects is it necessary to implement the simulation
method described in Fig. 15 for interacting defects.

DEFECT INTERACTION SIMULATIONS

Computer experiment simulations indicate that ceramic precipitates
and preprecipitation structures should exert a strong influence on
the nucleation, growth and stability of voids in reactor fuel
element cladding alloys. The content of the present section is
centered on this topic and is based on the central results of
collision cascade and defect annealing simulations.

Cascade simulation shows that vacancy aggregates initially
appear as stringy, non-compact forms[7] such as those shown in Fig. 16a.
Annealing simulation indicates that these noncompact forms are
subsequently collected together and transformed into compact void
nuclei such as that shown in Fig. 16b. These compact nuclei have

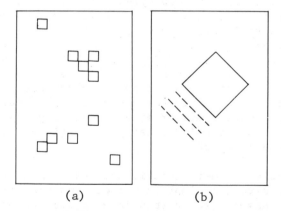

(a)   (b)

Fig. 16. Comparison of vacancy aggregates (a) stringy structure of
         vacancy aggregates produced directly by a cascade (See also
         Figs. 9-11); (b) the compact void structure of vacancy
         aggregates which form during annealing. Void facets lie
         on crystallographic planes; the parallel dashed lines indi-
         cate atom planes in the matrix around the void.

low index faceted surfaces and are formed as a result of a few highly
correlated vacancy migration steps. Other regular and stable vacancy
aggregates are also formed but are not of immediate concern. These
regular forms are linear arrays of divancies along <100> and <1000>
lines in bcc and hcp systems, respectively; and vacancy platelets
on (110), (111) and (10$\bar{1}$0) planes in bcc, fcc and hcp systems
respectively. In irradiated material, the quickest void nucleation
mode is the collision cascade production mode.[20]

In order to survive, a void nucleus must be stable against
thermal dissociation and dissociation by collision cascade overlap.
At a sufficiently high temperature all void nuclei will dissociate
thermally, in a pure metal. The stability of a void nucleus against
destruction by a collision cascade, in a pure material, depends
simply on the nucleus size being larger than the size of any cascade
initiated nearby. Computer simulation demonstrates that void nuclei
sufficiently large to be immune to destruction by collision cascade
encounters are, paradoxically, easily produced by collision cascade
overlap.[20] In fact, simulation shows that void nucleation occurs
easily whenever locally saturated or supersaturated defect densities
are attained. In this context, the occurrence of a locally supersaturated defect density automatically infers the presence of void
nuclei because these densities cannot exit unless large vacancy
clusters are produced either by cascade overlap or by the successive
superposition of individual Frenkel pairs. A supersaturated defect
density is one which is greater than the reciprocal of the recombination region volume.

Figure 17a indicates the deployment of cascade regions in α-iron
just before the radiation exposure is sufficient to produce frequent
overlap of cascade regions. Figure 17b depicts the beginnings of
overlap. Because of the Swiss-cheese structure of a cascade region,
it takes about three-fold overlap of cascades to displace each
atom in the crystal at least once.[20] Purely as a statistical result,
some extant vacancy clusters in a previously damaged region are
increased in size by the overlap of a new cascade. Specifically,
new large vacancy clusters are created immediately adjacent to pre-
existing clusters. Conversely, other pre-existing vacancy clusters
are erased by the overlap of a new cascade.

The introduction of metalloid impurities in computer simulation
experiments affects void nucleation, growth and stability in a
striking manner. Results for carbon in α-Fe will be used as examples.
In this instance, the key feature is the tetragonal strain field
about C in α-Fe and how it is accommodated at a void (free) surface.[21]
Figure 18 describes the strain field in bulk α-Fe. The two important
distance parameters are the tetragonal distance D_T and the ring
distance D_R. Normally D_T is 1.22 hlc and D_R is 0.93d_1 where d_1 is
the interatomic spacing and hlc is the half lattice constant. At a

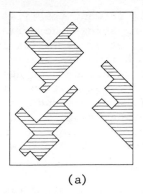
(a)

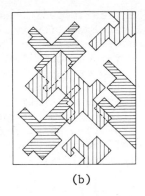
(b)

Fig. 17. Comparison of damage region distributions (a) below and (b)
at defect density saturation. Each cascade produces a
damaged region with a recognizable crystallographic orientation due to quasi-channeling.   Each cross hatched form in
(a) represents a damaged region produced by a single cascade; the deployment of damaged regions is characteristic
of an irradiation exposure below that required for saturation
of the defect density. Damaged regions overlap after prolonged irradiation and this can cause instantaneous production of void nuclei sufficiently large to be immune to
destruction by a subsequent cascade; (b) represents damage
region deployment at the onset of defect density saturation.

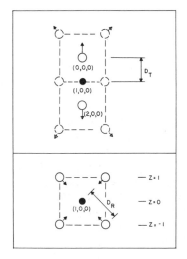

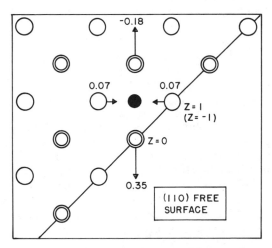

Fig. 18. Atom sites and characteristic distances pertinent
to the discussion of C in
α-Fe. The filled circle
represents the C atom.

Fig. 19. The most stable carbon
atom position near a
(110) free surface in
α-Fe. Strain is given as
a fraction of hlc.

free surface (Fig. 19),C positions itself between the first two atom planes with the tetragonality axis running across the surface "plane". In this position D_R is not much changed, but D_T is radically adjusted. Addition of more carbon atoms leads to a strong pre-precipitate Fe-C compound at the void surface which is stronger than the host metal. This stabilizes the void against thermal decomposition. It also serves to repel mobile interstitial configurations and contributes a stabilizing influence against annihilation events on the part of both mobile interstitial point defect configurations and interstitial loops by virtue of the repulsive image forces associated with the harder pre-precipitate layer.

Any type of hard precipitate can serve as a void nucleation agent if the interface between it and the host metal contains pits, valleys or steps. Computer simulation of defect accumulation at the pitted interface between a hard precipitate (ceramic) and soft matrix (metallic alloy) shows that void nuclei form preferentially in the pits and valleys. In the simulations this is due purely to the fact that the displacement energy in the precipitate exceeds that in the metal and that one or more sides of the void are shielded. This effect is shown in Fig. 20. A large supersaturation of vacancies occurs in such regions. It exceeds the theoretical average saturated state density and approaches the highest densities produced by cascades. In Fig. 20, for example, the theoretical average number of vacancies would be 37 per 1000 sites. In the simulation the density is 68 per 1000 sites. In this example each atom has been displaced 17 times. Irradiation-induced re-solution of metalloid atoms occurs continually at the precipitate. This occurrence provides agents for impurity stabilization of the void nuclei.

The vacancy and interstitial cluster distributions in Fig. 20 were obtained by random introduction of individual Frenkel pairs into a 1000-atom crystal. Each time a Frenkel pair was introduced, each component (vacancy and interstitial) defect was tested for stability with the extant defect distribution. This simulation corresponds to intense electron irradiation at 0°K, in that no defect migration was allowed. Extrapolation of the plot of "free" space for the introduction of new defects as a function of the number of extant stable Frenkel pairs (N_F) suggests that a little over twice the theoretical average defect density at saturation can be produced in a local region. In the case of the example given in Fig. 20 this means that a maximum of 74 defects could be achieved per 1000 atom sites. The state shown (68 defects) is the most dense state observed in the simulations thus far.

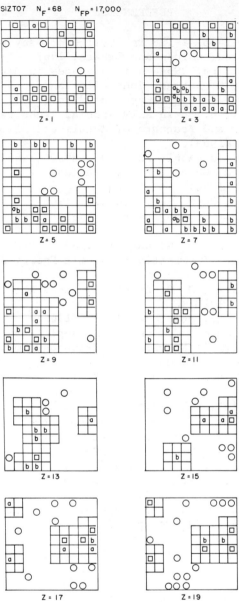

Fig. 20. Distribution of vacancies (small squares) and interstitials (open circles) in a 1000-atom crystallite after 17,000 Frenkel pairs (N_{FP}=17,000) had been produced. There are 10 z-planes in the crystallite, e.g., z=1, 3, ..., 19, where the unit of length is a half-lattice constant. Each plane contains 100 sites. The large squares represent sites covered by the recombination region about each vacancy on which interstitial atom occupancy is forbidden. The totality of large-square sites represents the vacancy cluster region. A void nucleus is represented by the maps for planes Z=1 - 11.

REFERENCES

1. J. R. Beeler, Jr.; pp. 1-128 in Physics of Many Particle Systems.
   Ed. by E. Meeron. Gordon and Breach, 1967.
2. J. R. Beeler, Jr.; pp. 295-476 in Advances in Materials Research,
   Vol. 4. Ed. by H. Herman. John Wiley and Son, 1970.
3. J. R. Beeler, Jr., J. Appl. Phys. 35 2226 (1964).
4. J. R. Beeler, Jr.; p. 86 in ASTM Special Technical Publication
   No. 380. Flow and Fracture of Metals and Alloys in Nuclear
   Environments, (1965).
5. J. R. Beeler, Jr., J. Appl. Phys. 37 3000 (1966).
6. J. B. Gibson, A. N. Goland, M. Milgram and G. H. Vineyard, Phys.
   Rev. 120 1229 (1960).
7. J. R. Beeler, Jr., Phys. Rev. 150 470 (1966).
8. C. Erginsoy, G. H. Vineyard and A. Englert, Phys. Rev. 133 A595
   (1964).
9. J. R. Beeler, Jr. and D. G. Besco; p. 43 in Radiation Damage in
   Solids, Vol. 1. International Atomic Energy Agency, Vienna,
   1962.
10. J. R. Beeler, Jr. and D. G. Besco, J. Phys. Soc. Japan 18, Supple-
    ment III, 159 (1963).
11. J. R. Beeler, Jr., J. Nuc. Mat. 15 1 (1965).
12. J. R. Beeler, Jr. and D. G. Besco, J. Appl. Phys. 34 2873 (1963).
13. D. N. Seidman, Cornell University (Private Communication).
14. M. T. Robinson and O. S. Oen, Phys. Rev. 132 2385 (1963).
15. W. P. Chun, M. F. Beeler and J. R. Beeler, Jr., (Unpublished
    Work, 1970).
16. D. G. Doran, Radiation Effects 2 249 (1970).
17. R. Perrin and R. Bullough; p. 52 in The Nature of Small Defect
    Clusters, Vol. 1 (AERE-R5269). Ed. by M. J. Makin, (1966).
18. R. A. Johnson, Phil. Mag. 16 553 (1967).
19. J. R. Beeler, Jr. and R. A. Johnson, Phys. Rev. 156 677 (1967).
20. J. R. Beeler, Jr.; pp. 621-652 in Lattice Defects and Their
    Interactions. Ed. by R. R. Hasiguti. Gordon and Breach, 1967.
21. J. R. Beeler, Jr., M. F. Beeler and J. C. W. Hsu, AEC Quarterly
    Report ORO-3912-2 (1969), p. 20.

FAST NEUTRON EFFECTS IN GLASS-CERAMICS AND AMORPHOUS SEMICONDUCTORS

L. L. Hench, W. D. Tuohig and A. E. Clark

University of Florida

Gainesville, Florida

ABSTRACT

A review of radiation damage mechanisms in insulators and semi-conductors is presented. Recent results on fast neutron effects on the ac and dc electrical properties of insulating glass-ceramics, amorphous semiconductors, and semiconducting glass-ceramics, including damage thresholds of crystallite sizes, are discussed and interpreted in terms of amorphous band structure models and heterogeneous dielectric theory.

INTRODUCTION

The unique range of properties exhibited by both insulating glass-ceramics and amorphous semiconductors makes them potentially useful in a variety of applications which may involve exposure to nuclear irradiation. Although the neutron radiation sensitivity of insulating glasses[1] and various crystalline ceramic materials[2-4] have been investigated in some detail, relatively little is known about neutron damage in glass-ceramics or semiconducting glasses.

Several recent reports have been made concerning neutron and γ-ray effects on the behavior of switching devices made of chalcogenide glasses;[5,6] they have shown device insensitivity from fluences of 10^{16} nvt fast neutrons and 10^{11} rads/sec of γ-rays. The electrical conductivity of liquid chalcogenides has been found[7] to be unaffected by fast neutron dosages of as much as 1.8×10^{20} nvt. Considering that the measurements were made on liquids, this result is not surprising.

Results from this laboratory[8] have shown that the electrical conductivity of solid semiconducting glasses in the $V_2O_5-P_2O_5$ system are relatively unaffected by fast neutron fluences of up to 4×10^{17} nvt. However, an approximately 10% increase in electrical conductivity was observed after a dose of 1.25×10^8 rads of Co^{60} γ-ray irradiation. This behavior was attributed to an increase in the concentration of quasi-free charge carriers due to Compton scattering of electrons. Additional evidence in support of this argument will be presented in this paper.

Studies of the influence of fast neutrons on the stability of the electrical properties of heterogeneous $KPO_3-V_2O_5$ semiconducting glasses have been recently reported by one of the authors.[9] An important observation in the study was the degradation of electrical characteristics associated with $100 - 200$ Å ordered heterogeneities within the glass matrix. After a 2.7×10^{17} nvt fluence the electrical conductivity had decreased by a factor of seven, ac conductivity characteristics of the disordered glass had appeared, and a large dielectric loss peak was destroyed. An objective of the present paper is to discuss the importance of crystallite size on the fast neutron damage threshold in heterogeneous amorphous semiconductors.

Ionically conducting Li_2O-SiO_2 glasses containing submicron crystalline regions have also been shown by the authors to be strongly influenced by fast neutron exposure.[10] A fluence of 1×10^{17} nvt was sufficient to decrease the magnitude of the dielectric loss peak exhibited by the material. The radiation exposure also accelerated the sequence of crystallization reactions occurring during thermal treatment of the glass. The dissolution rate of lithium metasilicate crystals which appear as a precursor to equilibrium crystallization was enhanced by the fast neutron damage.

A final objective of this paper is to report the effect of increasing the cumulative fast neutron dosage on the electrical properties of the nucleated Li_2O-SiO_2 glasses.

EXPERIMENTAL PROCEDURE

All three glasses discussed were melted in electric muffle furnaces in covered Pt crucibles and formed into specimens, ~1.8 cm dia., 0.5 cm thick, by quenching into steel molds. The compositions studied, melting and annealing schedules are given in Table 1.

Electrical measurements were made in vacuum on polished samples with vacuum evaporated gold electrodes in a double guard ring configuration using an apparatus previously described.[8,10,11]

Table 1.   Compositions and Thermal Histories of Glasses

Glass Composition	Melting Temp. and Time	Annealing Temp. and Time
80 mol. % V_2O_5- 20 mol. % P_2O_5	966°C/10 hr.	300°C/30 min.
33 mol. % KPO_3- 67 mol. % V_2O_5	550°C/8 hr.	200°C/2 hr.
33 mol. % Li_2O - 67 mol. % SiO_2	1350°C/24 hr.	300°C/1 hr.

Gamma-ray irradiation experiments were performed using the 30,000 Ci ^{60}Co source at the University of Florida.   The γ-ray dose was determined by using the photometric $Fe^{2+} \rightarrow Fe^{3+}$ reaction as detailed in ASTM procedure D-1671-63.[12]   Specimen temperature during γ-radiation was approximately 25°C.

Both the V_2O_5-P_2O_5 and Li_2O-SiO_2 glasses were given neutron irradiation exposures in the Wright-Patterson Air Force Base test reactor which has a fast neutron flux capability of 1.5 x 10^{13} n/cm^2 (>0.1 mev). The fast neutron flux was measured using a ^{58}Ni[n,p] ^{59}Co reaction with a 2.9 mev threshold energy. Fluxes reported are based on activations measured 48 hrs after removal of the samples from the reactor. Cadmium wrapping was employed to protect the gold electrodes. Temperature monitoring of the reactor indicated that sample temperatures were in the range of 50°C throughout the exposures. Approximately four weeks lapse at 25°C between exposure and measurement was necessary to reduce activity to a tolerable level.

V_2O_5-P_2O_5 Amorphous Semiconductors

Previous reactor exposures of 80 mol.% V_2O_5-20 mol.% P_2O_5 glasses resulted in a slight increase in the electrical conductivity.[8] Separate ^{60}Co γ-ray experiments indicated this behavior was due to γ-ray ionization of charge carriers. The net effect was only a 10% increase in conductivity. However, additional data presented in Fig. 1 shows that the γ-ray exposure strongly influences the dielectric losses in the glasses.

The dielectric loss angle, tan δ, plotted as a function of log measuring frequency, Hz, increases due to the 6.8 x 10^7 rads γ-ray exposure. Well resolved dielectric loss peaks also appear. The frequency location of the loss peaks and their magnitude both suggest that appreciable interfacial polarization is occurring in the irradiated glasses.[13] Analysis of the temperature dependence of the frequency location of the dielectric loss peak,

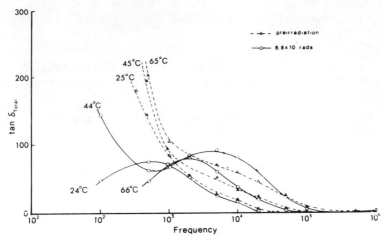

Fig. 1.   Development of dielectric loss peaks in an 80 mol.% V_2O_5–
20 mol.% P_2O_5 semiconducting glass with 6.8×10^7 rads
^{60}Co γ-ray exposure.

$$f_{max} = f_o \exp[-Q/RT], \qquad\qquad (1)$$

results in a measured activation energy for the loss process of 0.6 ev.
This is the same value calculated for the temperature dependence
of the dc conductivity of this glass.[9]   Therefore, it is reasonable
to assume that similar electronic carriers are involved in the relaxation process as contribute to the dc conductivity.

Since the γ-irradiation should not produce structural heterogeneities in the glass, the interfacial barriers involved in the
relaxation process must be associated with heterogeneities already
present or with the sample electrodes.   As discussed in a recent
paper concerning the theory of heterogeneous semiconductors,[14] either
internal heterogeneities or partially blocking electrodes can give
rise to dielectric loss peaks in these materials.

A γ-ray dosage of 1.25×10^8 rads produced an unusual low frequency resonance-like response which also appears to be an interfacial phenomena.   Figure 2 summarizes a series of measurements
of the frequency dependent capacitance of the 80/20 V_2O_5–P_2O_5 glass
exposed to 1.25×10^8 rads.   After irradiation an appreciable increase in the capacitance is observed even at 23°C.   At 40°C (post
radiation), a large increase in capacitance occurs, followed at
65°C by a wide low frequency resonance-like effect.   The capacitance
of the sample is negative over the range from 2.7×10^3 hz to
4.5×10^5 Hz, as measured by the –C scale on the Wayne Kerr B-221
bridge and the L scale on the B-601 bridge.   From 3×10^4 Hz to
1×10^5 Hz the negative C values are too large to be measured.

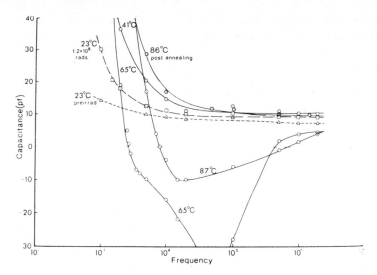

Fig. 2. Sequence of capacitance behavior of our 80/20 V_2O_5-P_2O_5
semiconducting glass after 1.2×10^8 rads ^{60}Co γ-ray irradiation. Sequence: 23°C pre-rad; 23°C post-rad; 65°C post-rad,
87°C post-rad; 86°C post-rad, post 119°C annealing; 41°C
post-rad, post annealing.

At a higher temperature, 87°C, the resonance-like behavior
appears to be damped and is measurable over the entire frequency
range. At 114°C the negative capacitance behavior disappeared
during the measurements, indicating a rapid annealing process
at this temperature. Post annealing data are shown in (Fig. 2) at
a measurement temperature of 86°C. After annealing, capacitance
values are nearly equivalent to the pre-radiation values for the
glass and are not shown for the sake of clarity.

It seems that the data of Figs. 1 and 2 can be interpreted in
one of two ways. Either (1) the ionizing radiation produces extra
charge carriers which are trapped at previously existing barriers
in the glass or at the electrode-glass interface, or (2) the radiation could create traps at interfacial regions which are populated
with "normal" carriers, the traps being created by ionization of
deep-lying electrons. The fact that the activation energy for the
relaxation loss process is the same as for dc conductions would
appear to favor the latter alternative. One would expect a lower
activation energy for a large number of ionized electrons. Also,
if pre-existing traps at interfaces were present, the relaxation
process should occur with "normal" carriers as well.

Creation of a wide distribution of deep-lying traps at interfacial regions also provides a reasonable explanation for the observed resonance behavior at 1.25×10^8 rads. In such a model, the

resonance-like behavior is associated with thermally activated and
field directed oscillations between trapping sites. At low temperatures, the oscillations simply contribute an additional interfacial
polarization to the capacitance. At sufficiently high temperatures,
resonance oscillations through the barriers become possible. A
portion of the population obtains irreversible sites on one side of
the barrier, thus producing a temperature dependent annealing of
the phenomena.

Partially Crystallized Semiconducting Glasses

Previously reported investigations of fast neutron effects in
heterogeneous 33 mol.% KPO_3-67 mol.% V_2O_5 semiconducting glasses
involved crystallites in the 100-200 Å size range.[4,15] A heat treatment at 288°C of glasses rapidly quenched from 800°C was employed
in the nucleation and growth of the crystallites. The studies
showed that fast neutron fluences of $<1.0 \times 10^{17}$ were required to
destroy dielectric loss peaks associated with the high conductivity
crystallites.

It is also possible to produce a small volume fraction, <1%,
of submicron crystals in the KPO_3-V_2O_5 glasses by casting from
temperatures of 550°C or below.[9] Dielectric loss peaks are exhibited
by such materials as shown in Fig. 3. The loss peaks have been
attributed to high conductivity PV_2 crystals giving rise to
Maxwell-Wagner-Sillars (MWS) interfacial polarization.[16] Similar
behavior has been observed in the FeO-P_2O_5 system.[17] The size of
the crystals in the sample of Fig. 3 are in the range of 0.1 μm, a
factor of 10 larger than in the previously reported study.[9]

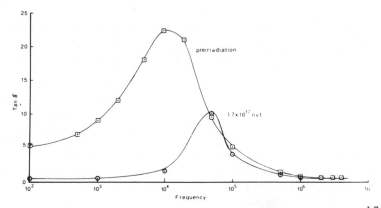

Fig. 3.   Dielectric loss peaks, before and after 1.7×10^{17} nvt
           irradiation, exhibited by a 33 mol.% KPO -67 mol.% V_2O_5
           semiconducting glass-ceramic.

The effect of the exposure of the heterogeneous glass to 1.7×10^{17} nvt fast neutrons is also shown in Fig. 3. Two major changes are readily apparent: the magnitude of the loss peak is markedly reduced and the location of the loss peak is shifted to a higher frequency. In terms of the MWS interfacial polarization model, the decrease in loss peak height indicates that either the volume fraction of crystallites has decreased or the conductivity difference between the phases is reduced, or the phase boundary has been degraded sufficiently that it will not sustain a space charge.

The frequency location of the loss peak is controlled by the morphology of the dispersed phase and consequently, a shift in frequency is evidence that morphological changes have occurred. Consequently, the loss behavior observed indicates that there is cumulative disordering of the dispersed submicron crystals within the glass matrix from the fast neutron irradiation. The damage threshold for the larger crystals appears to be in the range of an order of magnitude greater than the 100-200 Å crystals. Consequently, these results would suggest that the damage threshold for crystals dispersed in a glass matrix may be linearly proportional to the size of the crystals. Additional studies to extend the range of validity of this conclusion are in progress.

$Li_2O-2SiO_2$ Glass-Ceramics

Previous investigations have established that heat treating 33 mol.% Li_2O-SiO_2 glasses at 500°C for 5-6 hr precipitates metastable lithium metasilicate crystals that are ~200 Å long and

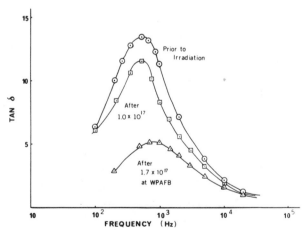

Fig. 4. Influence of cumulative fast neutron exposure on the dielectric loss peak of a 500°C/6 hr. heat treated $Li_2O-2SiO_2$ glass.

50 Å wide.[18,19] A MWS dielectric loss peak appears in the glass
concurrent with the presence of the crystals. Additional heat
treatment causes the metastable crystals to resorb and nucleate
the equilibrium lithium disilicate crystal phase in the process.
The dielectric loss peak decreases in magnitude and shifts to a
higher frequency as a result of the resorbtion.

A recent paper by the authors showed that the metastable
nucleation process was unaltered by a 1.0×10^{17} nvt irradiation of
the glass prior to heat treatment.[10] However, the metasilicate
dissolution process was shown to be significantly accelerated by
the fast neutron irradiation. The explanation proposed was that
the Li$^+$ mobility was enhanced by the irradiation, making the
structure more susceptible to thermal alteration. An enhanced
mobility of the Li ions would also decrease the conductivity
difference between the matrix and the crystals, thereby reducing
the magnitude of the loss peak, as shown in Fig. 4. The absence
of a shift in frequency of the peak with the 1.0×10^{17} nvt
exposure indicates that appreciable morphological changes did not
occur.

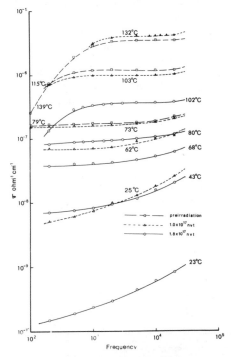

Fig. 5.   Change in electrical conductivity-temperature-frequency
behavior of a nucleated $Li_2O-2SiO_2$ glass with cumulative
fast neutron exposure.

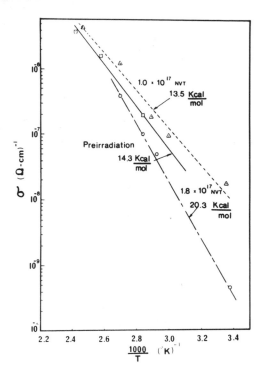

Fig. 6.   Fast neutron dependent changes in the conductivity—
          temperature dependence of a 500°C/6 hr heat treated
          $Li_2O-2SiO_2$ glass. (σ measured at 5 x 10^3 Hz.)

     Additional evidence to support this conclusion of enhanced Li^+
ion mobility is presented in Fig. 5.   The ac conductivity of the
Li_2O-SiO_2 glasses after a 500°C/6 hr heat treatment is plotted as
function of measuring frequency at various measuring temperatures.
The conductivity prior to irradiation is lower than the conductivity
after the 1.0 x 10^{17} exposure.   The radiation enhancement of the
conductivity appears to be due to an increase in the pre-exponential
term,

$$\sigma = \sigma_o \exp - [Q/RT] \qquad\qquad (2)$$

rather than a decrease in activation energy for the ionic conduction.
Figure 6 shows that the activation energy for the glass before and
after the 1·.0 x 10^{17} nvt radiation remains at approximately
15 kcal/mole, which is equivalent to that previously reported for
this material.

     Increasing the fast neutron fluence to 1.7 x 10^{17} continues
the degradation of the dielectric loss peak (Fig. 4) to one-third

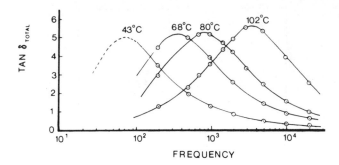

Fig. 7.   Temperature dependent variation in dielectric loss peak
          location for a $Li_2O-2SiO_2$ nucleated glass after
          1.8×10^{17} nvt irradiation.

the magnitude of the preradiation peak.   In addition, the location
of the peak is shifted to higher frequencies.

It can be seen in Figs.5 and 6 that the magnitude of the glass
conductivity is decreased by the additional neutron exposure, primarily by an increase in the activation energy to ~20 kcal/mole.
Such a decrease in activation energy may be a result of densification
of the glass structure such as observed in SiO_2 even at 10^{17} nvt.[20]
It is also reasonable that the Li^+ ions would be appreciably
influenced by densification because of their low bonding energy.

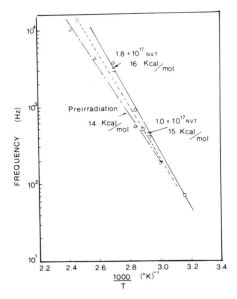

Fig. 8.   Fast neutron effects on the temperature dependence of the
          dielectric loss peak location of a nucleated $Li_2O-2SiO_2$
          glass.

Thermal variation of the loss peak location for the 1.8×10^{17} irradiated sample is given in Fig. 7. A plot of the logarithm of the frequency of maximum loss vs. 1/T (Eq. 1) yields the activation energy for the relaxation process. As shown in Fig. 8, the activation energy is not appreciably changed by the radiation. Because of the decrease in bulk conductivity of the 1.8×10^{17} nvt sample, a consistent explanation for the destruction of the loss peak requires a degradation of the metasilicate crystals that give rise to the peak. The shift in the frequency of the loss peak suggests that the axial ratio of the metasilicate crystals is reduced by the radiation in a like manner to that produced by additional 500°C thermal treatments.[10,18] It can be concluded that the change in the electrical properties of the nucleated glasses is a result of enhanced ionic mobility at low dosages (1.0×10^{17} nvt of less) followed by destruction of the submicron crystal nuclei at larger dosages.

CONCLUSIONS

The major point to be re-emphasized is that electrical properties of heterogeneous glasses that are governed by submicron crystals are susceptible to fast neutron damage at fluences of $1-2 \times 10^{17}$ nvt. The size of the crystals affects the damage threshold. Ionizing radiation produces electronic complications in addition to the neutron related structural changes for semiconducting glasses.

ACKNOWLEDGMENTS

The authors gratefully acknowledge the financial support of the Advanced Research Projects Agency of the Department of Defense monitored by the AFCRL under Contract No. F-19628-68-8-0058 and AROD Contract No. DAHC04-70-C-0024. The assistance of Lt. B. Wilson, Major F. Buoni, and Mr. A. Bauer of the Wright-Patterson Reactor Facility is also acknowledged.

REFERENCES

1. E. Lell, N. J. Kreidl and J. R. Hensler; p. 73 in Progress in Ceramic Science, Vol. 4. Ed. by J. Burke, Pergamon Press, 1966.
2. W. Primak and R. Kampworth, J. Appl. Phys. 39 (12) 5651–5658 (1968).
3. M. Nachman, L. Cojocaru and L. Ribco, P. 1 in Nukleonik, 10 Band, 1. Heft (1967).
4. D. J. Barber and N. J. Tighe, J. Amer. Ceram. Soc. 51 (11) 611 (1968).
5. (a) S. R. Ovshinsky, E. J. Evans, D. L. Nelson, J. Fritzsche, IEEE Trans. on Nuclear Science, Dec. 1968.
   (b) R. R. Shanks, J. H. Helbers and D. L. Nelson, "Ovonic Computer Circuits Development", Technical Rept. AFAL-TR-69-309, June 1970.
   (c) R. R. Shanks, D. L. Nelson, R. L. Rowler, H. C. Chambers and D. J. Niehaus, "Radiation Hardening Circuitry Using New Devices", Technical Report AFAL-TR-70-15, March 1970.

6. E. J. Evans, "A Feasibility Study of the Applications of Amorphous
   Semiconductors to Radiation Hardening of Electronic Devices",
   Picatinny Arsenal Technical Report 3698,
7. J. T. Edmond, J. C. Male and P. F. Chester, J. Sci. Inst. (J. of
   Physics E) 1 [2] 373 (1968).
8. L. L. Hench and G. A. Daughenbaugh, J. Nuclear Materials 25 58-63
   (1968).
9. L. L. Hench, J. Non-crystalline Solids, 2 250-277 (1970).
10. W. D. Tuohig and L. L. Hench, J. Nuclear Materials 31 86-92 (1969).
11. L. L. Hench, "Dielectric Relaxation in Materials Analysis", Society
    of Aerospace Materials and Process Engineers, Proceedings of
    the 14th Annual Symposium, Cocoa Beach, Florida, November 1968
    (Reprinted as Technical Paper 428, Engineering Progress at
    the University of Florida, Gainesville, Florida.)
12. Test for absorbed gamma radiation dose in the Fricke dosimeter,
    ASTM Standards 29 719 (1963).
13. (a) L. K. H. von Beek; p. 69 in Progress in Dielectrics, Vol. 7
        CRC Press.
    (b) L. L. Hench and H. F. Schaake, "Electrical Properties of Glass",
        in Introduction to Glass Science. Eds. D. Pye and H. Simpson,
        Plenum Press, New York (in press).
14. H. F. Schaake, "Theory of Heterogeneous Semiconductors", in
    Physics of Electronic Ceramics. Ed. by L. L. Hench and D. B.
    Dove, M. Dekker, Inc. (1971), (in press).
15. L. L. Hench, A. E. Clark and D. L. Kinser, "Neutron Irradiation
    Effects in Partially Crystallized Semiconducting Glasses",
    submitted to J. Non-crystalline Solids.
16. A. Fuwa, "Electrical Properties of Glasses and Crystals in the
    $K_2O-V_2O_5-P_2O_5$ System", M. S. Thesis, University of Florida, 1970.
17. D. L. Kinser, J. Electrochem. Soc. 117 (4) 546 (1970).
18. D. L. Kinser and L. L. Hench, J. Amer. Ceram. Soc., 52 445 (1968).
19. D. L. Kinser and L. L. Hench, J. Materials Sci. 5 369 (1970).
20. R. E. Jaeger, J. Amer. Ceram. Soc. 51 57 (1970).

DISCUSSION

R. H. Doremus (G. E. Research and Development Center): The conductivity
of a lithium silicate glass drops sharply when a continuous path of
glass between electrodes is blocked by surface or internal
crystallization. Can a layer of low conductivity material form in the
lithium disilicate glass at high irradiation, causing the observed
drop in conductivity?
Authors: The formation of a low conductivity layer could cause the
drop in conductivity observed but the change in loss angle can be
attributed only to bulk damage. Since the samples were irradiated
over all sides bulk damage seems more consistent.

FABRICATION AND IRRADIATION BEHAVIOR OF URANIUM MONONITRIDE

V. J. Tennery, T. N. Washburn and J. L. Scott

Oak Ridge National Laboratory

Oak Ridge, Tennessee

ABSTRACT

Uranium mononitride is a candidate fuel for nuclear reactors for applications in space because of its high thermal conductivity, high fissile atom density, and dimensional stability at high temperature. High purity UN ceramics fabricated from nitride powders by uniaxial or isostatic pressing and sintered in nitrogen at temperatures >2000°C, were irradiated at fuel temperatures to 1500°C and 2 w/o burnup. Results indicate no chemical reaction zone between fuel and metallic cladding, fuel swelling less than 2.8% ΔV/V per atomic percent burnup, and maximum fission gas release of 7% of that generated.

INTRODUCTION

Several kilowatts of electrical power will be required for future planned space facilities such as the Manned Orbiting Laboratory, the Moon Station and the Mars Mission Vehicle. When these relatively large quantities of electrical power are required, a compact fast reactor offers a decided advantage over fuel cells, solar energy, or isotopic power. Conversion of thermal to electrical power generation will be by either a liquid metal Rankine cycle or by a Brayton cycle with inert gas as the working fluid.

The two limiting design criteria for the compact fast reactor are weight and high operating temperatures. The weight is strongly affected by the amount of shielding, and since the shield thickness is fixed by allowable dose rates to people and equipment,[1] the

587

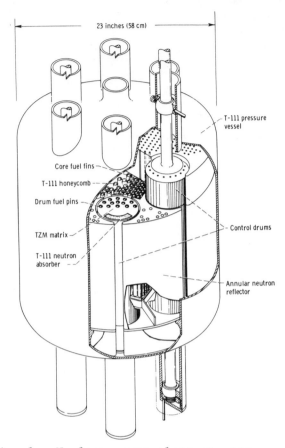

Fig. 1. Nuclear power plant reactor.

shielding weight is a function of the cube of its diameter. The need
to reduce this shield weight in a space vehicle is obvious. The
operating temperature of the reactor is governed by the radiator
surface area, since heat rejection into space is by radiation. An
attractive peak coolant temperature is about 1000°C; therefore,
refractory metal components are required for the system.

A typical core design is shown in Fig. 1.[2] The fuel pins, which
are cooled with ^{7}Li, are surrounded by a TZM (Mo-0.5% Ti-0.08% Zr)
reflector and T-111 (Ta-8% W-2% Hf) pressure vessel. Reactor control
is achieved by use of six rotating drums containing fuel pins on one
side and T-111 neutron absorber on the other. Fission heat is trans-
ferred from this core to an intermediate heat exchanger by use of
lithium and thence to the working fluid.

SELECTION OF THE FUEL

Because of the high operating temperatures, a bulk ceramic
fuel or a ceramic fuel dispersed in a ceramic or refractory metal
matrix is required for the application. A list of candidate fuels
is shown in Table 1. Uranium nitride has the most attractive combination of properties of the materials shown. The nearest competitor,
UC, has the disadvantage of reacting with all refractory metals,
whereas UN is compatible with some, including W and W-Re alloys.
Little is known about the irradiation properties of US or the borides.
Dispersion fuels, such as UO_2-W, have a much lower U density and
the core diameter is therefore much greater for a specific system
design than with bulk UN. Uranium dioxide has the disadvantages
of a significantly lower uranium density than UN and a much lower
thermal conductivity than UN or UC, as is indicated in Fig. 2.
Uranium metal exhibits severe irradiation swelling and has a low
melting temperature which removes it from the list of fuel candidates
for this application.

Table 1. Properties of Candidate Fuels for Space Reactor Applications.

Fuel	Matrix Crystal Structure	Density (g/cm^3) Matrix	Uranium	Liquidus Temperature (°C)	Mean Matrix Thermal Conductivity, K (w^{-1} cm^{-1} $°C^{-1}$)
UO_2	fcc, CaF_2	10.97	9.6	2800	0.03
UN	fcc, NaCl	14.32	13.6	2850	0.22
UC	fcc, NaCl	13.63	12.97	2400	0.21
UB_4	Tetragonal	9.3	8.58	2550	0.268
US	fcc	10.87	10.19	2460	0.12
UB_2	Hexagonal	12.87	12.35	2440	
UO_2-BeO	Hexagonal	3.01		2150	0.188
UO_2-W	bcc	19.30		2750	1.46
UO_2-Mo	bcc	10.22		2610	1.42

FABRICATION OF UN

The UN powders used in this experiment were produced by the
hydride-dehydride-nitride process using uranium metal as the
starting material. Reaction (Eq. 1) results in fragmentation of
the bulk U since the UH_3 has a lower density than the metal. The
hydride is easily decomposed

$$2 \; U + 3 \; H_2 \rightarrow 2 \; UH_3 \qquad\qquad (1)$$

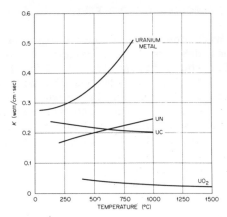

Fig. 2   Thermal conductivity of reactor fuel materials.

by heating at 300 to 350°C with finely divided U being the product.
The U is then reacted with N_2 (Eq. 2) at temperatures of about
300°C and a pressure of 760 torr of N_2

$$4 \ U + 3 \ N_2 \rightarrow 2 \ U_2N_3. \tag{2}$$

The sesquinitride (U_2N_3) is decomposed at 900°C in vacuum according
(Eq. 3) to produce fine particles of UN powder.

$$2 \ U_2N_3 \rightarrow 4 \ UN + N_2. \tag{3}$$

    The powder is gray-bronze in appearance and inert-gas-fusion
analysis shows a typical contamination of 150 ppm of oxygen.   The
powder is handled at all times in a purified Ar atmosphere since
the fine UN powders are extremely reactive with O_2.   The particle
size distribution of the powders is determined by means of a gas
sedimentation method and the distribution of some of the powders
are shown in Fig. 3a.

    The nitride powders are fabricated into both annular and solid
cylindrical fuel specimens for the irradiation experiments.   Uranium
nitride powders are very difficult to fabricate using the uniaxial
pressing technique.   However, this method was used for preparing
the fuel for this work because of the existing technology and
equipment.   A scanning electron microscope is employed to determine
the morphology of the UN particles, and a typical result is shown
in Fig. 3b.

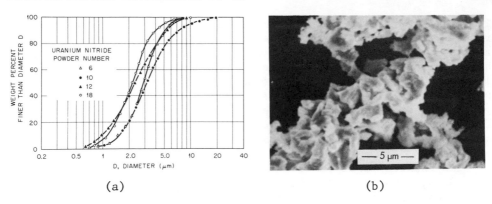

(a)   (b)

Fig. 3.   Characterization of UN powders (a) particle size distribution
          curves as determined by a gas sedimentation method; (b)
          angular particles as revealed by scanning electron microscopy.

        The sharp angular character of the nitride particles is the
apparent reason for the difficulty encountered when fabricating by
uniaxial pressing.   To aid in lubricating the particles during the
pressing process, they are coated with camphor as shown in the process

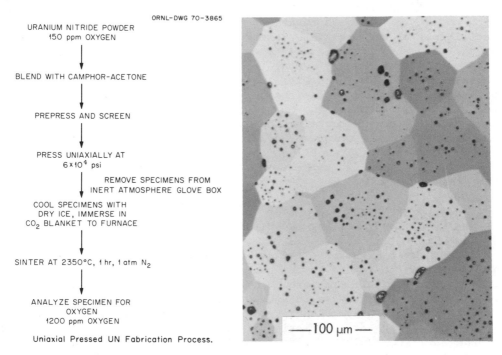

Fig. 4. Process flow diagram for   Fig. 5. Microstructure of unir-
        preparation of UN for   radiated UN.   Black areas
        irradiation experiments.   are pores; density was
                                              95% of theoretical.

flow sheet in Fig. 4. The specimens are cooled with dry ice during
transfer to the sintering furnace due to the extreme pyrophoric
nature of this material. The camphor is completely removed from
the UN during the sintering process. The microstructure of the
sintered UN produced by this process is shown in Fig. 5. The fuel
specimens were ground on the outside surface to close tolerances
by centerless grinding, whereas end grinding and final machining
of the inside surface of the annular pellets were accomplished by
the use of diamond grinding tools.

The chemical analyses for major elements are typically 94.40%
U, 5.40% N, 0.03% C, and 0.12% O by weight. Total trace metallic
impurities are about 300 ppm with the majority being comprised of
Al, Fe, Ni, and Cr.

IRRADIATION BEHAVIOR OF UN

The objective of the ORNL program for irradiation testing of
UN is to obtain basic information on the irradiation performance
of fuel pins at fuel temperatures from 1000 to 1500°C, cladding
outside surface temperatures of 900 to 1400°C, and linear heat ratings
of 5 to 10 kw/ft. These experiments are conducted in the poolside
facilities[3] of the Oak Ridge Research Reactor (ORR). The position
of the test capsule, relative to the reactor face, is adjustable
to permit operation over a range of selected conditions. The
capsules are instrumented with thermocouples on the cladding surface
of each fuel pin and at the fuel center line of one pin.[4] The
temperature gradient of the outer capsule wall is measured and used
as a calorimeter to determine the heat output of each fuel pin.
The irradiation performance characteristics of most interest in
this program are (1) swelling, (2) release of fission gases
from the fuel and (3) compatibility of the fuel and cladding
material.

Fuel Swelling and Fission-Gas Release

Most crystalline materials tend to swell in the presence of
high neutron fluxes due to the generation of interstitial atoms or
ions in the crystal structure. In a fuel material where fissioning
is also occurring, the solid and gaseous fission products also contribute significantly to the swelling. The swelling is a function
of the percentage of the fissionable atoms that are actually fissioned
in the fuel (commonly referred to as burnup), the burnup rate, the
temperature of the fuel during the fissioning, the neutron energies
and flux in the fuel during fissioning, the mechanical properties
of the fuel, and the amount of restraining force applied on the fuel
by its containment or cladding.

In the irradiation experiments we have conducted, the refractory metal claddings have been generally successful in restraining swelling of the UN fuel. In a fuel pin the initial swelling of fuel fills the space between the fuel and cladding, and subsequent swelling stresses the cladding. This latter situation is undersirable because most cladding metals tend to become weakened and somewhat embrittled in a neutron flux, and the forces from fuel swelling may cause the cladding to fail. An ideal fuel from this standpoint, then, would swell very slowly under its design irradiation conditions. If the fuel swelling rate is appreciable, high strength claddings must be used to restrain the increase in fuel volume. Swelling of the UN is restrained in the axial direction only by the binding forces of the intimate contact between fuel and cladding. The end pellets of the fuel column normally have higher swelling rates and are observed to swell a few thousandths of an inch axially, while the length of the other pellets remains relatively unchanged. Total swelling of the fuel in our tests is less than 5%, with about 3% being diametral increase to fill the gap between fuel and cladding. Any remaining swelling is either an increase in fuel length or an increase of the fuel pin diameter. In our tests the pin diameter increases[5] have been less than 0.5%

The gases, Xe and Kr, are produced during the fissioning of ^{235}U and, ideally, these gases should be completely retained in the solid fuel. Complete retention is desirable if it does not lead to unacceptable swelling rates as discussed previously. Extensive fission-gas release from the fuel is undesirable because of the high gas pressures that can be generated in the fuel pin, causing high cladding stresses. High gas release is also undesirable because a cladding failure of any kind is then accompanied by extensive radioactive contamination of the primary coolant.

In an actual ceramic fuel, complete retention of fission gas is not observed and an important objective of fuel irradiation experiments is to establish both the amount of fission-gas release under a given set of experimental conditions and the mechanism by which fission gas is retained. Generally, fission-gas release is a function of the temperature of the fuel during irradiation, the microstructure (grain size), the physical state of the fuel (i.e., whether it is badly cracked due to thermal cycling, etc.), and the extent of heavy metal burnup.

The fission gases are generated as individual atoms, and condensation of the gas atoms forms bubbles in a size range up to about 800 Å in the UN grains. However, relatively little coalescence of the fission-gas bubbles takes place within the grain. The small fission gas bubbles migrate to the grain boundaries under the influence of temperature gradients and surface tension forces. The equilibrium between the internal gas pressure and the surface tension of the bubbles changes upon reaching a grain boundary; consequently, coalescence and bubble expansion take place.

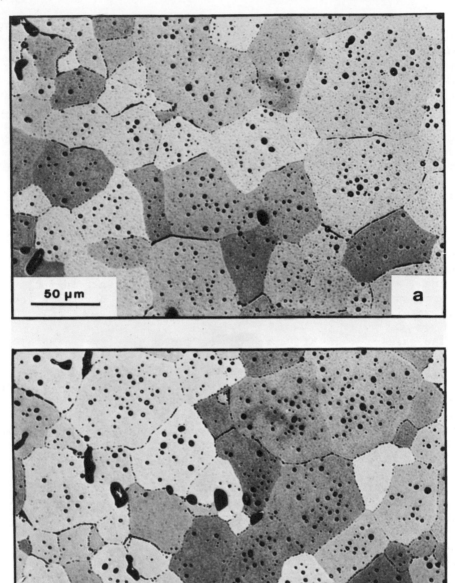

Fig. 6.   Microstructures of irradiated UN fuel (a) outside of surface
          region of fuel pellet, (b) midradius region.

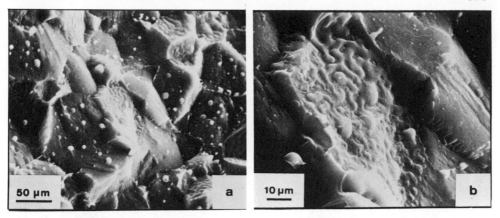

Fig. 7. Scanning electron fractographs of irradiated UN fuel pellet,
         (a) grain boundary and small fission-gas bubbles in the bulk
         UN, (b) grain boundary region containing a group of fission-
         gas bubbles which have coalesced together.

A typical microstructure of irradiated UN at the pellet mid-
radius and outside surface is shown in Fig. 6. This specimen was
irradiated to an average burnup of about 1.75 a/o in 5800 hr at
1500°C. The grain boundaries are all delineated by collections of
fission-gas bubbles, whereas the intergranular porosity is primarily
due to the fabrication process. At the outer edge of the fuel,
accumulation of fission gases at the grain boundaries is more severe
and often leads to complete separation of grains. This condition is
more severe at the outer surface because the fissioning rate is
higher in this region and the temperature gradient which is the
driving force for bubble migration is most severe in this area.

Accumulation of fission gases at the grain boundaries is vividly
portrayed by examination of replicas of the fuel by electron microscopy. Figure 7 illustrates the appearance of a fractured surface
of the fuel as revealed by examining a shadowed replica of the
surface by electron microscopy. Note that very small fission-gas
bubbles are evident in the transgranular fractures as in Fig. 7a,
while at the grain boundaries the bubbles have coalesced and grown
to very large sizes as in Fig. 7b, which is a portion of the (a)
field at higher magnification. These same conditions are also
revealed by examining carbon replicas of the fractured UN surface
by means of transmission electron microscopy in Fig. 8. Figure 8a
illustrates the condition of a grain boundary region, while Fig. 8b
shows very small, discrete fission-gas bubbles within a UN grain
along with the much larger fabrication porosity.

It is apparent from these micrographs that the accumulation of
fission gases at the grain boundaries of UN causes separation of
the grains, resulting in a significant increase in volume of the

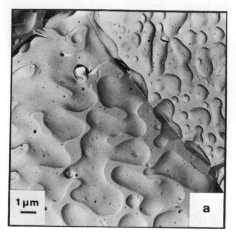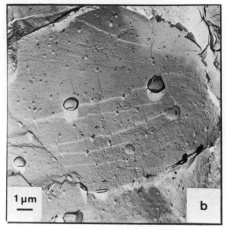

Fig. 8.   Transmission electron fractographs of irradiated UN fuel
          pellet (a) intergranular fracture region showing large
          concentration of fission-gas bubbles, (b) transgranular
          fracture region showing discrete fission-gas bubbles in
          the bulk UN.

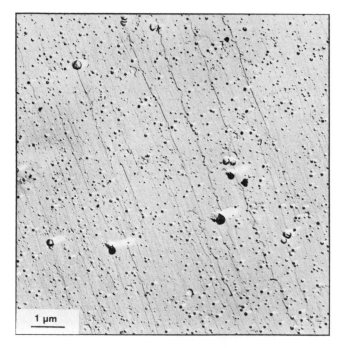

Fig. 9.   Electron micrograph of replica made from cleaved surface
          of irradiated single crystal of UN.

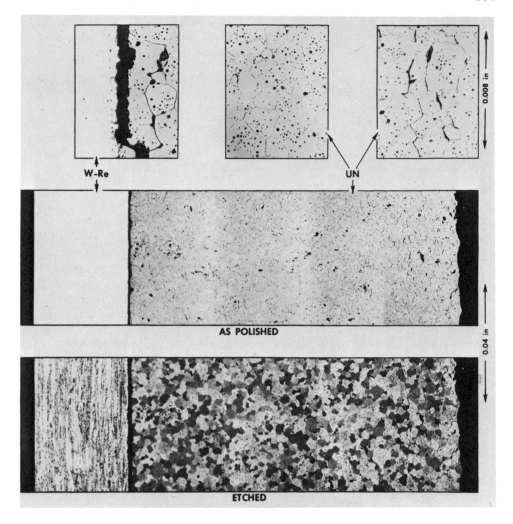

Fig. 10.   Longitudinal section through top fuel pin of capsule UN-3.
           Pin cladding was W-Re alloy.

fuel body. One potential method of reducing swelling, therefore,
would be to eliminate the grain boundaries as sites of fission-gas
coalescence. We have irradiated single crystal UN under conditions
similar to those noted previously, and fuel swelling was markedly
decreased.[5]   Figure 9 shows very small fission-gas bubbles in the
single crystal UN; no gross voids like those in Fig. 6 were observed.
The absence of grain boundary sites in this specimen virtually
eliminated fission-gas bubble coalescence. The difficulty in
fabricating single crystals of UN limits the practical application
of this approach to reducing fuel swelling. It is obvious, however,
that the elimination of grain boundaries could significantly reduce

the mechanical property requirements of the cladding for restraining
fuel swelling in this material. A practical compromise solution
would probably be to use large-grain polycrystalline UN.[6]

It is significant to note that even with the massive coalescence
of fission-gas bubbles at the grain boundaries, as shown in
Figs. 6-8, the UN fuel body at temperatures below 1500°C normally
releases 1% of the fission gases generated. However, at a fuel
temperature of 1700°C, as much as 97% of the fission gases produced
are released.[7] This infers a "breakaway" phenomenon in the release
of fission gases from the UN body between 1500 and 1700°C.

Compatibility with Claddings

The experimental evidence indicates that UN is compatible with
W and the W-Re alloys. Figure 10 illustrates a typical condition
in a UN fuel pin after irradiation testing for 5800 hr with the
W-Re cladding at 1300°C. There is no significant chemical reaction
between the UN and the W-Re. As noted in the top-left insert of
Fig. 10, there is a penetration of UN into the grain boundaries of
the 0.030 in. thick W-Re cladding to a depth of about 20 μm. We have
also noted similar penetrations of W-Re into the UN, but neither
case is considered to limit the use of these materials from a compatibility standpoint. Another alloy of interest as cladding for.UN
is T-111 (Ta-8%W-2%Hf). This alloy is not compatible with UN, and a
W liner, either vapor deposited[8] on the T-111 tube inside surface or
as a mechanically inserted liner, is used to separate the UN fuel
from the T-111 tube. Our investigations on use of these material
combinations to date are inconclusive.[9]

SUMMARY

Uranium nitride powders have been synthesized by the hydride-
dehydride-nitride process. The morphology of the particles was
determined by means of scanning electron microscopy. The particle
size distributions were measured by means of a gas sedimentation
method. The UN irradiation specimens were fabricated and sintered
such that they contained only 1200 ppm O_2 as an impurity.

The major effect of the stated irradiation conditions on UN
was to generate fission-gas bubbles which were observed to accumulate
at the grain boundaries of the polycrystalline UN with fuel temperatures to 1500°C. These bubbles caused the UN to swell until it
touched the cladding, with the cladding then restraining further
swelling. The compatibility of the UN with W and W-25% Re under
the stated irradiation conditions was excellent at cladding surface
temperatures of 1300 to 1350°C.

ACKNOWLEDGMENTS

The authors acknowledge the contributions of many associates in the Metals and Ceramics Division toward this work. The work of E. L. Long, Jr., and his associates in the Hot Cell Laboratory is particularly appreciated. Also, the contributions of S. C. Weaver and R. B. Fitts were very helpful.

This research was sponsored by the U. S. Atomic Energy Commission under contract with the Union Carbide Corporation.

REFERENCES

1. H. W. Davidson, Preliminary Analysis of Accidents in a Lithium-Cooled Space Nuclear Powerplant, NASA-TM S-1937, National Aeronautics and Space Administration (January 1970).

2. G. P. Lahte *et al.*, Preliminary Considerations for Fast Spectrum, Liquid-Metal Cooled Nuclear Reactor Program for Space-Power Applications, NASA TND-4315, National Aeronautics and Space Administration (March 1968).

3. D. B. Trauger, Some Major Fuel Irradiation Test Facilities of the Oak Ridge National Laboratory, ORNL-3574, Oak Ridge National Laboratory (April 1964).

4. V. A. DeCarlo *et al.*, Design of a Capsule for Irradiation Testing of Uranium Nitride Fuel, ORNL-TM-2363, Oak Ridge National Laboratory (February 1969).

5. S. C. Weaver, K. R. Thoms, and V. A. DeCarlo, Trans. Am. Nucl. Soc. 12 (2) 547 (1969).

6. R. F. Hilbert, V. W. Storhok, and W. Chubb, Trans. Am. Nucl. Soc. 13 102 (1970).

7. Richard F. Hilbert, Victor W. Storhok, and W. Chubb, Trans. Am. Nucl. Soc. 12 547 (1969).

8. R. L. Heestand, J. I.Federer, and C. F. Leitten, Jr., Preparation and Evaluation of Vapor Deposited Tungsten, ORNL-3662, Oak Ridge National Laboratory (August 1964).

9. T. N. Washburn, K. R. Thoms, S. C. Weaver, D. R. Cuneo, and E. L. Long, Jr., Trans. Am. Nucl. Soc. 13 101 (1970).

DISCUSSION

S. C. Carniglia (Kaiser Aluminum and Chemical Corp): Topography of the grain boundaries suggest viscous (i.e., liquid) behavior at 1500°C in a material nominally melting at 2850°C. Is this interpretation correct?

Authors: UN becomes very plastic at the irradiation temperature and we suspect that the UN has deformed plastically near the grain boundaries where bubble coalesence occurred.

H. Palmour III (N. C. State University):   Could the 1200 ppm O_2
content be contributing to enhanced grain boundary diffusion, and
hence to the localization of fission gas bubbles?   If so, would not
the newer, low O_2 material show greater stability?
Authors: It could perhaps be dependent upon the oxygen content.
The answer to this question is not presently available.   Experiments
will be started shortly which will provide an answer.

R. F. Stoops (N. C. State University):   Does porosity such as that
in your specimens help to prevent swelling?
Authors: There is no evidence that such closed porosity has any
effect on swelling.   We are using 85% dense specimens in some current
work to determine whether open porosity has an effect on swelling.